High-κ Gate Dielectrics

Series in Materials Science and Engineering

Series Editors: **B Cantor**, University of York, UK
M J Goringe, School of Mechanical and Materials
Engineering, University of Surrey, UK

Other titles in the series

Microelectronic Materials
C R M Grovenor

Aerospace Materials
B Cantor, H Assender and P Grant

Solidification and Casting
B Cantor and K O'Reilly

Topics in the Theory of Solid Materials
J M Vail

Fundamentals of Ceramics
M Barsoum

High Pressure Surface Science and Engineering
Y Gogotsi and V Domnich

Forthcoming titles in the series

Physical Methods for Materials Characterisation – Second Edition
P E J Flewitt and R K Wild

Metal and Ceramic Matrix Composites
B Cantor, F P E Dunne and I C Stone

Computer Modelling of Heat, Fluid Flow and Mass Transfer in
Materials Processing
C-P Hong

Other titles of interest

Applications of Silicon–Germanium Heterostructure Devices
C K Maiti and G A Armstrong

Series in Materials Science and Engineering

High-κ Gate Dielectrics

Edited by

Michel Houssa
Laboratoire Materiaux et Microelectronique de Provence,
Universite de Provence, France
Silicon Processing and Device Technology Division,
IMEC, Belgium

Institute of Physics Publishing
Bristol and Philadelphia

British Library Cataloguing-in-Publication Data

A catalogue record for this book is available from the British Library.

ISBN 0 7503 0906 7

Library of Congress Cataloging-in-Publication Data are available

Series Editors: **B Cantor and M J Goringe**

Commissioning Editor: Tom Spicer
Production Editor: Simon Laurenson
Production Control: Sarah Plenty and Leah Fielding
Cover Design: Victoria Le Billon
Marketing: Nicola Newey and Verity Cooke

Published by Institute of Physics Publishing, wholly owned by the Institute of Physics, London

Institute of Physics Publishing, Dirac House, Temple Back, Bristol BS1 6BE, UK

US Office: Institute of Physics Publishing, The Public Ledger Building, Suite 929, 150 South Independence Mall West, Philadelphia, PA 19106, USA

Typeset by Alden Bookset, Northampton, UK
Printed in the UK by MPG Books Ltd, Bodmin, Cornwall

Contents

Foreword

The success of the semiconductor industry relies on the continuous improvement of integrated circuits performances. This improvement is achieved by reducing the dimensions of the key component of these circuits: the metal–oxide–semiconductor field effect transistor (MOSFET). Indeed, the reduction of device dimensions allows the integration of a higher number of transistors on a chip, enabling higher speed and reduced costs. One of the key elements that allowed the successful scaling of silicon-based MOSFETs is certainly the superb material and electrical properties of the gate dielectric so far used in these devices, namely silicon dioxide. This material presents indeed several important features that allowed its use as a gate insulator. First of all, an amorphous silicon dioxide layer can be thermally grown on silicon with excellent control in thickness and uniformity, and naturally forms a very stable interface with the silicon substrate, with a low density of intrinsic interface defects. Secondly, silicon dioxide presents an excellent thermal and chemical stability, which is required for the fabrication of transistors, which includes annealing steps at high temperatures (up to 1000 °C). Next, the energy bandgap of silicon dioxide is quite large, about 9 eV, which confers excellent electrical isolation properties to this material, like its large energy band offsets with the conduction and valence bands of silicon and high breakdown fields, of the order of 13 MV/cm. Finally, the use of poly-silicon as gate electrode in a self-aligned CMOS (complementary metal–oxide–semiconductor) technology is also a determining factor in the scaling of the transistor structures.

All these superior properties allowed the fabrication of properly working MOSFETs with silicon dioxide gate layers as thin as 1.5 nm. However, further scaling down of the silicon dioxide gate layer thickness, required for the future CMOS technologies, is problematic. Indeed, the leakage current flowing through the transistors, arising from the direct tunnelling of charge carriers, exceeds 100 A/cm^2, which lies well above the specifications given by the International Technology Roadmap for Semiconductors, especially for low operating power and low standby power technologies. In addition, the reliability of ultrathin silicon dioxide layers becomes also an issue, namely the device lifetime, based on time-dependent gate dielectric breakdown, is not expected to reach 10 years at device operating conditions.

An alternative way of decreasing the silicon dioxide thickness in aggressively scaled MOSFET is to use a gate insulator with a higher relative dielectric constant κ than silicon dioxide (3.9). One could then use a physically thicker gate layer, yet with the same electrical thickness than sub-one nanometer silicon dioxide layers. This could potentially reduce the leakage current flowing through the transistors and also improve the reliability of the gate dielectrics. Consequently, tremendous worldwide research efforts have been focused in recent years to the investigation of so-called high-κ gate dielectrics for the potential replacement of silicon dioxide in advanced CMOS technologies.

The purpose of this book is to give a state-of-the art overview of high-κ gate dielectrics. The book consists of several contributions from internationally recognized experts in the field, collected into five different sections. The first section gives a brief introduction to the field, recalling the issues related to aggressive silicon dioxide thickness scaling and describing the requirements of alternative gate dielectrics. Section 2 is devoted to three major deposition techniques of high-κ dielectrics, i.e. atomic layer deposition, chemical vapour deposition and pulsed laser deposition. The physical, chemical and electrical characterization of high-κ gate dielectrics is covered in section 3. Section 4 deals with important theoretical investigations of high-κ dielectrics. The last section is devoted to technological aspects, i.e. to the behaviour and integration of high-κ dielectrics into modern CMOS technologies.

This book should serve as a valuable reference for researchers and engineers working in the field, as well as a good introductory book for PhD students who wish to perform research on high-κ dielectrics and advanced CMOS technologies. It should also give a good introduction to the field of advanced gate dielectrics for professors and students involved in graduate and postgraduate courses on modern semiconductor devices and technology of advanced integrated circuits. It is worth mentioning that this book is one of the first to review the field of high-κ gate dielectrics.

Acknowledgments

I would like to express my deepest gratitude to all the colleagues who have accepted to contribute to this book. This book is clearly the outcome of their willingness to share their great expertise in the field of high-κ gate dielectrics, resulting in chapters of very high scientific quality. I want to acknowledge them all for the time and effort they have spent in preparing their contributed chapters. At the Institute of Physics Publishing, I am indebted to Tom Spicer, Senior Commissioning Editor, and to Simon Laurenson, Production Manager, as well as their coworkers. Thanks to them, this review book project became a reality. My colleagues at IMEC and the University of

Provence are also gratefully acknowledged for the stimulating discussions we shared. Last but not least, I wish to thank my wife, Nathalie, and my son, Arnaud. Their encouragements and continuous support helped me considerably in preparing this monograph.

Michel Houssa
Leuven, Belgium
October 2003

SECTION 1

INTRODUCTION

Chapter 1.1

High-κ gate dielectrics: why do we need them?

M Houssa and M M Heyns

Introduction

The success of the semiconductor industry relies on the continuous improvement of integrated circuit performance. This improvement is achieved by reducing the dimensions of the key component of these circuits: the MOSFET (metal–oxide–semiconductor field effect transistor). Indeed, the reduction of device dimensions allows the integration of a higher number of transistors on a chip, enabling higher speed and reduced costs. The scaling of MOSFETs follows the famous Moore's law, which predicts the exponential increase in the number of transistors integrated on a chip [1]. This law is shown in figure 1.1.1, where the number of devices integrated in the different generations of Intel's microprocessors is presented as a function of the production year of these circuits [2]. For the sake of comparison, the 4004 processor, manufactured in 1971, integrated 2250 transistors from the 10 μm technology node (channel length of the devices) and was running at 108 kHz. The latest Intel Pentium®4 processor, introduced in 2002, integrates about 53 million transistors from the 0.13 μm technology, and demonstrates a clock frequency up to 2.8 GHz (at the time of writing).

One of the key elements that allowed the successful scaling of silicon-based MOSFETs is certainly the excellent material and electrical properties of the gate dielectric so far used in these devices: SiO_2. This material indeed presents several important features that have allowed its use as gate insulator. First of all, amorphous SiO_2 can be thermally grown on silicon with excellent control in thickness and uniformity, and naturally forms a very stable interface with the silicon substrate, with a low density of intrinsic interface defects. Besides, one of the most important defects at the

3

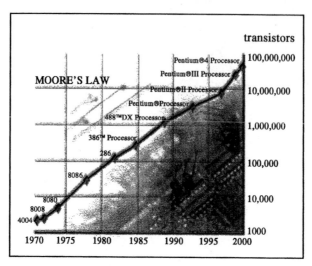

Figure 1.1.1. Illustration of Moore's law: number of transistors integrated in the different generations of Intel's microprocessors vs. the production year of these circuits [2].

(100)Si/SiO_2 interface, the P_{b0} centre (trivalent Si dangling bond), can be passivated very efficiently after post-metallization anneals performed in a hydrogen containing ambient. Secondly, SiO_2 presents an excellent thermal and chemical stability, which is required for the fabrication of transistors that includes annealing steps at high temperatures (up to 1000°C). Last but not least, the band gap of SiO_2 is quite large (about 9 eV), which confers excellent electrical isolation properties to this material, like its large energy band offsets with the conduction and valence bands of silicon and high breakdown fields, of the order of 13 MV cm^{-1}. It should be pointed out that the use of poly-silicon as gate electrode in a self-aligned CMOS technology was also a determining factor in the scaling of the transistor structures.

All these superior properties allowed the fabrication of properly working MOSFETs with SiO_2 gate layers as thin as 1.5 nm [3, 4]. However, as will be argued below, further scaling of the SiO_2 gate layer thickness is problematic. As illustrated in table 1.1.1, which is extracted from the specifications of the latest International Technology Roadmap for Semiconductors (ITRS) [5], the next generations of Si-based MOSFETs will require gate dielectrics with thicknesses below 1.5 nm, both for the high performance logic applications (like microprocessors for personal computers and workstations) and low operating power logic applications (like wireless applications). From a fundamental point of view, let us recall that the limit for SiO_2 thickness scaling is about 7 Å [6], below which the full band gap of the (bulk) insulator is not formed.

Table 1.1.1. EOT for the future generations of Si-based MOSFET technologies, including high performance logic applications and low operating power applications.

Technology (nm)	Production year	EOT (nm)	
		High performance logic	Low operating power logic
150	2001	1.3–1.6	2.0–2.4
130	2002	1.2–1.5	1.8–2.2
107	2003	1.1–1.4	1.6–2.0
90	2004	0.9–1.4	1.4–1.8
80	2005	0.8–1.3	1.2–1.6
70	2006	0.7–1.2	1.1–1.5
65	2007	0.6–1.1	1.0–1.4
50	2010	0.5–0.8	0.8–1.2
25	2016	0.4–0.5	0.6–1.0

This table is extracted from the ITRS (2001 version) [5].

Limits to SiO$_2$ scaling

The first problem arising from the scaling of the SiO$_2$ layer thickness concerns the leakage current flowing through the metal–oxide–semiconductor structure. As a matter of fact, in ultrathin SiO$_2$ gate layers (thickness typically below 3 nm), charge carriers can flow through the gate dielectric by a quantum mechanical tunnelling mechanism [7, 8]. This mechanism involves the tunnelling of charge carriers through a trapezoidal energy barrier (as illustrated in figure 1.1.2), the so-called direct tunnelling process [9]. From a simple Wentzel–Kramers–Brillouin (WKB) approach [10], it can be shown that the tunnelling probability increases exponentially as the thickness of the SiO$_2$ layer decreases [7, 9]. This results in a large increase of the leakage current flowing through the device as the SiO$_2$ gate layer thickness decreases, as shown in figure 1.1.2, the leakage current density exceeding $100 \, \text{A cm}^{-2}$ at $V_{\text{ox}} = 1$ V in a 1 nm thick SiO$_2$ layer (V_{ox} being the potential drop across the dielectric layer). The leakage current specifications from the ITRS are represented in figure 1.1.2 by shaded areas, both for high performance logic and low operating power logic applications [5]. It is clear from this figure that the SiO$_2$ layer thickness scaling is limited by the leakage current specifications, typically to 2.5–2.2 nm for low operating power circuits and 1.6–1.4 nm for high performance circuits. Comparing these figures to those given in table 1.1.1, it appears that SiO$_2$ will most probably not be used as gate dielectric for the 80 nm (and below) technologies. Actually, in the most 'aggressive' high performance technologies, nitrided silicon dioxide is used as gate dielectric. In this case, layers as thin as 1.2 nm with acceptable leakage

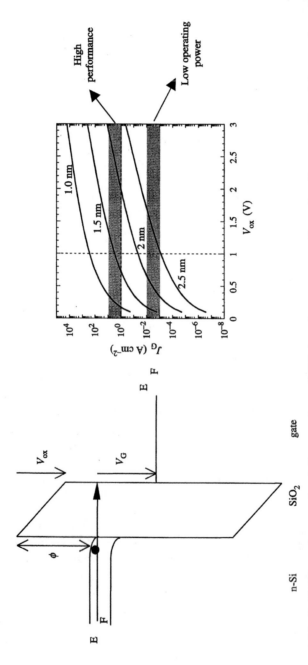

Figure 1.1.2. (Left) Schematic energy band diagram of an n-Si/SiO$_2$/metal gate structure, illustrating direct tunnelling of electrons from the Si substrate to the gate. ϕ is the energy barrier height at the Si/SiO$_2$ interface, V_{ox} the potential drop in the SiO$_2$ layer and V_G the applied gate voltage. (Right) Simulated tunnelling current through a metal–oxide–semiconductor as a function of the potential drop in the gate oxide, V_{ox}, for different SiO$_2$ gate layer thicknesses. Shaded areas represent the maximum leakage current specified by the ITRS for high performance and low operating power applications, respectively.

current can be used, making them viable gate dielectrics for the 70 nm high performance technologies.

Another issue related to SiO_2 thickness scaling concerns reliability aspects. During the operation of MOSFETs in integrated circuits, charge carriers flow through the device, resulting in the generation of defects in the SiO_2 gate layer and at the Si/SiO_2 interface [11–13]. When a critical density of defects is reached, breakdown (or quasi-breakdown) of the gate layer occurs, resulting in the failure of the device [14–16]. It was shown by Degraeve *et al.* [14] that the time-to-breakdown distributions of ultrathin SiO_2 layers could be quite well reproduced by a percolation approach, assuming that breakdown occurred via the formation of a percolation path between defects generated during the electrical stress.

The maximum gate voltage $V_{G,max}$ that can be applied to a MOSFET is presented in figure 1.1.3 as a function of the SiO_2 gate layer thickness at different temperatures and for the following specifications [17]: gate dielectric lifetime fixed at 10 years, failure rate fixed at 0.01% and chip area of 0.1 cm². These curves (solid lines in figure 1.1.3) were obtained from extrapolations of time-dependent dielectric breakdown measurements performed at high stress voltages, according to gate voltage, failure percentile and area scaling laws [14, 17]. It is evident that $V_{G,max}$ decreases when t_{ox} decreases and the temperature is raised, which results from the decrease in the critical density of defects necessary to trigger breakdown [14, 16]. The dotted curves in figure 1.1.3 correspond to the specifications of the ITRS for the different

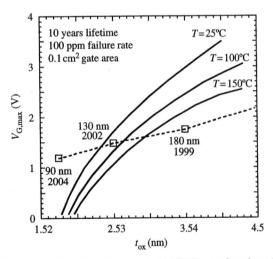

Figure 1.1.3. Maximum gate voltage $V_{G,max}$ of MOSFETs as a function of SiO_2 gate layer thickness at different temperatures and corresponding to the following specifications [16]: gate dielectric lifetime fixed at 10 years, failure rate fixed at 0.01% and chip area of 0.1 cm². Solid lines correspond to extrapolations from time-dependent dielectric breakdown measurements and the dashed line corresponds to the ITRS specifications.

generations of low operating power circuits [5]. The SiO_2 thickness limit is fixed at the crossover between the extrapolated curves and the technological specifications. This limit is found to be about 2.2 nm at room temperature and 2.8 nm at 150°C. Reliability requirements thus appear even more severe than leakage current requirements with respect to the scaling of the SiO_2 layer thickness.

Alternative gate dielectrics

From an electrical point of view, the metal–oxide–semiconductor structure behaves like a parallel plate capacitor: when a gate voltage V_G is applied to the gate, charges on the metal are compensated by opposite charges in the semiconductor, these latter charges forming the channel connecting the source and the drain of the transistor, as illustrated in figure 1.1.4. The capacitance C of this parallel plate capacitor is given by (in accumulation and inversion, neglecting poly-Si depletion effects)

$$C = \frac{A\varepsilon_r\varepsilon_0}{t_{ox}} \tag{1.1.1}$$

where A is the capacitor area, ε_r the relative dielectric constant of the material (3.9 for SiO_2), ε_0 the permittivity of free space ($8.85 \times 10^{-12}\,\mathrm{F\,m^{-1}}$) and t_{ox} the gate oxide thickness. From equation (1.1.1), it appears that decreasing t_{ox} allows us to increase the capacitance of the structure, and hence the increase in the number of charges in the channel for a fixed value of V_G. However, as pointed out above, the SiO_2 layer thickness approaches its limits. An alternative way of increasing the capacitance is to use an insulator with a higher relative dielectric constant than SiO_2 (it should be noticed that the

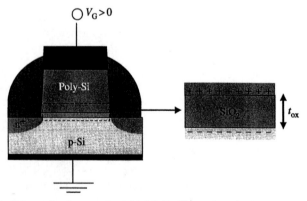

Figure 1.1.4. Schematic picture of a MOSFET, illustrating the behaviour of the MOS structure when a voltage V_G is applied to the gate, which is similar to a parallel plate capacitor.

relative dielectric constant is also represented by the letter κ, and one speaks about high-κ materials). One could then use a thicker gate layer and, hopefully, reduce the leakage current flowing through the structure and also improve the reliability of the gate dielectric.

The equivalent oxide thickness (t_{eq}) of a material is defined as the thickness of the SiO$_2$ layer that would be required to achieve the same capacitance density as the high-κ material in consideration. According to equation (1.1.1), t_{eq} is thus given by

$$\frac{t_{eq}}{\varepsilon_{r,\mathrm{SiO_2}}} = \frac{t_{\mathrm{high-}\kappa}}{\varepsilon_{r,\mathrm{high-}\kappa}} \qquad (1.1.2)$$

where $t_{\mathrm{high-}\kappa}$ and $\varepsilon_{r,\mathrm{high-}\kappa}$ are the thickness and relative dielectric constant of the high-κ material, respectively. As an example, using ZrO$_2$ as gate dielectric ($\varepsilon_r \approx 20$) would allow us to use a 5.1 nm thick layer in order to achieve a capacitance equivalent to a 1 nm thick SiO$_2$ layer; the equivalent oxide thickness of this ZrO$_2$ layer is thus 1 nm.

Actually, when a high-κ metal oxide like ZrO$_2$ or HfO$_2$ is deposited on an Si substrate, an ultrathin low-κ interfacial layer, either SiO$_x$ or SiM$_y$O$_x$ (where M is Zr or Hf) forms at the silicon interface, as illustrated in figure 1.1.5. This interfacial layer either grows during the deposition of the high-κ dielectric or during post-deposition anneal processes. It should be noticed that another low-κ layer can also form at the high-κ dielectric/metal gate interface.

The capacitance of the gate stack, C_{tot}, then results from the combination in series of the low-κ and high-κ dielectric layer capacitances, i.e.

$$\frac{1}{C_{tot}} = \frac{1}{C_{\mathrm{low-}\kappa}} + \frac{1}{C_{\mathrm{high-}\kappa}} \qquad (1.1.3)$$

The equivalent oxide thickness then reads

$$t_{eq} = \left(\frac{\varepsilon_{r,\mathrm{SiO_2}}}{\varepsilon_{r,\mathrm{low-}\kappa}}\right) t_{\mathrm{low-}\kappa} + \left(\frac{\varepsilon_{r,\mathrm{SiO_2}}}{\varepsilon_{r,\mathrm{high-}\kappa}}\right) t_{\mathrm{high-}\kappa} \qquad (1.1.4)$$

The presence of the low-κ interfacial layer increases the equivalent oxide thickness of the gate stack, and should thus be as thin as possible to achieve the equivalent oxide thickness (EOT) required by the ITRS. For example,

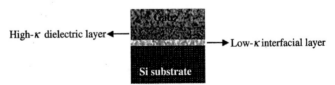

Figure 1.1.5. Schematic illustration of a MOS structure with a high-κ gate stack, formed by a low-κ interfacial layer and a high-κ dielectric layer.

a gate stack formed by a 7 Å SiO_x interfacial layer and a 5.1 nm ZrO_2 layer results in an EOT of 1.7 nm (assuming $\varepsilon_{r,low-\kappa} = 3.9$), as compared to an EOT of 1 nm if the interfacial layer was not present.

A lot of research efforts have been focused recently on the investigation of high-κ gate dielectrics [18–32], for the potential replacement of SiO_2 in advanced CMOS technologies. A list of materials studied in the literature is given in table 1.1.2, together with their relative dielectric constants.

The benefit of using high-κ materials as gate dielectric in the next generations of MOSFETs is illustrated in figure 1.1.6. This figure presents the high frequency capacitance–voltage and current–voltage characteristics of a MOS capacitor with a 4 nm $Al_xZr_{1-x}O_2$ gate layer. Comparing the capacitance–voltage characteristics of this device with simulated curves (taking into account quantum mechanical effects) allows us to extract the equivalent oxide thickness of the $Al_xZr_{1-x}O_2$ layer, which is found to be 1.1 nm. The leakage current of the capacitor is then compared to the simulated tunnelling current through a 1.1 nm SiO_2 layer. It is evident that the leakage current flowing through the device can be reduced by several orders of magnitude by using an $Al_xZr_{1-x}O_2$ gate layer.

However, the material that could potentially replace SiO_2 as gate dielectric in advanced CMOS technologies should also satisfy a long list of other requirements [27], e.g.:

- good thermal stability in contact with Si, preventing the formation of a thick SiO_x interfacial layer and the formation of silicide layers;
- low density of intrinsic defects at the Si/dielectric interface and in the bulk of the material, providing high mobility of charge carriers in the channel and sufficient gate dielectric lifetime;

Table 1.1.2. Examples of high-κ materials studied in the literature for the potential replacement of SiO_2 as advanced gate dielectrics.

Material	ε_r	Material	ε_r
Al_2O_3	9–11	$Al_xZr_{1-x}O_2$	9–25, 14[a]
Gd_2O_3	9–14	La_2O_3	21–30
Yb_2O_3	10–12	ZrO_2	14–25
Dy_2O_3	11–13	HfO_2	15–26
Nb_2O_5	11–14	Ta_2O_5	25–26
Y_2O_3	12–18	TiO_2	50–80
$Hf_xSi_{1-x}O_y$	3.9–26, 11[b]	$SrTiO_3$	200
$Zr_xSi_{1-x}O_y$	3.9–25, 12[b]	$Ba_xSr_{1-x}TiO_3$	200–300

[a] Typical value for $x = 0.5$.
[b] Typical values for Hf and Zr silicates corresponding to $x = 0.35$.

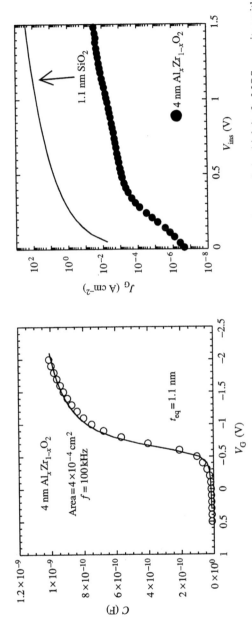

Figure 1.1.6. High frequency capacitance–voltage characteristics (left) and current–voltage characteristics (right) of a MOS capacitor with a 4 nm $Al_xZr_{1-x}O_2$ gate layer. Solid lines in these figures are calculated characteristics for a 1.1 nm SiO_2 gate layer.

- a sufficiently large energy band gap, providing high energy barriers at the Si/dielectric and metal gate/dielectric interfaces, in order to reduce the leakage current flowing through the structure;
- material's compatibility with CMOS processing, like, for example, high thermal budgets.

The potential high-κ gate dielectrics should thus meet most of these requirements; these latter requirements will be discussed in detail in the following chapters.

References

[1] Moore G E 1965 *Electronics* **38** 114–117
[2] See Moore's law at http://www.intel.com/research/silicon/mooreslaw.htm
[3] Timp G *et al* 1999 *Tech. Dig. Int. Electron Devices Meet.* (Piscataway: IEEE) p 55
[4] Chau R, Kavalieros J, Roberds B, Schenker R, Lionberger D, Barlage D, Doyle B, Arghavani R, Murthy A and Dewey G 2000 *Tech. Dig. Int. Electron Devices Meet.* (Piscataway: IEEE) p 45
[5] See the International Technology Roadmap for Semiconductors (2001 edn) at http://public.itrs.net
[6] Muller D A, Sorsch T, Moccio S, Baumann F H, Evans-Lutterodt K and Timp G 1999 *Nature* **399** 758
[7] Depas M, Vermeire B, Mertens P W, Van Meirhaeghe R L and Heyns M M 1995 *Solid-State Electron.* **38** 1465
[8] Lo S H, Buchanan D A, Taur Y and Wang W 1997 *IEEE Electron Dev. Lett.* **18** 209
[9] Fromhold A T 1981 *Quantum Mechanics for Applied Physics and Engineering* (New York: Dover)
[10] Merzbacher E 1998 *Quantum Mechanics* (New York: Wiley)
[11] Harari E 1978 *J. Appl. Phys.* **49** 2478
[12] DiMaria D J, Cartier E and Arnold D 1993 *J. Appl. Phys.* **73** 3367
[13] DiMaria D J and Cartier E 1995 *J. Appl. Phys.* **78** 3883
[14] Degraeve R, Groeseneken G, Bellens R, Ogier J L, Depas M, Roussel P J and Maes H 1998 *IEEE Trans. Electron Devices* **45** 904
[15] Houssa M, Nigam T, Mertens P W and Heyns M M 1998 *J. Appl. Phys.* **84** 4351
[16] Stathis J H 1999 *J. Appl. Phys.* **86** 5757
[17] Degraeve R, Pangon N, Kaczer B, Nigam T, Groeseneken G and Naem A 1999 *Tech. Dig. VLSI Symposium* (Piscataway: IEEE) p 59
[18] Kim H S, Campbell S A and Gilmer D C 1997 *IEEE Electron Dev. Lett.* **18** 465
[19] Chaneliere C, Autran J L, Devine R A B and Balland B 1998 *Mater. Sci. Eng.* **R22** 269
[20] Houssa M, Degraeve R, Mertens P W, Heyns M M, Jeon J S, Halliyal A and Ogle B 1999 *J. Appl. Phys.* **86** 6462
[21] Kwo J, Hong M, Kortan A R, Queeney K T, Chabal Y J, Mannaerts J P, Boone T, Krajewski J J, Sergent A M and Rosamilia J M 2000 *Appl. Phys. Lett.* **77** 130
[22] Wilk G D, Wallace R M and Anthony J M 2000 *J. Appl. Phys.* **87** 484
[23] Eisenbeiser K *et al* 2000 *Appl. Phys. Lett.* **76** 1324
[24] Copel M, Gribelyuk M and Gusev E 2000 *Appl. Phys. Lett.* **76** 436

[25] Park D G, Cho H J, Lim K Y, Lim C, Yeo I S, Roh J S and Park J W 2001 *J. Appl. Phys.* **89** 6275

[26] Morais J, da Rosa E B O, Miotti L, Pezzi R P, Baumvol I J R, Rotondaro A L P, Bevan M J and Colombo L 2001 *Appl. Phys. Lett.* **78** 2446

[27] Wilk G D, Wallace R M and Anthony J M 2001 *J. Appl. Phys.* **89** 5243

[28] Houssa M, Afanas'ev V V, Stesmans A and Heyns M M 2001 *Appl. Phys. Lett.* **79** 3134

[29] Copel M, Cartier E and Ross F M 2001 *Appl. Phys. Lett.* **78** 1607

[30] Lucovsky G 2001 *J. Vac. Sci. Technol.* **A19** 1353

[31] Misra V, Heuss G P and Zhong H 2001 *Appl. Phys. Lett.* **78** 4166

[32] Xu Z, Houssa M, De Gendt S and Heyns M M 2002 *Appl. Phys. Lett.* **80** 1975

SECTION 2

DEPOSITION TECHNIQUES

Chapter 2.1

Atomic layer deposition

Mikko Ritala

Introduction

An inherent consequence of the replacement of SiO_2 and SiO_xN_y as the gate dielectrics in the future generations of CMOS devices is that a new method for the dielectric deposition will be needed too. Among the various candidates, atomic layer deposition (ALD, also known by the names atomic layer epitaxy (ALE) and atomic layer chemical vapour deposition (ALCVD), for example) offers certain important characteristics like excellent large area uniformity, outstanding conformality, and atomic level control of film composition and thickness [1–4]. As a consequence, ALD of high-κ oxides has been a subject of increasingly intense research during the past few years. On the other hand, while new in the gate oxide application, the ALD method has been examined and developed continuously since its introduction in the late 1970s [5–8]. Most importantly, in thin film electroluminescent (TFEL) display devices [1], which were the original motivation for developing ALD and remained the only industrial application of ALD for a long time, one uses insulator films that are required to have quite similar properties as the high-κ gate dielectrics. Therefore, ALD processes for many potential high-κ gate oxide materials were developed and characterized already before the current high-κ research began [1].

This chapter makes an overview of ALD and, in particular, its application to high-κ gate oxide deposition. The chapter begins with the introduction of the basic principle of ALD, followed by a discussion on its advantages and limitations. Precursor chemistry and ALD reactors, the two important requirements for a successful utilization of ALD, complete the introductory part. Next, the special issues related to the high-κ gate oxide deposition on silicon by ALD are discussed. The rest of the chapter is devoted to a survey of high-κ ALD oxide processes, mainly focusing on those processes that have dominated the research but also introducing alternative processes available. Because device integration issues are discussed elsewhere in the book (see the chapter by E. Young and V. Kaushik), they are covered in this chapter only when directly related to ALD, the most important one being the compatibility of the ALD oxide processes with silicon.

Atomic layer deposition

Basic principle

Like CVD, ALD is a chemical gas phase thin film deposition method. The distinct feature of ALD is that the film is grown through sequential saturative surface reactions that are realized by pulsing the two (or more) precursors into the reactor alternately, one at a time, separated by purging or evacuation steps.

As an example, figure 2.1.1 demonstrates one ALD cycle in deposition of ZrO_2 from $ZrCl_4$ and H_2O, which is one of the most extensively examined high-κ gate oxide ALD processes. The film surface left from the previous cycle is terminated with hydroxyl groups. When dosed upon this surface, $ZrCl_4$ reacts with the hydroxyl groups forming $-O-ZrCl_3$ and/or $(-O-)_2ZrCl_2$ surface groups. Under ideal ALD conditions, each reaction step is saturative, which in this case means that all the hydroxyl groups, and any other possibly existing adsorption sites like bare oxide ions, become consumed from the surface. The following purge (or evacuation) step removes the excess $ZrCl_4$ molecules and HCl formed as a reaction

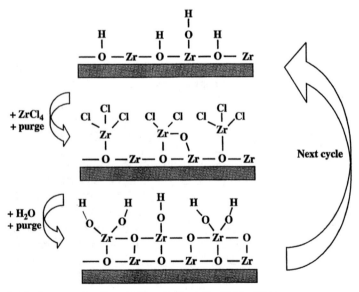

Figure 2.1.1. An example of an ALD cycle from the $ZrCl_4-H_2O$ process.

by-product. When dosed upon this surface, water reacts with the $-ZrCl_x$ species converting them into ZrO_2 covered by hydroxyl groups. Again the reaction is saturative so that all the chlorides are removed. Finally, the second purge/evacuation step completes the cycle leaving the surface covered by the same number of hydroxyl groups as at the beginning of the cycle. The outcome of this one ALD cycle is a deposition of a fixed amount, most often a part of a monolayer, of ZrO_2 on the surface. In order to grow a thicker film, the ALD cycle is repeated as many times as it is necessary to reach the target thickness.

Advantages and limitations of ALD

The distinctive feature of ALD is that the reactions are *saturative* (figure 2.1.2). This makes the film growth *self-limiting* that, in turn, gives the method a number of advantages as summarized in table 2.1.1. The self-limiting growth ensures that each cycle (possibly excluding the very first cycles, however, see the 'Special issues related to ALD of high-κ gate oxides' section) deposits the same amount of material on all surfaces independent of the precursor dose received as long as the dose is high enough to saturate the reactions. As a consequence, the ALD method offers excellent large area uniformity and conformality (figure 2.1.3). In addition, film thicknesses are accurately controlled simply by the number of deposition cycles applied (figure 2.1.4). This makes it also straightforward to tailor film composition at an atomic layer level (figure 2.1.5). Preparation of multicomponent and multilayer materials is further facilitated by the fact that process temperature windows are often reasonably wide so that binary processes are easy to combine.

The self-limitation of the film growth eliminates adverse effects of precursor dose variation not only spatially but also in time scale, providing

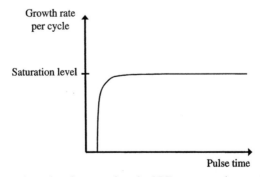

Figure 2.1.2. Saturation of surface reactions in ALD processes is experimentally verified by observing that the deposition rate per cycle stabilizes to a constant level with increasing precursor pulse time or dose.

Table 2.1.1. Relationships between characteristics and advantages of ALD.

Characteristic feature of ALD	Implication for film deposition	Practical advantage
Self-limiting growth process	Film thickness is dependent only on the number of deposition cycles	Accurate and simple thickness control
	No need for reactant flux homogeneity	Large area capability Large batch capability Excellent conformality No problems with inconstant vaporization rates of solid precursors Good reproducibility Straightforward scale-up
	Atomic level control of material composition	Capability to produce sharp interfaces, nanolaminates and superlattices Possibility to interface modification
Separate dosing of reactants	No gas phase reactions	Allows a use of precursors highly reactive towards each other, thereby enabling effective precursor utilization and short cycle times
	Sufficient time is given to complete each reaction step	High quality materials are obtained at low deposition temperatures
Processing temperature windows are often wide	Processing conditions of different materials are readily matched	Straightforward preparation of multilayer structures in a continuous process

again of course that the doses remain high enough to saturate the reactions. Therefore, solid sources are easier to use in ALD than in CVD where inconstant precursor fluxes, caused by surface area decrease through sintering, may cause problems to the process control.

The alternate supply of the precursors in well-separated pulses ensures that the precursors never meet in the gas phase. This eliminates risks of gas phase reactions with possible detrimental consequences such as particle formation.

(*a*)

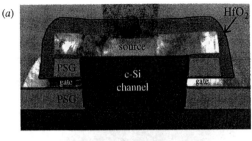

(*b*)

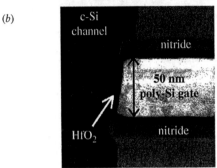

Figure 2.1.3. (*a*) TEM image of a vertical replacement gate MOSFET with the highly conformal ALD HfO$_2$ as the gate dielectric. (*b*) Higher magnification view of the active area. (Reprinted with permission from Hergenrother J M *et al* 2001 50 nm vertical replacement-gate (VRG) nMOSFETs with ALD HfO$_2$ and Al$_2$O$_3$ gate dielectrics *Tech. Dig. Int. Electron Devices Meeting (IEDM)*. Copyright 2001 IEEE.)

The major drawback of ALD is the low deposition rate which is a direct consequence of the stepwise film growth where in most processes only a fraction of a monolayer is deposited in one cycle. Deposition rates are typically in the range of 100–300 nm h^{-1}. Luckily, the low film thicknesses make the high-κ gate oxides an application where the deposition rate

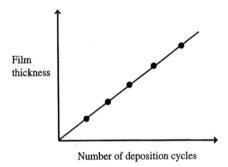

Figure 2.1.4. Schematic of the ideal linear film thickness correlation with the number of deposition cycles applied.

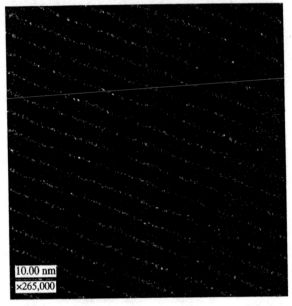

Figure 2.1.5. Cross-sectional TEM image of a $Zr_xSi_yO_z$–$Zr_xTi_yO_z$ nanolaminate.

requirements are somewhat more relaxed than in most other areas. The throughputs quoted by commercial reactor manufacturers are commonly in a range of 10–20 wafers per hour per reactor module.

In other applications, particularly in the TFEL flat panel display production, the low growth rate of ALD is effectively compensated by large batch processing. Again, the self-limiting growth mechanism gives ALD an advantage allowing one to process tens of substrates in a single compact reactor without sacrificing too much in the time required to complete one ALD cycle [1]. Obviously, the throughput increasing effect of large batch processing could be substantial also in the gate oxide application but until now this possibility appears to have remained quite unexplored while the single wafer reactors have dominated on the market. However, steps in this direction are the mini-batch flow-type reactor processing either four 200 mm wafers or three 300 mm wafers at the same time [10] and the less common fill-in–hold–pump-down type reactor with up to 26 wafers [11].

Precursor chemistry

Successful utilization of ALD is dependent on two factors: proper precursors and fast and efficient reactors. In this section the basic issues on ALD precursors are explored. The later sections will review the chemistry used in ALD of high-κ oxides. A more thorough survey of ALD chemistry, including also materials other than high-κ oxides as well as characterization of the ALD reactions, can be found in [1].

Table 2.1.2 summarizes the requirements for ALD precursors. First of all, the precursors must be volatile enough to ensure efficient transportation for saturating the surface reactions. Gases and high vapour pressure liquids are the preferred choices because they are easy to handle in external cylinders and they allow a supply of high doses in short time period. Representative examples of such precursors for high-κ oxide deposition are trimethyl aluminium and water. However, as already noted, solids can rather easily be used in ALD as well because the self-limiting growth mechanism takes care of small variations in precursor fluxes. A typical requirement to the vapour pressure in the source is about 0.1 torr but with larger substrate areas higher values become preferable. A major concern related to solids is transportation of fine particles from the source to the substrate by the carrier gas. Obviously, this problem is the most severe for the very fine particle size solids, as unfortunately is the case for the two widely used high-κ oxide precursors $ZrCl_4$ and $HfCl_4$. However, with a careful source design this problem seems to have been eliminated.

As the ALD method relies on saturative, self-limiting surface reactions, it is of utmost importance that the precursors and, as importantly, the surface species formed thereof do not decompose thermally on their own. Any decomposition reaction that would lead to a film deposition in excess of that arising from the surface exchange reactions must be avoided in an effort to maintain atomic level accuracy in film thickness control and uniformity. In this respect the thermally stable metal halides are the best choices while the use of metal compounds with organic ligands is often limited to temperatures below 300°C or even lower.

To achieve fast saturation in each reaction step, the precursor dosed onto the substrate should react rapidly with the surface species left from the previous precursor pulse. This calls for precursors that react aggressively with each other, though one should recognize that what eventually matters is

Table 2.1.2. Summary of ALD precursor requirements.

Volatility
Aggressive and complete reactions
No self-decomposition
No etching of the film or substrate material
No dissolution into the film or substrate
Sufficient purity
Unreactive volatile by-products
Inexpensive
Easy to synthesize and handle
Nontoxic and environmentally friendly

the reactivity towards the surface intermediates instead of their parent molecule. Another related requirement is that the reactions should be complete to provide high film purity. While looking for appropriate precursor combinations, thermodynamic calculations are valuable whenever the required data are available: the more negative the Gibbs free energy change is for a given net reaction, the more likely this combination of precursors is suitable for use in ALD. Also any information referring to high reactivity, like remarks on moisture sensitivity or gas phase reactions encountered in CVD experiments, give useful hints for choosing precursors to ALD. In this respect the $ZrCl_4$–H_2O process serves as a good example: in CVD this is a problematic precursor combination as they react easily in gas phase and thus water is often formed *in situ* from CO_2 and H_2, but in ALD where the gas phase reactions are inherently eliminated this has proven a good process.

Precursors should not etch the material formed during the deposition process, but in a few cases this has been found to occur. An example related to the high-κ gate oxides is etching of niobium oxide by niobium chloride through formation of niobium oxychloride:

$$Nb_2O_5(s) + 3NbCl_5(g) \rightarrow 5NbOCl_3(g)$$

As a consequence, it is impossible to deposit Nb_2O_5 from $NbCl_5$ [12]. A similar etching reaction takes place also between Ta_2O_5 and $TaCl_5$ but only above 300°C so that the $TaCl_5$–H_2O process may be used below this temperature [13, 14]. A somewhat different and less detrimental side reaction occasionally found in deposition of multicomponent materials is an exchange of cations between the metal precursor and the oxide upon which it is dosed [15]:

$$3TiO_2(s) + 4AlCl_3(g) \rightarrow 2Al_2O_3(s) + 3TiCl_4(g)$$

Quite often such a reaction self-terminates after a few cycles and may thus be quite difficult to observe, but it can complicate the control of composition of multicomponent films.

Ideally, the by-products formed in the surface exchange reactions should be unreactive so that they can be easily purged away from the reactor. This argument favours $Al(CH_3)_3$ over $AlCl_3$, for example, because methane is less reactive than HCl. While HCl is not expected to cause detrimental corrosion reactions in deposition of high-κ oxides, it may still readsorb on the surface and thereby block adsorption sites from the precursor molecules. Furthermore, if the adsorption site blocking is not uniform across the substrate, it may cause some thickness nonuniformity.

The other requirements listed in table 2.1.2 are obvious and self-explanatory. It is worth noting, however, that it is often hard to fulfil all the requirements simultaneously and then the lower ones in table 2.1.2 are those which must be sacrificed while developing new ALD processes. For example, $Al(CH_3)_3$ is a pyrophoric compound and thus needs special care in handling, but at the same time this high reactivity is one of the key factors in making the $Al(CH_3)_3$–H_2O precursor combination perhaps the best ALD process developed so far.

ALD reactors

Nearly all the commercial ALD reactors appear to be of the flow type where the pressure is most commonly 1–10 torr and inert gas is used for purging the reactor between the precursor pulses. This is because purging is faster than evacuating the reactor. In a well-designed reactor, one ALD cycle can be completed in less than a second [16], though cycle times of a couple of seconds seem to be more common. Because of their convenience and importance in production, flow-type reactors are extensively used also in research, but there high vacuum systems are used too, especially while performing detailed mechanism studies on surface reactions [1]. In the following, the crucial parts of the flow-type ALD reactors are examined. More detailed descriptions of the ALD reactors can be found in [1, 7, 8, 17–19], for example.

Inert gas supply. The most commonly used inert gases are nitrogen and argon. Besides purging the excess precursor molecules and by-products out of the reactor, inert gases are also used for transporting the precursors. Purity is the key property because the inert gas supply serves as the largest source of impurities, well exceeding any minor leakage possibly left after careful helium leak testing [1]. Reactive residuals, particularly moisture, must be at a ppm level, or lower. For the best combination of speed and utilization efficiency of precursors and inert gas, the reactors are typically run at pressures of 1–10 torr, and the inert gas consumption varies from 0.5 slm in small research reactors to a few tens of slm in production reactors.

Precursor sources with sequencing control. Depending on the vapour pressure of the precursor in comparison with the reactor pressure, the sources can be divided into two groups. The high vapour pressure sources are either gases or liquids which have vapour pressures higher than the total pressure in the reactor. These precursors are easily pulsed into the reactor with the aid of fast valves. Transport gases are not necessarily needed for the introduction but are quite often used with liquids to ensure fast delivery of the required doses.

Low vapour pressure compounds need to be heated to obtain the required vapour pressures, typically 0.1–1 torr. The heated sources can be either inside the reactor in separately controlled temperature zones, or they can be external vessels. In either case the source lines and any valving elements must be heated to higher temperatures to avoid condensation.

A clever way of pulsing low vapour pressure sources is the inert gas valving system [1, 7, 8, 17] which has no temperature limitations. An alternative way for feeding low vapour pressure sources is liquid injection delivery where a solution of the precursor in an appropriate inert solvent is injected into a vaporization chamber having a high enough temperature for complete volatilization.

To avoid film deposition onto the source line walls and the consequent need of frequent cleaning, separate source lines are often used for the different precursors. These lines merge only in the reaction chamber right before the substrates.

Reaction chamber. Desorption of one precursor from the reaction chamber walls during the pulse sequence of the other precursor would of course be highly disturbing to the ALD process. On the other hand, complete removal of physisorbed molecules from the reaction chamber walls is time consuming and could easily require excessively long purge periods. Therefore, the ALD reaction chamber walls should be kept hot enough—often they are at the same temperature as the substrate—so that film is deposited there too, even if this calls for regular cleaning of the chamber. On the other hand, if so desired, for example, to get more freedom in the reaction chamber configuration, the reaction chamber and its heating system may be enclosed within a cold wall vacuum chamber [19]. Then the cold walls of the vacuum chamber and the hot walls of the reaction chamber limit an intermediate space that must be completely isolated from the source lines and the reaction chamber to avoid water contamination from the cold walls.

The flow-type ALD reactors can be classified further into basically two groups depending on how the precursor flow is directed to the substrate surface (figure 2.1.6): the flow channel type (often called also a travelling-wave reactor) where the precursors flow across the substrate, and the top injection type where the precursor flow is directed perpendicular to the substrate surface, often being first dispersed uniformly with the aid of a shower head. The latter are familiar from CVD reactors while the flow channel reactors are more specific to ALD and better address the key requirements for the high throughput, i.e. fast saturation of the surface reactions and fast switching from one precursor to the other.

In the flow channel reactors the precursors are transported along a narrow flow channel lined by the substrate and its holder on one side, and the reactor wall in close proximity on the other side (in batch reactors another substrate replaces the reactor wall as the other lining of the flow channel [1]). When the precursor molecules flow through the channel, they make multiple hits with the substrate (and the wall). The multiple hit conditions increase the probability of finding an open adsorption site and thereby promote the utilization efficiency of the precursor. Another benefit is that the surface becomes rapidly saturated, thus allowing use of subsecond pulse times. The pluglike flow conditions in the flow channel ensure that the precursor

FLOW CHANNEL REACTOR

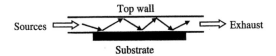

TOP INJECTION REACTOR

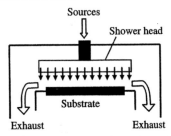

Figure 2.1.6. Schematics of the two common configurations of flow-type ALD reactors. In both cases the sources include appropriate valving systems, and the two precursors are usually led into the reaction chamber through separate lines. In the top injection reactor the showerhead may be replaced also by other kinds of flow dispersing device.

pulses retain their shapes without extensive diffusion broadening. As a result, the reactor is rapidly purged. The flow channel reactors are also straight-forward to scale up simply by adding several, as many as tens of flow channels parallel to each other as done in the TFEL display production, for example [1].

One concern related to the flow channel reactors is the possibility of decreased growth rate and thickness nonuniformity arising from reaction by-product readsorption [1, 20, 21]. The by-product adsorption is likely to be pronounced in the flow channel reactors because there the by-products travel in front of the precursor pulse and thus can first adsorb without competition with the precursor molecules. If firmly adsorbed, the by-products block adsorption sites from the metal precursors and thereby decrease the growth rate. Further, thickness nonuniformity may develop because the number of by-product molecules in the gas increases on the way from the leading edge of the substrate to the trailing edge. Nevertheless, even with the chloride-based processes, uniformities better than 2% have been achieved across 300 mm wafers. Rotation of the substrate should also minimize this effect.

In properly designed top injection (showerhead) reactors, each site on the substrate should receive the precursor pulse front at about the same time without being first exposed to the by-products. Therefore, this reactor configuration should be less vulnerable to the adverse effects of by-product readsorption. On the other hand, to incorporate the showerhead, the chamber must be more open than in the flow channel type reactors and therefore it is more slowly purged; one study reported 5 s purge times for the showerhead reactor [22]. One might try to overcome this problem by completing the purging only in the important volume between the showerhead and the substrate but then the residual CVD reactions outside this volume, though downstream, would cause a serious risk of particle contamination. Finally, the top injection reactor configuration can hardly be applied for processing batches larger than a few wafers.

For plasma-enhanced ALD, two options exist for positioning the substrate in relation to the plasma source: the substrate may either be immersed in the plasma or it may be downstream of the plasma so that only radicals but not ions and electrons flow onto the substrate. While the plasma immersion configuration ensures high density of radicals, it also contains a severe risk of plasma damage, like creation of trapped charge. This risk is minimized in the downstream configuration that is therefore preferable, even if the radical flux onto the substrate is smaller because of radical recombination between the plasma source and the substrate. However, because oxidation of the high-κ oxide–silicon interface is a serious concern, it is quite doubtful if oxygen radicals can be used at all in the gate oxide deposition (cf the 'Special issues related to ALD of high-κ gate oxides' section).

In addition to the above described reactor configurations, there is also a third kind of flow-type reactor where the substrates are placed on a holder which rotates so that the substrates are alternately exposed to two continuous precursor flows [1, 23]. However, this reactor type has only been used in compound semiconductor ALD research and there are no commercial reactors available.

Vacuum pumps and related exhaust equipment. As the flow-type reactors operate at 1–10 torr pressures with reasonably high inert gas flow rates, Roots blowers and mechanical pumps are the common choices for pumping. No high vacuum pumps are needed because the purity of the ALD system is based on efficient purging with high purity inert gas, rather than ultimately low base pressure. This is the most evident in the case of oxide ALD with water as an oxygen source. As it is well known, water as a polar molecule is the most difficult residual gas to remove while pumping for high vacuum, but in the ALD oxide processes water is repeatedly pulsed into the reactor. It is clear, therefore, that the water residual level during the metal precursor pulse soon after the water pulse is determined primarily by the purging efficiency and the purity of the inert gas, not the base pressure attained before the process.

The exhaust system may be equipped with various traps, afterburners and related exhaust management devices, depending on the precursors being employed. These are common to CVD systems and will not be dealt with in more detail here. It must be noted, however, that when large overdoses are used, the excess precursors and possibly also by-products are likely to react or condense in some part of the exhaust line producing significant amounts of solid material to be taken care of.

Connections to other tools. Integration of the ALD oxide processes to the overall CMOS process flow requires that the reactors are equipped with the standard connections for attaching to cluster tools or automatic wafer loaders. This allows also various pre- and post-deposition treatments to be done without exposing the wafers to air in between.

Because the beauty of the ALD method is that a good process control is attained simply by relying on the self-limiting growth reactions, separate process control tools are quite rarely used. Techniques like quadrupole mass spectrometry and quartz crystal microbalance have therefore been used only in basic research on reaction mechanisms [1]. Nevertheless, it still might be worth consideration of equipping also the production and development reactors with these tools as they would provide valuable information about the vacuum condition and the success of precursor dosing.

While most of the commercial reactors are of the flow type, there is at least one exception that relies on the fill-in–hold–pump-down principle [11]. As noted, pumping down the reactor is more time consuming than purging and the minimum cycle times have been 12–15 s. To compensate for the longer cycle times, this particular reactor is designed for batch processing of up to 26 wafers at a time.

Special issues related to ALD of high-κ gate oxides

Deposition of high-κ gate oxides for the future generation CMOS devices is a very demanding task. The special requirements for the deposition of any high-κ oxide by any method for this application go back to the fact that the high-κ oxides will be needed in devices where the thickness of the so far dominant SiO_2 gate dielectric should be scaled to 1.0 nm and less. In other words, the equivalent oxide thickness (EOT) of the gate stack, including both the high-κ oxide and any interface layers, should be less than 1.0 nm. However, as silicon is readily oxidized at its surface, a thin interface layer of SiO_2 or a mixture of SiO_2 and the high-κ oxide, i.e. the so-called metal silicate, is easily formed. In such a case, there are two capacitors in series and the total EOT is a sum of the EOTs of the high-κ oxide and the interface layer:

$$EOT_{tot} = EOT_{high\text{-}κ} + EOT_{interface}$$

If the interface layer is pure SiO_2, $EOT_{interface}$ may be replaced by the physical thickness of this layer, d_{SiO_2},

$$EOT_{tot} = EOT_{high\text{-}κ} + d_{SiO_2}$$

From the above it is clear that to achieve $EOT_{tot} \leq 1.0$ nm, the interfacial SiO_2 must be thinner than 1.0 nm, and to gain a reasonable physical thickness increment from the high-κ oxide, the SiO_2 layer should preferably be less than 0.5 nm. Even if it appears that often the interface layer is not pure SiO_2 but also contains the high-κ oxide, though to a lesser extent [24, 25], the thickness and EOT of this layer are still of concern because the permittivity of the SiO_2-rich silicate is closer to SiO_2 than the high-κ oxide.

On the other hand, it is well recognized that the $Si-SiO_2$ interface is a nearly ideal semiconductor–insulator interface with a very low density of interface traps, while the high-κ oxides have a tendency to form much more defective interfaces. Therefore, for low interface trap densities, it would be beneficial to have SiO_2 at the interface with silicon. These contradictory requirements lead to the special challenge for the MOSFET high-κ gate oxide processing: how to control the high-κ oxide–Si interface so that it would contain an acceptably low density of traps while at the same time keeping the interfacial SiO_2 or silicate layer thin enough to reach the targeted EOT_{tot} values.

The first step in addressing these requirements is to choose the high-κ oxide material so that it will not react with silicon during the deposition or the following annealing treatments. This issue is not discussed here, except just to note that this stability requirement has focused the research mainly on oxides of Al, Zr, Hf, and the rare earths, especially La and Y.

The second issue to be considered is how the silicon surface is treated prior to the high-κ oxide deposition. The starting surfaces for ALD of high-κ gate oxides have been either ultrathin oxide or nitride layers or hydrogen terminated silicon. The oxide and nitride layers are commonly formed by first etching away the native oxide layer followed by controlled oxidation or nitridation. The oxide layer may be formed by thermal oxidation or by low temperature treatment with aqueous solutions containing ozone or hydrogen peroxide, for example. The latter are called chemical oxides and they differ from the thermal oxides somewhat in their chemical nature, one important difference apparently being the high hydroxyl coverage on the chemical oxides [26]. Nitridation is commonly done by high temperature annealing in the presence of ammonia. Hydrogen terminated silicon is, in turn, formed by the HF-last etching procedure which removes the native oxide and passivates the surface of the abruptly terminating silicon crystal with Si–H bonds, though small amounts of fluorine and hydrogen residues may remain too [27]. X-ray photoelectron spectroscopy (XPS) studies showed that when exposed for 20 min to water vapour at partial pressures typical to ALD oxide processes, the hydrogen terminated Si(100) surface is stable up to 300°C; only fluorine residues and silicon dangling bonds were thought to be hydroxylated and defect sites possibly oxidized [27]. A similar water vapour exposure at 350 and 400°C caused a formation of about 6 Å oxide layer, and hydrogen desorption was proposed to be the limiting step in the oxidation by water vapour [27]. The stability against water vapour at low temperatures suggests that the hydrogen terminated surface could enable the deposition of the high-κ oxide on silicon without a creation of a SiO_2 interface layer. It must be noted, however, that when the first metal precursor molecules have adsorbed on the silicon surface, its resistance against water weakens [28]. Furthermore, as oxygen insertion from molecular oxygen into the surface Si–Si bonds is possible without removal of the surface hydrogen [29], even trace levels of O_2 may have a significant effect.

In principle, at least, starting with the hydrogen terminated silicon could allow one to deposit the high-κ oxide directly on Si without any low permittivity interface layer. Unfortunately, it has been found that getting an ALD process started on hydrogen terminated silicon is always more difficult than on oxide or nitride surfaces [26, 30]. Figure 2.1.7 compares the initial stages of HfO_2 growth from $HfCl_4$ and H_2O on hydrogen terminated silicon, silicon covered by chemical oxide formed using ozone-based chemistry, and thermal SiO_2 and Si–O–N [26]. The growth on chemical oxide is linear from the very first cycles but on thermal SiO_2 and Si–O–N it takes 15–25 cycles and on the hydrogen terminated surface 30–50 cycles to reach the linear growth. These differences are understood in terms of different reactive site densities on the starting surfaces: the chemical oxide surface contains the highest density of hydroxyl groups, thereby providing the easiest nucleation. The other extreme is the hydrogen terminated surface where the Si–H groups are quite inert and prevent a rapid conversion of the surface into hydroxyl group terminated. As compared with the other surfaces, the chemical oxide was also concluded to ensure the most two-dimensional growth resulting in the best uniformity. It was also demonstrated that a chemical oxide as thin as 0.17 nm was enough to ensure linear growth from the very beginning of the growth, though in this case the deposition rate was lower and the resulting film was less dense than on 0.5–0.7 nm chemical oxides. This difference was related to an incomplete coverage of silicon with the nominally 0.17 nm thick oxide layer.

Though the hydrogen terminated silicon has been found to be a problematic starting surface for all the ALD high-κ oxide processes, there are differences in the lengths of the initial nonlinear regime (figure 2.1.8) [30, 31]. The reason for the $Al(CH_3)_3$–H_2O process showing the fastest nucleation is apparently the capability of $Al(CH_3)_3$ to react with the Si–H surface groups [28]. Such a high reactivity seems to be missing from $ZrCl_4$ and $HfCl_4$.

Besides retarding the start of the film growth, the poor nucleation of ZrO_2 and HfO_2 on the hydrogen terminated silicon also causes remarkable microscopic level nonuniformity to the films (figure 2.1.9) [30, 32]. As a consequence, the electrical properties of these films tend to be inferior to those deposited on alternative starting surfaces. The more readily starting $Al(CH_3)_3$–H_2O process results in better morphology, however (figure 2.1.10) [30, 33]. In addition, no interface layer is observable with TEM or XPS.

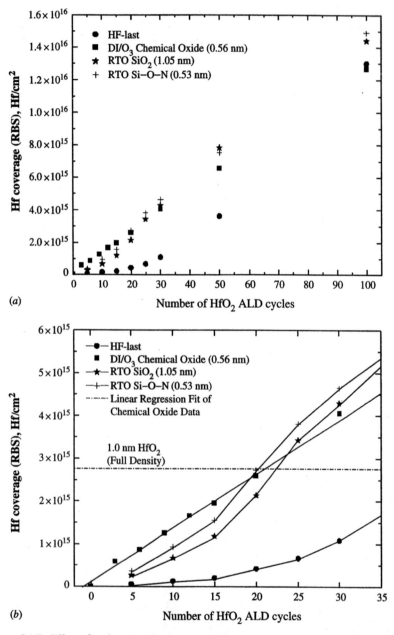

Figure 2.1.7. Effect of various starting layers on Hf coverage measured with RBS vs. the number of $HfCl_4$–H_2O ALD cycles. (*b*) shows a subset of data of (*a*). (Reprinted from Green M L, Ho M -Y, Busch B, Wilk G D, Sorsch T, Vonard T, Brijis B, Vandervorst W, Räisänen P I, Muller D, Bude M and Grazul J 2002 *J. Appl. Phys.* **92** 7168. Copyright 2002 American Institute of Physics.)

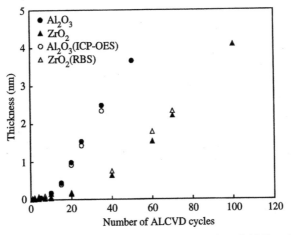

Figure 2.1.8. Thickness of Al_2O_3 and ZrO_2 films as a function of ALD cycles applied on hydrogen terminated silicon. (Reprinted from Nohira H *et al* 2002 Characterization of ALCVD-Al_2O_3 and ZrO_2 layer using x-ray photoelectron spectroscopy *J. Non-Cryst. Solids* **303** 83–87. Copyright 2002, with permission from Elsevier Science.)

Therefore, one quite common approach has been to start the gate oxide deposition with a thin layer of Al_2O_3 followed by another (most often ZrO_2 and HfO_2) high-κ oxide having a permittivity higher than that of Al_2O_3 (see 'Mixed oxides and nanolaminates' section). Similarly, an ultrathin (~0.5 nm) silicon nitride layer deposited by ALD was recently found to promote uniform nucleation in the $Zr(O^tBu)_4$–H_2O process while at the same time also suppressing the oxidation of silicon [34].

The third critical issue in the high-κ gate oxide deposition is how oxidative the ALD process is towards silicon. The different oxygen sources used in ALD oxide processes can be arranged in the following order

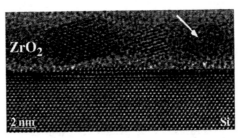

Figure 2.1.9. Cross-sectional TEM image of a ZrO_2 film deposited from $ZrCl_4$ and H_2O on hydrogen terminated silicon. The arrow points to an amorphous region separating the ZrO_2 crystallites. (Reprinted from Copel M, Gribelyuk M and Gusev E 2000 *Appl. Phys. Lett.* **76** 436. Copyright 2000 American Institute of Physics.)

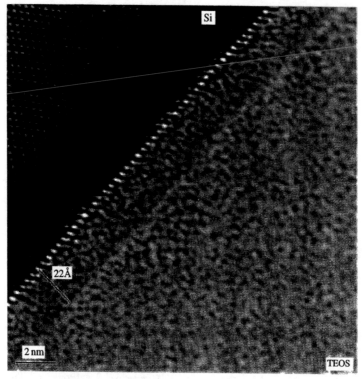

Figure 2.1.10. Cross-sectional TEM image of an Al_2O_3 film deposited from $Al(CH_3)_3$ and H_2O on hydrogen terminated silicon. (Reprinted from Gusev E P, Copel M, Cartier E, Baumvol I J R, Krug C and Gribelyuk M A 2000 *Appl. Phys. Lett.* **76** 176. Copyright 2000 American Institute of Physics.)

corresponding to their expected oxidation power: oxygen radicals > ozone > hydrogen peroxide > water > alkoxides of metals with highly stable oxides. The last ones were chosen as new oxygen sources for ALD with just the MOSFET high-κ gate oxides in mind [35]. The central idea is that in the alkoxides of those metals that form more stable oxides than silicon, the oxygen is so tightly bound to the metal that it cannot oxidize silicon. This was indeed the case at least in the $AlCl_3$–$Al(O^iPr)_3$ process (figure 2.1.11). Besides serving as oxygen sources, the metal alkoxides also deposit their metals into the films.

Figures 2.1.9–2.1.13 compare silicon–high-κ oxide interfaces obtained by various ALD processes on hydrogen terminated silicon. The effect of water is found to depend on the metal precursor: the $Al(CH_3)_3$–H_2O process produces an abrupt Si–Al_2O_3 interface [33] (figure 2.1.10) whereas in the $ZrCl_4$–H_2O (figure 2.1.9) [30, 32, 36] and $HfCl_4$–H_2O [37] processes an interface layer is formed. This leads to a speculation that in the ZrO_2 and

Figure 2.1.11. Cross-sectional TEM image of an Al$_2$O$_3$ film deposited on hydrogen terminated silicon from AlCl$_3$ and Al(O^{i}Pr)$_3$. (Reprinted from Ritala M, Kukli K, Rahtu A, Räisänen P I, Leskelä M, Sajavaara T and Keinonen J 2000 Atomic layer deposition of oxide thin films with metal alkoxides as oxygen sources *Science* **288** 319. Copyright 2000 American Association for the Advancement of Science.)

HfO$_2$ processes the poor nucleation enhances silicon oxidation. Indeed, XPS studies indicated that after ten ZrCl$_4$–H$_2$O cycles a thin interfacial oxide had formed but no significant ZrO$_2$ growth was observable on hydrogen terminated silicon [30]. This suggests that ZrO$_2$ deposition can start only after the surface has become oxidized and/or hydroxylated. Consequently, nonuniform oxidation of silicon leads to nonuniform nucleation of ZrO$_2$ which eventually leads to nonuniform film morphology.

The Y(thd)$_3$–O$_3$ process (thd = 2,2,6,6-tetramethyl-3,5-heptanedione) forms an interface layer thicker than 1 nm (figure 2.1.12) [38, 39]. Though not yet clearly verified, this may be assumed characteristic to ozone, thereby ruling out ozone-based processes as candidates for high-κ gate oxides. Even thicker interfacial layers of 2–3 nm were found from the ZrO$_2$ films deposited by oxygen plasma enhanced ALD (figure 2.1.13) [40]. Nevertheless,

Figure 2.1.12. Cross-sectional TEM image of a Y_2O_3 film deposited from $Y(thd)_3$ and O_3 on hydrogen terminated silicon. (Reproduced by permission of The Electrochemical Society, Inc. from Gusev E P *et al* 2001 *Electrochem. Soc. Proc.* **2001-9** 189.)

remarkably low EOT of 1.4 nm was measured for a film with a physical thickness of 6.5 nm, thereby implying that the interface layer is not SiO_2 but a mixture of zirconium and silicon oxides.

At this point it must be emphasized that the examination of the interfaces formed in the high-κ oxide deposition processes is complicated because the interface layers may also grow in thickness afterwards when the samples are stored in air. For example, while absent right after the

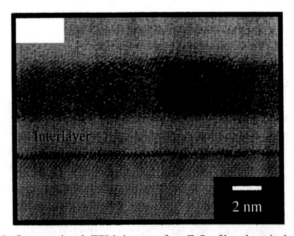

Figure 2.1.13. Cross-sectional TEM image of a ZrO_2 film deposited on hydrogen terminated silicon from $Zr(NEt_2)_4$ with oxygen plasma enhanced ALD. (Reprinted from Kim Y, Koo J, Han J, Choi S, Jeon H and Park C -G 2002 *J. Appl. Phys.* **92** 5443. Copyright 2002 American Institute of Physics.)

Al(CH$_3$)$_3$–H$_2$O process, a thin ($\sim$0.15 nm) interface layer was found after extended room temperature air exposure of thin ($<$4 nm) Al$_2$O$_3$ films [30]. This is not unique to Al$_2$O$_3$ or ALD but also other high-κ oxides are permeable to oxygen and water when thin enough. Therefore, to be able to draw definite conclusions about the interface layers, the samples should be capped *in situ*, either by appropriate oxygen diffusion preventing materials or just by making the films themselves thick enough.

ALD of high-κ gate oxide materials

Common remarks on ALD oxide processes

The ALD processes of the highest interest for the high-κ gate oxides employ water as an oxygen source. In these processes surface hydroxyl groups play an important role as reactive adsorption sites for the metal precursors as shown in figure 2.1.1. Hydroxyl groups that are repeatedly formed during each water pulse are not necessarily all stable, however, but they can combine with each other forming an oxide ion and a water molecule that may desorb from the surface. This process that leads to a loss of reactive hydroxyl groups from the surface is called dehydroxylation. Due to the heterogeneity of surface hydroxyl groups, they are lost over a range of temperatures. As a result, the density of hydroxyl groups remaining on the surface decreases quite smoothly with increasing temperature, and this is quite often reflected in a similarly decreasing oxide film deposition rate. On the other hand, by using large water doses the dehydroxylation may be compensated for to some extent with a consequent increase in the deposition rate [41, 42]. In this case hydroxyl groups are formed not only in reactions between water and the metal precursor but also by dissociative adsorption of water on coordinatively unsaturated surface sites, i.e. on bare metal and oxide ions on the surface.

Reaction mechanism studies have verified the mechanism presented in figure 2.1.1: at all the temperatures examined up to 500°C at least one chlorine atom was released, on average, per adsorbing zirconium atom during the ZrCl$_4$ pulse [43]. In line with the expected decrease of the hydroxyl group coverage, the number of chlorine atoms released during the ZrCl$_4$ pulse decreased from 2 to 1 with increasing temperature. On the other hand, in a similar TiCl$_4$–H$_2$O process the number of chlorine atoms released per adsorbing titanium atom decreased below 1 already when the temperature exceeded 250°C, thereby implying that part of the TiCl$_4$ molecules were adsorbing without undergoing exchange reactions with hydroxyl groups, i.e. either molecularly or dissociatively [44]. While the effect of HCl readsorption cannot fully be excluded, this observation implies that also in the absence of hydroxyl groups the metal precursors could adsorb on the oxide surface firmly enough to survive the following purge period.

In most of the ALD oxide processes the deposition rates have been in a range of $0.3–1.0\,\text{Å/cycle}$. This means that most often only a fraction of a monolayer is deposited in each cycle. Steric hindrance between the adsorbed metal species and low density of reactive $-OH$ sites are usually considered the limiting factors for the deposition rate. It is quite hard to distinguish which one of these is dominating because fewer hydroxyl groups mean also that more ligands are left bound to the metal precursor after its adsorption, thereby causing more steric hindrance. Anyhow, the outcome is the same: the number of metal atoms incorporated into the metal precursor adsorption layer is too low to form a full atomic layer of the oxide after the water pulse.

Since ALD employs chemical compounds as precursors, there is always a possibility of film contamination by residues from these compounds. In the most often examined ALD high-κ oxide films the residual contents are typically in a range of a few tenths of atomic per cent, but films deposited outside the optimized process conditions may contain rather more impurities. The most common reason for the films containing residues is that they are deposited at too low temperatures where the reactions remain incomplete; especially hydroxyl groups are left in increasing amounts when the deposition temperature is lowered. With organometallic precursors contamination may also be found at high temperatures where these compounds start to decompose. On the other hand, oxidation states of the metals in the high-κ oxide films always seem to be correct, i.e., there are no lower oxidation states present. Metal–silicon bonding has not been observed either.

Density of the ALD oxides, in comparison to the bulk material, depends in the first place on the structure of the films. Polycrystalline materials are reported to have quite high densities, exceeding 90% of the bulk values. Amorphous films, by contrast, may have much lower densities as is understandable from their disordered structure. For example, Al_2O_3 films have a density of about $3.0\,\text{g cm}^{-3}$ even after annealing at 800°C while the bulk value is $3.97\,\text{g cm}^{-3}$ [30, 45]. There are no reports on epitaxial growth of oxides on silicon by ALD.

Al_2O_3

$Al(CH_3)_3–H_2O$ process

Basically all the ALD Al_2O_3 studies on gate oxides have used the $Al(CH_3)_3–H_2O$ process. This precursor combination is often considered as the most ideal ALD process developed so far: both precursors are volatile liquids and thus easy to handle, they show high reactivity towards each other, they are relatively inexpensive and have high enough thermal stability to allow deposition temperatures of 400°C and even higher, and the reaction by-product methane is unreactive. The only major concern is the pyrophoric nature of trimethyl aluminium.

The $Al(CH_3)_3$–H_2O process was demonstrated for the first time by Higashi and Flemming [46] and already that study proved the high purity, conformality, and good electrical characteristics of the films. The breakdown field strength was as high as $8\,MV\,cm^{-1}$ and the interface trap density was relatively low, about $10^{11}\,eV^{-1}\,cm^{-2}$ at midgap. On the other hand, the presence of negative charge with a density of $6 \times 10^{11}\,cm^{-2}$ and relatively low permittivity of 7 were also reported. Later, it was shown that higher permittivities of 8–9 [47], and even 11 after annealing at 1000°C [39, 48], are attainable, but the problem of negative charge seems to remain.

The $Al(CH_3)_3$–H_2O process has been thoroughly examined. The most typically reported deposition rate is 1.1 Å/cycle [41, 46, 49] but the rate is temperature dependent decreasing with increasing temperature, apparently because of the increasing dehydroxylation of the surface, and thus the rate of 1.1 Å/cycle appears to be an upper limit [41, 42, 49–51]. Various reaction mechanism studies [42, 49, 51] agree that the film growth proceeds through alternate surface reactions where $Al(CH_3)_3$ adsorbs by undergoing exchange reactions with the surface hydroxyl groups, and the resulting $-Al(CH_3)_{3-x}$ surface species serve as reactive sites for water in the next step of the process which converts the surface back to hydroxyl group terminated. Water can also adsorb on coordinatively unsaturated surface sites producing more hydroxyl groups. The retarded start of growth on hydrogen terminated silicon (figure 2.1.8) has already been commented on.

Usually no carbon has been observed in the films with techniques having a detection limit of 0.1 at.% or above. A hydrogen content of 1 at.% was measured for a film deposited at 250°C, independently of the water dose applied [41], but as common to ALD oxides, the hydrogen contents increase towards low temperatures. For example, about 5 at.% hydrogen was found from the films grown at 200°C [22]. In general, an increase in deposition temperature increases both permittivity and refractive index [47, 50], apparently because of the lowered hydrogen content.

The ALD deposited Al_2O_3 films are amorphous and smooth, and they form an abrupt and interface layer free contact to silicon (figure 2.1.10) [33]. The amorphous state may also be retained after high temperature anneal at 900°C, especially if capped by the gate electrode prior to annealing [52]. Without capping the films may crystallize more easily [52, 53]. Overall, the crystallization is affected by several factors: annealing temperature, annealing time, thickness of the film, silicon surface pretreatment, and presence or absence of capping layers. As estimated from refractive index after 30 s anneal, a structural transformation of 50 nm films occurs between 800 and 900°C [39]. On the other hand, when annealed for longer times (10 min), the films started to transform into γ-Al_2O_3 already at 800°C with a concomitant reduction in thickness and widening of the band gap [54]. While partial crystallization was found to deteriorate leakage properties [54], more completed crystallization and the associated densification decreased leakage currents by several orders of magnitude [53, 54]. Finally, it has been noted that while the Al_2O_3 films deposited onto hydrogen terminated silicon show crystallization onset at around 800°C, those deposited onto oxide surface remain amorphous even after 1100°C spike annealing [55].

Electron spin resonance studies have indicated that the Al_2O_3–Si interface contains dangling bond type defects similar to the SiO_2–Si interface [56]. Thus, the Al_2O_3–Si was concluded to be basically of the SiO_2–Si type, though under enhanced stress. The properties of the standard thermal SiO_2–Si interface were approached after annealing (≥ 650°C) under oxygen but not in vacuum, thereby implying that the initial abruptness of the interface prevents thermal adaptation from occurring unless an additional SiO_x interlayer is grown. In the presence of oxygen such an interface layer is readily formed, however. When 2.5 nm Al_2O_3 films on silicon were annealed for 30 min at 600°C under variable oxygen pressures, an originally missing interfacial SiO_2 layer was formed and its thickness increased logarithmically with oxygen pressure to roughly 0.8 nm for atmospheric pressure oxidation [57]. Comparison with bare silicon samples showed that such thin Al_2O_3 layers have in essence no barrier effect against oxygen diffusion. And as already noted, even storage in air at room temperature is enough to cause a growth of about 0.15 nm of interface layer [30]. On the other hand, if the films are annealed in the absence of oxygen, they start to degrade at temperatures exceeding 900°C through desorption of SiO and some oxygen-deficient aluminium species [57]. However, as the following example implies, in the conventional CMOS process flow the annealing-induced degradation of the gate oxide appears to be effectively retarded by the overlying gate electrode. Anyhow, an accurate control of oxygen partial pressure during anneals is of utmost importance for controlling the Al_2O_3–Si interface.

Encouraging device characteristics were obtained when ALD-deposited Al_2O_3 was integrated into a sub 0.1 μm n-MOS process with polycrystalline silicon gates and 1000°C rapid thermal anneal [48]. The uniformity across a 200 mm wafer was excellent, and the leakage current reduction was more than 100-fold compared to SiO_2 in an EOT range of 1.0–1.5 nm. The reliability was estimated to exceed that of SiO_2-based devices. From a plot of EOT vs. physical thickness a permittivity as high as 11 and an interface layer EOT of 0.75 nm were evaluated [39, 48]. Reduction in channel mobility and charging were identified as the main concerns of the Al_2O_3 gate dielectrics.

Promising MOS characteristics were reported also in another study, particularly after dielectric improvement anneals in O_2 or ultraviolet O_3 ambient [58]. With n^+-poly-Si/Al_2O_3/p-Si MOS capacitors, interface densities in the range of 8×10^{10} $eV^{-1} cm^{-2}$ near the Si midgap were measured, while frequency dispersion was as small as 20 mV (1 kHz–1 MHz) and hysteresis was 15 mV under an electric field of $8\,MV\,cm^{-1}$. At a gate voltage of $-2.5\,V$, Al_2O_3 with an EOT of 3.6 nm had a leakage current of about $5\,nA\,cm^{-2}$ which is about three orders of magnitude lower than that of a thermal SiO_2 with a similar EOT. Also the breakdown and reliability characteristics were excellent. Again a quite high negative fixed charge density of $2 \times 10^{12}\,cm^{-2}$ with an associated shift in the flatband voltage and a control of interfacial oxide were identified as the main challenges for the future development.

One concern related to Al_2O_3, and to other high-κ oxides too, is severe boron diffusion from the p^+-polysilicon gate through the gate oxide into the silicon substrate during high temperature anneals with the consequence of a large flatband voltage shift [52]. This may be suppressed, however, either by a thin ($\sim 0.5\,nm$) silicon oxynitride layer between Al_2O_3 and Si [52] or by remote plasma nitridation of Al_2O_3 [59], both being effective up to 850°C. As the effectiveness of the latter procedure was attributed to the formation of AlN, it would be interesting to examine if a thin AlN capping layer deposited by ALD [60–66] could also suppress boron diffusion.

The $Al(CH_3)_3$–H_2O process has been successfully applied also in the vertical replacement gate MOSFET devices, the geometry of which sets very high demands on the conformality of the gate oxide film deposition process (cf figure 2.1.3) [9]. Using an upper limit of $1\,A\,cm^{-2}$ for the gate leakage current density, Al_2O_3 was concluded to be scalable down to EOT $\leq 1.3\,nm$.

Alternative Al_2O_3 processes

While the $Al(CH_3)_3$–H_2O process has dominated in the gate oxide studies, there are several alternative chemistries for ALD of Al_2O_3. One of the first ALD processes ever developed was the $AlCl_3$–H_2O combination [5–8, 15, 67–72] which was originally developed for the TFEL display insulators [1]. The process was successful in this application and has been in industrial use for nearly 20 years. Thus, even if the solid nature of $AlCl_3$ calls for some extra effort in source design and daily operation, it does not form any major obstacle to the production capability. For the gate oxide research it would be especially interesting to know what kind of an interface the $AlCl_3$–H_2O process would form with silicon because that would add to our understanding on the origin of the difference between the $Al(CH_3)_3$–H_2O and the $ZrCl_4/HfCl_4$–H_2O processes, i.e. whether the absence of the interface layer in the $Al(CH_3)_3$–H_2O process is characteristic of the Al_2O_3 material itself or the $Al(CH_3)_3$ precursor.

Other aluminium precursors used together with water are $Al(CH_3)_2Cl$ [73], $Al(OC_2H_5)_3$ [68], and $Al(OCH_2CH_2CH_3)_3$ [68]. On the other hand, besides water, also alcohols [68, 74], N_2O [75], NO_2 [76], O_3 [77], and aluminium alkoxides [35, 78] have been used as oxygen precursors, the last ones serving also as the metal source and known to produce sharp interfaces with silicon (figure 2.1.11). Characteristic to most of the Al_2O_3 films deposited by ALD has been stoichiometric composition, high purity, except at very low temperatures of 200°C and below where especially hydrogen residues become substantial, permittivity of about 8–9, and high breakdown fields. Deposition rates have typically been in the range of 0.5–1.0 Å/cycle, decreasing somewhat with increasing temperature.

Also oxygen plasma-assisted processes have been reported [79] but these are of limited interest for gate oxides because of the obvious risk of oxidizing the silicon surface.

ZrO_2 and HfO_2

$ZrCl_4$–H_2O and $HfCl_4$–H_2O processes

From the ALD and precursor chemistry points of view ZrO_2 and HfO_2 are very similar and are thus dealt with together. The first ALD processes for these oxides were based on the corresponding tetrachlorides ($ZrCl_4$ and $HfCl_4$) and water [80, 81], and these processes have been the most extensively used also in the gate oxide research. Reaction mechanism studies using QMS and QCM have verified that the ZrO_2 growth proceeds as sketched in figure 2.1.1 [43, 82] and the same applies to the HfO_2 process too [83].

There are two main concerns related to the $ZrCl_4$- and $HfCl_4$-based processes. The first one is particle transportation from the sources to the films as these solids consist of very fine particles. However, with careful source design this problem can be solved. The second concern is the already discussed poor nucleation on hydrogen terminated silicon (figure 2.1.9). A common way around this problem is to use thin silicon oxide (chemical or thermal) or aluminium oxide made by ALD as a starting layer. Occasionally also chlorine residues left from the precursors are considered as a potential problem but their effects on the MOSFET characteristics remain still to be verified. Anyhow, it is known that the chlorine residues are preferentially located at the ZrO_2–Si interface when the film is deposited onto hydrogen terminated silicon [30] and at the ZrO_2–SiO_2 interface when deposited on the native oxide [84].

As common to the ALD oxide processes, the deposition rates in these ZrO_2 and HfO_2 ALD processes have been found to decrease with increasing temperature. Typically values of 0.5–1.0 Å/cycle have been reported, though occasionally also higher rates have been obtained with different reactors.

The impurity contents in the ZrO_2 and HfO_2 films are temperature dependent. In films deposited at 500°C no chlorine was detected with RBS, and there was less than 0.4 at.% hydrogen as measured with nuclear reaction analysis [80, 81]. However, at 300°C which seems to be the most often used deposition temperature in the gate oxide studies, about 0.6–0.8 at.% chlorine and 1.5 at.% hydrogen were left in the ZrO_2 films [82, 85], and 0.4 at.%

chlorine and 1 at.% hydrogen in the HfO_2 films [86]. HfO_2 films deposited at 225°C contained 4 at.% chlorine residues [83] while there were 4–5 and 5–6 at.% chlorine and hydrogen, respectively, in the ZrO_2 films deposited at 180°C [85].

Post-deposition annealing has been found to decrease the chlorine residue contents in the ZrO_2 and HfO_2 films (figures 2.1.14 and 2.1.15) [84]. Annealing of ZrO_2 films at 1000°C for 60 s decreases the chlorine level by about an order of magnitude, and at 1050°C chlorine disappears almost completely (figure 2.1.14(A)). Chlorine residing at the ZrO_2–SiO_2 interface is quite efficiently eliminated already at 900°C. By contrast, in the HfO_2 films chlorine is remarkably more stable: the bulk content is only slightly affected even after annealing at 1050°C, and while the interfacial chlorine can be quite efficiently removed (figure 2.1.15(A)), this requires higher temperatures than with ZrO_2. With both oxides, the annealing time does not have any significant effect (figures 2.1.14(B) and 2.1.15(B)) as compared with the

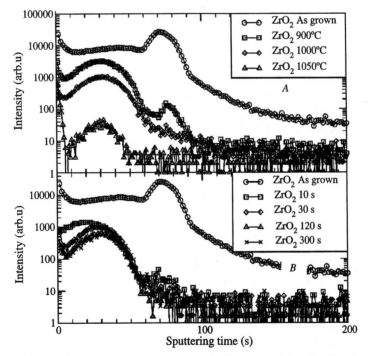

Figure 2.1.14. TOF-SIMS depth profiles of chlorine in 10 nm thick ZrO_2 films deposited on native oxide at 300°C and post-deposition annealed in nitrogen at different temperatures for 60 s (A), and at 1000°C for different times (B). (Reprinted from Ferrari S, Scarel G, Wiernier C and Fanciulli M 2002 Chlorine mobility during annealing in N_2 in ZrO_2 and HfO_2 films grown by atomic layer deposition *J. Appl. Phys.* **92** 7675. Copyright 2002 American Institute of Physics.)

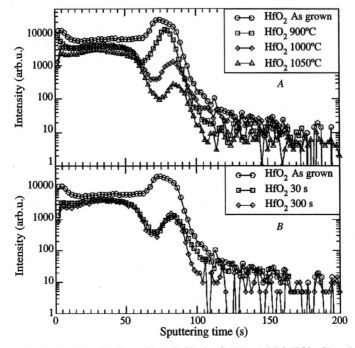

Figure 2.1.15. TOF-SIMS depth profiles of chlorine in 10 nm thick HfO$_2$ films deposited on native oxide at 300°C and post-deposition annealed in nitrogen at different temperatures for 60 s (*A*), and at 1000°C for different times (*B*). (Reprinted from Ferrari S, Scarel G, Wiernier C and Fanciulli M 2002 Chlorine mobility during annealing in N$_2$ in ZrO$_2$ and HfO$_2$ films grown by atomic layer deposition *J. Appl. Phys.* **92** 7675. Copyright 2002 American Institute of Physics.)

annealing temperature. It was also shown that chlorine removal from the interface is well correlated with the growth of the interfacial silicon oxide. It must be noted that these films were annealed uncapped and any overlayer, like a polysilicon gate, is likely to retard the chlorine removal.

XRD studies have revealed that the crystal structure of ZrO$_2$ films depends on deposition temperature and film thickness. In thicker films ($\geq$100 nm) monoclinic and/or tetragonal phases were detected when deposited at 500°C [80] but at about 200°C the films were cubic [82]. Thinner (20–40 nm) films deposited at 300°C contained tetragonal ZrO$_2$ as the dominant phase whereas those deposited at 210°C were amorphous [85, 87]. TEM and electron diffraction investigations have indicated that already a few nanometres thick ZrO$_2$ films deposited at 300°C are crystalline with a cubic structure [36], but also an orthorhombic phase has been reported [88]. Overall, the trend appears quite clear; the film structure develops from an amorphous phase through the metastable tetragonal and/or cubic phases to the stable monoclinic phase as the deposition temperature and/or film thickness increase. In other words, the higher the temperature, the lower the thicknesses where the first crystallites are observed and where the monoclinic phase starts to dominate. The same applies also to the crystallization of the originally amorphous films by post-deposition annealing: the thinner the film, the higher the temperature needed for its crystallization, e.g., 7.0 nm film crystallizes at 400°C but 2.7 nm film requires 500°C [89]. On the other hand, when laminated as 1 nm thick layers confined by amorphous Al_2O_3 layers, ZrO_2 stays amorphous above 900°C [30].

HfO_2 films are predominantly monoclinic but tetragonal phase is often present too [24, 81, 86]. Again the thinnest films (~5 nm and below) may be amorphous as deposited but they crystallize during annealing. For example, a few nanometres thick HfO_2 films deposited on chemical silicon oxide at 300°C were amorphous as examined with TEM [26]. Anyhow, they still had a density of 90% of the crystalline HfO_2.

Thermal stability is an important property for integration of the high-κ oxides into the overall CMOS process. Annealing of a ZrO_2/SiO_2 stack in oxygen at 930°C did not cause intermixing of the oxide layers but significant oxygen indiffusion occurred increasing the SiO_2 layer thickness [32]. In vacuum an uncapped ZrO_2/SiO_2 stack was stable at 900°C but at 1000°C it decomposed into zirconium silicides and SiO which sublimed [32, 88]. Capping with a CVD poly-Si gate did not make the structure survive 1000°C annealing but now the reactions started at the ZrO_2–poly-Si interface [88]. On the other hand, it was observed that while poly-Si deposited by CVD reacted with ZrO_2 at temperatures above 750°C, sputtered poly-Si was stable even at 1050°C [90]. This difference was related to the oxygen vacancies created in ZrO_2 during the poly-Si CVD. Indeed, it was later shown that by using an appropriately modified poly-Si CVD process the detrimental reactions between ZrO_2 and CVD-poly-Si may be avoided [91]. These results shed new light on the common argument that ZrO_2 could not be used together with poly-Si gates.

Quite wide ranges of permittivities of 15–23 and 12–22 have been reported for the ALD deposited ZrO_2 and HfO_2, respectively [1, 24, 38, 39, 85–87, 92]. Permittivities measured from thicker films, which thereby perhaps better represent the film bulk properties, have typically been about 20 for ZrO_2 and 16 for HfO_2. When deposited on thin SiO_2 interface layers both ZrO_2 and HfO_2 have provided four to five orders of magnitude reduction in leakage current as compared with SiO_2 with a similar EOT of 10–20 Å (see, e.g. [39]). However, when deposited on hydrogen terminated silicon, poor dielectric characteristics have been obtained because of the poor nucleation (figure 2.1.9). Thus, there remains a challenge how to pretreat the interface to ensure uniform nucleation, EOT below 1.0 nm and low leakage.

Whether the interface layer between ZrO_2 and silicon is silicon nitride or oxide has a quite remarkable effect on the electrical properties of the gate stack. With a 1.3 nm nitride interface layer 5.0 nm ZrO_2 films resulted in an

EOT of 1.4 nm, approximately 10^{-5} A cm^{-2} leakage at 1 V from flatband, and interface density of about 1×10^{12} cm^{-2} eV^{-1} [36]. With a chemical oxide interface significantly lower interface trap density of 3×10^{11} cm^{-2} eV^{-1} was achieved but the EOT was somewhat higher because of lower permittivity and larger physical thickness of the interface layer. The amorphous layer left in between ZrO$_2$ and silicon was found to have a moderate permittivity of 6–7, thereby suggesting intermixing of ZrO$_2$ and the chemical oxide. No frequency dispersion was observed in any of these films. Hysteresis of the C–V curves was voltage dependent, and also the surface pretreatment had a marked effect: for ± 2 V sweeps the hysteresis was 8–10 mV for the chemical oxide but 130 mV for the nitrided surface. This was related to elevated trapping caused by the large nitrogen content close to the dielectric–silicon interface. In another study, a 7.4 nm thick ZrO$_2$ film deposited onto 1.0–1.5 nm chemical oxide resulted in an EOT of 2.8 nm [92]. From the position of the flatband voltage the effective density of fixed negative charge was estimated to be 5×10^{12} cm^{-2} while the hysteresis of the flatband voltage gave 1.5×10^{12} cm^{-2} as the density of trapped electrons.

Studies on TiN/ZrO$_2$ (5 nm)/SiO$_2$ (0.7 nm)/p-Si MOS capacitors (EOT $\sim$ 3.0 nm) where ZrO$_2$ was deposited on a thermal oxide revealed that the interface trap density may depend also on the method used for the gate electrode deposition [93]. When TiN was deposited by ALD from TiCl$_4$ and NH$_3$ at 450°C, the interface trap density was 1.8×10^{11} cm^{-2} eV^{-1} whereas sputtered TiN resulted in a three times higher trap density of 4.7×10^{11} cm^{-2} eV^{-1}. Clearly, a damage-free process should be used for the gate metal deposition.

Comparison of ZrO$_2$ and HfO$_2$ deposited on a chemical oxide as gate dielectrics in conventional self-aligned MOSFETs with polysilicon gate electrodes showed HfO$_2$ to give better results than ZrO$_2$ [94]. The reasons for the difference remained somewhat unsolved but appeared to be related more to the integration issues rather than the quality of the dielectric film itself. After 10 s annealing at 1000°C the transistors with HfO$_2$ gate dielectrics showed promising characteristics, though with about 15% channel mobility degradation as compared with SiO$_2$ control devices. The EOT was quite strongly dependent on the annealing procedure: EOT = 3.0 nm for 30 s at 850°C, and EOT = 1.7 nm for 10 s at 1000°C. The gate leakage increased accordingly but was still acceptable, e.g. 35 mA cm^{-2} for the 1000°C anneal. The threshold voltage was quite high referring to negative charge of about 2×10^{12} cm^{-2}. If the annealing was done at lower temperatures, 850 and 900°C, the density of negative charge was higher, about 5×10^{12} cm^{-2}.

A major concern related to the high-κ gate oxide integration is high charge concentration in the films causing largely shifted and unstable threshold voltages as well as carrier scattering and thereby reduced mobility.

Addressing this issue Wilk *et al* [95] showed that when HfO_2 is deposited on a 0.5 nm chemical oxide and annealed at 900°C for 30 s under nitrogen, negative fixed charge densities below $2 \times 10^{11}\,cm^{-2}$ can be reached, representing a tenfold decrease from the generally reported values. Such an annealing increased the EOT by about 0.3 nm, however, thereby suggesting that a small increment in the interface oxide may be necessary to minimize fixed charge.

Very encouraging results were recently reported on employing the mobility enhancement in strained silicon to compensate the channel mobility degradation caused by the high-κ gate oxide [96]. This was demonstrated with a gate stack (EOT = 2.8 nm) consisting of HfO_2 deposited by ALD on thin (<1 nm) interfacial silicon oxynitride layer. The mobility measured for the strained silicon NMOSFET was 60% higher than in an unstrained silicon device with a similar HfO_2/SiO_xN_y gate dielectric, and 30% higher than in the conventional NMOSFET with unstrained silicon and SiO_2 gate oxide. Of course a device with strained silicon and SiO_2 resulted in the highest mobility, but with 1000 times higher gate leakage than the device with the HfO_2/SiO_xN_y gate dielectric.

After gate metal deposition the MOSFET devices are commonly annealed in forming gas to decrease the interface trap density by passivating dangling bonds at the interface. However, it has recently been recognized that annealing in D_2 rather than H_2 improves the reliability characteristics of the devices. This was shown to be the case also with the Pt/HfO_2 (3.0 nm)/SiO_2 (1.0 nm)/Si MOS capacitors where the HfO_2 layer was deposited by ALD on thermal SiO_2 [97]. Annealing at 450°C for 30 min in D_2 instead of H_2 resulted in a lower degree of charge trapping, less generation of interface states, larger charge-to-breakdown and longer time-dependent dielectric breakdown characteristics under electrical stress. The initial C–V and I–V characteristics were not affected. SIMS depth profiles revealed that the major difference between the films annealed in D_2 and H_2 was the incorporation of deuterium at the SiO_2/Si interface while in the rest of the dielectric stack no differences were observable. Therefore, the improvement in the reliability characteristics was ascribed to the heavy mass effect of deuterium on the Si–D bond strength at the SiO_2/Si interface, and consequently this improvement should be applicable to other high-κ oxides as well.

The HfO_2 process has been integrated also to the VRG-MOSFET geometry where it showed excellent conformality (figure 2.1.3) and very good intra-wafer uniformity across a 200 mm wafer [9]. The films showed low gate leakage current and it was estimated that the material should be scalable below EOT = 1.0 nm. On the other hand, the negative charge density estimated from the threshold voltage was quite high, $\geq 3 \times 10^{12}\,cm^{-2}$. Nevertheless, high drive currents with good short channel performance were achieved.

Alternative ZrO$_2$ and HfO$_2$ processes

The concerns related to ZrCl$_4$ and HfCl$_4$ being fine particle size solids with a modest volatility, poor nucleation on hydrogen terminated silicon and chlorine residues left in the films have motivated studies on alternative zirconium and hafnium precursors. Among these ZrI$_4$ and HfI$_4$ are otherwise quite similar to the chlorides but the weaker metal–iodine bond strength and larger size of iodide as compared to chloride may lead to lower amounts of halide residues: in the ZrO$_2$ films deposited at 375°C and above iodine content was below the detection limit of XPS (0.1–0.2 at.%) but at 250–350°C about 1 at.% iodine was left in the films [98–101]. The oxygen source in these studies was aqueous H$_2$O$_2$ but this hardly has any major effect as compared with using just water. The iodine contents in HfO$_2$ films deposited from HfI$_4$ and H$_2$O were 0.9 and 0.4 at.% at 225 and 300°C, respectively [86, 102]. Thus, the differences in the halide contents compared to the chloride precursors are quite minimal at 300°C and above 500°C but at the lowest temperatures and around 400°C the differences increase somewhat. Furthermore, high temperature annealing seems to remove iodine more efficiently than chlorine [103].

Apparently because of the weaker bonding between the metals and iodine, saturation of the surface reactions is not fully complete even at 300°C [98, 102]. It is thought that desorption of iodine from the surface opens new adsorption sites and thereby causes a slow increase in deposition rate with increasing metal iodide pulse time. Anyhow, this process appears to be slow enough so that the uniformity of the films deposited in small research reactors, at least, is not deteriorated as compared with the chloride processes. Since also the iodides are fine particle size solids, the particle problem still remains. A formation of a 2 nm thick interface layer has also been found [100, 101]. Further, though the dielectric constant of ZrO$_2$ grown from ZrI$_4$ was above 20 when deposited at 250–325°C, it exhibited quite a substantial decrease at higher deposition temperatures of 400°C and above [99]. In the case of HfO$_2$, the major benefits of HfI$_4$ over HfCl$_4$ in respect of the electrical properties are the lower oxide charge trap density and higher breakdown field [86] but it remains still to be explored if the differences survive high temperature annealing.

Unlike most other metal precursors used in ALD of oxides, iodides react also with molecular oxygen at reasonably low temperatures, thereby offering a hydrogen-free alternative to the oxide deposition. HfO$_2$ films have been deposited with a HfI$_4$–O$_2$ process at 570–755°C [104]. The iodine content of these films was below the detection limit of TOF-ERDA, about 0.1 at.%. However, also this process resulted in a 1.5–2.0 nm interface layer.

Alkoxides of zirconium and hafnium show unfortunately poorer thermal stability than those of titanium, the third metal in the same group in the periodic table. Of the basic alkoxides, only zirconium *tert*-butoxide, Zr(O^{t}Bu)$_4$, has been studied in more detail but it was found to decompose quite substantially and the films were of a lower quality than those obtained from the halides [105]. On the other hand, Chang *et al* [106, 107] managed to deposit highly uniform films with an alternate supply of $Zr(O^tBu)_4$ and O_2 above 300°C, but this process was shown to be based on a decomposition of $Zr(O^tBu)_4$. Thus, despite the good cycle-to-cycle repeatability, the process was not self-limiting ALD but well-controlled pulsed CVD and lower temperatures must be used to achieve the self-limiting growth. Indeed, at 200°C a self-limiting growth with a rate of 2 Å/cycle was demonstrated [34]. Even at this low temperature, about 1.2 nm interfacial layer was formed on hydrogen terminated silicon. On the other hand, an about 0.5 nm silicon nitride layer deposited by ALD successfully suppressed the oxidation. After 3 min annealing at 850°C an EOT of 1.6 nm was measured for a 4.2 nm ZrO_2–0.5 nm silicon nitride stack. The interface trap density was estimated to be about $5 \times 10^{12}\,cm^{-2}\,eV^{-1}$ and the positive fixed charge density was $5 \times 10^{12}\,cm^{-2}$.

Replacement of two *tert*-butoxide ligands in $Zr(O^tBu)_4$ by bidentate donor-functionalized dimethylaminoethoxide (dmae, $-OCH_2CH_2NMe_2$) ligands gives coordinatively better saturated $Zr(O^tBu)_2(dmae)_2$ with improved thermal stability, but still the ALD reactions were not entirely saturative [108]. Nevertheless, with short exposure times and low deposition temperatures below 300°C, reasonably well-controlled growth could be achieved. The films contained rather high hydrogen residue contents (>8 at.%), however, and also some carbon and nitrogen residues were left in the films.

Another donor-functionalized alkoxide examined is the 1-methoxy-2-methyl-2-propanolate complex of hafnium: $Hf(OCMe_2CH_2OMe)_4$ ($Hf(mmp)_4$) [109]. Below 300°C the growth rate was very low below 0.2 Å/cycle. At higher temperatures, in turn, $Hf(mmp)_4$ decomposition contributed to the film growth. Thus, it appears that when the alkoxide becomes functionalized well enough for enhanced stability through good shielding of the metal centre, the reactivity towards water becomes adversely affected and approaches that of β-diketonates. The films deposited at 300°C where the decomposition was still quite minimal contained 4–6 at.% carbon and as much as 14–18 at.% hydrogen but annealing at 850°C decreased them to about 2 at.% each. Accordingly, *C–V* curves of the as-deposited films showed substantial hysteresis but in the annealed films the hysteresis width was considerably decreased.

A metal alkylamide hafnium tetrakis(ethylmethylamide), $Hf(NEtMe)_4$, was recently found to be a quite promising precursor for HfO_2 [110]. The compound is volatile liquid (evaporation temperature 60°C). Reactions with water were reasonably well saturative at 250°C though separate experiments indicated weak decomposition. The carbon and nitrogen contents were below 0.5 at.% but the films contained about 2 at.% hydrogen. The permittivity was about 14 as measured from films thicker than 100 nm.

In another study, properties of six alkylamides ($M(NMe_2)_4$, $M(NEtMe)_4$, $M(NEt_2)_4$ with M = Zr or Hf) were compared as ALD oxide precursors, again with water as the oxygen source [111]. All these compounds are liquids under vaporization conditions and thus avoid problems related to solid precursors. The upper limit for the applicable deposition temperature was set by the metal precursor decomposition, increasing in steps of 50°C in the order of $M(NMe_2)_4 < M(NEtMe)_4 < M(NEt_2)_4$. The zirconium compounds were limited to 100°C lower temperatures than the hafnium compounds with the corresponding ligands. Below their decomposition temperatures all six precursors were reported to produce uniform and pure films with less than 1 at.% carbon and 0.25 at.% nitrogen.

$Zr(O^tBu)_4$ and $Zr(NEt_2)_4$ have also been used in oxygen plasma enhanced ALD processes [40]. Despite the relatively thick 2.0–3.0 nm interface layers seen with TEM (figure 2.1.13), EOT as low as 1.4 nm was measured for a 6.5 nm thick film deposited from $Zr(NEt_2)_4$. The leakage current of this film was also very low, $3 \times 10^{-9}\,\text{A cm}^{-2}$.

ZrO_2 films have also been deposited from $Zr(thd)_4$, Cp_2ZrCl_2 and $Cp_2Zr(CH_3)_2$ (Cp = cyclopentadienyl) with ozone as an oxygen source [112]. The bulky $Zr(thd)_4$ resulted in a modest deposition rate of 0.24 Å/cycle while the cyclopentadienyl compounds gave a higher rate of 0.55 Å/cycle. Carbon and hydrogen residue levels were below 0.5 at.% in the films deposited at 300–400°C. No results on issues specific to the gate oxide application were reported but it is very probable that also in these processes ozone oxidizes silicon too extensively. On the other hand, most recently, it was demonstrated that $Cp_2Zr(CH_3)_2$ may be used also with water, thereby offering an interesting halide-free alternative to ALD of ZrO_2 [113]. The hydrogen and carbon contents were also very low, below 0.1 at.%, in the films deposited at 350°C.

Finally, also hafnium nitrate $Hf(NO_3)_4$ has been used as a precursor in ALD of HfO_2 films [114]. The easy decomposition of this compound, and also other metal nitrates like $Zr(NO_3)_4$, has made them attractive precursors for metal oxide CVD, but this property makes it quite doubtful to use them in ALD. Indeed, although it was first checked that the decomposition begins at 180°C and therefore the ALD experiments were made at 160°C, the deposition rate was still noted to be highly sensitive to the condition of the deposition chamber, history of quartz ware and precursor purity. Nevertheless, EOT of 2.1 nm was reported for a 5.7 nm thick HfO_2 film.

Sc_2O_3, Y_2O_3 and lanthanide oxides

The early work on the ALD of oxides of scandium [115], yttrium [39, 116, 117], lanthanum [118–120] and cerium [121, 122] was based on processes employing metal β-diketonate complexes and ozone as precursors. Typically 3–6 at.% carbon residues are left in the films [39, 120]. But as figure 2.1.12

shows, ozone seems to be too oxidizing towards silicon causing a formation of an interface layer thicker than 1.0 nm [38, 39]. As a consequence, these processes are hardly usable for gate dielectrics with EOT far below 2.0 nm. Anyhow, in the EOT range of 2.0–2.5 nm the leakage current reduction is 4–5 orders of magnitude as compared with SiO_2 [39].

Quite recently, it was shown that Sc_2O_3 can be deposited from a scandium cyclopentadienyl $ScCp_3$ and water [115]. This process gave a higher deposition rate than the $Sc(thd)_3$ process (0.75 vs. 0.13 Å/cycle). On the basis of what is known from the other high-κ oxide ALD studies, the oxidation of silicon should be less important when using water compared to ozone, but the $ScCp_3$–H_2O and $Sc(thd)_3$–O_3 processes were not compared in this respect. Permittivities were not reported either. It has also been briefly reported that La_2O_3 may be deposited from lanthanum tris(bis(trimethylsilyl)amide) ($La[N(SiMe_3)_2]_3$) and water [123]. As volatile cyclopentadienyl and silylamide compounds are known for also other group 3 metals and lanthanides, these precursor groups may open new ways for ALD of high-κ oxides. It appears, however, that poor thermal stability of some of these compounds will limit their use in ALD.

$LaAlO_3$

$LaAlO_3$ is considered as one of the most potential ternary oxides for the gate dielectrics because of its stability in contact with silicon. On the other hand, as compared with binaries, ternary oxides are more challenging materials to be deposited because a strict control of the ratio of the two cations is required. $LaAlO_3$ films with less than 2 at.% carbon residues have been deposited by ALD from $La(thd)_3$, $Al(acac)_3$ and ozone [124]. The as-deposited films were amorphous but annealing at 900°C crystallized them. However, the above-considered limitations of ozone-based process obviously apply also in this case.

SiO_2 and Si_3N_4

SiO_2 is of course not a high-κ alternative to itself but is still important to consider here because mixing of SiO_2 with high-κ oxides is known to stabilize an amorphous structure and improve the interface characteristics. In addition, SiO_2 may also be employed as a very thin and controlled interface layer. The same applies to Si_3N_4 which is therefore also covered here.

ALD of SiO_2 has been examined using the following precursor combinations: $SiCl_4$–H_2O [125–129], $Si(NCO)_4$–H_2O [130], $Si(NCO)_4$–$N(C_2H_5)_3$ [131] and $CH_3OSi(NCO)_3$–H_2O_2 [132, 133]. These SiO_2 processes have been found very different from the metal oxide ALD processes: the SiO_2 processes require long, tens of seconds, exposure times to saturate the surface reactions, to be compared with subsecond exposures in metal oxide processes. As a result, the effective deposition rates are much lower for SiO_2. On the other hand, as only a monolayer or two of binary SiO_2 is needed at the interface, the low deposition rate is not that serious a concern. In addition, it has been observed that when mixed with some metal oxides, SiO_2 depositing reactions become more feasible than in the binary ALD SiO_2 processes.

One interesting observation on SiO_2 ALD is that the $SiCl_4$–H_2O process may be catalysed with ammonia and pyridine so that it proceeds at room temperature [127–129]. The catalytic effect decreases with increasing temperature above room temperature, however, thus making it difficult to combine this process with metal oxide processes that require higher temperatures.

The slow reactions in SiO_2 ALD seem to be characteristic to silicon as similar problems have been met in Si_3N_4 ALD processes too. Nevertheless, Si_3N_4 films with reasonably good properties have been obtained using combinations of $SiCl_4$–NH_3 [134–137], SiH_2Cl_2–NH_3 plasma [138, 139], SiH_2Cl_2–thermally cracked NH_3 [140] and Si_2Cl_6–N_2H_4 [141]. Nakajima *et al* [136] used thermal cycling where the $SiCl_4$ exposure (170 torr) was done at 340–375°C and the NH_3 exposure (300 torr) at 550°C but even in this process the exposure times were 300 and 90 s, respectively. Nevertheless, when a 2.2 nm thick film deposited with 20 cycles was post-deposition annealed for 90 min at 550°C, the permittivity increased from 5.7 to 7.2, giving an EOT = 1.2 nm [137]. The improvement in leakage current as compared with SiO_2 with a similar EOT was 10- to 100-fold at 1 V and below but the difference decreased towards higher voltages. The interface trap density measured for a film with somewhat higher EOT of 2.4 nm was found to be nearly the same as that of SiO_2 but the film contained positive fixed charge at a density of 7.5×10^{11} cm^{-2} [135]. An ultrathin (~ 0.5 nm) layer made with this silicon nitride ALD process was found to prevent silicon oxidation in a $Zr(O^tBu)_4$–H_2O process at 200°C [34]. It has also been shown that an about two monolayers (~ 0.4 nm) thick Si_3N_4 layer deposited on top of thermal SiO_2 is effective in suppressing boron penetration through the gate dielectric [136].

Other high-κ oxides

There are several oxide materials that possess attractively high permittivity but which are generally ruled out as the primary candidates for gate dielectrics because of their thermodynamically favourable reactions with silicon. Nevertheless, also these oxides are briefly surveyed here because there is a possibility that they could be mixed with the stable oxides. This is known to increase the permittivity from that possessed by the stable oxides whereas it remains to be explored how the stabilities with silicon would be affected in the solid solutions. In addition, a change of CMOS process flow towards a replacement gate approach could make these materials interesting also in themselves.

Ta$_2$O$_5$ and Nb$_2$O$_5$ can be deposited in a well-controlled manner by ALD from the corresponding metal ethoxides and water [142, 143]. For tantalum, TaCl$_5$ can be used as well, though only below the onset temperature of the etching reaction, about 300°C [13, 14, 144–146]. By contrast, with NbCl$_5$ the etching reaction prevents the deposition at all temperatures [12]. Ta$_2$O$_5$ has been deposited also from TaI$_5$ and H$_2$O$_2$ and this process leaves no iodine residues detectable with XPS [147]. Also in this case etching of the film by the tantalum precursor was observed, starting at 350°C. Other approaches to ALD of Ta$_2$O$_5$ have been the use of oxygen plasma for oxidizing Ta(OEt)$_5$ [148] and reactions between TaCl$_5$ and Ta(OEt)$_5$ [149]. In general, as-deposited Ta$_2$O$_5$ films are amorphous but from the halides also crystalline films have been obtained, though only in a temperature range where the etching reaction starts to play a role. The structural outcome of Nb$_2$O$_5$ is dependent on the underlying material: on glass and amorphous Ta$_2$O$_5$ film Nb$_2$O$_5$ is amorphous but on polycrystalline indium–tin oxide it is polycrystalline.

Typical permittivities of 21–25 have been measured for Ta$_2$O$_5$. When deposited from Ta(OEt)$_5$ and H$_2$O on hydrogen terminated silicon, EOT of 1.2 nm was achieved with a film having a physical thickness of 3.0 nm [150]. An EOT vs. physical thickness plot gave an estimation of 0.8 nm for the interface layer. On the other hand, the films deposited on Si with the plasma enhanced ALD process had permittivities as high as 40 which is a very high value for amorphous Ta$_2$O$_5$ [148]. The film itself as well as the Ta$_x$Si$_y$O$_z$ interface layer was thought to be densified by the oxygen plasma, thereby causing the high permittivities. The EOT (1.7 nm) of the Ta$_2$O$_5$ film–interface layer stack was almost equal to the physical thickness of the interface layer (1.8 nm). Nb$_2$O$_5$, in turn, is very leaky, complicating permittivity determination, but values of 50 have been estimated [151]. By mixing with other oxides the leakage current can be significantly decreased, but this decreases the permittivity too.

TiO$_2$ films have been deposited by ALD from TiCl$_4$ [152–157], TiI$_4$ [158–160], Ti(OMe)$_4$ [161], Ti(OEt)$_4$ [162, 163] and Ti(O^{i}Pr)$_4$ [164, 165] as titanium precursors and water as the oxygen source, though occasionally also hydrogen peroxide has been used. In addition, with TiI$_4$ molecular oxygen may be used too [159]. The resulting TiO$_2$ films are quite poor insulators because, as characteristic to this material, they are oxygen deficient and the oxygen vacancies serve as donors. In fact, some of these films are conductive enough to be measured with a four-point probe, giving resistivities in the range of 1 Ω cm. This conductivity has prevented a reliable determination of the permittivity. However, when mixed with Al$_2$O$_3$, TiO$_2$ forms good insulators which are used in the commercial TFEL displays [1, 15].

SrTiO$_3$ and BaTiO$_3$ films have been deposited by ALD using strontium and barium cyclopentadienyl compounds, Ti(O^{i}Pr)$_4$ and water as precursors

[166, 167]. Because of the precursor decomposition, the films were deposited at 325°C or below. After annealing at 500°C a permittivity of 180 was measured for SrTiO$_3$ and 165 for BaTiO$_3$ when the film thickness was more than 200 nm, but when the film thickness was decreased to 50 nm, the permittivity decreased to 100. The reason for using cyclopentadienyl compounds for Sr and Ba is that the more common β-diketonates do not react with water at temperatures low enough to avoid precursor decomposition. On the other hand, alternative chemistries examined for SrTiO$_3$ deposition include a use of Sr(thd)$_2$–O$_3$–Ti(O^{i}Pr)$_4$–H$_2$O [168], Sr(thd)$_2$–oxygen plasma–Ti(O^{i}Pr)$_4$–oxygen plasma [169], and Sr(methd)$_2$– oxygen plasma–Ti(O^{i}Pr)$_4$–oxygen plasma, where methd = 1-(2-methoxy- ethoxy)-2,2,6,6-tetramethyl-3,5-heptanedionate [170]. It appears that lower carbon residue levels are achieved with the plasma enhanced processes than with the ozone based process.

SrTa$_2$O$_6$ is another multicomponent high-κ oxide deposited by ALD. In this case a single source precursor SrTa$_2$(OEt)$_{10}$(dmae)$_2$ has been used together with either water [171] or oxygen plasma [172], the former approach better transferring the correct Sr/Ta ratio from the precursor into the films. Permittivities ranging from 30 to 90 were measured depending on the post- deposition annealing temperature.

Mixed oxides and nanolaminates

The atomic level accuracy in controlling film thickness and composition makes ALD a nearly ideal method for controlled deposition of mixed oxides and nanolaminates, the latter consisting of stacks of layers, each layer being typically 1–15 nm in thickness. Mixed oxides and nanolaminates are deposited simply by combining binary oxide ALD cycles in an appropriate manner, i.e. as thoroughly mixed as possible for the mixed oxides, or separated into subsequences each depositing discrete layers for the nanolaminates. Another way of making mixed oxides is to supply two metal precursors in one cycle so that one of the metal compounds serves not only as a metal but also as an oxygen source, e.g. ZrCl$_4$ and Si(OEt)$_4$ for Zr–Si–O [35] and Hf(NMe$_2$)$_4$ and (^{t}BuO)$_3$SiOH for Hf–Si–O [123].

Mixed oxides and nanolaminates have been quite extensively examined for thicker (50–200 nm) insulators, like those used in TFEL displays and integrated passive capacitors [1, 69, 70, 72, 151, 173–177]. By combining two high-κ insulators into a mixed oxide or, perhaps preferably, a nanolaminate stack, the leakage current properties can be significantly improved as compared with the constituent binaries, while the permittivities show inter- mediate values. Sometimes, under favourable conditions like the stabilization of the tetragonal ZrO$_2$ phase in the nanolaminates, the permittivities may even exceed those of the binaries [72]. However, as the future gate oxides must have much higher capacitances than the insulators in the above applications, thick nanolaminate stacks cannot be applied but only a few layers can be afforded.

A great deal of the ALD high-κ gate oxide research has focused on some sort of mixed oxide and nanolaminate. The motivations for this are basically threefold:

1. control of the interfaces with the silicon substrate and the gate electrode;
2. stabilization of amorphous structure;
3. optimization of the composition and structure for the best combination of EOT and leakage current.

However, as it becomes evident below, the third issue is actually a consequence of the first two.

As noted earlier, the poor nucleation of ZrO_2 and HfO_2 on hydrogen terminated silicon leads to nonuniform film microstructure (figure 2.1.9) and may even prevent closure of the film at very low thicknesses. By contrast, Al_2O_3 forms a uniform film without interface layer (figure 2.1.10), though also in this case the start of the growth is somewhat retarded (figure 2.1.8). On the other hand, while about twice the permittivity of SiO_2, the permittivity of Al_2O_3 is still much lower than those of ZrO_2 and HfO_2. Therefore, it has been quite obvious to combine Al_2O_3 and ZrO_2 or HfO_2 so that a thin Al_2O_3 is deposited first at the interface and then the growth is continued with the higher permittivity oxides which nucleate uniformly on Al_2O_3.

For example, $Al_2O_3/ZrO_2/Al_2O_3$ and $Al_2O_3/Al_xZr_yO_z/Al_2O_3$ three-layer structures with a total thickness of 4.7–5.7 nm deposited onto hydrogen terminated silicon resulted in EOT values of 1.1–1.7 nm [30]. Large hysteresis of 400 mV was observed before annealing, while annealing at 600°C decreased the hysteresis to 15–40 mV which still indicates rather high bulk trap density. The interface state density near the midgap was 4–6×10^{12} cm^{-2}eV^{-1}. Further annealing decreased the interface trap density to 1×10^{11} cm^{-2}eV^{-1} but at the same time the EOT increased, apparently indicating formation of a thin interfacial oxide. Leakage currents at 1 V were in the 10^{-6}–10^{-7} A cm^{-2} range for the stacks with 1.0 nm Al_2O_3 interface layers and about 10^{-3} A cm^{-2} for 0.5 nm Al_2O_3 interface layers.

In another similar study, an EOT of 1.0 nm was reported for an $Al_2O_3/ZrO_2/Al_2O_3$ laminate structure with Pt gate [178]. The best fit to an MEIS (medium energy ion scattering) measurement indicated that the laminate stack had a structure of Si–0.5 nm $Al_{0.1}Si_{0.3}O_{0.6}$–0.5 nm $Zr_{0.33}Al_{0.1}O_{0.57}$–3.0 nm ZrO_2–0.5 nm Al_2O_3. Leakage current of 3×10^{-4} A cm^{-2} was measured at 1 V from flatband voltage and the interface state density was 2×10^{11} cm^{-2}eV^{-1}. The capacitance measured with the poly-Si gate was significantly lower than that measured with the Pt gate. This difference was explained by the growth of an interfacial oxide layer during the poly-Si deposition and its activation anneal at 700°C. Plasma nitridation of the nanolaminate improved its barrier properties against oxygen diffusion and thereby diminished the capacitance degradation by the poly-Si gate, but still the capacitance did not reach that obtained with the Pt gate.

Integration of the high-κ oxides in the CMOS devices calls for attention not only to the lower interface with the silicon substrate but also to the top one between the high-κ oxide and the gate electrode. Any low permittivity layer formed at the top interface has a detrimental effect on the total capacitance, similarly to the Si substrate/dielectric interface. Also other kinds of detrimental interaction have been observed at the upper interface, the most serious ones being the destructive reactions between polysilicon and the high-κ oxide, particularly ZrO_2 as discussed on p 45. With HfO_2 the interactions have not been as destructive but still leakage current increment and growth of a few detrimentally large polysilicon grains were observed after CVD of polysilicon gates [179]. While not directly related to each other, these two adverse effects were both attributed to a partial reduction of HfO_2 by silane used as a precursor in polysilicon CVD. They both were eliminated by capping the HfO_2 layer with ten ALD cycles of Al_2O_3 corresponding to about one monolayer: the leakage current was decreased by a factor greater than 10^4 and the extensive polysilicon grain growth was completely eliminated.

Grain boundaries are known to serve as preferential leakage channels in polycrystalline high-κ oxides, and they are also likely to act as easy diffusion paths for the dopants from the polysilicon gate into the silicon substrate. An illustrative verification of the grain boundary leakage was obtained with tunnelling AFM measurements on ALD HfO_2 [180]. Nevertheless, low leakage current densities have been achieved also with polycrystalline gate dielectrics [9]. Thus, it still remains an open question how harmful the polycrystalline structure of the gate dielectric eventually is for the MOSFET operation. Anyhow, several studies have focused on how to obtain at the same time high permittivity and amorphous film structure stabilized through all the annealing procedures. The most common approach with the ALD made high-κ gate oxides has been to mix Al_2O_3 into ZrO_2 or HfO_2. As the following examples will show, increasing the aluminium content in these solid solutions increases the resistance against crystallization, but at the expense of lowered permittivity.

Hf–Al–O mixtures deposited with a 1:1 Hf:Al cycle ratio (60% Al, $\kappa = 12$) retained amorphous structure during annealing at 900°C for 30 min but crystallized during 1 h annealing at the same temperature [95, 181]. When the Al content was increased to 81% by using a Hf:Al cycle ratio of 1:3, the film remained amorphous even when annealed at 1000°C for 30 s but the permittivity was only about 10. In addition, it has been shown that in the mixed oxides deposited with the Hf:Al cycle ratio of 1:3, the amount of negative fixed charge is nearly as high as in the binary Al_2O_3 ($-1.2 \times 10^{12}\,cm^{-2}$ vs. $-1.3 \times 10^{12}\,cm^{-2}$, respectively) [182]. This, together with the low permittivity led to a conclusion that there is little benefit of using such mixed oxides where the aluminium content is high. By contrast, with a 1:1 cycle ratio ($k \approx 16$) scaling of EOT down to about 1.3 nm was found to be possible [182]. Further decrease of the EOT towards 1 nm will be largely dependent on the capability of controlling the interface layer.

In a study focusing on the Hf-rich end of the solid solutions, a film with a composition of $(HfO_2)_{0.67}(Al_2O_3)_{0.33}$ remained amorphous after annealing at 900°C for 20 s while a film with a higher Hf content, $(HfO_2)_{0.85}(Al_2O_3)_{0.15}$, was crystallized [183]. Silicon oxide formation at the interface decreased with increasing Al content and this was related to the smaller oxygen diffusion coefficient of Al_2O_3 and less grain boundary diffusion because of the increased crystallization temperature. Unfortunately, no electrical data were reported.

Also nanolaminate structures are effective in stabilizing amorphous structure, though apparently not as effective as the solid solutions with high Al content. A nanolaminate with a structure of 2.0 nm HfO_2–1.0 nm Al_2O_3– 2.0 nm HfO_2–0.5 nm Al_2O_3–2.0 nm HfO_2 grown in this respective order on a 1.3 nm SiO_2 layer on silicon was found to retain its amorphous and laminated structure even after annealing at 870°C for 5 min in nitrogen [184]. Only annealing at 920°C caused a disappearance of the laminate structure and crystallization of HfO_2. The interface layer stayed unchanged. An increase of the permittivity from 10 for the as-deposited nanolaminate to 17 after annealing at 920°C was related to the mixing of HfO_2 and Al_2O_3.

Zhao *et al* [185] studied thermal stability of a 77 nm thick multilayer structure made by applying 15 cycles Al_2O_3 + 98 × (8 cycles ZrO_2 + 5 cycles Al_2O_3) + 15 cycles Al_2O_3, which resulted in a double layer thickness of 0.77 nm. The as-deposited structure was amorphous. When annealed at different temperatures for 1 min, the laminate structure was stable up to 600°C, above which mixing started to become completed at 800°C. Importantly, crystallization of tetragonal ZrO_2 occurred only after annealing at 900°C. A similar study on already originally well-mixed Zr–Al–O films with 40–60 mol% Al_2O_3 showed that the crystallization occurred only at 1000°C when the annealing time was 1 min [186]. The difference in the crystallization onset temperatures was related to the lower Al_2O_3 content (26 mol%) in the multilayer structure. In both cases the crystallization was observed at about 100°C lower temperatures when the annealing time was longer as it was in the *in situ* high temperature XRD measurements. These observations led to a conclusion that an amorphous mixture of ZrO_2 and Al_2O_3 is thermodynamically more stable than the two separated amorphous phases but crystallization makes the phase separation favourable.

As noted, the adverse effect of adding aluminium oxide to zirconium and hafnium oxides is the decrease of the permittivity. One may compensate for this decrease by adding some other oxide with a higher permittivity, like Nb_2O_5 or Ta_2O_5. For example, amorphous Hf–Al–Nb–O mixed oxide films (physical thickness $\approx 5\,nm$, EOT $\approx 2\,nm$) had essentially the same effective permittivity as binary HfO_2, whereas the permittivities of Hf–Al–O films were much lower [103]. In addition, the C–V curves of the Hf–Al–Nb–O films were essentially hysteresis free while the Hf–Al–O films showed considerable hysteresis. The negative effect of the niobium oxide addition was increased leakage. Similarly, addition of niobium to the Zr–Al–O films increased their permittivity so that a Zr–Al–Nb–O film with a physical thickness of 8 nm gave an EOT of 1.9 nm while the leakage current density was of the order of $10^{-5}\,A\,cm^{-2}$ at 1 V [150].

Summary

The atomic level accuracy in thickness and composition control, excellent large area uniformity and perfect conformality make ALD a very promising method for the high-κ gate oxide deposition. The research on using ALD for this application has largely focused on three processes: $Al(CH_3)_3$–H_2O, $ZrCl_4$–H_2O and $HfCl_4$–H_2O. Besides binary oxides, also solid solutions and nanolaminates have been extensively examined.

While many promising results have been obtained in the EOT range above 1 nm, a lot of challenges still remain. In particular, the requirements on having low EOT and good interface characteristics with silicon are largely contradictory. As a consequence, the future research for scaling EOT further down to 1 nm and beyond with acceptable interface and leakage characteristics must focus on the control of the interface with the silicon substrate. Fixed charge in the high-κ oxide films forms another issue affecting flatband voltages and channel mobility. In an effort to understand and decrease the charge densities, alternative ALD processes available for these oxides might be worth further evaluation.

References

[1] Ritala M and Leskelä M 2002 *Handbook of Thin Film Materials* vol 1 ed H S Nalwa (New York: Academic Press) p 103
[2] Leskelä M and Ritala M 2002 *Thin Solid Films* **409** 138
[3] Sneh O, Clark-Phelps R B, Londergan A R, Winkler J and Seidel T E 2002 *Thin Solid Films* **402** 248
[4] Leskelä M and Ritala M 1999 *J. Phys. IV* **9** Pr8–837
[5] Suntola T and Antson J 1977 *US Patent No* 4 058 430
[6] Suntola T, Antson J, Pakkala A and Lindfors S 1980 *SID 80 Digest* **11** 108
[7] Suntola T S, Pakkala A J and Lindfors S G 1983 *US Patent No* 4 389 973
[8] Suntola T S, Pakkala A J and Lindfors S G 1983 *US Patent No* 4 413 022
[9] Hergenrother J M *et al* 2001 *Tech. Dig. Int. Electron Devices Meeting* p 51
[10] www.moo-han.com

[11] McDougall B A, Zhang K Z, Vereb W W and Paranjpe A 2002 *ALD 2002—AVS Topical Conference on Atomic Layer Deposition, Seoul*

[12] Elers K-E, Ritala M, Leskelä M and Rauhala E 1994 *Appl. Surf. Sci.* **82/83** 468

[13] Aarik J, Aidla A, Kukli K and Uustare T 1994 *J. Cryst. Growth* **144** 116

[14] Kukli K, Ritala M, Matero R and Leskelä M 2000 *J. Cryst. Growth* **212** 459

[15] Skarp J 1984 *US Patent No* 4 486 487

[16] Elers K-E, Saanila V, Soininen P J, Li W-M, Kostamo J T, Haukka S, Juhanoja J and Besling W F A 2002 *Chem. Vap. Deposition* **8** 149

[17] Suntola T 1994 *Handbook of Crystal Growth* vol 3b ed D T J Hurle (Amsterdam: Elsevier) p 601

[18] Ylilammi M 1995 *J. Electrochem. Soc.* **142** 2474

[19] Skarp J I, Soininen P J and Soininen P T 1997 *Appl. Surf. Sci.* **112** 251

[20] Siimon H, Aarik J and Uustare T 1997 *Electrochem. Soc. Proc.* **97-25** 131

[21] Siimon H and Aarik J 1997 *J. Phys. D: Appl. Phys.* **30** 1725

[22] Paranjpe A, Gopinath S, Omstead T and Bubber R 2001 *J. Electrochem. Soc.* **148** G465

[23] Tischler M A and Bedair S M 1986 *Appl. Phys. Lett.* **48** 1681

[24] Cho M-H, Roh Y S, Whang C N, Jeong K, Nahm S W, Ko D-H, Lee J H, Lee N I and Fujihara K 2002 *Appl. Phys. Lett.* **81** 472

[25] Renault O, Samour D, Damlencourt J-F, Blin D, Martin F, Marthon S, Barrett N T and Besson P 2002 *Appl. Phys. Lett.* **81** 3627

[26] Green M L *et al* 2002 *J. Appl. Phys.* **92** 7168

[27] Kim Y B, Tuominen M, Raaijmakers I, de Blank R, Wilhelm R and Haukka S 2000 *Electrochem. Solid-State Lett.* **3** 346

[28] Frank M M, Chabal Y J and Wilk G D 2003 *Appl. Phys. Lett.* **82** 4758

[29] Zhang X, Garfunkel E, Chabal Y J, Christman S B and Chaban E E 2001 *Appl. Phys. Lett.* **79** 4051

[30] Besling W F A *et al* 2002 *J. Non-Cryst. Solids* **303** 123

[31] Nohira H *et al* 2002 *J. Non-Cryst. Solids* **303** 83

[32] Copel M, Gibelyuk M and Gusev E 2000 *Appl. Phys. Lett.* **76** 436

[33] Gusev E P, Copel M, Cartier E, Baumvol I J R, Krug C and Gribelyuk M A 2000 *Appl. Phys. Lett.* **76** 176

[34] Nakajima A, Kidera T, Ishii H and Yokoyama S 2002 *Appl. Phys. Lett.* **81** 2824

[35] Ritala M, Kukli K, Rahtu A, Räisänen P I, Leskelä M, Sajavaara T and Keinonen J 2000 *Science* **288** 319

[36] Perkins C M, Triplett B B, McIntyre P C, Saraswat K C, Haukka S and Tuominen M 2001 *Appl. Phys. Lett.* **78** 2357

[37] Cho M, Park J, Park H B, Hwang C S, Jeong J and Hyun K S 2002 *Appl. Phys. Lett.* **81** 334

[38] Gusev E P *et al* 2001 *Electrochem. Soc. Proc.* **2001-9** 189

[39] Gusev E P, Cartier E, Buchanan D A, Gribelyuk M, Copel M, Okorn-Schmidt H and D'Emic C 2001 *Microelectron. Eng.* **59** 341

[40] Kim Y, Koo J, Han J, Choi S, Jeon H and Park C-G 2002 *J. Appl. Phys.* **92** 5443

[41] Matero R, Rahtu A, Ritala M, Leskelä M and Sajavaara T 2000 *Thin Solid Films* **368** 1

[42] Rahtu A, Alaranta T and Ritala M 2001 *Langmuir* **17** 6506

[43] Rahtu A and Ritala M 2002 *J. Mater. Chem.* **12** 1484

[44] Matero R, Rahtu A and Ritala M 2001 *Chem. Mater.* **13** 4506

[45] Busch B W, Pluchery O, Chabal Y J, Muller D A, Opila R L, Kwo J R and Garfunkel E 2002 *MRS Bull.* **27** 206

[46] Higashi G S and Flemming C G 1989 *Appl. Phys. Lett.* **55** 1963

[47] Fan J-F and Toyoda K 1993 *Japan. J. Appl. Phys.* **32** L1349

[48] Buchanan D A *et al* 2000 *Tech. Dig. Int. Electron Devices Meeting* p 223

[49] Ott A W, Klaus J W, Johnson J M and George S M 1997 *Thin Solid Films* **292** 135

[50] Yun S J, Lee K-H, Skarp J, Kim H R and Nam K-S 1997 *J. Vac. Sci. Technol. A* **15** 2993

[51] Juppo M, Rahtu A, Ritala M and Leskelä M 2000 *Langmuir* **16** 4034

[52] Park D-G, Cho H-J, Yeo I-S, Roh J-S and Hwang J-M 2000 *Appl. Phys. Lett.* **77** 2207

[53] Ericsson P, Begtsson S and Skarp J 1997 *Microelectron. Eng.* **36** 91

[54] Afanas'ev V V, Stesmans A, Mrstik B J and Zhao C 2002 *Appl. Phys. Lett.* **81** 1678

[55] Zhao C *et al* 2003 *Mater. Res. Soc. Symp. Proc.* **745** 9

[56] Stesmans A and Afanas'ev V V 2002 *Appl. Phys. Lett.* **80** 1957

[57] Copel M, Cartier E, Gusev E P, Guha S, Bojarczuk N and Poppeller M 2001 *Appl. Phys. Lett.* **78** 2670

[58] Park D-G, Cho H-J, Lim K-Y, Lim C, Yeo I-S, Roh J-S and Park J W 2001 *J. Appl. Phys.* **89** 6275

[59] Cho H-J, Park D-G, Lim K-Y, Ko J-K, Yeo I-S, Park J W and Roh J-S 2002 *Appl. Phys. Lett.* **80** 3177

[60] Elers K-E, Ritala M, Leskelä M and Johansson L-S 1995 *J. Phys. IV* **5** C5 1021

[61] Mayer T M, Rogers J W Jr and Michalske T A 1993 *Chem. Mater.* **3** 641

[62] Bartram M E, Michalske T A, Rogers J W Jr and Paine R T 1993 *Chem. Mater.* **5** 1424

[63] Liu H, Bertolet D C and Rogers J W Jr 1995 *Surf. Sci.* **340** 88

[64] Riihelä D, Ritala M, Matero R, Leskelä M, Jokinen J and Haussalo P 1996 *Chem. Vap. Deposition* **2** 277

[65] Asif Khan M, Kuznia J N, Skogman R A, Olson D T, MacMillan M and Choyke W J 1992 *Appl. Phys. Lett.* **61** 2539

[66] Asif Khan M, Kuznia J N, Olson D T, George T and Pike W T 1993 *Appl. Phys. Lett.* **63** 3470

[67] Aarik J, Aidla A, Jaek A, Kiisler A-A and Tammik A-A 1990 *Acta Polytech. Scand. Chem. Technol. Metall. Ser.* **195** 201

[68] Hiltunen L, Kattelus H, Leskelä M, Mäkelä M, Niinistö L, Nykänen E, Soininen P and Tiitta M 1991 *Mater. Chem. Phys.* **28** 379

[69] Kattelus H, Ylilammi M, Saarilahti J, Antson J and Lindfors S 1993 *Thin Solid Films* **225** 296

[70] Kattelus H, Ylilammi M, Salmi J, Ranta-aho T, Nykänen E and Suni I 1993 *Mater. Res. Soc. Symp. Proc.* **284** 511

[71] Ritala M, Saloniemi H, Leskelä M, Prohaska T, Friedbacher G and Grasserbauer M 1996 *Thin Solid Films* **286** 54

[72] Kukli K, Ritala M and Leskelä M 1997 *J. Electrochem. Soc.* **144** 300

[73] Kukli K, Ritala M, Leskelä M and Jokinen J 1997 *J. Vac. Sci. Technol. A* **15** 2214

[74] Jeon W-S, Yang S, Lee C and Kang S-W 2002 *J. Electrochem. Soc.* **149** C306

[75] Kumagai H, Toyoda K, Matsumoto M and Obara M 1993 *Japan. J. Appl. Phys.* **32** 6137

[76] Drozd V E, Baraban A P and Nikiforova I O 1994 *Appl. Surf. Sci.* **82/83** 583

[77] Kim J B, Kwon D R, Chakrabarti K, Lee C, Oh K Y and Lee J H 2002 *J. Appl. Phys.* **92** 6739

[78] Räisänen P I, Ritala M and Leskelä M 2002 *J. Mater. Chem.* **12** 1415

[79] Jeong C-W, Lee J-S and Joo S-K 2001 *Japan. J. Appl. Phys.* **40** 285

[80] Ritala M and Leskelä M 1994 *Appl. Surf. Sci.* **75** 333

[81] Ritala M, Leskelä M, Niinistö L, Prohaska T, Friedbacher G and Grasserbauer M 1994 *Thin Solid Films* **250** 72

[82] Aarik J, Aidla A, Mändar H, Uustare T and Sammelselg V 2002 *Thin Solid Films* **408** 97

[83] Aarik J, Aidla A, Kiisler A-A, Uustare T and Sammelselg V 1999 *Thin Solid Films* **340** 110

[84] Ferrari S, Scarel G, Wiernier C and Fanciulli M 2002 *J. Appl. Phys.* **92** 7675

[85] Kukli K, Ritala M, Aarik J, Uustare T and Leskelä M 2002 *J. Appl. Phys.* **92** 1833

[86] Kukli K, Ritala M, Sajavaara T, Keinonen J and Leskelä M 2002 *Thin Solid Films* **416** 72

[87] Ferrari S, Dekadjevi D T, Spiga S, Tallarida G, Wiemer C and Fanciulli M 2002 *J. Non-Cryst. Solids* **303** 29

[88] Gribelyuk M A, Callegari A, Gusev E P, Copel M and Buchanan A 2002 *J. Appl. Phys.* **92** 1232

[89] Zhao C, Roebben G, Bender H, Young E, Haukka S, Houssa M, Naili M, De Gendt S, Heyns M and Van Der Biest O 2001 *Microelectron. Rel.* **41** 995

[90] Perkins C M, Triplett B B, McIntyre P C, Saraswat K C and Shero E 2002 *Appl. Phys. Lett.* **81** 1417

[91] Callegari A, Gusev E, Zabel T, Lacey D, Gribelyuk M and Jamison P 2002 *Appl. Phys. Lett.* **81** 4157

[92] Houssa M, Tuominen M, Naili M, Afanas'ev V, Stesmans A, Haukka S and Heyns M M 2000 *J. Appl. Phys.* **87** 8615

[93] Park D-G, Lim K-Y, Cho H-J, Cha T-H, Yeo I-S, Roh J-S and Park J W 2002 *Appl. Phys. Lett.* **80** 2514

[94] Kim Y *et al* 2001 *Tech. Dig. Int. Electron Devices Meeting* p 455

[95] Wilk G D *et al* 2002 *Tech. Dig. VLSI Meeting, Honolulu* p 88

[96] Rim K *et al* 2002 *Tech. Dig. VLSI Meeting, Honolulu* p 12

[97] Sim H and Hwang H 2002 *Appl. Phys. Lett.* **81** 4036

[98] Kukli K, Forsgren K, Aarik J, Uustare T, Aidla A, Niskanen A, Ritala M, Leskelä M and Hårsta A 2001 *J. Cryst. Growth* **231** 262

[99] Kukli K, Forsgren K, Ritala M, Leskelä M, Aarik J and Hårsta A 2001 *J. Electrochem. Soc.* **148** F227

[100] Kukli K, Ritala M, Uustare T, Aarik J, Forsgren K, Sajavaara T, Leskelä M and Hårsta A 2002 *Thin Solid Films* **410** 53

[101] Forsgren K, Westlinder J, Lu J, Olsson J and Hårsta A 2002 *Chem. Vap. Deposition* **8** 105

[102] Forsgren K, Hårsta A, Aarik J, Aidla A, Westlinder J and Olsson J 2002 *J. Electrochem. Soc.* **149** F139

[103] Kukli K, Ritala M, Leskelä M, Sajavaara T, Keinonen J, Gilmer D C, Hegde R, Rai R and Prabhu L 2003 *J. Mater.Sci.; Mater. Electron.* **14** 361

[104] Kukli K, Ritala M, Sundqvist J, Aarik J, Lu J, Sajavaara T, Leskelä M and Hårsta A 2002 *J. Appl. Phys.* **92** 5698

[105] Kukli K, Ritala M and Leskelä M 2000 *Chem. Vap. Deposition* **6** 297

[106] Chang J P and Lin Y-S 2001 *J. Appl. Phys.* **90** 2964

[107] Chang J P, Lin Y-S and Chu K 2001 *J. Vac. Sci. Technol. B* **19** 1782

[108] Matero R, Ritala M, Leskelä M, Jones A C, William P A, Bickley J F, Steiner A, Leedham T J and Davies H O 2002 *J. Non-Cryst. Solids* **303** 24

[109] Kukli K, Ritala M, Leskelä M, Sajavaara T, Keinonen J, Jones A C and Roberts J L *Chem. Mater.* submitted

[110] Kukli K, Ritala M, Sajavaara T, Keinonen J and Leskelä M 2002 *Chem. Vap. Deposition* **8** 199

[111] Hausmann D M, Kim E, Becker J and Gordon R G 2002 *Chem. Mater.* **14** 4350

[112] Putkonen M and Niinistö L 2001 *J. Mater. Chem.* **11** 3141

[113] Putkonen M, Niinistö J, Kukli K, Sajavaara T, Karppinen M, Yamauchi H and Niinistö L 2003 *Chem. Vap. Deposition* **9** 207

[114] Conley J F Jr, Ono Y, Zhuang W, Tweet D J, Gao W, Mohammed S K and Solanki R 2002 *Electrochem. Solid-State Lett.* **5** C57

[115] Putkonen M, Nieminen M, Niinistö J and Niinistö L 2001 *Chem. Mater.* **13** 4701

[116] Mölsä H, Niinistö L and Utriainen M 1994 *Adv. Mater. Opt. Electron.* **4** 389

[117] Putkonen M, Sajavaara T, Johansson L-S and Niinistö L 2001 *Chem. Vap. Deposition* **7** 44

[118] Seim H, Nieminen M, Niinistö L, Fjellvåg H and Johansson L-S 1997 *Appl. Surf. Sci.* **112** 243

[119] Seim H, Mölsä H, Nieminen M, Fjellvåg H and Niinistö L 1997 *J. Mater. Chem.* **7** 449

[120] Nieminen M, Putkonen M and Niinistö L 2001 *Appl. Surf. Sci.* **174** 155

[121] Mölsä H and Niinistö L 1994 *Mater. Res. Soc. Symp. Proc.* **335** 341

[122] Päiväsaari J, Putkonen M and Niinistö L 2002 *J. Mater. Chem.* **12** 1828

[123] Gordon R G, Becker J, Hausmann D and Suh S 2001 *Chem. Mater.* **13** 2463

[124] Nieminen M, Sajavaara T, Rauhala E, Putkonen M and Niinistö L 2001 *J. Mater. Chem.* **11** 2340

[125] Sneh O, Wise M L, Ott A W, Okada L A and George S M 1995 *Surf. Sci.* **334** 135

[126] Klaus J W, Ott A W, Johnson J M and George S M 1997 *Appl. Phys. Lett.* **70** 1092

[127] Klaus J W, Sneh O and George S M 1997 *Science* **278** 1934

[128] Klaus J W, Sneh O, Ott A W and George S M 1999 *Surf. Rev. Lett.* **6** 435

[129] Klaus J W and George S M 2000 *Surf. Sci.* **447** 81

[130] Gasser W, Uchida Y and Matsumura M 1994 *Thin Solid Films* **250** 213

[131] Yamaguchi K, Imai S, Ishitobi N, Takemoto M, Miki H and Matsumura M 1998 *Appl. Surf. Sci.* **130–132** 202

[132] Morishita S, Gasser W, Usami K and Matsumura M 1995 *J. Non-Cryst. Solids* **187** 66

[133] Morishita S, Uchida Y and Matsumura M 1995 *Japan. J. Appl. Phys.* **34** 5738

[134] Klaus J W, Ott A W, Dillon A C and George S M 1998 *Surf. Sci.* **418** L14

[135] Nakajima A, Yoshimoto T, Kidera T and Yokoyama S 2001 *Appl. Phys. Lett.* **79** 665

[136] Nakajima A, Yoshimoto T, Kidera T, Obata K, Yokoyama S, Sunami H and Hirose M 2001 *J. Vac. Sci. Technol. B* **19** 1138

[137] Nakajima A, Khosru Q D M, Yoshimoto T, Kidera T and Yokoyama S 2002 *Appl. Phys. Lett.* **80** 1252

[138] Goto H, Shibahara K and Yokoyama S 1996 *Appl. Phys. Lett.* **68** 3257

[139] Yokoyama S, Goto H, Miyamoto T, Ikeda N and Shibahara K 1997 *Appl. Surf. Sci.* **112** 75

[140] Yokoyama S, Ikeda N, Kajikawa K and Nakashima Y 1998 *Appl. Surf. Sci.* **130–132** 352

[141] Morishita S, Sugahara S and Matsumura M 1997 *Appl. Surf. Sci.* **112** 198

[142] Kukli K, Ritala M and Leskelä M 1995 *J. Electrochem. Soc.* **142** 1670

[143] Kukli K, Ritala M, Leskelä M and Lappalainen R 1998 *Chem. Vap. Deposition* **4** 29

[144] Aarik J, Aidla A, Kukli K and Uustare T 1994 *J. Cryst. Growth* **144** 116

[145] Kukli K, Ritala M, Matero R and Leskelä M 2000 *J. Cryst. Growth* **212** 459

[146] Kukli K, Aarik J, Aidla A, Kohan O, Uustare T and Sammelselg V 1995 *Thin Solid Films* **260** 135

[147] Kukli K, Aarik J, Aidla A, Forsgren K, Sundqvist J, Hårsta A, Uustare T, Mändar H and Kiisler A-A 2001 *Chem. Mater.* **13** 122

[148] Song H-J, Lee C-S and Kang S-W 2001 *Electrochem. Solid-State Lett.* **4** F13

[149] Kukli K, Ritala M and Leskelä M 2000 *Chem. Mater.* **12** 1914

[150] Kukli K, Ritala M, Leskelä M, Sajavaara T, Keinonen J, Gilmer D, Bagchi S and Prabhu L 2002 *J. Non-Cryst. Solids* **303** 35

[151] Kukli K, Ritala M and Leskelä M 1997 *Nanostruct. Mater.* **8** 785

[152] Ritala M, Leskelä M, Nykänen E, Soininen P and Niinistö L 1993 *Thin Solid Films* **225** 288

[153] Ritala M, Leskelä M, Johansson L-S and Niinistö L 1993 *Thin Solid Films* **228** 32

[154] Aarik J, Aidla A, Uustare T and Sammelselg V 1995 *J. Cryst. Growth* **148** 268

[155] Aarik J, Aidla A, Sammelselg V, Siimon H and Uustare T 1996 *J. Cryst. Growth* **169** 496

[156] Aarik J, Aidla A, Kiisler A-A, Uustare T and Sammelselg V 1997 *Thin Solid Films* **305** 270

[157] Kumagai H, Matsumoto M, Toyoda K, Obara M and Suzuki M 1995 *Thin Solid Films* **263** 47

[158] Kukli K, Ritala M, Schuisky M, Leskelä M, Sajavaara T, Uustare T and Hårsta A 2000 *Chem. Vap. Deposition* **6** 303

[159] Schuisky M, Aarik J, Kukli K, Aidla A and Hårsta A 2001 *Langmuir* **17** 5508

[160] Aarik J, Aidla A, Uustare T, Kukli K, Sammelselg V, Ritala M and Leskelä M 2002 *Appl. Surf. Sci.* **193** 277

[161] Pore V, Rahtu A, Leskelä M, Ritala M, Sajavaara T and Keinonen J *Chem. Vap. Deposition* at press

[162] Ritala M, Leskelä M and Rauhala E 1994 *Chem. Mater.* **6** 556

[163] Aarik J, Karlis J, Mändar H, Uustare T and Sammelselg V 2001 *Appl. Surf. Sci.* **181** 339

[164] Döring H, Hashimoto K and Fujishima A 1992 *Ber. Bunsenges. Phys. Chem.* **96** 620

[165] Ritala M, Leskelä M, Niinistö L and Haussalo P 1993 *Chem. Mater.* **5** 1174

[166] Vehkamäki M, Hatanpää T, Hänninen T, Ritala M and Leskelä M 1999 *Electrochem. Solid-State Lett.* **2** 504

[167] Vehkamäki M, Hänninen T, Ritala M, Leskelä M, Sajavaara T, Rauhala E and Keinonen J 2001 *Chem. Vap. Deposition* **7** 75

[168] Kosola A, Putkonen M, Johansson L-S and Niinistö L 2003 *Appl. Surf. Sci.* **211** 102

[169] Lee J H, Cho Y J, Min Y S, Kim D and Rhee S W 2002 *J. Vac. Sci. Technol. A* **20** 1828

[170] Kil D-S, Lee J-M and Roh J-S 2002 *Chem. Vap. Deposition* **8** 195

[171] Vehkamäki M, Ritala M, Leskelä M, Jones A C, Davies H O, Sajavaara T and Rauhala E *J. Electrochem. Soc.* submitted

[172] Lee W J, You I-K, Ryu S-O, Yu B-G, Cho K-I, Yoon S-G and Lee C-S 2001 *Japan. J. Appl. Phys.* **40** 6941

[173] Kukli K, Ihanus J, Ritala M and Leskelä M 1996 *Appl. Phys. Lett.* **68** 3737

[174] Kukli K, Ritala M and Leskelä M 2001 *J. Electrochem. Soc.* **148** F35

[175] Kukli K, Ritala M and Leskelä M 1999 *J. Appl. Phys.* **86** 5656

[176] Kattelus H, Ronkainen H and Riihisaari T 1999 *Int. J. Microcircuits Electron. Packag.* **22** 254

[177] Kattelus H, Ronkainen H, Kanniainen T and Skarp J 1998 *Proceedings of 28th European Solid-State Device Res. Conf., ESSDERC'98, Bordeaux* p 444

[178] Jeon S, Yang H, Chang H S, Park D G and Hwang H 2002 *J. Vac. Sci. Technol. B* **20** 1143

[179] Gilmer D C *et al* 2002 *Appl. Phys. Lett.* **81** 1288

[180] Lysaght P S, Chen P J, Bergmann R, Messina T, Murto R W and Huff H R 2002 *J. Non-Cryst. Solids* **303** 54

[181] Ho M Y *et al* 2002 *Appl. Phys. Lett.* **81** 4218

[182] Carter R J, Tsai W, Young E, Caymax M, Maes J W, Chen P J, Delabie A, Zhao C, DeGendt S and Heyns M 2003 *Mater. Res. Soc. Symp. Proc.* **745** 35

[183] Yu H Y *et al* 2002 *Appl. Phys. Lett.* **81** 3618

[184] Cho M-H, Roh Y S, Whang C N, Jeong K, Choi H J, Nam S W, Ko D-H, Lee J H, Lee N I and Fujihara K 2002 *Appl. Phys. Lett.* **81** 1071

[185] Zhao C, Richard O, Bender H, Caymax M, De Gendt S, Heyns M, Young E, Roebben G, Van Der Biest O and Haukka S 2002 *Appl. Phys. Lett.* **80** 2374

[186] Zhao C *et al* 2002 *J. Non-Cryst. Solids* **303** 144

Chapter 2.2

Chemical vapour deposition

S A Campbell and R C Smith

Introduction

Chemical vapour deposition is a process by which gaseous molecular precursors are converted to solid-state materials, usually in the form of a thin film, on a heated surface [1, 2]. In figure 2.2.1 some of the fundamental processes accessible to the precursor molecules are illustrated. The gaseous molecules are introduced into the chamber. Some gas-phase reactions may occur. Normally this leads to more reactive daughter products. These daughter products, along with unreacted precursor molecules, are transported to the vicinity of the wafer where they may adsorb onto the surface. In many cases, the fluence of these unreacted species is higher due to their higher concentrations in the gas phase; however, due to their lower reactivity, they are less likely to adhere on the surface of the wafer. It is also typically true that the daughter by-products are more easily transported to the wafer surface due to their lighter mass (and therefore higher diffusivity). On the substrate surface the adsorbed precursor molecules and daughter molecules may: (1) desorb from the surface and re-enter the transport flow, (2) diffuse along the substrate surface, and/or (3) react to form a solid deposit. Once reaction has occurred, the by-products desorb from the surface and are removed from the reaction chamber via the transport flow.

Advantages and limitations of CVD

Many methods are available for depositing metal oxide thin films. Some of these are reviewed in the other sections of this chapter. Chemical vapour deposition (CVD) has been used in a number of industries (electronics, optics, tool strengthening, etc) to deposit a wide variety of materials [3, 4]. CVD is extremely attractive to the electronics industry. Unlike ALD and

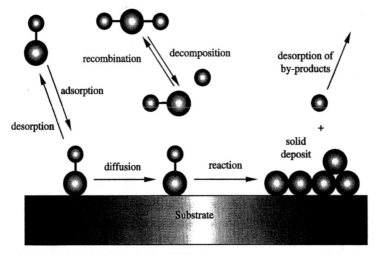

Figure 2.2.1. Fundamental steps in the CVD process.

MBE, it is widely used in the manufacture of integrated circuits. Most semiconductor and insulating films used in microelectronics are deposited by CVD. The reasons for this acceptance include ease of manufacture, deposition rates that can be controlled from 1 to $1000 \, \text{nm} \, \text{min}^{-1}$, good film conformality, and the ability to control composition relatively easily. The other widely used thin film deposition technique is physical vapour deposition (PVD). PVD techniques often are limited to line-of-sight deposition, requiring that the substrate be directly opposite to the source in the deposition system, unless the substrate is adequately heated to provide a surface diffusion length comparable to the feature size/depth. However, for planar structures such as the gate insulator of conventional MOSFETs, surface coverage is not a serious problem. One of the largest concerns for using PVD to manufacture gate insulators is the effect of plasma damage. Although ion bombardment of the substrate can be minimized in a properly selected plasma geometry, it cannot be eliminated. Furthermore, plasma processes such as the sputtering of metal targets are a relatively rich source of x-rays that can further damage the deposited films. Surface damage can be problematic when surface roughening or interdiffusion of layers that have already been deposited must be avoided. This is particularly true for gate insulators where both the bulk of the film and the interface must be perfect (i.e., have only satisfied bonds or have defect levels well outside the silicon bandgap) to the part per million level. While plasma damage may be reduced by post-deposition annealing, there has been considerable reluctance to accept PVD as a viable technique for depositing thin insulating layers. CVD can be performed at much higher deposition pressures (up to atmospheric pressure) than PVD techniques. PVD methods usually require high vacuum (often 10^{-8} torr or below) to minimize incorporation of contaminants into the film. The higher operating pressures available for CVD allow more efficient processing with less time to pump to a base vacuum to remove contaminants from the deposition system.

Though CVD holds many advantages over PVD methods, currently, CVD also has several important concerns. While developing a PVD process is often quite straightforward, CVD processes may involve complex chemistries that often are not well understood. Since CVD processes involve the decomposition of a molecular precursor, the use of the proper precursor(s) for a given material system is vital. Poorly chosen chemistries can lead to the presence of unacceptable levels of residual impurities in the film, which may act as trap sites. This is a particular concern in applications such as gate insulators in which one might reasonably expect the layers to experience charge injection under normal device operation. Gases may be introduced along with the deposition precursor to volatize the reaction by-products on the surface of the wafer to reduce chemical contamination in the film. Excessive reactions in the gas phase can also lead to particle formation (homogeneous nucleation). Particles incorporated into the film often produce a poor film morphology. The large surface area associated with an aerosol may also scavenge the reactive species from the chamber, dramatically slowing the film deposition process, particularly if the particles are swept from the reactor in the gas stream rather than depositing on the wafer.

Another consideration for CVD processes is the deposition temperature. For most CVD processes, thermal energy is needed to decompose precursors to begin the deposition process. The temperature required to drive this process at a usable rate may restrict the grain size or phase of the material that can be deposited unless a post-deposition anneal (PDA) is used. Thus, successful development of a CVD process often requires substantial optimization. In some cases, an acceptable precursor is simply not available. However, new precursors and more elegant CVD methods are continuously being introduced.

Reactors for CVD

CVD is nearly always performed in a vacuum or inert atmosphere to prevent incorporation of unwanted matter during deposition [1]. Although a variety of reactor designs can be used, each consists of several basic components. A diagram of a typical CVD reactor is shown in figure 2.2.2. The basic components of a CVD system include: (1) a method for introduction of precursors into the reactor, commonly either a mass flow controller for gaseous precursors or a carrier gas for vapours from liquids and sublimed solids. In some cases liquid droplets may be injected into the reactor if

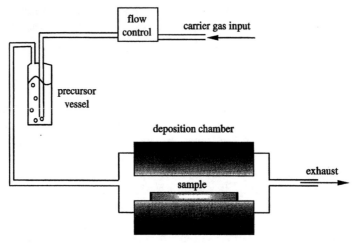

Figure 2.2.2. Schematic diagram of the basic components of a typical CVD reactor.

vapours are not available; (2) an apparatus for supporting substrates within the reactor; (3) a method for heating the substrates such as a resistive heater, a high intensity lamp, or RF coils; and (4) an outlet to the pumping system for the removal of gaseous reaction by-products and unreacted gases. The effectiveness of a reactor design depends on the gas flow dynamics within the system and upon its ability to effectively transport precursor molecules to the substrate surface while removing the volatile reaction products. Poor flow dynamics may lead to the accumulation of gas-phase products near the surface where they can become incorporated in the film as impurities.

Generally, for the conditions used in typical CVD processes, the flow of gases within a reactor is laminar. However, at a surface the velocity of the gaseous molecules is necessarily zero. As a result, a boundary layer exists where the gas velocity increases from zero to that of the bulk flow. Near the reactor inlet the gradient is small but the thickness of the boundary layer increases gradually along the flow direction. At some point downstream, the flow through the reactor stabilizes and the boundary layer assumes a more constant thickness. The presence of a boundary layer complicates the picture of the CVD process. Precursor molecules and/or their reactive daughter molecules must diffuse through the boundary layer prior to adsorbing onto the substrate surface and reacting [1]. Similarly, before reaction by-products can be removed from the system they must pass through the boundary layer into the bulk gas flow. This is particularly a concern for the high mass precursors often used to deposit high-κ materials. In mixed gas systems the diffusion coefficient of the heavy element may be substantially less than that of the lighter gases, making diffusion through the stagnant layer difficult. Since the boundary layer thickness decreases as the square root of the bulk gas velocity, these complications can be reduced either by increasing the flow of reactant gases into the chamber or by reducing the overall pressure of the system.

Thermal or concentration boundary layers may also develop under certain conditions within the reactor, which further complicates the uniformity of precursor transport throughout the reactor and may result in poor film uniformity. The gases located near hot surfaces, such as the substrate or reactor walls, are heated much more rapidly than those farther away. Thus, a temperature boundary layer that is similar in shape to the velocity boundary layer exists above the surface of a substrate. In general, higher and more uniform temperatures are located downstream from the reactor inlet, where the thermal gradient is not as steep.

Over the course of a deposition, the overall concentration of reactants decreases as precursor molecules are consumed in reactions at the substrate surface and as the boundary layer becomes more concentrated in reaction by-products. As a result, the number of molecules reaching the substrate will be larger for regions that are closer to the inlet. Depletion of precursor can cause nonuniform deposition across the substrate. In some geometries simply tilting the substrate so that the flow of precursor molecules is not parallel to the surface can improve the film uniformity.

Deposition behaviour in CVD

Conventional CVD is a thermally activated process. Therefore, the deposition temperature is potentially the most important variable for deposition control. Temperature influences the reaction rate of the precursors at the surface and thereby the overall deposition rate. The deposition temperature also partly controls the film composition, microstructure, and morphology. Because materials properties are controlled by these film characteristics, it is vital to understand how the deposition conditions affect them.

A generic illustration of the dependence of growth rate on temperature for a typical CVD process is shown in figure 2.2.3. At low deposition temperatures, the growth rate exhibits Arrhenius-type kinetics, i.e., the deposition rate increases as $\exp(-E_A/kT)$, where E_A is an activation energy, k is Boltzman's constant, and T is the temperature (K). Deposition in this region can often be extremely conformal, even for high aspect ratio structures, as long as the temperature is uniform since the deposition rate is largely independent of gas flow patterns. Reactors designed to operate in this regime are often multiwafer systems with excellent thermal uniformity. At intermediate temperatures, the rate of chemical reaction is faster than the rate of precursor transport to the substrate surface. In this region the deposition rate is independent of deposition temperature and is controlled by the precursor flux. Reactors designed to run in this regime are typically single

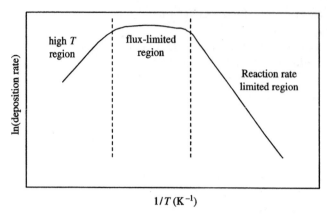

Figure 2.2.3. Generic plot showing the temperature-dependent behaviour of the CVD process.

wafer reactors with carefully controlled flow dynamics. For higher deposition temperatures the deposition rate is often observed to decrease. This behaviour is usually attributed to depletion of the precursor in the gas phase by particle formation. However, the decrease in deposition rate has also been observed in systems where the deposition pressures are low enough to prevent molecules from interacting in the gas phase prior to arriving at the substrate surface and thereby eliminate gas-phase particle formation [3, 5]. It has been suggested that at high deposition temperatures the rate of precursor desorption becomes significant in comparison to the rate of reaction, which leads to the slower observed deposition rates [5]. Since exposing the substrate to unnecessarily large thermal cycles is normally undesirable, few processes are run in this regime.

For most films, the film composition is difficult to control by simple temperature variation. However, in cases where two materials are deposited simultaneously from different precursors, each with its own rate versus temperature profile, the deposition temperature can play a larger role in determining the final stoichiometry of the film.

Materials deposited by CVD display a range of morphologies depending upon the deposition conditions [2]. As grown, CVD films may be amorphous, polycrystalline, or single crystalline. Amorphous films often result at low deposition temperatures where the mobility of the deposited material on the surface is very limited. Under such conditions there is little reorganization of the material to form crystalline deposits. Crystalline films are observed for depositions performed at higher temperatures and for some films that were amorphous as deposited but then subjected to a high temperature PDA. Molecules in a crystalline film exhibit long-range order. Polycrystalline films, in which each grain is a single crystal, are often observed in the materials of interest here. However, grain boundaries can sometimes cause problems

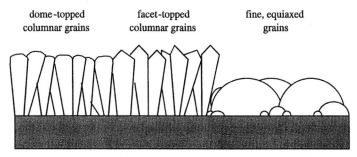

Figure 2.2.4. Diagram of the common grain structures observed in thin films deposited by CVD.

including the potential for increased electrical conduction, increased diffusion of impurities, inhomogeneous film properties, and surface roughening.

Three types of grain structure generally result for polycrystalline films grown by CVD: (1) columnar grains with rounded tops, (2) columnar grains with faceted tops, or (3) fine equiaxed grains [1]. These structures are illustrated in figure 2.2.4. Fine-grained films are usually preferred over columnar films due to their superior mechanical properties. Films of the columnar type tend to contain more void space between neighbouring grains. Also, the columnar structure can lead to anisotropic behaviour in the properties of the material. Fine-grained structures are often obtained under conditions where the rate is governed by surface kinetics. Therefore, films deposited at either low pressure or temperature and at high precursor concentrations will tend to form fine grains. For depositions at higher substrate temperature grain growth is fast and columnar structures will often develop. Columnar grains are also often observed in thicker films due to increased grain growth in one direction.

Precursor considerations

Precursors for CVD of metal oxides generally fall into one of three classifications: organometallic, metalorganic, or inorganic compounds. In organometallic compounds the ligands are coordinated to the metal centre via a direct metal–carbon bond, while for metalorganic compounds there is no direct metal–carbon bonding. Examples of organometallic precursors include $AlMe_3$ and $(C_5H_4Me)Pt(Me)_3$. Common metalorganic precursors include $Ti(OCH(CH)_3)_4$ and zirconium(IV) acetylacetonate. Inorganic compounds contain no organic ligands and include metal tetrahalides such as $TiCl_4$ and ZrI_4, and metal hydrides such as SiH_4 [6].

Although precursors of many types exist, CVD precursors must possess the same general properties. Ideal precursors are either a room temperature gas or possess a high vapour pressure (>0.1 torr) at reasonable source temperatures to allow for adequate mass transfer into the deposition chamber. Liquids and gaseous precursors are preferred because the vapour pressure is generally stable and well-established delivery methods, such as direct liquid injection and mass flow controllers, are available. The use of solid precursors is complicated by the dependence of the evaporation rate on the source surface area. Because the surface area of the particles that make up a typical source changes over time, the volatility of the precursor and therefore mass transfer into the deposition system can be inconsistent. This problem can be at least partially overcome by exposing fresh particle surfaces by grinding, stirring, or some other method. Alternatively, some solid sources can be dissolved in an inert solvent for delivery. Techniques such as atomic layer CVD, where alternating monolayers of precursor are dosed onto the substrate surface, and direct liquid injection CVD, where the precursor is dissolved in an appropriate solvent and introduced as a solution, do not rely on a continuous flux of precursor to the substrate.

Another important consideration is the temperature necessary for deposition to occur. Because the dimensions in advanced microelectronic devices are extremely small, processing often must be performed at low temperature ($<700°C$) to prevent interdiffusion of layers that have already been processed. Requirement of additional reagents in a CVD process may also complicate the process design. This is especially problematic when depositing oxides onto silicon from precursors that require oxygen or H_2O as a secondary reactant. For such a system the silicon substrate can readily be oxidized. For this reason single source precursors are more desirable than those requiring additional reagents. Precursors should be stable during storage. If the shelf life of a precursor is short the deposition rate may change due to the presence of impurities or to decomposition of the precursor within the vessel. Finally, one needs to consider the environmental, health, and safety factors associated with a precursor. One must be aware of, and make efforts to mitigate, potential problems associated with toxicity, flammability, and explosion hazards, as well as knowing how the precursor and its reaction by-products will affect the environment.

The vapour pressure is the limiting factor encountered in the search for CVD precursors for many materials. This is especially the case for the deposition of materials containing elements with high atomic number. In general, as molecular mass increases the vapour pressure decreases and the compound will more likely be a solid at room temperature. Highly fluorinated ligands may be substituted to increase precursor volatility. However, while the fluorinated ligands increase the precursor volatility, they may also lead to fluorine incorporation in the film, which can dramatically change the properties of a material. Furthermore, materials of complex composition, e.g. $BaTiO_3$ and $PbZr_xTi_{1-x}O_3$, often preclude the use of single source precursors. In such cases more sophisticated deposition strategies using a combination of two or more precursors may be necessary.

In an interesting paper, Taylor and co-workers compared the deposition of a high-κ material, TiO_2, with various precursors [5]. They found that the choice of precursor had a strong effect on the film microstructure when the deposition was run in the reaction rate limited regime. At higher temperatures the film morphologies merged. They suggest that the precursor surface coverage and diffusion are important determinants of film morphology at low temperature. No effort was made to study the electrical properties. It is anticipated that these properties, which are sensitive to ppm levels of electrically active defects, may depend much more strongly on the deposition chemistry in all temperature regimes.

CVD precursors to group IV metal oxides

The early work on high permittivity metal oxide insulators focused primarily on Ta_2O_5 and simple group IVB binary oxides. Precursors used to deposit

Table 2.2.1. Precursors for CVD of group IV oxides. References for each process are listed in the text.

Precursor	b.p. (°C)	T_b (°C)	Minimum T_d (°C)
$TiCl_4$	136.4	n.r.	200
$Ti(O-i-Pr)_4$	232	80–120	250
$[Ti(\mu-ONep)(ONep)_3]_2$	n.r.	120	300
$Ti(NO_3)_4$	40 (subl.)/0.2 mm	22–40	230
$ZrCl_4$	331 (subl.)	158–200	800
$Zr(O-t-Bu)_4$	81/3 mm	50–70	300
$[Zr(\mu-ONep)(ONep)_3]_2$	n.r.	170	300
$Zr(acac)_4$	171–173 (m.p.)	150–220	350
$Zr(tfac)_4$	130 (subl.)/0.005 mm	130–180	300
$Zr(thd)_4$	180 (subl.)/0.1 mm	200–230	500
$Zr(acac)_2(hfip)_2$	Dist. 80°/10^{-3} mm	77.5	350
$Zr(O-i-Pr)_2(thd)_2$	110–111 (m.p.)	195	350
$Zr(O-t-Bu)_2(thd)_2$	261–263 (m.p.)	195	350
$Zr_2(O-i-Pr)_6(thd)_2$	160–163 (m.p.)	200	350
$Zr(NO_3)_4$	95 (subl.)/0.2 mm	60–85	285
$HfCl_4$	320 (subl.)	n.r.	n.r.
$Hf(O-t-Bu)_4$	90/5 mm	n.r.	n.r.
$Hf(tfac)_4$	n.r.	190	300
$Hf(NO_3)_4$	95 (subl.)/0.2 mm	60–85	250

T_b, bubbler temperature; T_d, deposition temperature; n.r., not reported.

TiO_2, ZrO_2, and HfO_2 are listed in table 2.2.1. Metal chlorides were among the first compounds in CVD processing [7–13]. Liquid $TiCl_4$ has a relatively low boiling point (136.4°C), while $ZrCl_4$ and $HfCl_4$ are solids that sublime readily at temperatures near 300°C. Oxide films can be prepared from the chlorides either by hydrolysis, equation (2.2.1), or via direct oxidation, equation (2.2.2). In the case of deposition from zirconium and hafnium chloride, temperatures greater than 800°C are often necessary and may lead to significant oxidation of the silicon substrate by the reactant gases [7, 9–11].

$$TiCl_4 + 2H_2O \xrightarrow{\Delta} TiO_2 + 4HCl \qquad (2.2.1)$$

$$TiCl_4 + O_2 \xrightarrow{\Delta} TiO_2 + 2Cl_2 \qquad (2.2.2)$$

For these reasons there has been limited interest in using group IV chlorides for the CVD of metal oxides for microelectronic applications. Chlorides and iodides are widely used to deposit MO_2 films by atomic layer deposition (ALD) using the chemistry shown in equation (2.2.1) [8] as described in another chapter of this book. The key to these ALD processes is to operate at temperatures where the CVD deposition rate is negligible. Thus, one either uses very high temperature sources such as halides and operates at moderate temperatures, or one uses the MO or OM sources at very low substrate temperatures.

Metal alkoxides, such as those shown in figure 2.2.5, are attractive as precursors due to their high vapour pressures and lower deposition temperatures. Titanium isopropoxide, $Ti(OCH(CH_3)_2)_4$, has been the most widely used precursor for the deposition of titanium dioxide thin films [14–21]. High-quality films can be deposited from $Ti(OCH(CH_3)_2)_4$ either by direct thermolysis, equations (2.2.3) and (2.2.4), or by hydrolysis, equation (2.2.5).

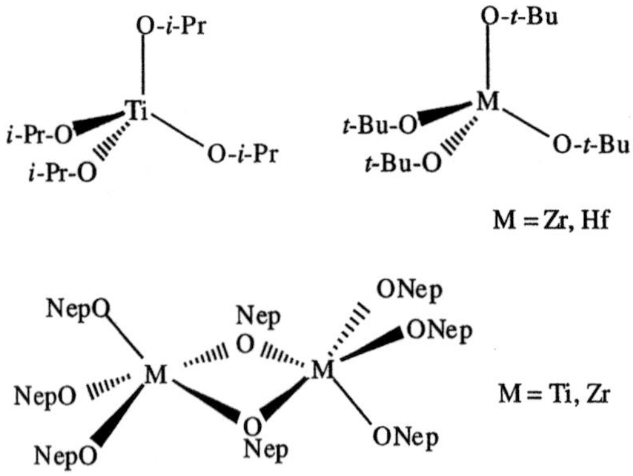

Figure 2.2.5. Schematic drawing of group IV alkoxide precursors: $M(O\text{-}t\text{-}Bu)_4$, $Ti(O\text{-}i\text{-}Pr)_4$, and $[M(\mu\text{-}ONep)(ONep)_3]_2$.

$$Ti[OCH(CH_3)_2]_4 \xrightarrow{\Delta} TiO_2 + 2CH_3CHCH_2 + 2(CH_3)_2CHOH \qquad (2.2.3)$$

$$Ti[OCH(CH_3)_2]_4 \xrightarrow{\Delta} TiO_2 + 4CH_3CHCH_2 + 2H_2O \qquad (2.2.4)$$

$$Ti[OCH(CH_3)_2]_4 + 2H_2O \xrightarrow{\Delta} TiO_2 + 4(CH_3)_2CHOH \qquad (2.2.5)$$

Similarly, $Zr(OC(CH_3)_3)_4$ [7, 22–25] and $Hf(OC(CH_3)_3)_4$ [7, 26] have been used to deposit zirconium and hafnium dioxide, respectively. The vapour pressure of zirconium tetra-*tert*-butoxide (ZTB), e.g., is the highest of all of the commonly used precursors for ZrO_2. It has a sufficient vapour pressure to even allow a reasonable deposition rate with a room temperature bubbler [23, 24, 27, 28]. Most metal alkoxides are sensitive to moisture and must be handled under inert atmosphere conditions.

In β-diketonate complexes, illustrated in figure 2.2.6, the metal centre is surrounded by eight oxygen atoms, making them generally more stable against hydrolysis than alkoxides. The zirconium and hafnium acetylacetonates, $Zr(acac)_4$ and $Hf(acac)_4$, decompose to give ZrO_2 and HfO_2 at 350°C but require source temperatures around 200°C for sufficient vaporization [29–32]. The tri- and per-fluorinated analogues, $M(tfac)_4$ and $M(hfac)_4$, are more volatile but the increased risk of fluorine contamination is undesirable [29, 30, 32–36]. Tetramethylheptanedionato zirconium, $Zr(thd)_4$ (this precursor can also be abbreviated $Zr(dpm)_4$ where dpm stands for dipivaloylmethanato), is especially stable, can be handled in air, and has been shown to deposit high-quality films [32, 37–40]. Unfortunately, $Zr(thd)_4$ requires source temperatures between 200 and 230°C and substrate temperatures greater than 500°C. Furthermore, the addition of O_2 is often necessary to help reduce the amount of residual carbon in films deposited from β-diketonate compounds.

The recent development of mixed ligand precursors, shown in figure 2.2.7, attempts to combine the favourable properties of alkoxide and

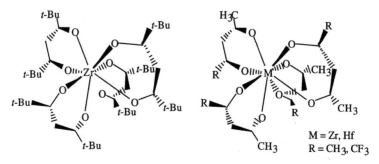

Figure 2.2.6. Schematic drawing of group IV β-diketonate precursors: $Zr(thd)_4$ and $M(acac)_4$.

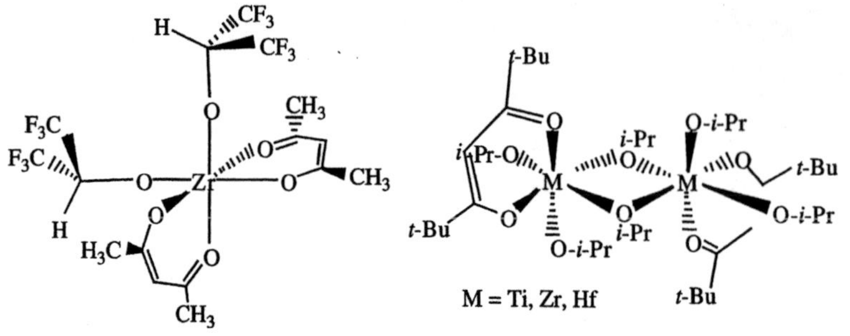

Figure 2.2.7. Schematic drawing of group IV mixed ligand precursors: *cis*-Zr(acac)$_2$(hfip)$_2$ and M$_2$(O-*i*-Pr)$_6$(thd)$_2$.

β-diketonate complexes [41−45]. The liquid compound Zr(acac)$_2$(hfip)$_2$, where hfip = OCH(CF$_3$)$_2$, has been used to deposit ZrO$_2$ films as low as 350°C [44, 45]. The compound is reported to be more stable towards hydrolysis than Zr[OC(CH$_3$)$_3$]$_4$ and the vapour pressure lies between the zirconium β-diketonates and *tert*-butoxide. The mixed tetramethylheptanedionato-alcoholato complexes, Zr(OCH(CH$_3$)$_2$)$_2$(thd)$_2$ and [Zr(OCH(CH$_3$)$_2$)$_3$(thd)]$_2$, have been used in direct liquid injection CVD [41, 42, 46]. The volatilities of Zr(OCH(CH$_3$)$_2$)$_2$(thd)$_2$ and [Zr(OCH(CH$_3$)$_2$)$_3$(thd)]$_2$ do not appear to be a significant improvement over the acetylacetonates, as evaporator temperatures for the depositions were 195°C. Zirconium dioxide films could be deposited, however, between 300 and 600°C from 0.1 M THF solutions of the compounds. [Zr(OCH(CH$_3$)$_2$)$_3$(thd)]$_2$ and Ti(OCH(CH$_3$)$_2$)$_2$(thd)$_2$ have also been used to deposit lead zirconate titanate (PZT) using direct liquid injection CVD [42].

The anhydrous metal nitrates, M(NO$_3$)$_4$, are the only single source precursors that are completely carbon, hydrogen, and halogen free [5, 47−51]. The volatility of the anhydrous nitrates arises from the covalent nature of the interaction between the metal centre and the oxygen atoms of the chelating nitrato ligands (figure 2.2.8). Typical source temperatures range from 40°C for titanium nitrate to 80°C for zirconium and hafnium nitrate. No additional oxygen is required for deposition, and high-quality films of TiO$_2$, ZrO$_2$, and HfO$_2$ have been deposited at temperatures below 250°C.

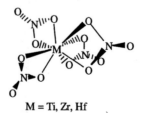

M = Ti, Zr, Hf

Figure 2.2.8. Schematic drawing of group IV anhydrous metal nitrate precursors.

Because the nitrate ligands themselves are strongly oxidizing, the anhydrous group IV nitrates should not be mixed with liquid or solid organic compounds or other readily oxidized materials. However, the reactivity of the ligands in these compounds makes them attractive candidates for ALD. In pure form we have seen no evidence that anhydrous metal nitrates are shock sensitive or subject to violent decomposition upon heating.

Impurities in group IVB CVD films

One of the issues in the CVD of metal oxides from MO sources is the inclusion of impurities such as carbon and hydrogen in the films. These impurities may arise from an incomplete scission of one of the organic ligands, or, more likely, due to the incorporation of by-products into the film. These could be gaseous by-products that impinge on the surface, or they may be species produced by the surface decomposition of the adsorbed molecules. At this writing little is known about the effect of these impurities on the electrical properties of the film; however, it is reasonable to expect that they will affect leakage currents, charge trapping, and the reliability of the films. One significant caution in this area is the meaning of 'low-carbon' and 'carbon-free' claims that one often sees in the literature [52]. Due to its light mass, the detection limit of carbon is typically about 1% using RBS, EDX, and other common thin-film techniques. Devices, however, are typically sensitive to electrical active defects of well below 1 ppm. Naturally, hydrogen is often even more difficult to detect and is seldom measured.

One can argue, from the effects of these impurities in SiO_2, that the effects of these impurities in the metal oxides may be significant. The effect of carbon in SiO_2 has been most heavily studied as it relates to the thermal oxidation of SiC. Charge centres in SiO_2 on SiC produces low inversion layer mobility. Excess carbon in the oxide is believed to form defect clusters with an energy about 3 eV above the valence band edge of SiC [53]. Research suggests that these defects are $\sim 2\,nm$ from the interface and that they are the cause of the slow or 'near-interface' states, typically seen in the large hysteresis loops in 4H material [54]. Various other studies have shown that removal of carbon from the surface, either by pre-oxidation treatments, or by post-oxidation anneals, reduces the film charge, while increasing the carbon by techniques such as CO_2 annealing increases the film charge. Post-deposition annealing of many of the metal oxide high-κ films using water vapour or other oxidizing ambients is also found to significantly decrease the leakage current, although this effect is not necessarily due to volatilization of carbon. Other possible explanations include repairing oxygen vacancies and thickening of the interfacial layer.

Fredriksson and Forsgren studied the deposition of ZrO_2 from several of the β-diketonates with various oxygen sources. They mapped out regions of low impurity concentrations in the films [55]. One of the most extensive studies of ZrO_2 CVD from ZTB was done by Cameron and George. They found that depositing ZrO_2 from 50 mtorr of ZTB in a sealed, nonflowing reactor produced carbon concentrations as high as 14%, and that the carbon content increased with increasing temperature. Chen *et al* [52] investigated the use of a novel precursor, $Zr(O\text{-}i\text{-}Pr)_2(thd)_2$, for the deposition of ZrO_2. They were able to obtain films with carbon content less than 1%, but the equivalent oxide thickness of these films was rather high, suggesting that the interfacial layer may be thick. Ohshita *et al* used tetrakisdiethylamidohafnium $(Hf(NEt_2)_4)$ with O_2 as a source for depositing HfO_2. They found that increasing the deposition temperature decreased the carbon content [56]. Harsta has studied metal iodides as carbon-free and hydrogen-free sources for both CVD and ALD of binary group IVB oxides [57]. A group at Minnesota has developed the use of anhydrous metal nitrates of the form $M(NO_3)_4$, where M = Ti, Zr, Sn, and Hf as carbon-free and hydrogen-free single sources for forming high-κ layers [47]. As with all oxygen-rich sources, however, the key to the successful use of these precursors is the control of the interfacial layer formation. This will be discussed in a later section.

The addition of water during the deposition or during PDAs significantly decreases the carbon content, but may increase the growth of the interfacial layer. Based on infrared and kinetic measurements, Bradley [22] proposed that below 300°C, the hydrolysis of ZTB proceeds in a two-step process to form methylpropene as:

$$Zr[OC(CH_3)_3]_4 + 2H_2O \xrightarrow{\Delta} ZrO_2 + 4C(CH_3)_3OH \qquad (2.2.6)$$

$$4C(CH_3)_3OH \longrightarrow 4CH_2{=}C(CH_3)_2 + 4H_2O \qquad (2.2.7)$$

Recently, the stable, solid *neo*-pentoxides of titanium and zirconium were used to deposit the corresponding oxides in an MOCVD process [58].

The effect of hydrogen in high-κ materials is, again, relatively unknown. The effect of this impurity in SiO_2, however, is instructive. This has been studied by Afanas'ev and Stesmans. Their hydrogen transport model [59] suggests that impacting electrons release H^+ ions near the Si/SiO_2 interface which then randomly hop around in the gate dielectric stack, where they can be trapped and form hydrogen-induced defects including Si–O bond rupture. This process can ultimately lead to electrical breakdown in SiO_2. Houssa has studied similar effects in $Si/SiO_2/ZrO_2/TiN$ structures and found that the results obeyed this model. Neutral defects in ZrO_2 were believed to be ZrOH centres [60]. This would suggest that reducing the hydrogen content in the high-κ films will improve their reliability and, perhaps, reduce their charge density.

Morphology and structure of IVB films

For application as a planar FET gate insulator, any of the proposed techniques would have sufficient step coverage. For nonplanar structures, such as trench capacitors, wrap-around gates and FINFETs, however, the step coverage of the film is an extremely important consideration. Generally, one finds that step coverage improves when going from PVD to CVD to ALD. Given the desirability of CVD for manufacturing, and given the impurity and charge problems seen in ALD, producing highly conformal CVD coatings is of considerable interest. Early work in this area was done by Gilmer [61]. Using anhydrous titanium nitrate as a precursor, they found that the deposition uniformity on 10 μm deep, 2 μm wide trenches increased dramatically as the deposition temperature was lowered from 490 to 170°C. Such a temperature reduction would move the reaction away from any flux-limited region (figure 2.2.3). The deposition rate was found to decrease when the process was operated in a reaction rate limited regime, where no nonuniformity could be observed in the trench. Obviously, a trade-off would have to be made in a manufacturing environment between uniformity and deposition rate.

Akiyama *et al* have studied this problem numerically [62]. They found that microscale trenches and holes can be occluded near their openings with thinner film grown on top of the substrate leaving a wider void deep in the hole by a set of CVD operations at two temperature levels, i.e., a film is grown first at a higher temperature and then at a lower temperature. This operation was effectively verified by low pressure metalorganic CVD experiments with ZrO_2 films.

The crystallinity of the film plays an important role in its performance as a high-κ material. It is now well known that the permittivity is a strong function of crystallinity [63]. It is also likely, but not proven, that crystallinity will affect thermal stability, leakage current, impurity diffusion, and reliability. All of the group IVB binary metal oxides are found to be amorphous when deposited at sufficiently low temperature. In some cases, the deposition rate at these temperatures is too low to be useful. A similar effect is seen in plasma-enhanced CVD processes, where the amorphous to polycrystalline transformation occurs at sufficiently high power densities [64]. At moderate (300–700°C) deposition and/or anneal temperatures, however, all of the simple binary IVB oxides undergo an amorphous to polycrystalline phase transformation. Thus, for device application, the films are all poly-crystalline. The crystal phase of HfO_2 and ZrO_2 deposited appears to depend on the film thickness [65]. Thin films appear to favour the lower symmetry phases. ZrO_2 shows the more pronounced effect, going from cubic (fluoride) or tetragonal for films less than 30 nm to monoclinic for thicker films. In HfO_2 very thin films (~6 nm) show some tetragonal phase, while thicker films are exclusively Baddelleyite (monoclinic).

Film roughness can have a significant effect on the performance of high-κ films. Because a rough film has thin spots and because the leakage current decreases exponentially with film thickness in the direct tunnelling regime, it is possible to show that [66]

$$J_{rough} = J_{smooth} \exp[\sigma^2 / 2t_0^2] \qquad (2.2.8)$$

where J_{rough} and J_{smooth} are the leakage current densities of rough and smooth films, respectively, σ is the standard deviation of the film roughness, and t_0 is a parameter that is material dependent. For polycrystalline films, lowering the deposition temperature decreases the roughness. This is not surprising since reducing the deposition temperature reduces the adspecies surface mobility. Thus, molecules on the surface are much less likely to have sufficient energy to diffuse into kink sites where they may be incorporated into growing facets. Interestingly, for HfO_2 it has been shown [66] that post-deposition annealing at temperatures up to 700°C does not increase the roughness.

Other high-κ metal oxides

Ta_2O_5 was one of the first materials studied as a gate insulator. It has a dielectric constant of 20–26, a band gap of 4.5 eV, but appears to have a low band offset to Si. It is also not thermodynamically stable on silicon. The precursors used in the CVD process include $Ta(N(CH_3)_2)_5$ [67, 68], $Ta(O(C_2H_5)_5$ [69–72], and $TaC_{12}H_{30}O_5N$ [73]. Each is typically used with an oxidant such as NO, N_2O, O_2, or H_2O. Like the group IV metal oxides, films deposited by these precursors have a significant carbon and hydrogen impurity concentration unless the oxidant partial pressure in the reactor is high. High temperature post-deposition oxidant ambient annealing was often required to reduce the leakage current density and improve the film density and charge trap density. This produces additional interfacial layer growth, as discussed earlier.

Studies have also shown that the incorporation of impurities into Ta_2O_5 affects the film's properties greatly; however, the results are not consistent. Cava *et al* [74] found that substitution of low valence elements such as N and Zr into the O or Ta position would reduce the leakage current of the films. This was contradicted by Jung's results [71] that instead indicated the opposite trend.

Al_2O_3 has also received considerable attention. It has a relatively modest dielectric constant (~8), but has a bandgap of 8.7 eV. More importantly it is thermodynamically stable on silicon. Al_2O_3 can be deposited by CVD [75], although much more work has been done using ALD. When depositing from the acacs, particle formation and inclusion into the film are possible [76].

Al_2O_3 films have 10^{11} interface states/cm^2, but often have high bulk trap densities. Manchanda [77] found that the interface state density and traps in the films can be greatly suppressed by adding small amounts of a dopant such as Si and Zr without sacrificing the dielectric constant. An interface state density as low as 3×10^{10} can be reached.

Pseudobinary alloys

As discussed, simple binary metal oxides have severe problems related to crystallization. For these reasons, a number of groups have investigated alloying binary oxides with silicon, aluminium, and nitrogen to form amorphous films. The amorphous films tend to have lower permittivities, both due to the phase transformation [63] and due to the chemical composition of the film [79]. Generally, one trades off increased thermal stability for lower permittivity. In some cases, however, the bandgaps are expected to increase with increasing levels of alloying. This will reduce the leakage current, which may allow thinner films that can offset the loss of permittivity. This section will focus on the silicates since almost all of the work on aluminates has been done by PVD or ALD, and the work on nitrided high-κ films has not yet been published.

The silicates of the group IVB metal oxides have been the most heavily studied pseudobinaries. Zr and Hf are predicted [80] to be thermodynamically stable on silicon from MO_2 to $MSiO_4$ to SiO_2. Higher concentrations of oxygen are generally considered desirable as this reduces the likelihood of forming silicide bonds in the material [81]. Generally, a Si-rich mixture is desired to make the film stable to high enough temperatures for dopant activation anneals which are done at approximately 1000°C. For low concentration pseudobinary mixtures, a glass transition temperature exists between 800 and 900°C at which phase separation into the end-member components, SiO_2 and ZrO_2, occurs [82]. Higher Si concentrations are stable to higher temperatures.

Most studies in this area have involved separate source gases for the metal and the silicon, with the oxidant as a third gas. One exception is the work by Zurcher *et al.* They report two single-source zirconium silicate precursors, $Zr(acac)_2(OSiMe_3)_2$ and $Zr(acac)_2[OSi(t\text{-}Bu)Me_2]_2$, which they used to deposit films of $Zr_{1-x}Si_xO_2$ in the temperature range 400–700°C. Silicon concentrations in the range 0.05–0.25 were obtained [83]. For three-gas CVD work, silicon sources have included silane, disilane, TEOS [84], and tris-diethylamino-silane $(HSi(NEt_2)_3)$ [85].

The deposition of aluminium silicates has also been studied. Kuo and Chuang have shown that both tetraethylorthosilicate (TEOS) and hexamethyldisilazane (HMDSN) can be used with aluminium tri-*sec*-butoxide (ATSB) to form aluminium silicates with variable silicon concentrations [86].

Interfacial layers

It is generally found that the deposition of metal oxide films on silicon forms an interfacial layer. Since this layer is in series with the deposited material, its equivalent oxide thickness must be added to that of the high-κ layer. Keeping this layer as thin as possible is essential to obtaining usable high-κ stack structures.

The group IVB metal oxides seem to nucleate readily on Si from all of the MO sources. Unfortunately, they also produce an amorphous zirconium silicate interfacial layer [78, 87, 88]. The CVD of Al_2O_3 from the acac also shows the formation of a mixed aluminate interfacial layer [89]. It is interesting to speculate on the source of these layers. The transition layer of lanthanum oxide film on Si as prepared by metalorganic CVD has been studied by x-ray photoelectron spectroscopy, cross-sectional scanning transmission electron microscopy, and energy dispersive x-ray analysis. It was revealed that the diffusion of silicon into the lanthanum oxide occurs during the film deposition and post-annealing, and consequently, a lanthanum silicate is formed. The composition of lanthanum and silicon in the silicate is nonstoichiometric and gradually changes in the direction of the film thickness. These results show that the suppression of the silicon diffusion is essential in controlling the properties of the dielectric films [90]. Similarly, work by Harada *et al* has shown that the out-diffusion of silicon into HfO_2 leads to an unintentional silicate interfacial layer, and that the thicker this interfacial layer becomes, the lower the Weibull slope for the reliability. On the other hand, an intentionally grown silicate had an improved reliability, suggesting that the interface between the binary and the pseudobinary layers may play an important role in the breakdown process [91]. A number of groups have shown that nitridation of the surface of the silicon before high-κ deposition will reduce the thickness of the interfacial layer, but most nitridations also increase the fixed charge density of the stack. Recently, work has begun on the deposition of high-κ materials on very thin ($\sim$0.5 nm) layers of SiO_xN_y grown thermally from NO.

When using the nitrato sources, however, the interfacial layer is not the silicate, but instead is SiO_xN_y when the deposition process was run at 350°C. Interestingly, Park *et al* found that this precursor appeared to form a silicate interfacial layer when deposited at 200°C, but this layer appeared to phase segregate into SiO_2 and HfO_2 when post-deposition annealed at temperatures above 500°C [92]. It may be that the nitrogen content of these films is sufficient to suppress the updiffusion of silicon, while the typical MO sources do not form an oxynitride at the interface. The thickness of the interfacial oxynitride was found to be independent of the deposition temperature over the range 250–500°C, but longer deposition times produced a greater interfacial layer thickness. It has been suggested that this oxidation proceeds via the Mott–Cabrera mechanism, which involves the drift of dissolved oxidants through the interfacial layer [93, 94]. By reducing the deposition time, the thickness of the interfacial layer was reduced to 0.6 nm [66].

Summary

CVD is the preferred technique for the manufacture of devices with high-κ gate insulators. The outcome of the CVD process depends heavily on the reactor, the precursors, the deposition process, and the post-deposition thermal cycles. Research on the binary metal oxides has now been joined by research on silicates, aluminates, and nitrided metal oxides. Control of the interfacial layer is key to developing a successful high-κ stack.

References

[1] *Handbook of Chemical Vapor Deposition* 1992 ed H O Pierson (Park Ridge, NJ: Noyes Publications)

[2] Sherman A 1987 *Chemical Vapor Deposition for Microelectronics* (Park Ridge, NJ: Noyes Publications)

[3] *The Chemistry of Metal CVD* 1994 eds T T Kodas and M J Hampden-Smith (Weinheim: VCH)

[4] *CVD of Nonmetals* 1996 ed W S J Rees (Weinheim: VCH)

[5] Taylor C J *et al* 1999 Does chemistry really matter in the chemical vapor deposition of titanium dioxide? Precursor and kinetic effects on the microstructure of polycrystalline films *J. Am. Chem. Soc.* **121** 5220–5229

[6] Forsgren K and Harsta A 1999 CVD of ZrO_2 using ZrI_4 as metal precursor *J. de Physique IV* **9 pt 1** Pr8-487–Pr8-491

[7] Powell C F 1966 Chemically deposited nonmetals *Vapor Deposition* ed J M Blocher (New York: Wiley) pp 343–420

[8] Copel M, Gribelyuk M and Gusev E 2000 Structure and stability of ultrathin zirconium oxide layers on Si(001) *Appl. Phys. Lett.* **76** 436–438

[9] Yamane H and Hirai T 1987 Preparation of ZrO_2-film by oxidation of $ZrCl_4$ *J. Mater. Sci. Lett.* **6** 1229–1230

[10] Choi J-H, Kim H-G and Yoon S-G 1992 Effects of the reaction parameters on the deposition characteristics in ZrO_2 CVD *J. Mater. Sci.: Mater. Electron.* **3** 87–92

[11] Tauber R N, Dumbri A C and Caffrey R E 1971 Preparation and properties of pyrolytic zirconium dioxide films *J. Electrochem. Soc.* **118** 747–754

[12] Ghoshtagore R N and Norieka A J 1970 Growth characteristics of rutile film by chemical vapor deposition *J. Electrochem. Soc.* **117** 1310–1314

[13] Yeung K S and Lam Y W 1983 A simple chemical vapour deposition method for depositing thin TiO_2 films *Thin Solid Films* **109** 169–179

[14] Fictorie C P, Evans J F and Gladfelter W L 1994 Kinetic and mechanistic study of the chemical vapor deposition of titanium dioxide thin films using tetrakis-(isopropoxo)-titanium(IV) *J. Vac. Sci. Technol. A* **12** 1108–1113

[15] Kim T W *et al* 1994 Optical and electronic properties of titanium dioxide films with a high magnitude dielectric constant grown on p-Si by metalorganic chemical vapor deposition at low temperature *Appl. Phys. Lett.* **64** 1407

[16] Lu J, Wang J and Raj R 1991 Solution precursor chemical vapor deposition of titanium oxide thin films *Thin Solid Films* **204** L13–L17

[17] Rausch N and Burte E P 1992 Thin high-dielectric TiO_2 films prepared by low pressure MOCVD *Microelectron. Eng.* **19** 725–728

[18] Rausch N and Burte E P 1993 Thin TiO_2 films prepared by low pressure chemical vapor deposition *J. Electrochem. Soc.* **140** 145–149

[19] Siefering K L and Griffin G L 1990 *J. Electrochem. Soc.* **137** 814–818

[20] Yan J *et al* 1996 Structural and electrical characterization of TiO_2 grown from titanium tetrakis-isopropoxide (TTIP) and $TTIP/H_2O$ ambients *J. Vac. Sci. Technol. B* **14** 1706–1711

[21] Yoon Y S *et al* 1994 Structural properties of titanium dioxide films grown on p-Si by metal-organic chemical vapor deposition at low temperature *Thin Solid Films* **238** 12–14

[22] Bradley D C 1989 Metal alkoxides as precursors for electronic and ceramic materials *Chem. Rev.* **89** 1317–1322

[23] Gould B J *et al* 1994 Chemical vapour deposition of ZrO_2 thin films monitored by IR spectroscopy *J. Mater. Chem.* **4** 1815–1819

[24] Takahashi Y, Kawae T and Nasu M 1986 Chemical vapour deposition of undoped and spinel-doped cubic zirconia film using organometallic process *J. Cryst. Growth* **74** 409–415

[25] Xue Z *et al* 1992 Chemical vapor deposition of cubic-zirconia thin films from zirconium alkoxide complexes *Eur. J. Solid State Inorg. Chem.* **29** 213–225

[26] Pakswer S and Skoug P 1970 Thin dielectric oxide films made by oxygen assisted pyrolysis of alkoxides *Second International Conference on Chemical Vapor Deposition* (Los Angeles, CA: The Electrochemical Society)

[27] Cho B O, Lao S, Sha L and Chang J P 2001 *J. Vac. Sci. Technol. A* **19** 2751–2761

[28] Kukli K, Ritala M and Leskela M 2000 *Chem. Vap. Deposition* **6** 297–302

[29] Desu S B, Shi T and Kwok C K 1990 Structure, composition and properties of MOCVD ZrO_2 thin films *Chemical Vapor Deposition of Refractory Metals and Ceramics* ed B M Gallois (Pittsburgh, PA: Materials Research Society) pp 349–355

[30] Balog M *et al* 1972 Thin films of metal oxides on silicon by chemical vapor deposition with organometallic compounds. I *J. Cryst. Growth* **17** 298–301

[31] Kim J S, Marzouk H A and Reucroft P J 1995 Deposition and structural characterization of ZrO_2 and yttria-stabilized ZrO_2 films by chemical vapor deposition *Thin Solid Films* **254** 33–38

[32] Balog M *et al* 1977 The chemical vapour deposition and characterization of ZrO_2 films from organometallic compounds *Thin Solid Films* **47** 109–120

[33] Balog M *et al* 1979 The characteristics of growth of films of zirconium and hafnium oxides (ZrO_2, HfO_2) by thermal decomposition of zirconium and hafnium

β-diketonate complexes in the presence and absence of oxygen *J. Electrochem. Soc.* **126** 1203–1207

[34] Hwang C S and Kim H J 1993 Deposition and characterization of ZrO_2 thin films on silicon substrate by CVD *J. Mater. Res.* **8** 1361–1367

[35] Balog M *et al* 1977 Chemical vapor deposition and characterization of HfO_2 films from organo-hafnium compounds *Thin Solid Films* **41** 247–259

[36] Kim E-T and Yoon S-G 1993 Characterization of zirconium dioxide film formed by plasma enhanced metal-organic chemical vapor deposition *Thin Solid Films* **227** 7–12

[37] Si J, Peng C H and Desu S B 1992 Deposition and characterization of metalorganic chemical vapor deposition ZrO_2 thin films using $Zr(thd)_4$ *Mater. Res. Soc. Symp. Proc.* **250** 323–329

[38] Si J, Desu S B and Tsai C-Y 1994 Metal–organic chemical vapor deposition of ZrO_2 films using $Zr(thd)_4$ as precursors *J. Mater. Res.* **9** 1721–1727

[39] Peng C H and Desu S B 1994 Metalorganic chemical vapor deposition of ferroelectric $Pb(Zr,Ti)O_3$ thin films *J. Am. Ceram. Soc.* **77** 1799–1812

[40] Akiyama Y, Sato T and Imaishi N 1995 Reaction analysis of ZrO_2 and Y_2O_3 thin film growth by low-pressure metalorganic chemical vapor deposition using β-diketonate complexes *J. Cryst. Growth* **147** 130–146

[41] Jones A C *et al* 1998 Liquid injection MOCVD of zirconium dioxide using a novel mixed ligand zirconium precursor *Chem. Vap. Deposition* **4** 197–201

[42] Jones A C *et al* 1999 Metalorganic chemical vapor deposition (MOCVD) of zirconia and lead zirconate titanate using a novel zirconium precursor *J. Eur. Ceram. Soc.* **19** 1431–1434

[43] Kim D-Y, Lee C-H and Park S J 1996 Preparation of zirconia thin films by metalorganic chemical vapor deposition using ultrasonic nebulization *J. Mater. Res.* **11** 2583–2587

[44] Morstein M, Pozsgai I and Spencer N D 1999 Composition and microstructure of zirconia films obtained by MOCVD with a new, liquid, mixed acetylacetonato-alcoholato precursor *Chem. Vap. Deposition* **5** 151–158

[45] Morstein M 1999 Volatile zirconium bis(acetylacetonato)bis(alcoholato) complexes containing heterosubstituted alcoholato ligands *Inorg. Chem.* **38** 125–131

[46] Jones A C *et al* 1998 MOCVD of zirconia thin films by direct liquid injection using a new class of zirconium precursor *Chem. Vap. Deposition* **4** 46–49

[47] Colombo D G *et al* 1998 Anhydrous metal nitrates as volatile single source precursors for the CVD of metal oxide films *Chem. Vap. Deposition* **4** 220–222

[48] Gilmer D C *et al* 1998 Low temperature CVD of crystalline titanium dioxide films using tetranitratotitanium(IV) *Chem. Vap. Deposition* **4** 9–11

[49] Smith R C *et al* 2000 Low temperature chemical vapor deposition of ZrO_2 on Si(100) using anhydrous zirconium(IV) nitrate *J. Electrochem. Soc.* **147** 3472–3476

[50] Smith R C *et al* 2000 Chemical vapour deposition of the oxides of titanium, zirconium and hafnium for use as high *k* materials in microelectronics devices. A carbon-free precursor for the synthesis of hafnium dioxide *Adv. Mater. Opt. Electron.* **10** 105–114

[51] Park J *et al* 2002 Chemical vapor deposition of HfO_2 thin films using a novel carbon-free precursor *J. Electrochem. Soc.* **149** G89–G94

[52] Chen H W, Landheer D, Wu X, Moisa S, Sproule G I, Chao T S and Huang T Y 2002 Characterization of thin ZrO_2 films deposited using $Zr(Oi-Pr)_2(thd)_2$ and O_2 on Si(100) *J. Vac. Sci. Technol. A—Vac. Surf. Films* **20** 1145–1148

[53] Bassler M, Afanas'ev V V and Pensl G 1998 Interface state density at implanted 6H SiC/SiO_2 MOS structures *Mater. Sci. Forum* **264–268** 861–864

[54] Pensl G, Bassler M, Ciobanu F, Afanas'ev V, Yano H, Kimoto T and Matsunami H 2000 Traps at the SiC/SiO_2 interface *Mater. Res. Soc. Proc.* **640** H3.2.1–H3.2.11

[55] Fredriksson E and Forsgren K 1997 Thermodynamic modelling of MOCVD of ZrO_2 from beta-diketonates and different oxygen sources *Surf. Coatings Technol.* **88** 255–263

[56] Ohshita Y *et al* 2002 Using tetrakis-diethylamido-hafnium for HfO_2 thin-film growth in low-pressure chemical vapor deposition *Thin Solid Films* **406** 215–218

[57] Harsta A 2001 Halide CVD of dielectric and ferroelectric oxides *13th European Conference on Chemical Vapor Deposition (EUROCVD 13)* (Athens: Journal de Physique IV)

[58] Gallegos J J *et al* 2000 Neo-pentoxide precursors for MOCVD thin films of TiO_2 and ZrO_2 *Chem. Vap. Deposition* **6** 21–26

[59] Afanas'ev V V and Stesmans A 1999 Hydrogen-related leakage currents induced in ultrathin SiO_2/Si structures by vacuum ultraviolet radiation *J. Electrochem. Soc.* **146** 3409–3414; Houssa M, Stesmans A, Carter R and Heyns M M 2001 Stress-induced leakage current in ultrathin SiO_2 layers and the hydrogen dispersive transport model *Appl. Phys. Lett.* **78** 3289–3291

[60] Houssa M, Afanas'ev V V, Stesmans A and Heyns M M 2001 Defect generation in Si/SiO_2/ZrO_2/TiN structures: the possible role of hydrogen *Semicond. Sci. Technol.* **16** L93–L96

[61] Gilmer D C 1998 Chemical vapor deposition and characterization of titanium dioxide thin films *PhD Thesis Chemistry* (Minneapolis, MN: Minnesota Press)

[62] Akiyama Y, Sato T and Imaishi N 1995 Occlusion of micro-size trench and hole by thin film growth via CVD—3-dimensional Monte Carlo simulation *Proceedings of the 1995 ASME/JSME Thermal Engineering Joint Conference* Part 2 (of 4): ASME/JSME Thermal Engineering Joint Conference—Proceedings. v 2 1995 (New York: ASME)

[63] Forsgren K *et al* 2002 Deposition of HfO_2 thin films in HfI_4-based processes *J. Electrochem. Soc.* **149** F139–F144

[64] Colpo P *et al* 2000 Characterization of zirconia coatings deposited by inductively coupled plasma assisted chemical vapor deposition *J. Vac. Sci. Technol. A—Vac. Surf. Films* **18** 1096–1101

[65] Liu R *et al* 2001 Materials and physical properties of novel high-κ and medium-k gate dielectrics *Gate Stack and Silicide Issues in Silicon Processing II* (San Francisco, CA: MRS)

[66] Chen F, Smith R, Campbell S A and Gladfelter W L 2002 Effect of HfO_2 deposition on interfacial layer thickness and roughness, Proceedings of the ECS Meeting, Spring 2002 *Spring Meeting of the Electrochemical Society* (Philadelphia, PA: ECS)

[67] Son K A *et al* 1998 Ultrathin Ta_2O_5 film growth by chemical vapor deposition of $Ta(N(CH3)_2)_5$ and O_2 on bare and SiO_xN_y-passivated Si(100) for gate dielectric applications *J. Vac. Sci. Technol. A* **16** 1670–1675

[68] Son K A *et al* 1998 Deposition and annealing of ultrathin Ta_2O_5 films on nitrogen passivated Si(100) *Electrochem. Solid-State Lett.* **1** 178–180

[69] Kim S O *et al* 1994 Fabrication of n-metal–oxide–semiconductor field effect transistor with Ta_2O_5 gate prepared by plasma enhanced metalorganic chemical vapor deposition *J. Vac. Sci. Technol. B* **12** 3006

[70] Park D *et al* 1998 Transistor characteristics with Ta_2O_5 gate dielectric *IEEE Electron Device Lett.* **19** 441

[71] Jung H *et al* 2000 Electrical characteristics of an ultrathin (1.6 nm) TaO_xN_y gate dielectric *Appl. Phys. Lett.* **76** 3630

[72] Lu Q *et al* 1998 Leakage current composition between ultrathin Ta_2O_5 films and conventional gate dielectrics *IEEE Electron Device Lett.* **19** 341

[73] Luan H F 1998 Ultrathin high quality Ta_2O_5 gate dielectric prepared by in-situ rapid thermal processing *International Electron Devices Meeting* (New York: IEEE)

[74] Cava R F 1995 *Nature* **377** 215

[75] Temple D *et al* 1990 Formation of aluminum oxide films from aluminum hexa-fluoroacetonate at 350–450°C *J. Electron. Mater.* **19** 995

[76] Pulver M, Nemetz W and Wahl G 2000 CVD of ZrO_2, Al_2O_3 and Y_2O_3 from metalorganic compounds in different reactors *Surf. Coatings Technol.* **125** 400–406

[77] Manchanda L *et al* 1998 Gate quality doped high *k* films for CMOS beyond 100 nm: 3–10 nm Al_2O_3 with low leakage and low interface states *International Electron Devices Meeting* (New York: IEEE)

[78] Kawamoto A, Jameson J, Griffin P, Cho K and Dutton R 2001 Atomic scale effects of zirconium and hafnium incorporation at a model silicon/silicate interface by first principles calculations *IEEE Electron Device Lett.* **22** 14–16

[79] Kato H *et al* 2001 Fabrication of hafnium silicate films by plasma-enhanced chemical vapor deposition *Proceedings of 2001 International Symp. on Electrical Insulating Materials (ISEIM 2001)/2001 Asian Conf: Proceedings of the International Symposium on Electrical Insulating Materials*

[80] Wang S Q and Mayer J W 1988 Reactions of Zr thin films with SiO_2 substrates *J. Appl. Phys.* **64** 4711

[81] Wilk G D, Wallace R M and Anthony J M 2000 Hafnium and zirconium silicates for advanced gate dielectrics *J. Appl. Phys.* **87** 484–492

[82] Rayner G Jr, Therrien R and Lucovsky G 2001 The structure of plasma-deposited and annealed pseudo-binary ZrO_2–SiO_2 alloys *Gate Stack and Silicide Issues in Silicon Processing, San Francisco, CA: Materials Research Society Symposium—Proceedings*

[83] Zurcher S, Morstein M, Spencer N D, Lemberger M and Bauer A 2002 New single-source precursors for the MOCVD of high-kappa dielectric zirconium silicates to replace SiO_2 in semiconducting devices *Adv. Mater.* **14** 171–177

[84] Gladfelter W L, Smith R C, Taylor C J, Roberts J T, Hoilien N, Campbell S A, Tiner M, Hegde R and Hobbs C 2001 Chemical vapor deposition of amorphous films of titania, zirconia or hafnia alloyed with silica *Gate Stacks and Silicides II* (Mater. Res. Soc.)

[85] Ishikawa M *et al* 2001 HfO_2 and $Hf_{1-x}Si_xO_2$ deposition by MOCVD using TDEAH *Advanced Metallization Conference (AMC 2001), Montreal: Advanced Metallization Conference (AMC)*

[86] Kuo D H and Chuang P Y 2002 Growth behaviors of low-pressure metalorganic chemical vapor deposition aluminum silicate films deposited with two kinds of silicon sources: hexamethyldisilazane and tetraethyl orthosilicate *J. Vac. Sci. Technol. A—Vac. Surf. Films* **20** 1511–1516

[87] Chen H-W, Huang T-Y, Landheer D, Wu X, Moisa S, Sproule G I and Chao T-S 2002 Physical and electrical characterization of ZrO_2 gate insulators deposited on Si(100) using $Zr(Oi-Pr)_2(thd)_2$ and O_2 *J. Electrochem. Soc.* **149**(6) F49–F55

[88] Chang J P and Lin Y-S 2001 Ultra-thin zirconium oxide films deposited by rapid thermal chemical vapor deposition (RT-CVD) as alternative gate dielectric *Gate Stack and Silicide Issues in Silicon Processing II, San Francisco, CA: Materials Research Society Symposium—Proceedings*

[89] Klein T M *et al* 1999 Evidence of aluminum silicate formation during chemical vapor deposition of amorphous Al_2O_3 thin films on Si(100) *Appl. Phys. Lett.* **75** 4001–4003

[90] Yamada H, Shimizu T and Suzuki E 2002 Interface reaction of a silicon substrate and lanthanum oxide films deposited by metalorganic chemical vapor deposition *Japan. J. Appl. Phys. Part 2—Lett.* **41** L368–L370

[91] Harada Y *et al* 2002 Specific structural factors influencing on reliability of CVD-HfO_2 *2002 Symposium on VLSI Technology Digest of Technical Papers, Honolulu, HI: IEEE Symposium on VLSI Circuits, Digest of Technical Papers 2002*

[92] Park J *et al* 2002 Chemical vapor deposition of HfO_2 thin films using a novel carbon-free precursor: characterization of the interface with the silicon substrate *J. Electrochem. Soc.* **149** G89–G94

[93] Mott F 1947 *Trans. Faraday Soc.* **43** 429

[94] Cabrera N and Mott N F 1948 *Rep. Prog. Phys.* **12** 163

Chapter 2.3

Pulsed laser deposition of dielectrics

Dave H A Blank, Lianne M Doeswijk,
Koray Karakaya, Gertjan Koster and Guus Rijnders

Introduction

The flexibility of the pulsed laser deposition (PLD) set-up offers the researcher a large freedom of choice in target material, ablation characteristics, target–substrate geometry, ambient gas and its pressure, and substrate temperature. All these parameters influence plasma formation, film growth, and film properties. Deposition parameters can be varied over several orders of magnitude. In particular, the ambient pressure is known for its large dynamic range of more than seven orders of magnitude. Furthermore, the kinetic energies of species arriving at the substrate surface can be tuned over three orders of magnitude (0.1–400 eV). The large dynamic ranges of deposition parameters result in different regimes of ablation and deposition. Due to powerful lasers with short pulse lengths, high laser power densities can be realized. This makes the ablation of a wide range of materials possible.

PLD is attractive for research on novel dielectrics because it is fast and one can easily investigate a wide range of different materials and compositions. Currently, a major issue in the growth of oxide materials with PLD is the control of the surface morphology. For high-κ materials it is necessary to control the thickness and roughness of the thin films down to an atomic scale. Such well-controlled growth can also be used to manufacture artificially layered structures of different materials with different properties. In this way, it is possible to create a whole new class of materials. It would be possible to create materials tailor made to applications. Such materials are also ideal for the purpose of understanding the physics and the search for materials with as yet unknown properties.

PLD has become the workhorse for the development of complex oxide materials. Much effort is put into the deposition of excellent textured layers without grain boundaries. In general, the properties of highly oriented films approximate the properties of single crystals. Single or multi-layer structures require a well-conditioned process technique. The deposited layers must have a large homogeneity with well-defined material properties, smooth surfaces, and, in the case of oxides, the correct oxygen stoichiometry.

In 1988, several groups have succeeded in growing, with PLD, the high temperature superconducting $YBa_2Cu_3O_7$ layers with the orthorhombic phase and correct oxygen stoichiometry, without a further anneal step [1]. Simultaneously, the effect of the partial oxygen pressure during deposition became clear. Borman and Hammond [2, 3] showed the relation between the oxygen partial pressure during deposition and the oxygen concentration in the deposited layer. Stable orthorhombic phases could be obtained at a deposition temperature of about 700°C and an oxygen pressure of 0.5 mbar. The success of using PLD became clear because of the *in situ* growth process that could be used at high (oxygen) pressure. Therefore, up until now this deposition method is the most used technique for the deposition of $YBa_2Cu_3O_7$ and related complex oxide thin films.

Nowadays, a variety of laser types are applied in this field (see table 2.3.2 and the references therein). Moreover, it is feasible to combine PLD with standard *in situ* diagnostic techniques, and growth monitoring became possible even at relatively high deposition pressures using ellipsometry and the so-called *high pressure* reflecting high energy electron diffraction (RHEED). These developments have helped to make PLD a mature technique to fabricate complex materials and structures.

Basic principles

The technique of PLD is similar to other vapour deposition techniques. It requires a vacuum chamber, a temperature controllable substrate holder, and source material. However, in contrast to all other evaporation techniques, the power source, a laser, is placed outside the vacuum chamber. Lenses focus the laser beam in order to get a sufficiently large energy density to ablate the source material. The vaporized material forms a plume and, if a substrate is placed inside or near the plume, part of the evaporated material will sublimate on it.

The first application of laser radiation as external energy source to vaporize materials and to deposit thin films in a vacuum chamber was reported by H.M. Smith and A.F. Turner, using a ruby-laser [4]. Systematic investigations were performed in the late 1970s followed in the late 1980s by deposition equipment for highly crystalline dielectric thin films [5] and semiconductor epitaxial layers for bandgap engineering [6] and the breakthrough by fabricating the high-T_c superconductors [7]. However, despite the recent progress, the total effort is still small compared to, e.g., MBE, which in fact was discovered at the same time! The reason for this can be found in the so-called splashing effect (droplet formation) and the lack of common interest before the perovskite materials became interesting. Figure 2.3.1 shows a schematic view of a PLD set-up.

The deposition technique has had a number of different names, like laser-assisted sputtering, laser-assisted deposition and annealing (LADA), pulsed laser evaporation (PLE), laser molecular beam epitaxy (LMBE), laser-induced flash evaporation (LIFE), laser ablation and finally, the most accepted name nowadays—PLD. A first extensive overview about thin-film deposition by PLD for a variety of materials can be found in a paper by Cheung and Sankur [8]. Additional information about PLD can be found in the book of D.B. Chrisey and G.K. Hubler [9].

A variety of lasers with wavelengths in the range of $10\,\mu m - 100\,nm$ are used in thin-film technology. The choice of laser depends on a number of conditions, which originate from the laser or from the target material. Typical laser characteristics are wavelength, laser energy, pulse width, and repetition frequency. Important target parameters are reflectivity, which is dependent on the wavelength, thermal conductivity, heat capacity, and density. The most applied laser types used for the evaporation of metals are given in table 2.3.1.

Normal pulsed ruby and Nd:glass lasers are in general not suitable for the thin-film process because of the long pulse length (small power density). The CO_2 and Nd:YAG lasers are the most commonly used lasers for the deposition of all kinds of material. In the case of oxides, the excimer laser has some advantages, due to the short wavelength. As an example, in table 2.3.2, laser types used for the fabrication of $YBa_2Cu_3O_7$ thin films are given.

The features and advantages of PLD particularly in the case of the fabrication of complex oxides are:

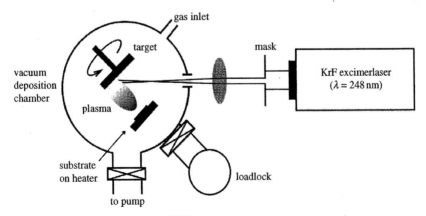

Figure 2.3.1. Schematic view of a PLD set-up.

Table 2.3.1. Lasers applied for the deposition of thin films.

Laser	Mode of operation	Wavelength (μm)	Pulse length (ns)
CO_2(CW)	Continuous (CW)	10.6	
CO_2(TEA)	Q-switched	10.6	200
Nd:YAG	Q-switched	1.06	200
Excimer	Xe, Kr, Ar	UV	15–200

Table 2.3.2. Laser types used for the deposition of $YBa_2Cu_3O_7$ thin films. The references give the first characteristic information about the deposition process.

Laser type	Wavelength	References
Pulsed CO_2	10.6 μm	[10, 11]
Pulsed Nd:YAG	1.06 μm	[12–15]
Excimer ArF	193 nm	[16, 17]
Excimer KrF	248 nm	[18–22]
Excimer XeCl	308 nm	[23–26]

- The power source is placed outside the vacuum chamber, which reduces the chamber dimensions.
- Deposition can take place in an oxygen environment, which makes **PLD** suitable for *in situ* preparation.
- Pressure during deposition can be significantly higher compared to standard deposition techniques, which makes it possible to grow compounds that are unstable at low pressures.
- Process parameters, like substrate temperature, oxygen pressure, energy density of the laser beam, and distance variation between target and substrate, are independent variables, which have collective and individual influence on the ultimate result. Therefore, the optimum preparation conditions can be found relatively simply and fast.
- The plasma consists of highly energetic particles, like ions and excited neutrals.
- The stoichiometry can be maintained during the evaporation, which makes an adjustment of the elements in the target unnecessary.
- The ablation equipment is relatively simple and flexible. The number of targets can be extended (for multilayers) and, if necessary, material from these targets can be evaporated simultaneously with the use of a beam splitter.

- The high deposition rates make PLD convenient for the deposition on substrates that are known to be sensitive for diffusion.
- The PLD process can also be applied for structuring deposited single layers, which makes it very attractive for industrial applications.

The physical processes involved in PLD are rather complicated despite its ease of operation. A pulsed laser beam is focused on a target that generates a dense layer of vapour on the target surface. The vapour absorbs most of the energy in the laser pulse. This happens in time scales of typically 20 ns and the vapour is transformed to an energetic plasma. The plasma then starts to expand into the surrounding ambient gas driven by the pressure gradient. For low background pressures the initial expansion of the plume agrees well with a classical drag-force model. The expansion front of the plasma at higher pressures can be well described with the shock propagation model [27]. While expanding, the thermal energy will be transformed into kinetic energy and the plasma will 'freeze'. At this point the plasma has obtained a high forward velocity with a certain angular distribution. The angular distribution of material and velocity depends on a number of parameters, such as laser fluence on the target and size, and geometry of the laser spot. Next, the plasma will decelerate due to the interaction with the background gas. At a characteristic distance, the forward velocity of the plasma is equal to the average thermal velocity of the gas. The particles in the plasma will then quickly adopt the thermal energy of the ambient gas and have thus become thermalized.

From the above, four aspects can be distinguished at the thin-film fabrication with PLD (see also figure 2.3.2):

(a) interaction between the laser beam and the target;
(b) interaction of evaporated material with the laser beam;
(c) adiabatic expansion of the plasma;
(d) growth of the film.

Figure 2.3.2. Plasma due to laser interaction with target (1), with evaporated material (2), expansion of the plasma (3), and film growth (4).

The decoupling of the vacuum hardware and the energy source (laser) makes PLD easily adaptable to different operational modes without the constraints imposed by the use of internally powered evaporation sources. This makes the dynamic range possible for deposition parameters large, a unique feature of PLD. Furthermore, PLD distinguishes itself from other thin-film techniques by a relatively high deposition rate during the laser pulse and the high kinetic energies possible for species arriving at the substrate surface.

The most important deposition parameters concern the laser characteristics (e.g. wavelength, pulse duration, and beam profile), target properties (e.g. density, absorption coefficient, thermal conductivity, and melting temperature), target–laser interaction (e.g. ablation rate and target surface morphology), substrate properties (e.g. material, crystallinity, and temperature), target–substrate geometry, and deposition ambient (e.g. gas present and its pressure).

Target–laser interaction

The PLD process starts with the ablation of the target material. A couple of thousand laser pulses irradiate the target during the deposition of a 100–300 nm thin film. If the target surface were left unchanged after each pulse, every consecutive pulse would result in the same target–laser interaction and could be described by the same characteristics. Laser pulses, however, rarely remove material in a smooth layer-by-layer fashion. Instead, the laser-irradiated target surface becomes altered, both physically and chemically.

Photonic energy is coupled into the target *via* electronic processes [28, 29]. Photons interact with the outermost (bound or free) valence electrons of atoms. Metals are characterized by their loosely bound outermost electrons, which are essentially free to travel around from atom to atom. As a consequence of this behaviour, the optical properties of most metals can be quite successfully described within the framework of the free electron gas model. For irradiation of metals with light of frequencies much lower than the plasma frequency, absorption is controlled by the free carriers in the material. Electromagnetic radiation with such large wavelengths is, however, strongly reflected, requiring large laser beam intensities to compensate for the large reflection losses. As the wavelength of the incident radiation decreases, the absorption and the reflectivity decrease. For even smaller wavelengths, the exact dependence of the optical parameters on wavelength depends on the specific band structure of the individual materials.

Unlike metals, the outermost electrons of insulators completely fill their shell. Insulators, though, can be characterized by their band gap E_g. For radiation with a photon energy greater than E_g, electronic transitions

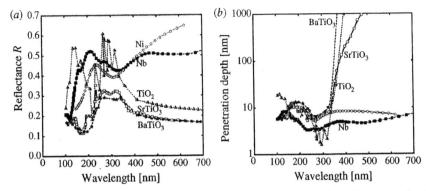

Figure 2.3.3. (*a*) Reflectance and (*b*) penetration depth as a function of wavelength of incident radiation for several target materials. The optical data (refractive index *n* and extinction coefficient *k*) were taken from [30] and [31].

between the valence and conduction bands are induced. The strongly absorbing region of the spectrum coincides with the high-reflectivity region because of the inherently large refractive indices for photon energies larger than the band gap. At energies below the band gap, the optical properties of insulators are determined by the much weaker intraband electronic transitions. Overall, insulators strongly resemble metals at wavelengths above their individual band edges as indicated by the reflectance[1] and penetration depth[2] data given in figure 2.3.3. The optical data (refractive index *n* and extinction coefficient *k*) were taken from [30] and [31].

Upon high-density irradiation ($>10^8 \, \mathrm{W \, cm^{-2}}$), materials quickly pass into an absorbing state because of optical breakdown, reducing the difference between opaque and transparent materials [32, 33]. Basically, electrons that are either initially present in the conduction band or generated by single- or multiphoton absorption are accelerated by the electric field of the laser radiation, gaining energy. When the electrons gain sufficient energy, they can create additional free electrons by impact ionization: upon colliding with an atom an energetic carrier is slowed down while creating an additional low-energy carrier pair. All carriers are accelerated again in the electric field present until the process repeats, and a carrier avalanche develops. This process can rapidly bring the electron density to $10^{18} \, \mathrm{cm^{-3}}$, which makes a material essentially opaque. The electric field amplitude of an electromagnetic wave is given by

---

[1]$R = [(n-1)^2 + k^2]/[(n+1)^2 + k^2]$ at normal incidence.
[2]Penetration depth $= 1/\alpha = \lambda/4\phi nk$ with α the absorption coefficient and λ the wavelength of incident radiation.

$$E = \left(\frac{2\Phi}{cn\varepsilon_0}\right)^{0.5} \tag{2.3.1}$$

where Φ is the power density, ε_0 the permittivity of free space, c the velocity of light, and n the refractive index. A material with a refractive index of 2 irradiated with a laser pulse (duration 20 ns) of a fluence of 2.0 J cm^{-2} is subjected to a field strength of 1.9×10^5 V cm^{-1}, sufficient to cause breakdown in a variety of target materials.

Although the reflectance for the metals is still higher than for the oxides, the optical properties of the metals and oxides presented in figure 2.3.3 are quite similar in the case of high energy density nanosecond laser irradiation of 248 nm. Therefore, the difference between metal and oxide target material removal is largely determined by the thermal properties of the materials. An overview of some optical and thermal properties is given in table 2.3.3 for different target materials.

The ratio of the optical absorption depth (penetration depth) $1/\alpha$ and the thermal diffusion length determines the temperature profile at the target surface during the laser pulse. The thermal diffusion length is given by [8, 9, 39]:

$$(2D\tau_p)^{0.5} = \left(\frac{2\kappa\tau_p}{\rho C_p}\right)^{0.5} \tag{2.3.2}$$

where D is the thermal diffusivity, τ_p the laser pulse duration, κ the thermal conductivity, ρ the specific weight, and C_p the specific heat at constant pressure. For the materials presented in table 2.3.3, the thermal diffusivity for the metals is an average of 2.3×10^{-5} m^{2}s^{-1} and for the oxides a factor of 10 lower (3.2×10^{-6} m^2 s^{-1}). If the optical absorption depth is much smaller

Table 2.3.3. Overview of some optical and thermal properties for target materials used in this research [30, 31, 34–38].

Target material	Absorption coefficient ($\lambda = 248$ nm) (10^6 cm^{-1})	Reflectance ($\lambda = 248$ nm)	Thermal conductivity (300 K) (W m^{-1}K^{-1})	Specific heat (298 K) (J kg^{-1}K^{-1})	Melting temperature (1 atm) (K)
Ni	1.5	0.45	91	440	1728
Fe	1.3	0.43	80	450	1811
Nb	2.9	0.48	54	260	2750
TiO$_2$ (rutile)	1.5/1.7[a]	0.27/0.34	7/10	711	2095
SrTiO$_3$	1.6	0.29	11	540	2353
BaTiO$_3$	1.3	0.13	6	440	1898

[a] $E \perp c$ axis / $E \parallel c$ axis.

than the thermal diffusion length, the laser energy is absorbed in the target surface layer and the thermal diffusivity controls the heating characteristics. In the opposite case, the optical absorption depth mainly determines the ablation depth. The former is the case for the above-mentioned target materials for laser irradiation at 248 nm as shown in table 2.3.4. Although the temperature dependence of the thermal and optical properties during laser ablation was disregarded, the use of constant thermal properties was sufficient to obtain a first indication of the difference between metal and oxide ablation.

The higher ablation threshold of metals in comparison with oxides is mainly caused by the higher reflectance and thermal conductivity of metals. The former results in less photonic energy coupled into the target material and the latter results in a larger volume into which the absorbed energy diffuses. The effect of the laser beam energy profile becomes visible for material removal from metal targets close to their ablation threshold. The scanning electron microscope (SEM) image in figure 2.3.4(a) shows the Nb target morphology after 200 pulses at $3.5 \, \text{J cm}^{-2}$. The mask size and optical magnification used were $7 \times 21 \, \text{mm}^2$ and 0.11, respectively. Although the average energy density of $3.5 \, \text{J cm}^{-2}$ was not sufficient for homogeneous ablation of Nb, the higher intensity at the centre of the ablation spot did result in material removal. It demonstrates the importance of the use of a homogeneous laser spot, which can be obtained by the use of a smaller mask aperture: e.g., the energy profile of the laser beam caused a spatial variation

Table 2.3.4. Optical penetration depth, thermal diffusion length, their ratio, enthalpy of evaporation, and the ablation threshold for target materials used in this research.

Target material	Penetration depth ($\lambda = 248$ nm) (nm)	Thermal diffusion length (300 K) (μm)	Ratio (10^{-3})	Enthalpy of evaporation H_{ev} (10^5) (J mol^{-1})	Ablation threshold ($\lambda = 248$ nm) (J cm^{-2})
Ni	6.7	1.1	6.1	4.5[a]	0.85[b]–2.2 [40, 41]
Fe	7.8	1.1	7.1	3.5 [42]	2 [43, 44]
Nb	3.4	1.1	3.1	6.8 [42]	3.5
TiO$_2$	5.8	0.4	14.5	17[a]	< 0.5 [45]
SrTiO$_3$	6.3	0.4	15.8	30[a]	0.1 [46]
BaTiO$_3$	7.8	0.3	26.0	30[a]	—

[a] Indication of a value for H_{ev} based on given enthalpies in [42] for lower temperatures.

[b] Matthias *et al* [41] reported a threshold of $0.85 \, \text{J cm}^{-2}$ for visible damage at the target surface by melting and an ablation threshold of $2.2 \, \text{J cm}^{-2}$ for plasma formation. Hiroshima *et al* [40] reported a non-zero deposition rate at a fluence of $0.85 \, \text{J cm}^{-2}$.

Figure 2.3.4. Nb target morphology after 200 pulses as a function of fluence for (*a*) $3.5\,\mathrm{J\,cm^{-2}}$, (*b*) $4.5\,\mathrm{J\,cm^{-2}}$, and (*c*) $6.0\,\mathrm{J\,cm^{-2}}$. The mask used consisted of seven rectangular apertures of $7 \times 2\,\mathrm{mm^2}$.

in energy of 40% for a 7.0 mm wide mask, which can be reduced to 10% by the use of a 4.0 mm wide mask.

Influence of target morphology

For a specific laser and target, the fluence mainly determines the target morphology evolution under pulsed laser irradiation. The target surface characteristics can influence the ablation rate, the film stoichiometry, and its surface roughness. Depending on the exact target morphology, a rough target surface can either decrease or increase the ablation rate. The former is a result of an increased surface area and, therefore, reduced fluency, while the latter is the result of better light coupling into the target. If induced target surface structures are aligned in the direction of the laser beam, the plasma is also shifted in the direction of the laser beam. Exfoliation of fragile microstructures formed at the target surface results in rough film surfaces, because the plasma carries the loose debris towards the substrate, where it condenses onto the growing thin film. For multi-element materials, it is important to realize stoichiometric ablation [46, 47]. A first indication, though not sufficient, of stoichiometric ablation is a smooth target morphology, which also offers the possibility of a longer stable use of the target. These requirements result in a critical ablation threshold, for which a desirable target morphology is obtained.

Repetitive laser irradiation on a non-rotating target can induce periodic surface structures, which have been observed for a wide class of target materials such as metals [48], ceramics [49], polymers [50], and semiconductors [51, 52]. Sipe *et al* [53] considered that the periodic structures result from inhomogeneous energy profiles associated with the interference of the incident beam with a surface scattered field for fluences near the damage threshold ($<0.55\,\mathrm{J\,cm^{-2}}$). The pattern spacing for polarized light was determined to be $\lambda(1 \pm \sin\theta)^{-1}$ and $\lambda(\cos\theta)^{-1}$, where λ is the wavelength of the laser beam and θ the angle of incidence of the beam. Clark *et al* [54] discussed the possibility of interfering or beating waves, resulting in a larger spacing. They also attributed wavelike structures to frozen capillary waves. These waves can be generated as a result of the impulse that the target surface suffers when material is ablated from it. This impulse would seed capillary waves in much the same way that dropping a stone into a lake generates waves. The ripple spacing is determined primarily by how long the impulse lasts, which in turn depends largely on the time the surface stays molten. To create a periodic target structure, the target material must freeze before the waves decay away. Since short-wavelength structures have the fastest decay, they are more likely to decay away before the target surface is solid again. Finally, the wavelength of the light incident on the target surface can be much larger than the free-space wavelength, because of the presence of a plasma close to the target surface. The refractive index of the plasma can be smaller than one and, hence, the wavelength of the laser pulse as seen by the surface is increased due to passage through the plasma.

Figure 2.3.4 shows the target morphology after 200 pulses on a non-rotating Nb target as a function of the incident laser fluence. The target morphology shows a periodic structure for fluences of $4.5\,\mathrm{J\,cm^{-2}}$ and higher. The 'waves' are perpendicular to the direction of the incoming laser beam. In the case of figure 2.3.4(c), the laser beam came from the right side creating a steep, rougher side and a longer, smooth side of a wave.

In the case of ablation of non-rotating BaTiO$_3$ and SrTiO$_3$ targets, the desirable smooth target morphology could be achieved. Figure 2.3.5 shows the target morphology after 200 pulses on a SrTiO$_3$ target as a function of laser fluence. A periodic structure is visible for a fluence of $0.6\,\mathrm{J\,cm^{-2}}$, but for a fluence of $1.0\,\mathrm{J\,cm^{-2}}$ the periodicity disappears. The target morphology hardly changed between 1.0 and $2.5\,\mathrm{J\,cm^{-2}}$, and showed a flat surface with many cracks due to surface melting and subsequent fast solidification.

Figure 2.3.6 shows the target morphology after 200 pulses on a non-rotating TiO$_2$ target as a function of laser fluence. The SEM images are quite different from the former results. The target morphology changes more severely for the different fluences used. For a low fluence of $0.6\,\mathrm{J\,cm^{-2}}$, no periodicity perpendicular to the direction of the incoming laser beam is visible. A periodic structure is present for ablation at $1.5\,\mathrm{J\,cm^{-2}}$, though quite different from the case of metal or SrTiO$_3$ ablation. The target morphology shows a wavy pattern for higher fluences, even up to $6.0\,\mathrm{J\,cm^{-2}}$. No flat target

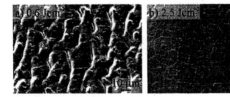

Figure 2.3.5. SrTiO$_3$ target morphology after 200 pulses at a fluence of (a) $0.6\,\mathrm{J\,cm^{-2}}$ and (b) $2.5\,\mathrm{J\,cm^{-2}}$.

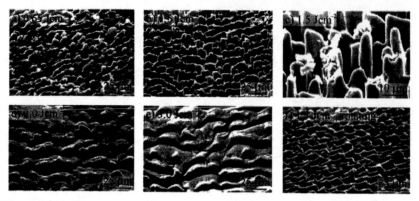

Figure 2.3.6. TiO_2 target morphology after 200 pulses as a function of fluence for (a) $0.65\,J\,cm^{-2}$, (b) $1.5\,J\,cm^{-2}$, (c) $1.5\,J\,cm^{-2}$ enlarged, (d) $2.0\,J\,cm^{-2}$, (e) $3.0\,J\,cm^{-2}$, and (f) $1.5\,J\,cm^{-2}$ on a rotating target.

morphology was achieved for TiO_2 ablation. The waves are again perpendicular to the direction of the incoming laser beam, independent of spot size and shape. For the spot sizes used ($5\,mm^2$), the wave period is between 15 and 20 μm, which is the same order of magnitude as in the case of metal ablation. In the case of TiO_2, a fluence of at least $2.0\,J\,cm^{-2}$ is recommended, since the target morphology contains loose debris for fluences under $2.0\,J\,cm^{-2}$ as shown in figure 2.3.6(c). The situation changes for a rotating target, because the ablation spots of consecutive pulses can partially overlap. Although overlap only occurred for a small range of angles of about $\pm 10°$ (with $0°$ referring to complete overlap[3]), the resulting target morphology changed drastically as shown in figure 2.3.6(f) in comparison with figure 2.3.6(b). It appeared that the target morphology was already debris free for a laser fluence of $1.5\,J\,cm^{-2}$, while for a laser fluence of $1.0\,J\,cm^{-2}$ the target morphology still showed fragile microstructures. Atomic force microscopy images of TiO_2 films deposited at $1.0\,J\,cm^{-2}$ showed that exfoliation, indeed, took place.

In the case of the oxide targets, the repetitive laser irradiation not only changed the target morphology, but also the target stoichiometry. The yellowish target colour changed into grey after ablation, indicative of a metal-rich layer. The electrical resistance of the TiO_2 target surface, determined by a two-probe measurement, decreased at least four orders of magnitude[4] after ablation. The electrical resistance of the $SrTiO_3$ target

---

[3]Overlap calculated for a target with a diameter of 2.54 cm, and a 3.0 mm long and 1.5 mm wide spot positioned at the edge of the bottom part of the target.

[4]The electrical resistance of the non-ablated oxide surface was higher than the maximum measurable resistance (20 MΩ).

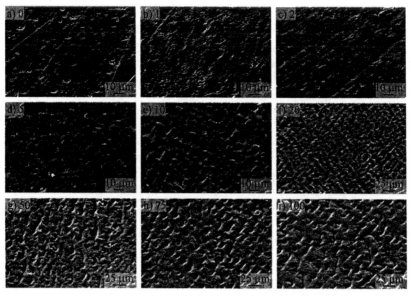

Figure 2.3.7. TiO$_2$ target morphology after ablation at a fluence of 2.0 J cm^{-2} as a function of number of pulses: (*a*) 0, (*b*) 1, (*c*) 2, (*d*) 5, (*e*) 10, (*f*) 20, (*g*) 50, (*h*) 75, and (*i*) 100.

surface decreased at least one order of magnitude. XPS on a TiO$_2$ target after ablation at 1.5 J cm^{-2} for 200 pulses showed, indeed, strong oxygen reduction, resulting in Ti-rich material.

An important question is if these target morphologies directly develop after one laser pulse or are more pulses needed before a steady-state morphology is induced? The evolution of the target morphology of a TiO$_2$ target under consecutive irradiation with laser pulses of 2.0 J cm^{-2} is presented in figure 2.3.7. The morphology after one laser pulse shows a lot of pits. Initial ablation of an SrTiO$_3$ target shows the same characteristic pits (see figure 2.3.8(*a*)). Since these pits are not present in the case of metal ablation, the pits are possibly induced by the release of oxygen (gas), which is in accordance with the former discussion about preferential ablation of oxygen from an oxide target. Consecutive pulses diminish the number of pits and induce the presence of cracks, indicative of solidification after melting. The first indication of the creation of periodic surface structures on the TiO$_2$ target is visible after 10 pulses and more clearly after 20 pulses. The steady-state regime is achieved between 50 and 75 pulses. In the case of SrTiO$_3$ ablation at 1.7 J cm^{-2}, steady state is already reached after 15 pulses.

Sipe *et al* [53] considered polarized laser radiation of fluences near the damage threshold in their theory. In this study, unpolarized light of higher

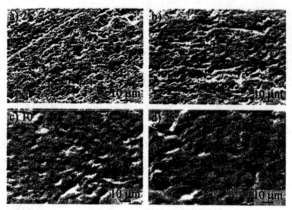

Figure 2.3.8. $SrTiO_3$ target morphology after ablation at a fluence of $1.7\,J\,cm^{-2}$ as a function of number of pulses: (*a*) 2, (*b*) 5, (*c*) 10, and (*d*) 25.

fluence was used, which induced a larger spacing as also reported by others [51, 55]. The desirable smooth surface after ablation was only achieved in the case of $SrTiO_3$ and $BaTiO_3$ ablation. It was not possible to induce the same smooth morphology in the case of TiO_2 ablation. Apparently, TiO_2 ablation characteristics fall in between those of the ablated metals and perovskites, not achieving a flat surface but showing some periodic surface structure, although not as strong as in the case of metal ablation. The metal and oxide properties differ mostly for the reflectance, thermal conductivity, and enthalpy of evaporation. Looking at the values of the physical parameters given in table 2.3.4, TiO_2 characterization falls, indeed, in between the values for metals and oxides in the case of the enthalpy of evaporation, H_{ev}. Its estimated H_{ev} is a factor of 2–5 higher than those of the metals, but a factor of 2 lower than those of the other oxides. The enthalpy of evaporation can be regarded as a cooling source at the target surface. A higher H_{ev} in combination with a lower κ for $SrTiO_3$ and $BaTiO_3$ confines the absorbed energy to the target surface. The lower H_{ev} in combination with the higher κ for the metals makes the energy absorption more a target volume process in comparison with $SrTiO_3$ and $BaTiO_3$. Furthermore, the Ti enrichment observed in this study changes the optical and thermal properties of the target surface during ablation. This affects the ablation characteristics as has been demonstrated for other materials. Span [56] investigated the dependence of surface target morphology and relative composition of a $La_{0.5}Sr_{0.5}CoO_3$ target after ablation as a function of fluence. For low fluences $(0.3–0.4\,J\,cm^{-2})$, large cones were formed at the target surface directed along the laser beam. The top of the cones consisted of refractory Co nuclei with a higher ablation threshold, shielding the stoichiometric target material from ablation.

Interaction of evaporated material with the laser beam

The evaporated material introduces a new aspect. The laser beam, which normally only hits the target material, is scattered by the evaporated material, resulting in an absorption of laser energy. This results in two effects. First, radiative heating of the free electrons takes place. These electrons strongly couple with the incident radiation and are accelerated. After the material has been emitted from the target it is accelerated during the whole laser pulse due to energy absorption. Collisions with the evaporated material increase the ionization rate. Second, the temperature of the evaporated material increases rapidly to extremely high values ($>10\,000\,K$). These high temperatures, close to the surface of the target, lead to thermionic emission of ions from the target, identical to other thermal processes (the temperature dependence of the degree of ionization is given by the Langmuir–Saha equation). On one hand, the evaporated material shields the target for the incident radiation, resulting in a decrease of the evaporation velocity. On the other hand, the evaporation velocity will increase due to the radiation energy of the high temperatures of the evaporated material.

Expansion plasmas

Material removed from the target expands into the vacuum chamber. In vacuum, the plasma angular distribution is determined by collisions of the ablated species among themselves. In the presence of an ambient gas, the plasma angular distribution is perturbed by additional collisions of the ablated species with the ambient gas atoms. In addition, the ambient gas can chemically react with the plume species and be incorporated into the growing film. In the case of deposition of oxide films from oxide targets, an additional oxygen flow into the vacuum chamber can be necessary to deposit stoichiometric films. Also, it gives the possibility to deposit metal oxide, nitride, or carbide films from metal targets in an oxygen, nitrogen, or methane ambient, respectively.

Most collisions of the ablated species among themselves occur while the plasma is small and close to the target. Since the pressure gradient in the vapour near the target surface is the largest perpendicular to the target surface, the axis of the laser-generated plasma is oriented along the target surface normal. If the emitted particle density is low enough, collisions between particles are limited. Provided the emission is truly thermal, the velocities are described by a half-range Maxwellian, i.e. a Maxwellian with only positive velocities normal to the target. More commonly in PLD, the emitted particle density is high enough that near-surface collisions occur. In a layer within a few mean free paths of the target surface, the Knudsen layer, the distribution function evolves to a full-range Maxwellian in a centre-of-mass coordinate system [57]. Therefore, a backflow of particles towards the target is created. Effectively, the exponential part of the velocity distribution for the velocity component normal to the target surface (v_x) changes from

$$\exp(-mv_x^2/2kT_s), \quad v_x \geq 0 \tag{2.3.3}$$

to

$$\exp(-m(v_x - u_K)^2/2kT_K), \quad -\infty < v_x < \infty \tag{2.3.4}$$

where m is the mass of the ablated species, k the Boltzmann constant, T_s the target surface temperature, u_K the centre-of-mass (or flow) velocity (similar to the velocity of sound), and T_K the temperature of the Knudsen layer. If the number density of ablated particles is very high, the downstream boundary of the Knudsen layer acts as the throat of a nozzle. The adiabatic expansion shows a flow velocity exceeding the velocity of sound.

The high kinetic energies possible for the ablated species are induced by the interaction of the laser beam with the plasma. Absorption of laser radiation increases the plasma temperature. The absorption would occur by an inverse Bremsstrahlung process, which involves the absorption of a photon by a free electron. Therefore, the plasma absorbs laser radiation only at distances close to the target, where the density of ionized species is high [58]. Introduction of an ambient gas into the vacuum chamber decreases the kinetic energies of the species. In the presence of an ambient gas, the plasma tends to confine because of additional collisions of the ablated species with the ambient gas atoms. So, an ambient gas present scatters, attenuates, and thermalizes the plasma species, changing their spatial distribution, deposition rate, and kinetic energy distribution. The ablated species can be regarded as an ensemble that experiences a viscous force proportional to its velocity through the ambient gas [59]. The equation of motion is $a = -\beta v$ with a the acceleration, v the velocity, and β the slowing coefficient, giving

$$x = (v_0/\beta)(1 - \exp(-\beta t)) \tag{2.3.5}$$

where x is the travelled distance from the target surface, v_0 the initial velocity, and t the travel time. This drag force model has shown good agreement at low pressures and short times. At higher pressures, the formation of a shock wave has been observed [60, 61]. The propagation of a spherical shock front is described by the following distance–time relation,

$$x = (k/p)^{0.2}t^{0.4} \tag{2.3.6}$$

where k denotes a constant proportional to the laser fluence and p the ambient pressure.

Influence of the ambient gas

In general, raising the ambient pressure results in an increase in fluorescence from all species due to collisions on the expansion front and subsequent collisions within the plasma. The plasma boundary sharpens, indicative of a shock front, resulting in spatial confinement of the plasma. The plasma will decelerate due to the interaction with the background gas. At a characteristic distance the forward velocity of the plasma is equal to the average thermal velocity of the gas. The particles in the plasma will then quickly adopt the thermal energy of the ambient gas and have thus become thermalized. Figure 2.3.2 shows the effect of the background pressure on the shape of the plasma. At low pressures the plasma 'fills' the whole chamber. At higher pressures the shape of the plasma is small, but very intense.

In a paper by Strikovski and Miller [62] an adiabatic thermalization model for PLD of oxides is discussed. Their assumptions are partly in agreement with experiments done with high pressure RHEED during deposition of $SrTiO_3$ (see later). Using this technique, one can measure the number of pulses needed to complete one unit cell layer of $SrTiO_3$ as a function of the (oxygen) background pressure. Two different regimes can be distinguished. At lower pressures (substrate is outside the plasma) the deposition is almost constant. At higher pressures, there is a clear pressure dependence.

To be able to adequately explain this characteristic behaviour, the changing plume shape has to be accounted for. During our experiments, we have observed the plume shape and indeed it changes dramatically as the pressure increases. In general, at low pressures (~ 0.01 mbar) a large plasma is observed (filling the whole chamber and hardly visible) with an intense plasma close to the target. At higher pressure (~ 0.05 mbar) the visible part of the plasma has a more spherical shape and at even higher pressures (approaching 0.1 mbar) this results in a plasma with three regimes: very intense spikelike plasma close to the target, a sphere-shaped plasma, and a larger soft (corona-like) plasma. Figure 2.3.2 shows the evolution of the plasma at different pressures. Here, clearly the difference in shape of the plasma can be seen.

There is some discussion in the literature about the extent of interaction of the ablated species with the ambient gas. Gonzalo *et al* [63] investigated the plasma expansion dynamics after ablation of a $BaTiO_3$ target at $2.0\,\mathrm{J\,cm^{-2}}$. The emission spectra of the plasma contained several lines, most of them related to neutrals (Ba, Ti) and ions (Ba^+, Ti^+). No evidence of oxidized species such as BaO and TiO was observed, even in an oxygen ambient. Kano *et al* [64] measured Mg^+, Mg_2^+, MgO^+, MgO_2^+, Mg_2O^+, and $Mg_2O_2^+$ ion intensities after ablation of MgO between 0.2 and $1.0\,\mathrm{J\,cm^{-2}}$. They concluded that the observed ions were directly emitted from the target surface in their ionic forms and were not modified by reactions in flight. This is in contradiction with the explanation given by Harilal *et al* [65], who concluded that species like Y and YO were generated directly from the $YBa_2Cu_3O_7$ target at $12.0\,J\,cm^{-2}$, while Y^+ was mainly produced just outside the target due to electron impact. As time evolved after a laser pulse, the Y^+ generation decreased and collisional recombinations resulted in enhanced spectral emission from Y. At a slightly later time, YO was generated from reactions between Y and O species. Tang *et al* [66] confirmed these conclusions by investigating the ablated species from a TiO_2 target at $0.5\,J\,cm^{-2}$ in vacuum. The ablated species included O, O_2, Ti, TiO, and TiO_2, also measured by Gibson [67] for low fluences ($<0.1\,J\,cm^{-2}$). Ablation in the presence of an O_2 beam resulted in an increased signal for TiO^+ and a lower signal for Ti^+ in comparison with ablation in vacuum. This suggests the reaction between ablated Ti atoms/ions and ambient oxygen gas atoms. Voevodin *et al* [68] found that an oxygen ambient led to a significant presence of atomic oxygen, and molecular ZrO and YO in the near-substrate region for ablation of a ZrO_2/Y_2O_3 target at $10.0\,J\,cm^{-2}$. These species existed in higher concentrations and for longer times after the laser pulse in an oxygen ambient in comparison with an argon ambient. The presence of an argon ambient promoted the generation of energetic ions. Also Dutouquet *et al* [69] determined that TiO was produced by chemical reactions between the metal Ti vapour and the oxygen gas in the case of ablation of a Ti target. Lichtenwalner *et al* [70] noted metal species interacting with oxygen gas atoms, increasing the presence of molecular oxide species. Furthermore, they noted that the centre of the plasma did not interact substantially with the oxygen gas. The ambient pressure influenced the degree of mixing between ablated species and ambient gas atoms. Although there is no complete consensus in the literature, oxidized species are often detected in plasmas in the case of ablation of oxide targets. The amount of oxidized species increases in the presence of oxygen gas. The influence of the ambient gas present shows to be strong.

Another approach is the mixture of reactive with a non-reactive gas. A small percentage of oxygen flow added to the argon flow has a dramatic influence on the plasma appearance. The exterior of the plasma seems to be more affected than the interior of the plasma, confirming the idea that the centre does not interact substantially with the oxygen gas.

Thin-film growth

The growth kinetic of the film prepared by PLD is different from the ones grown with other deposition techniques; e.g., using a multi-component target the plasma contains a diversity of particles, as in the case of $YBa_2Cu_3O_7$: Cu, Ba, Cu, O_2, Y, BaO, YO, and CuO [60] all in a vapour phase. There are no electric fields or accelerated noble gas ions present, and the substrate is bombarded in a pulsed way by large quantity of material. In particular, the high supersaturation of the deposited species and the enormous deposition rate during one laser pulse are typical for PLD. In general, the following growth processes are important for the formation of epitaxial thin films:

(a) The collision of an atom with the (heated) substrate, resulting in an adsorption or recoil from the substrate. The adsorption depends on the incident kinetic energy of the atom and varies for different particles.

(b) The adsorbed atom loses its excessive kinetic energy in about $2/u$ s, where u is the substrate lattice vibration frequency ($\sim 10^{14}\,\mathrm{s}^{-1}$). The atom, therefore, equilibrates thermally with the substrate in about $10^{-14}\,$s, having an energy kT, where T is the substrate temperature. At this stage, the atoms can desorb again resulting in a migration of the atoms over the substrate.

(c) Due to the collisions of the adsorbed atoms they will be bound to each other and to the substrate forming stable nuclei. This process strongly depends on the evaporation rate of the material, position of the substrate in the plasma and background pressure (note that the latter two are coupled). The PLD technique applying a pulsed laser results in heavily fluctuating evaporation rate. The effect of the pulsed method of deposition can be observed using the high pressure RHEED and will be discussed in the next section extensively.

(d) The coalescence of nuclei leads to the film growth. The coalescence process also strongly depends on the substrate temperature and position in the plasma. The mobility of the deposited material can be in the order of seconds [71], which is of the time scale of the pulse frequency of the laser (often 1 Hz is used). With high laser pulse frequencies the impinging atoms can hamper the coalescence.

(e) Due to the wide range of adjusting the deposition parameters, layer-by-layer or step-flow growth are feasible for many different materials

Besides the above-described processes, of course the type of substrate and its quality are of the utmost importance for the growth of (epitaxial) layers.

High-pressure reflected high-energy diffraction during PLD

Because there were almost no *in situ* diagnostic tools available to study the growth by PLD, not much was known about the pulsed method of deposition and the influence of the repetition rate of the laser and influence of background pressure. RHEED is often used for the analysis and monitoring of thin film growth in ultra-high vacuum deposition systems [72]. For PLD of oxide materials the diagnostics of the growing film surfaces by *in situ* RHEED is hampered by the relatively high oxygen pressure. Nevertheless, several groups have monitored the growth of complex oxides with RHEED and have shown intensity oscillations, by depositing under pressures compatible with their RHEED set-up. To incorporate oxygen in the as-grown films, different alternatives were used, e.g. low pressures (10^{-4}–1 Pa) of molecular oxygen [73, 74], NO_2 [75, 76] or O_3 [77], and alternatively pulsed oxygen sources [78]. A low deposition pressure during PLD, however, can lead to stress, usually compressive, in the film. As mentioned earlier, this is caused by the bombardment of the film during the deposition by highly energetic particles, originating from the plasma. Furthermore, some complex oxides, like high-T_c superconductors, are not stable in low oxygen pressure at high temperature and, therefore, must be deposited at high oxygen pressures of up to 30 Pa to avoid decomposition of the film [3].

We developed a RHEED system designed for growth-monitoring under high deposition pressures (up to 100 Pa) [79]. The main problem to be solved is the scattering loss at high pressure. In order to minimize the loss, the travelling path of the electrons in the high-pressure region has to be kept as short as possible. Furthermore, most of the commercial available electron sources use heated tungsten filaments to emit electrons. The oxygen pressure in the source should be very low ($<5 \times 10^{-4}$ Pa) to avoid short lifetimes of the filaments. The designed system satisfies these requirements, i.e., a low pressure in the electron gun and a high pressure in the deposition chamber [80].

Design of a PLD system for oxides

Excimer laser

The excimer laser belongs to the class of high-pressure gas lasers that operate in the ultraviolet spectral region. These lasers have in common that the laser medium consists of excimers: mostly a noble gas coupled with a halogen. Under normal circumstances the two atoms will be in the repulsive lower state and no excimer exists. If the noble atom is excited, it will be bound and form an excimer molecule with an energy state that is lower than the energy state of the single (excited) atom. The excimer drops to its ground state, emitting a photon. The excitation takes place in a high-voltage discharge between two electrodes. The distance between the two electrodes imposes the dimensions of the laser beam, typically 2 cm × 3 cm.

The pulse duration of the excimer laser is about 10–50 ns (although femtosecond lasers are also available nowadays, they are not widely used for PLD and therefore not discussed here) and its repetition frequency can go up to 1 kHz. Commercial systems with an output power of up to 500 W are available and the lifetime of a gas filling is about 10 million pulses. The wavelength depends on the inert gas/halogen combination used, (see table 2.3.5).

The most used lasers for deposition purposes are XeCl and KrF. Those lasers have a repetition rate ranging from 1 to 300 Hz and pulse duration of

Table 2.3.5. Most common gas/halogen mixtures, with their specific wavelength (nm).

	Argon	Krypton	Xenon
Fluorine	193	248	351
Chlorine	(175)[a]	222	308
Bromine	(161)[b]	(206)[a]	282

[a] Fluorescence.
[b] Predissociated.

20–30 ns. The pulse-to-pulse stability is better than 4% and the beam divergence of those excimer lasers is in the x and y directions of the beam, respectively, 1 and 3 mrad. Sometimes, an unstable resonator is used to decrease the divergence by a factor of 10. The laser beam is focused onto the target using a (projection) mask in combination with special optics. Here, the homogeneity of the beam can be increased using a homogenizer.

Deposition chamber

A schematic view of the deposition chamber, including the electron source assembly, is given in figure 2.3.9. Here, a (248 nm) laser beam is focused under 45° incidence onto a high density (>95%) target using a lens (focal

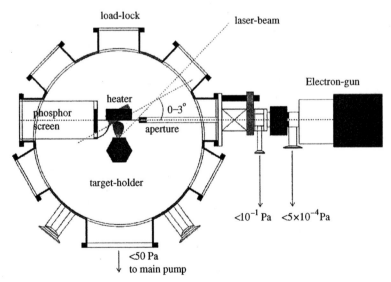

Figure 2.3.9. Schematic view of the deposition chamber, including the electron source assembly.

length ~ 500 mm). Both substrate holder including substrate holder and multi-target holder (which can hold up to five different targets) are mounted on a computer controlled *XYZ*-rotation stage and can be inserted *via* a load-lock system without breaking the vacuum. Selection of the targets, laser repetition, and the number of laser pulses per target material is fully automated.

The targets are mostly stoichiometric sintered pellets (Al_2O_3) or single crystals (e.g. $SrTiO_3$, SrO, BaO, CaO, HfO_2, TiO_2). In order to ablate from a 'fresh' part of the target with every pulse, the target must be pre-ablate.

For deposition at elevated temperatures, the substrate is mounted on a thermo-coax heater using silver paint allowing for good thermal contact. The temperature is measured inside the heater block with a K-type thermocouple and temperatures up to 950°C can be obtained using thermo-coax wires. The pressure is controlled by means of two mass flow controllers in combination with a pump restriction. The aperture of the restriction and the gas flow settings determine the final pressure. High pressures ($>10^3$ Pa), e.g., for annealing purposes are obtained by closing all valves to the pumps and flooding of the chamber with either oxygen or nitrogen. Oxygen gas (purity 4.5 or 6.0) can be used as well as inert gases (N_2, He, Ar, Ne).

RHEED

An electron source produces a beam with a minimum size of 100 μm, which in turn gives a spot size on the surface of $0.1 \times 3 - 0.1 \times 7$ mm^2 (at an angle of incidence roughly 0.8–2°). Nowadays, RHEED systems up to 35 keV are available. The heater has to be rotated in order to adjust the angle of incidence of the electron beam on the substrate. An additional rotation is used to change the azimuth.

The diffraction pattern on a fluorescent phosphor screen can be recorded by a computer controlled CCD camera. A frame grabber and software collect the images. For observing RHEED oscillations, a high speed camera can be used. Nowadays, several analysis softwares are available that enable one to analyse the diffraction data, i.e., full pattern as well as intensity variations of selected areas in the pattern. The on-chip integration option allows one to intensify reflections of weakly scattered beams, without changing the beam configuration.

High pressure RHEED

Figure 2.3.9 shows the complete high pressure RHEED set-up. A differential pumping unit is used to maintain a vacuum of better than 5×10^{-4} Pa in the electron source. The source is mounted on a flange connected to a stainless steel extension tube with an inner diameter of 8 mm. An aperture (diameter

250–500 μm) separates the tube from the deposition chamber. The pressure in the tube, which depends on the pump speed and the size of the aperture, is kept below 10^{-1} Pa. Using this two-stage pumping system, the pressure in the deposition chamber can be increased up to 100 Pa maintaining the low pressure in the electron source. The electron beam, which passes through the apertures inside the differential pumping unit and the tube, enters the deposition chamber near the substrate at a distance of about 50 mm. The *XY* deflection facility of the electron source is used to direct the electron beam through the aperture at the end of the tube. Many electron sources are also equipped with a beam rocking option, which in this case cannot be used. For this, a mechanical solution has to be employed by rocking the complete gun.

Small magnetic fields, like the earth magnetic field, can influence the electron beam and complicate the alignment of the beam through a small aperture. Therefore, special care has been taken to shield the electron beam from magnetic fields using μ-metal.

The fluorescent phosphor screen (diameter 50 mm) is mounted on a flange located near the substrate. The distance between the screen and substrate is 50 mm. The screen is shielded from the plasma in order to minimize contamination. The electron source, including the extension tube, is mounted on a *XYZ*-stage allowing us to adjust the distance between the substrate and end of the tube.

Although scattering of electrons in high oxygen pressure decreases the intensity of the electron beam, we have shown that growth monitoring of complex oxides at high oxygen pressures is feasible using RHEED.

Growth studies of SrTiO₃: pulsed laser *interval* deposition

As mentioned above, PLD has become an important technique to fabricate novel materials, including materials with high dielectric constants. Although there is the general impression that, due to the pulsed deposition, the growth mechanism differs partially from continuous physical and chemical deposition techniques, it has hardly been used. Here, we would like to introduce a growth method based on a periodic sequence: fast deposition of the amount of material needed to complete one monolayer followed by an interval in which no deposition takes place and the film can reorganize. This makes it possible to grow in a layer-by-layer fashion in a growth regime (temperature, pressure) where otherwise island formation would dominate the growth. This growth technique could be of importance if very thin layers are needed.

In order to be able to create a crystal structure by depositing consecutive unit cell layers of different materials, a layer-by-layer growth mode is a prerequisite: nucleation of each next layer may only occur after the previous layer is completed. Occasionally, the deposition conditions such as the substrate temperature and ambient gas pressure (oxygen in the case of oxide materials) can be optimized for true two-dimensional (2D) growth, e.g. homoepitaxy on $SrTiO_3(001)$.

In the case of homoepitaxy, kinetic factors will determine the growth mode, whereas in the case of heteroepitaxy also thermodynamic factors, e.g. misfit, are important to understand the growth mode. In fact, always layer-by-layer growth is predicted for homoepitaxy from a thermodynamic point of view [81]. However, during deposition of different kinds of material, i.e., metals, semiconductors, and insulators, independent of the deposition technique, a roughening of the surface is observed. Assuming only 2D nucleation, determined by the super-saturation [82], limited interlayer mass transport results in nucleation on top of 2D islands before completion of a unit cell layer. Still, one can speak of a 2D growth mode. However, nucleation and incorporation of ad-atoms at step edges is proceeding on an increasing number of unit cell levels, which is exhibited by damping of the RHEED intensity oscillations. In fact, an exponential decay of the amplitude is predicted assuming conventional MBE deposition conditions [83].

To understand the implications of the characteristics of PLD on growth, which are expected to be kinetic in origin, we used the following *in situ* RHEED studies. First, the transition of step flow growth to layer-by-layer growth on vicinal surfaces has been used to estimate diffusion parameters [84]. With the beam directed parallel to the terraces, intensity oscillation will disappear at a critical temperature, when all the material will be incorporated at the vicinal steps. The critical diffusion length is then of the order of the terrace width. This simple picture has been extended to include, for instance, non-linear effects [85, 86] and the fact that the transition is not sharp [87–89]. Secondly, another approach has been given by Vvedensky *et al* [90], by studying the relaxation of the RHEED intensity after interruption of the growth by MBE. Comparing the characteristic relaxation times observed during experiments with Monte Carlo simulations, the participating microscopic events can by identified (provided that all the relevant mechanisms are included in the simulation model). Note that this approach is closely connected with PLD, since after each laser pulse, relaxation of the RHEED intensity is observed and can be viewed as a kind of interrupted growth. Finally, the relaxation behaviour of the RHEED intensity after each laser pulse has been studied in case of PLD $SrTiO_3$ [71]. By time resolved RHEED, the relaxation times for different temperatures have been measured and an estimate of the diffusion barrier has been made.

Time resolved RHEED experiments

In figure 2.3.10, true 2D RHEED intensity oscillations are observed during deposition of $SrTiO_3$ with PLD at a temperature of 850°C and an oxygen pressure of 3 Pa. For these experiments single TiO_2 terminated $SrTiO_3$ single

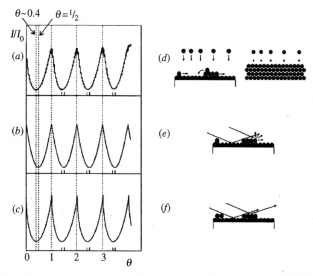

Figure 2.3.10. (*a*) Intensity oscillations during homo-epitaxial growth of SrTiO$_3$ at 850°C and 3 Pa, indicative of true layer-by-layer growth, (*d*). Calculated intensity oscillations using (*b*) a diffraction model of which a schematic representation is given in (*e*) and (*c*) a step density model of which a schematic representation is given in (*f*). The number of pulses needed to complete one unit cell layer is estimated to be 27.

crystals have been used [91, 92]. At lower temperatures (figure 2.3.11), 2D RHEED intensity oscillations are still visible. In this figure, another feature of PLD—the relaxation phenomenon—is exemplified. In the case of PLD, a typical value for the deposition rate within one pulse is of the order of

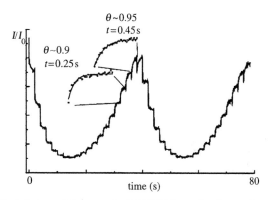

Figure 2.3.11. Modulation of the specular RHEED intensity due to the pulsed method of deposition ($T = 750$°C, $pO_2 = 3$ Pa); insets give enlarged intensity after one laser pulse plus fit to give characteristic relaxation times. Here, a deposition rate of 19 pulses per unit cell layer was extracted.

$10\,\mu\mathrm{m\,s^{-1}}$ [93, 94]. Therefore, a high supersaturation is expected when the plume is on and thus the number of 2D nuclei can be very high. Subsequently, when the plume is off, larger islands are formed through re-crystallization, exhibited by the typical relaxation of the RHEED intensity of the specular spot [73]. The insets in figure 2.3.11 are enlargements of the relaxation after a laser pulse. From this it is clear that the characteristic relaxation times depend, among other things, on the coverage during deposition. The relatively high temperature in combination with a low oxygen pressure enhances the mobility of the ad-atoms on the surface and, therefore, the probability of nucleation on top of a 2D island is minimized. The as-deposited ad-atoms can migrate to the step edges of 2D islands and nucleation only takes place on fully completed layers.

Depositing $SrTiO_3$ at a temperature of 800°C and an oxygen pressure of 10 Pa, with a continuous pulse frequency of 1 Hz (referred to as standard deposition conditions) the surface is transiting from a single level system to a multi-level system, as indicated by the strong damping of RHEED intensity oscillations in figure 2.3.12. A higher pressure causes the mobility of the particles on the surface to be lower. The probability of nucleation on top of a 2D island has increased, resulting in the observed roughening. To overcome the roughening at lower temperatures and higher pressures, several groups suggested the use of periodically interrupted growth [95, 79], leading to smoother surfaces. Annealing the surface will level off any roughness that has developed during deposition. This option is especially useful in the case of co-evaporated thin films. However, one still needs a considerable mobility on the surface, despite the longer waiting times. Sometimes, increasing the temperature to increase the activity is an option. However, for growing complex oxide materials, the regime of temperatures and pressures is limited

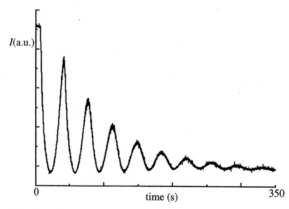

Figure 2.3.12. Intensity of the specular reflection during 'continuous' deposition of $SrTiO_3$ at 800°C and 10 Pa.

by the stability of the desired phases. Several groups have investigated the possibility of applying a form of growth manipulation to promote interlayer mass transport. Rosenfeld *et al* [96, 97] suggest periodically varying the temperature, varying the growth rate or applying ion bombardment to increase the number of nucleation sites in the initial stage of growth and thus decrease the average island size. Because of the phase stability mentioned above, changing the temperature (or oxygen pressure) is not an option. Periodically ion bombardment is very difficult to realize, maintaining the right stoichiometry in the case of oxide materials. In particular, in artificial layered structures mixing of succeeding layers is undesirable.

Growth rate manipulation to impose layer-by-layer growth could be a possibility, where fast deposition at the beginning of each unit cell layer is used to promote the formation of islands, thereby reducing their average size. However, the problem for PLD is not the formation of nuclei at the beginning of each unit cell layer, but lies more in the subsequent coalescence and ripening in between the laser pulses.

Since small islands promote interlayer mass transport, one can utilize the high supersaturation achieved by PLD by maintaining it for a longer time interval and suppress subsequent coarsening. Accordingly, to circumvent premature nucleation due to the limited mobility of the ad-atoms at a given pressure and temperature, causing a multi-level 2D growth mode, we introduce the possibility of interval deposition. Exactly one unit cell layer is deposited in a very short time interval, i.e. of the order of the characteristic relaxation times [71], followed by a much longer interval during which the deposited material can rearrange. During the short deposition intervals, only small islands will be formed due to the high super-saturation typical for PLD. The probability of nucleation on the islands increases with their average radius [82] and is, therefore, small in the case of fast deposition. The total number of pulses needed to complete one unit cell layer has to be as high as possible, to minimize the error introduced by the fact that only an integer number of pulses can be given. Both high deposition rate and sufficiently accurate deposition of one unit cell layer can be obtained by PLD using a high laser pulse frequency.

To prove this growth method, SrTiO₃ was deposited using the KrF excimer laser with a maximum repetition rate of 10 Hz as well as the XeCl excimer laser with a maximum repetition rate of 100 Hz. We used the same oxygen pressure and substrate temperature as in the case of continuous deposition where strong damping of the specular intensity is observed (figure 2.3.12). During deposition the growth was monitored using high pressure RHEED[5]. The incident angle of the 20 keV electrons was set at 1°. Wafers

---

[5]The deposition rate per pulse was estimated by depositing one unit cell layer at low frequencies followed by annealing at 850°C to restore the initial surface.

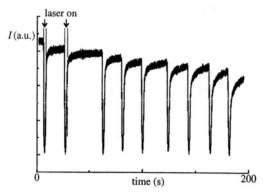

Figure 2.3.13. Intensity variations of the specular reflection during interval deposition using a laser reprate of 10 Hz, under identical conditions as figure 2.3.12.

with the smallest miscut angles ($<0.2°$) were selected for this study to exclude step-flow-like growth behaviour.

Figure 2.3.13 shows the RHEED intensity during ten cycles of deposition (at 10 Hz) and subsequently a period of no deposition, following the new approach. In this case, the number of pulses needed per unit cell layer was estimated to be about 27. In figure 2.3.14(*a*) the intensities at each maximum of both methods are compared. The decay of the intensity after each unit cell layer is significantly lower compared to the situation in figure 2.3.12. From this figure it can be seen that the recovery of the intensity after each deposition interval is fast at the beginning of deposition. The decrease in intensity can be ascribed to the fact that only an integer number of pulses can be given to complete a unit cell layer, besides nucleation on the next level. A slightly lower or higher coverage causes a change in recovery

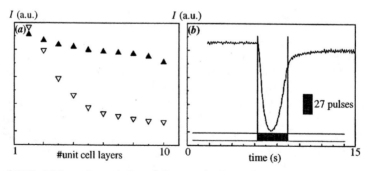

Figure 2.3.14. (*a*) Intensity variation of the specular beam during one deposition interval (10 Hz) and subsequent relaxation. (*b*) Intensities of the maxima during continuous deposition (inverted triangles) compared with the maxima during interval deposition at 10 Hz (upright triangles).

time. This situation will deteriorate with every subsequent unit cell layer, as follows from increasing relaxation times.

The intensity change during deposition of one unit cell layer at 10 Hz is shown in figure 2.3.14. The shape of the intensity curve at 10 Hz strongly resembles the calculated parabola from the intensity change of a two-level growth front with randomly distributed islands and island sizes [98]. From the shape of the curve it can be seen that the time needed to deposit one unit cell layer is still too long. This is because the deposition time interval of 2.7 s is longer than the characteristic relaxation time (~0.5 s). However, a significant suppression of the formation of a multilevel system has already been achieved at this point.

To avoid the above-mentioned situation, a similar experiment was performed using the XeCl laser with a pulse frequency of 100 Hz. The number of pulses needed to complete one unit cell layer was estimated to be 43 (i.e. deposition time of 0.43 s). Here, the increase of the RHEED intensity occurs after the deposition interval and recovers almost to the same level. The fact that the overall RHEED intensity slightly decreases is an indication that the number of 43 pulses is not exactly correct. If we continue the growth with 43 pulses per unit cell layer, the RHEED intensity will decrease continuously. Therefore, we periodically changed this number to 42 and observed, after an initial decrease of the interval maxima, that the intensity increased again, indicating that the surface becomes smoother. We repeated this procedure several times and in figure 2.3.15 an example of this sequence is given. Only the intensity change during the final 30 intervals of a total of 90 intervals (each constituting one unit cell layer) is depicted here. This led us to the possibility of, partly, correcting for the error due to the integer number of pulses: adjusting the amount of deposited material, by changing the number of pulses by one just after a decrease in maximum intensity. This way we can

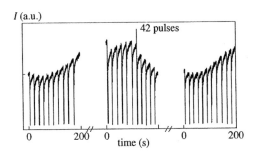

Figure 2.3.15. The intensity changes of the specular beam during interval deposition of SrTiO₃; the repetition rate used here is 100 Hz. The number of pulses needed to complete one unit cell layer is estimated to be 43. Sometimes, the number of pulses was changed by one, as indicated in the graph, to fine-tune the amount of deposited material.

maintain the level of RHEED intensity during deposition, suppressing the formation of a multi-level system. In fact, by doing this we proved the validity of this approach.

Conclusions

Three different stages of the PLD process were discussed: ablation, plasma expansion, and deposition of ablated species on a substrate with subsequent film growth. All stages are important to consider in understanding the relation between film deposition and the resulting film properties.

The influence of the laser fluence and the number of laser pulses on the resulting target morphology have been presented. Although different responses for metal and oxide targets to repetitive laser irradiation could be expected due to the different band structures of metals and oxides, the optical response is quite similar for 248 nm laser irradiation. Therefore, the difference in response is largely caused by differences in thermal properties such as thermal conductivity and enthalpy of evaporation. A flat target morphology after target ablation is desirable for a stable use of the target during thin-film deposition. However, this is not achievable for all materials. Metal targets show periodic structures of the order of μm after consecutive pulses of laser radiation, but the $SrTiO_3$ and $BaTiO_3$ targets do show a flat surface after ablation for relatively low fluences ($1.0\,J\,cm^{-2}$). The observed TiO_2 target ablation characteristics fall in between those of the ablated metals and perovskites, because ablation results in the presence of Ti, whose values for the thermal conductivity, specific heat, and the absorption coefficient are different from those of TiO_2. The straight ridges present in the TiO_2 target morphology after ablation at $1.5\,J\,cm^{-2}$ appears to be caused by the presence of Ti in their top layers, shielding the underlying stoichiometric target material from ablation. The final target morphology is dependent on fluence, number of pulses, and the movement of the target itself (rotating, scanning, or motionless). It can take between 15 and 75 pulses to reach a steady-state target morphology.

The consecutive plasma formation after ablation is highly dependent on ambient gas present and its pressure. The plasma appearance changes in colour and shape, resulting in different concentrations and angular distributions, respectively, of the ablated species. A small percentage of oxygen flow added to the argon flow has a dramatic influence on the plasma appearance in the case of $SrTiO_3$ ablation. The exterior of the plasma is more affected than the interior of the plasma, confirming the idea that the centre does not interact substantially with the oxygen gas.

For PLD, the high deposition rate and the energy of the particles at the substrate surface, both a function of the gas pressure, as well as the pulsed method of deposition, is expected to affect the growth of thin films. *In situ*

RHEED intensity monitoring of the specular reflection gives information about the nucleation of the deposited material. Moreover, by analysis of the RHEED pattern, both during growth as well as of any crystalline surface, one can judge the surface quality and detect, for example, the origination of precipitates.

During deposition of oxide materials, an oxide ambient is used and, although scattering of electrons in high oxygen pressure decreases the intensity of the electron beam, we have shown that growth monitoring at high oxygen pressures is feasible using RHEED. By two-stage pumping and enclosing the electron beam as long as possible in a vacuum tube, intensity losses due to scattering can be minimized. Introducing this technique has pushed the developments of the PLD technique enormously and the combination of the pulsed method of deposition with intensity monitoring results in the possibility of time resolved RHEED.

Besides the observed intensity oscillations in case of layer-by-layer growth, enabling accurate growth rate control, it became clear that intensity relaxation observed due to the typical pulsed method of deposition leads to a wealth of information about growth parameters.

The effect of the *pressure* and *temperature* on the diffusivity have been investigated using time resolved RHEED, by means of the relaxation behaviour after each laser pulse.

For successive deposition of unit cell layers of different targets to create artificially layered structures [99] or perfect multilayers, a true layer-by-layer mode is essential throughout the deposition of the whole film, since for every layer the starting surface has to be atomically flat again. A transition to multilevel has to be avoided as much as possible because the growth front may end up being thicker than each constituting layer. For many oxide materials one cannot freely choose the deposition temperature and pressure. Often, the desired phase is only stable in a limited regime of the deposition conditions, in combination with the substrate material. We have shown that it is possible with PLD to impose a single level 2D growth mode or layer-by-layer growth mode for $SrTiO_3$ despite unfavourable deposition conditions with respect to mobility. Depositing every unit cell layer at a very high deposition rate followed by a relaxation interval, we extend the typical high super-saturation for PLD keeping the average *island size* as small as possible. Therefore, the interlayer mass transport is strongly enhanced and the formation of a multi-level growth front does not occur. This technique, which we call pulsed laser *interval* deposition, is unique for PLD; no other technique has the possibility to combine very high deposition rates with intervals of no deposition in a fast periodic sequence. From the experiments, we conclude that for optimal results a high laser reprate and number of pulses per unit cell layer is desirable. The above-described deposition method shows very high potential in growing extremely thin and atomically smooth (metal) oxides, like materials with high dielectric constants.

References

[1] Roas B, Schultz L and Endres G 1988 *Appl. Phys. Lett.* **53** 1557

[2] Borman R and Nolting J 1989 *Appl. Phys. Lett.* **54** 2148

[3] Hammond R H and Borman R 1989 *Physica. C* **162–164** 703

[4] Smith H M and Turner A F 1965 *Appl. Opt.* **4** 147

[5] Sankur H and Cheung J T 1987 *J. Vac. Sci. Technol.* **A5** 2869

[6] Cheung J T and Sankur H 1987 *J. Vac. Sci. Technol.* **85** 705

[7] Dijkkamp D *et al* 1987 *Appl. Phys. Lett.* **51** 619

[8] Cheung J T and Sankur H 1988 *CRC Cnt. Rev. Solid State Mater. Sci.* **15** 63

[9] Chrisey D B and Hubler G K 1994 *Pulsed Laser Deposition of Thin Films* (New York: Wiley)

[10] Miura S, Yoshitake T, Satoh T, Miyasaka Y and Shohata N 1988 *Appl. Phys. Lett.* **52** 1008

[11] Tachikawa K, Ono M, Nakada K, Suzuki T and Kosuge S *High Tc superconducting films of $YBa_2Cu_3O_7$ prepared by a CO_2 laser beam evaporation* Private communication

[12] Koren G, Gupta A, Giess E A, Segmuller A and Laibowitz R B 1989 *Appl. Phys. Lett.* **54** 1054

[13] Lynds L, Weinberger B R, Potrepka D M, Peterson G G and Lindsay M P 1989 *Physica C* **159** 61

[14] Koren G, Baseman R J, Gupta A, Lutwyche M I and Laibowtz R B 1990 *Appl. Phys. Lett.* **56** 2144

[15] Lynds L, Weinberger B R, Peterson G G and Krasiski H A 1988 *Appl. Phys. Lett.* **52** 320

[16] Kwok H S, Zheng J P, Huang Z Q, Ying Q Y, Witanachchi S and Shaw D T 1989 *Proc. SERI Conf.*

[17] Fogarassy E, Fuchs C, Siffert P, Perriere J, Wang X Z and Rochet F 1988 *Solid State Commun.* **67** 975

[18] Dijkkamp A, Wu X D, Ogale S B, Inam A, Chase E W, Miceli P, Tarascon J M and Venkatesan T 1987 *Proc. Int. Workshop on Novel Mechanism of Superconductivity* 1

[19] Wu X D, Inam A, Venkatesan T, Chang C C, Chase E W, Barboux P, Tarascon J M and Wilkens B 1988 *Appl. Phys. Lett.* **52** 754

[20] Chang C C, Venkatesan T, Wu X D, Inam A, Chase E W, Hwang D M, Tarascon J M, Barboux P, England P and Wilkens B J 1989 *IEEE Trans. Magn.* **25** 2441

[21] Koren G, Polturak E, Fischer B, Cohen D and Kimel G 1988 *Appl. Phys. Lett.* **53** 2330

[22] Norton D P, Lowndes D H, Budai J D, Christen D K, Jones E C, McCamy J W, Ketcham Th D, St Julien D, Lay K W and Tkaczyk J E 1990 *J. Appl. Phys.* **68** 223

[23] Singh R K, Biunno N and Narayan J 1988 *Appl. Phys. Lett.* **53** 1013

[24] Singh R K, Narayan J, Singh A K and Krishnaswamy J 1989 *Appl. Phys. Lett.* **54** 2271

[25] Blank D H A, Adelerhof D J, Flokstra J and Rogalla H 1990 *Physica C* **167** 423

[26] Singh R K and Narayan J 1990 *J. Appl. Phys.* **67** 3785

[27] Liberman M A and Velikovich A L 1986 *Physics of Shock Waves in Gases and Plasmas* (Berlin: Springer)

[28] Boyd I W 1987 *Laser Processing of Thin Films and Microstructures* (Berlin: Springer)

[29] Willmott P R and Huber J R 2000 *Rev. Mod. Phys.* **72** 315

[30] Palik E D 1985 *Handbook of Optical Constants of Solids* (Orlando: Academic)

[31] Palik E D 1991 *Handbook of Optical Constants of Solids II* (San Diego: Academic)
[32] von Allmen M and Blatter A 1995 *Laser-Beam Interactions with Materials: Physical Principles and Applications* (Berlin: Springer)
[33] Wood R 1990 *Laser Damage in Optical Materials* (Bristol: Institute of Physics Publishing)
[34] http//www.webelements.com/
[35] Grimvall G 1999 *Thermophysical Properties of Materials* (Amsterdam: Elsevier)
[36] Karapet'yants M Kh and Karapet'yants M L 1970 *Thermodynamic Constants of Inorganic and Organic Compounds* (London: Ann Arbor-Humphrey Science Publishers)
[37] Lide D R 2000 *CRC Handbook of Chemistry and Physics* vol 81 (Boca Raton: CRC Press LLC)
[38] Cardarelli F 2000 *Materials Handbook* (London: Springer)
[39] Singh R K and Kumar D 1998 *Mater. Sci. Eng. R* **22** 113
[40] Hiroshima Y, Ishiguro T, Urata I, Makita H, Ohta H, Tohogi M and Ichinose Y 1996 *J. Appl. Phys.* **79** 3572
[41] Matthias E, Reichling M, Siegel J, Käding O W, Petzoldt S, Skurk H, Bizenberger P and Neske E 1994 *Appl. Phys. A* **58** 129
[42] Knacke B 1973 *Thermochemical Properties of Inorganic Substances* (Berlin: Springer)
[43] Fähler S and Krebs H U 1996 *Appl. Surf. Sci.* **96–98** 61
[44] Jordan R and Lunney J G 1998 *Appl. Surf. Sci.* **127–129** 968
[45] Durand H-A, Brimaud J-H, Hellman O, Shibata H, Sakuragi S, Makita Y, Gesbert D and Meyrueis P 1995 *Appl. Surf. Sci.* **86** 122
[46] Dam B, Rector J H, Johansson J, Kars S and Griessen R 1996 *Appl. Surf. Sci.* **96–98** 679
[47] Span E A F, Roesthuis F J G, Blank D H A and Rogalla H 1999 *Appl. Surf. Sci.* **150** 171
[48] Young J F, Preston J S, van Driel H M and Sipe J E 1983 *Phys. Rev. B* **27** 1155
[49] Doeswijk L M, de Moor H H C, Blank D H A and Rogalla H 1999 *Appl. Phys. A* **69** S409
[50] Milani P and Manfredini M 1996 *Appl. Phys. Lett.* **68** 1769
[51] Lowndes D H, Fowlkes J D and Pedraza A J 2000 *Appl. Surf. Sci.* **154–155** 647
[52] Young J F, Sipe J E and Driel H M 1984 *Phys. Rev. B* **30** 2001
[53] Sipe J E, Young J F, Preston J S and van Driel H M 1983 *Phys. Rev. B* **27** 1141
[54] Clark S E and Emmony D C 1989 *Phys. Rev. B* **40** 2031
[55] György E, Mihailescu I N, Serra P, Pérez del Pino A and Morenza J L 2002 *Surf. Coat. Tech.* **154** 63
[56] Span E A F 2001 Oxygen-permeable perovskite thin-film membranes by pulsed laser deposition *Ph.D. Thesis* University of Twente, The Netherlands
[57] Kelly R and Dreyfus R W 1988 *Surf. Sci.* **198** 263
[58] Singh R K, Holland O W and Narayan J 1990 *J. Appl. Phys.* **68** 233
[59] Geohegan D B 1992 *Appl. Phys. Lett.* **60** 2732
[60] Dyer P E, Issa A and Key P H 1990 *Appl. Phys. Lett.* **57** 186
[61] Ohkoshi M, Yoshitake T and Tsushima T 1994 *Appl. Phys. Lett.* **64** 3340
[62] Strikovski M and Miller J 1998 *Appl. Phys. Lett.* **73** 12
[63] Gonzalo J, Afonso C N and Ballesteros J M 1997 *Appl. Surf. Sci.* **109/110** 606
[64] Kano S, Langford S C and Dickinson J T 2001 *J. Appl. Phys.* **89** 2950

[65] Harilal S S, Radhakrishnan P, Nampoori V P N and Vallabhan C P G 1994 *Appl. Phys. Lett.* **64** 3377
[66] Tang Y, Han Z and Qin Q 1999 *Laser Chem.* **18** 99
[67] Gibson J K 1995 *J. Vac. Sci. Technol. A* **13** 1945
[68] Voevodin A A, Jones J G and Zabinski J S 2000 *J. Appl. Phys.* **88** 1088
[69] Dutouquet C and Hermann J 2001 *J. Phys. D* **34** 3356
[70] Lichtenwalner D J, Auciello O, Dat R and Kingon A I 1993 *J. Appl. Phys.* **74** 7497
[71] Blank D H A, Rijnders G, Koster G and Rogalla H 1999 *Appl. Surf. Sci.* **139** 17
[72] Bozovic I and Eckstein J N 1995 *MRS Bull.* **20** 32
[73] Karl H and Stritzker B 1992 *Phys. Rev. Lett.* **69** 2939
[74] Yoshimoto M, Ohkubo H, Kanda N, Koinuma H, Horiguchi K, Kumagai M and Hirai K 1992 *Appl. Phys. Lett.* **61** 2659
[75] Liu Z, Hanada T, Sekine R, Kawai M and Koinuma H 1994 *Appl. Phys. Lett.* **65** 1717
[76] Kanai M, Kawai T and Kawai S 1991 *Appl. Phys. Lett.* **58** 771
[77] Terashima T, Bando Y, Iijima K, Yamamoto K, Hirata K, Hayashi K, Kamigaki K and Terauchi H 1990 *Phys. Rev. Lett.* **65** 2684
[78] Shaw T M, Gupta A, Chern M Y, Batson P E, Laibowitz R B and Scott B A 1994 *J. Mater. Res.* **9** 2566
[79] Rijnders A J H M, Koster G, Blank D H A and Rogalla H 1997 *Appl. Phys. Lett.* **70** 1888
[80] For more information about high pressure RHEED, see: www.tsst.nl
[81] Rosenfeld G, Poelsema B and Comsa G 1997 *Growth and Properties of Ultrathin Epitaxial Layers* ed D A King and D P Woodruff (Amsterdam: Elsevier) chapter 3
[82] Markov I V 1995 *Crystal Growth for Beginners* (London: World Scientific) pp 81–86
[83] Yang H-N, Wang G-C and Lu T-M 1995 *Phys. Rev. B* **51** 17932
[84] Neave J H, Dobson P J, Joyce B A and Jing Zhang 1985 *Appl. Phys. Lett.* **47** 100
[85] Myers-Beahton A K and Vvedensky D D 1990 *Phys. Rev. B* **42** 5544
[86] Myers-Beahton A K and Vvedensky D D 1991 *Phys. Rev. A* **44** 2457
[87] Zandvliet H J W, Elswijk H B, Dijkkamp D, van Loenen E J and Dieleman J 1991 *J. Appl. Phys.* **70** 2614
[88] Shitara T, Vvedensky D D, Wilby M R, Zhang J, Neave J H and Joyce B A 1992 *Phys. Rev. B* **46** 6825
[89] Shitara T, Zhang J, Neave J H and Joyce B A 1992 *J. Appl. Phys.* **71** 4299
[90] Vvedensky D D and Clarke S 1990 *Surf. Sci.* **225** 373
[91] Koster G, Kropman B L, Rijnders A J H M, Blank D H A and Rogalla H 1998 *Appl. Phys. Lett.* **73** 2920
[92] Koster G, Rijnders G, Blank D H A and Rogalla H 2000 *Physica C* **339** 215–230
[93] Geohegan D B and Puretzky A A 1995 *Appl. Phys. Lett.* **67** 197
[94] Chern M Y, Gupta A and Hussey B W 1992 *Appl. Phys. Lett.* **60** 3045
[95] Gupta A, Chern M Y and Hussey B W 1993 *Physica C* **209** 175
[96] Rosenfeld G, Servaty R, Teichert C, Poelsema B and Comsa G 1993 *Phys. Rev. Lett.* **71** 895
[97] Rosenfeld G, Poelsema B and Comsa G 1995 *J. Cryst. Growth* **151** 230
[98] Lagally M G, Savage D E and Tringides M C 1989 *Reflection High Energy Electron Diffraction and Reflection Electron Imaging of Surfaces* ed P K Larsen and P J Dobson (London: Plenum) pp 139–174
[99] Koster G, Verbist K, Rijnders G, Rogalla H, van Tendeloo G and Blank D H A 2001 *Physica C* **353** 167–183

SECTION 3

CHARACTERIZATION

Chapter 3.1

Oxygen diffusion

R M C de Almeida and I J R Baumvol

Introduction

One of the strengths of amorphous (vitreous) SiO_2 as a gate dielectric material, when thermally grown directly on the active region of a single-crystalline Si semiconductor, is its thermal stability in further fabrication steps, in strict accordance with the phase diagram of the Si–O system. More specifically, this means that during further processing steps, usually performed at temperatures between 500 and 1100°C in different atmospheres, and therefore at different oxygen partial pressures, oxygen is adsorbed at the surface of SiO_2 films on Si and diffuses therein, having the following consequences:

(a) Atomic transport is limited to species coming from the gas phase. Oxygen, in particular, diffuses as molecular oxygen (O_2) in the films without interacting with the SiO_2 network to react in the near-interface region with non-fully oxidized silicon atoms (Si^{nfo}) or with Si atoms from the substrate. Alternatively, oxygen from the gas phase is exchanged for fixed oxygen from the solid phase in the near-surface region. Si remains essentially immobile. Although Si interstitials are produced as a result of oxygen arrival and reaction at the Si/SiO_2 interface, injected Si atoms are not observed in the bulk of the SiO_2 films after the thermal oxidation processing.

(b) Apart from additional, stoichiometric or sub-stoichiometric SiO_2 formation due to reaction of diffusing oxygen with substrate-Si, there are no other chemical reactions at the film interfaces with single-crystalline Si (active region of the transistor) or even with poly-crystalline Si (gate electrode), which is deposited on the SiO_2/Si structures.

(c) In the bulk of SiO_2 films on Si with thicknesses above 4–5 nm, chemical composition does not change. However, ultrathin silicon oxide films (thicknesses below 3–4 nm) on Si can be appreciably oxygen deficient and, in this case, thermal processing in O_2 completes the oxidation of Si^{nfo}, leading to a film with average stoichiometry closer to SiO_2.

The facts announced in (a)–(c) above were confirmed [1–3] using isotopic substitution and high resolution profiling methods.

SiO_xN_y films on Si thermally processed in oxygen-containing atmospheres in the same temperature range are essentially as stable as SiO_2. There are some minor differences resulting from the fact that O_2 is no longer the only diffusive species, as nitrogen may also migrate in different forms such as N, N_2, NO and others. In practice, the only noteworthy difference is the loss of part of the N content of SiO_xN_y films during further thermal processing steps.

While SiO_2 and SiO_xN_y films are chemically inert, all the alternative oxides of interest for high-κ dielectrics studied so far are reactive in O_2-containing atmospheres, even at O_2 partial pressures as low as 10^{-7} mbar. Furthermore, SiO_2 films are thermodynamically stable on Si, whereas most alternative oxides are not [4, 5].

In contrast to SiO_2 films, during annealing of alternative dielectrics of any thickness in oxygen-containing atmospheres at the same temperatures as mentioned above, oxygen may diffuse by strongly interacting with the three characteristic regions, namely surface, bulk, and interface, bringing about many consequences. Exchange and incorporation of oxygen from the gas phase for oxygen atoms previously existing in the film (solid phase) occurs throughout the whole high-κ film. When the propagating oxygen front arrives at the high-κ/Si interfacial region it reacts therein. Besides, there is an alternative, complementary channel of oxygen transport, which is analogous to that taking place in SiO_2, namely diffusion without interaction with the high-κ film network. This alternative channel also contributes to supply oxygen to the interface, producing SiO_2. One can summarize the effects as follows:

(a) Chemical reactions can take place in the bulk of high-κ films, including reaction of the high-κ oxide network with diffusing oxygen, changing its composition, which can then depart substantially from the original, as-deposited compositions. The electrical consequence is the modification of the dielectric constant of the high-κ film, which alters the flatband voltage of the associated capacitor and consequently the threshold voltage of the transistor.

(b) Oxygen arrival at the high-κ/Si interface, oxidizing substrate-Si and forming a SiO_2 layer as well as metal silicates, leading to high-κ/SiO_2 or high-κ/silicate stacks where the thickness of the SiO_2 or silicate interlayer will determine the lowering of the overall dielectric constant [6].

(c) Oxygen arrival to the high-κ/Si interface also triggers the injection of Si interstitials from the substrate due to the volume difference of Si atoms in Si and SiO$_2$. Injected interstitials are transported into the high-κ film and react therein, changing the composition of the high-κ film. Metal silicides can be formed by reaction between the metal gate and mobile Si that migrated to the dielectric/metal electrode interface, changing the Fermi level and the electrical conductivity, and even contributing with a lower-capacitance layer in series with the high-κ dielectric. Furthermore, Si precipitates and separate metal oxide phases can be formed through reduction of the oxide by the metal electrode film in the upper interface, since metal electrodes may become necessary.

(d) Finally, the transport of metallic species from the high-κ film into the channel region of the Si transistor, which will degrade the mobility of charge carriers in the transistor channel, has also been observed.

Reaction–diffusion in high-κ dielectrics on Si

In SiO$_2$ and SiO$_x$N$_y$ films on Si, which are essentially inert and where the oxidizing or oxynitriding species are the only mobile species, post-deposition thermal treatment of high-κ dielectrics may render mobile all the species present in the system, promoting a cascade of chemical reaction as well. Thus, while annealing in O$_2$ of SiO$_2$ films above a certain thickness (4–5 nm) on Si can be described as diffusion of O$_2$ followed by O$_2$–Si reaction at the interface, forming SiO$_2$, in high-κ dielectrics of any thickness a very typical reaction–diffusion process takes place, establishing a far more complicated physico-chemical situation with severe consequences on the electrical characteristics. This typical reaction–diffusion scenario significantly contributes to the complexity of replacing SiO$_2$ by high-κ oxides. Therefore, to control the integrity of the gate dielectric in ULSI processing—something which is more or less granted in the case of SiO$_2$ and SiO$_x$N$_y$ gate dielectrics—it is necessary to gather enough information on the stability of high-κ dielectrics on Si in oxygen-containing atmospheres, from extremely low (10^{-9} mbar) to rather high (10^2 mbar) O$_2$ partial pressures, and on the effects of different parameters like temperature, time, and partial pressures of the different gaseous components of the annealing atmospheres, as well as interlayer composition and thickness, in order to build up an adequate model of the system which is prone to be incorporated into circuit design codes. An annealing in oxygen-containing atmosphere will probably have to be intentionally performed in any high-κ dielectric into obtain leakage current and interface density of electronic states at low enough levels to fulfil the very strict requirements of technology. Thus, one can clearly see that our scientific, atomic scale understanding of all above-mentioned phenomena will have to be substantially increased before the desired control on high-κ dielectrics is reached, such that they can be comfortably incorporated into Si-based device fabrication technology.

Furthermore, it became clear from the very beginning that an extremely thin (typically 0.5 nm thick or less) interfacial layer of silicon oxide, nitride, or oxynitride, usually thermally grown or deposited on the surface of the active region of the silicon semiconductor substrate, before or after high-κ dielectric deposition, would be a convenient way of engineering the best replacement for silicon oxide. This interfacial layer can provide an SiO_2 (or SiO_xN_y)/Si-like interface, which is the one that presents the minimum density of electronic states (ideal passivation) on one hand, constituting on the other hand, when they contain nitrogen, a reaction–diffusion barrier to oxygen that limits further thickening of the interfacial layer, as well as a diffusion barrier to metallic species from the high-κ material that migrate into the transistor channel. However, since the essential need for an intermediate layer was only very recently made clear, one can still find more and more evidence appearing in the literature. A complete survey of the atomic scale mechanisms leading to improvement in stability by nitrogen-containing intermediate layers is beyond the scope of the present work. Here we will be limited to the presentation of the main empirical facts and their repercussions on stability.

Oxygen diffusion in aluminium oxide films on Si

Annealing of thin films in N_2 or in O_2

Al_2O_3 films with thickness in the range 5–45 nm, deposited by ALCVD on three different surfaces: (a) HF-stripped surface, (b) ultrathin (0.7 nm) chemically grown SiO_2 films on Si, and (c) ultrathin (1.2–2 nm), thermally grown SiO_2 films on Si, were investigated [7, 8]. The behaviour against a rapid thermal annealing (RTA) spike in N_2 at temperatures between 950 and 1050°C was addressed and concentration versus depth distributions of the different elements are shown in figure 3.1.1. One can see that the profiles of the different species change with annealing, indicating H exodiffusion, Si migration from the substrate into the Al_2O_3 layer, and Al and O loss. Some of these features have been reported earlier [9–11] for thinner aluminium oxide film samples annealed in O_2. Figure 3.1.2 shows Si migration into Al_2O_3 films on Si as observed by medium-energy ion scattering (MEIS).

Samples annealed in N_2 for 15 min or in UHV for 20 min at 800°C were characterized by x-ray photoelectron spectroscopy (XPS). Results indicate that: (a) the as-deposited aluminium film presents an abrupt

---

Figure 3.1.1. SIMS depth profiles for (a) as-deposited Al_2O_3 (12 nm)/SiO_2/Si and after spike annealings at (b) 950°C and (c) 1050°C (XRD results are given in the insets). From [7].

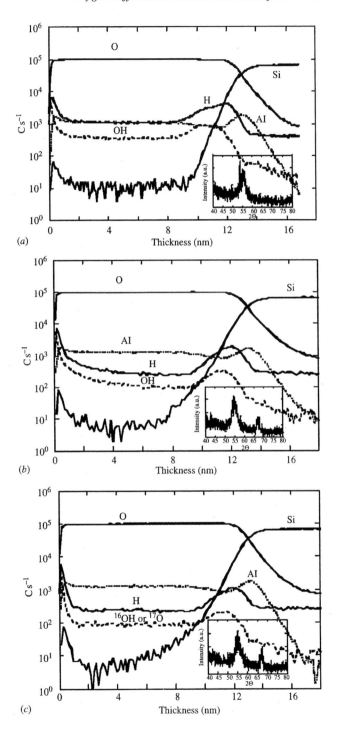

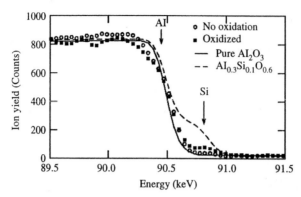

Figure 3.1.2. MEIS spectra taken with a normally incident beam of 100 keV protons, detected at a scattering angle of 110° before and after oxidation in O_2 at 800°C of 4 nm thick Al_2O_3 films deposited by ALCVD on Si and pre-annealed in vacuum at 600°C. From [11].

interface with the silicon substrate, almost without an SiO_2 intermediate layer; (b) annealing in N_2 or in UHV showed the non-negligible growth of a silicon oxide layer between the Al_2O_3 film and the Si substrate, the oxygen consumed for this oxidation of the Si substrate having been attributed to OH groups remaining from the sample preparation process.

Figure 3.1.3 shows ^{16}O and ^{18}O SIMS profiles for $Al_2{}^{16}O_3$ films annealed in O_2 isotopically enriched in the ^{18}O isotope ($^{18}O_2$). ^{18}O is seen to incorporate in the whole Al_2O_3 films, in exchange for ^{16}O atoms previously existing in the films ($^{16}O-^{18}O$ exchange). A pre-annealing in N_2 was seen to reduce considerably (but not eliminate) ^{18}O incorporation and $^{16}O-^{18}O$ exchange.

Thin aluminium oxide films deposited on SiO_2 (0.7 nm)/Si samples by ALCVD using TMA/H_2O cycles were also investigated. As-deposited Al_2O_3 films (6.5 nm thick) were subjected to RTA either in vacuum (5×10^{-7} mbar) or in 70 mbar of 98.5% ^{18}O-enriched O_2 ($^{18}O_2$), in order to distinguish between oxygen incorporated from the gas phase and that previously existing in the films.

Al profiles as determined by narrow resonance nuclear reaction profiling (NRP) [12, 13] are shown in figure 3.1.4 (top), indicating a

Figure 3.1.3. ^{18}O and ^{16}O SIMS depth profiles after oxidation in a closed Joule effect furnace (100 mbar, 800°C, 1 h) of 45 nm Al_2O_3/Si stacks either in $^{18}O_2$ alone or after a sequential treatment, a first $^{18}O_2$ oxidation followed by a second oxidation in $^{16}O_2$ ($^{18}O_2/^{16}O_2$) of (a) as-deposited layer, and N_2 annealed by RTP at (b) 800°C and (c) 900°C. The SIMS profiles are presented with a linear scale and the total oxygen profiles ($^{16}O + ^{18}O$) were normalized to 10^{15} C s^{-1}. From [7].

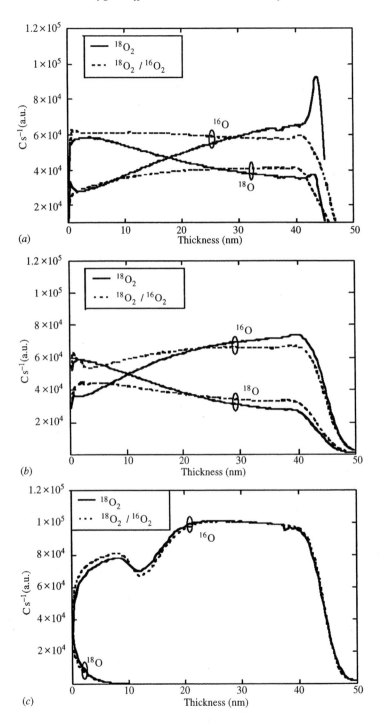

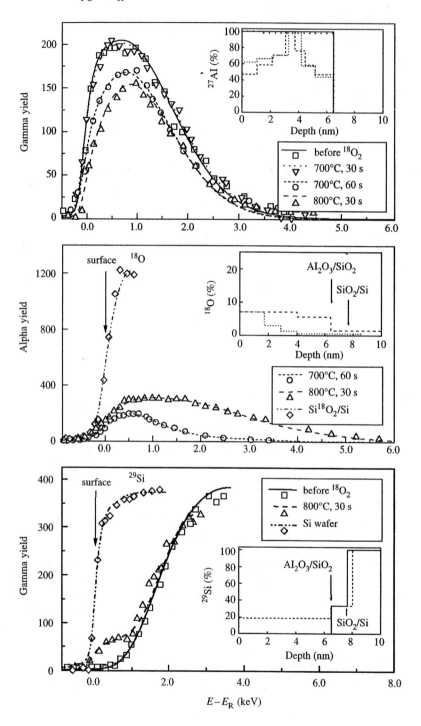

constant Al concentration in the as-deposited film, corresponding to stoichiometric Al_2O_3 and an abrupt interface with the substrate, within the accuracy and depth resolution of the technique (approximately 0.5 nm near the sample surface). This figure also shows a progressive loss of Al from the surface and interface regions as temperature and time of RTA in $^{18}O_2$ are increased, such that after annealing at 800°C for 30 s a substantial loss of Al atoms is observed. Figure 3.1.4 (centre) shows ^{18}O profiles also determined by NRP, indicating that for RTA at 700°C the incorporation of ^{18}O is mainly in near-surface regions ($\sim 7\%$), decreasing deeper into the Al_2O_3 films. RTA at higher temperatures also shows incorporation of ^{18}O in the near-interface region, indicating the growth of an SiO_2 interfacial layer. Finally, figure 3.1.4 (bottom) shows the ^{29}Si profiles, where Si migration into the Al_2O_3 films is seen. Si mobility was observed after RTA in O_2 at 700 and 800°C, whereas at 600°C Si movement was negligible or absent.

The areal densities of ^{18}O and ^{16}O after RTA in $^{18}O_2$ given in table 3.1.1 are evidence that ^{18}O is incorporated in exchange for ^{16}O atoms existent in the films. The areal densities of incorporated ^{18}O are in general slightly smaller than those of lost ^{16}O. The same has been observed for thicker aluminium oxide films subjected to longer annealing times.

The same $Al_2O_3/SiO_2/Si$ structures from figure 3.1.4 were submitted to RTA in high vacuum at the same temperatures and times. In all cases ^{16}O was not lost, and ^{27}Al and ^{29}Si profiles determined before and after annealing were superimposable, evidencing immobility of all species. Furthermore, annealing in a vacuum (700°C, 120 s) previous to RTA in $^{18}O_2$ has not altered the results for RTA in $^{18}O_2$ only (800°C, 30 s). It can be concluded then that oxygen from the gas phase plays an essential role in promoting the transport of the atomic species in the $Al_2O_3/SiO_2/Si$ structures.

So, for long enough times and/or high enough temperature and $^{18}O_2$ pressure, RTA of $Al_2O_3/SiO_2/Si$ structures leads to incorporation of ^{18}O mainly by $^{18}O-^{16}O$ exchange, in slightly smaller amounts than ^{16}O is lost. The silicon oxide buffer layer grows, aluminium moves towards the surface where it leaves the film, and silicon migrates towards the surface reacting in the region that was formerly an aluminium oxide film. Aluminium silicate,

Figure 3.1.4. Excitation curves of the $^{27}Al(p,\gamma)^{28}Si$ (top), $^{18}O(p,\alpha)^{15}N$ (centre), and $^{29}Si(p,\gamma)^{30}P$ (bottom) nuclear reactions around resonance energies, with the corresponding profiles in the insets for ultrathin $Al_2O_3/SiO_2/Si$ structures subjected to RTA in $^{18}O_2$. 100% of ^{27}Al, ^{18}O, and ^{29}Si correspond, respectively, to their concentrations in Al_2O_3, $Al_2^{18}O_3$, and Si. Excitation curves from a standard $Si^{18}O_2$ film on Si and from a virgin Si wafer are also shown, in order to determine the surface energy positions of ^{18}O and ^{29}Si, respectively. The arrows in the insets indicate the positions of the interfaces before RTA. From [9].

Table 3.1.1. Areal densities of ^{18}O incorporated in the $Al_2O_3/SiO_2/Si$ structures during rapid thermal annealings in $^{18}O_2$ and the corresponding areal densities of ^{16}O remaining in these structures. The areal density of ^{16}O in the as-deposited sample was $41.2 \times 10^{15}\,cm^{-2}$.

| | Areal density ($10^{15}\,cm^{-2}$) | | | |
| | ^{16}O | | ^{18}O | |
Temperature (°C)	30 s	60 s	30 s	60 s
600	40.6	39.8	0.17	0.40
700	39.6	38.1	0.62	1.36
800	36.5	–	3.65	–

or some other Si–Al–O compound, or SiO_2–Al_2O_3 phase mixture, were observed to form in the near interface region.

Thickness inhomogeneities in the Al_2O_3 films were observed for the higher annealing temperatures [14]. However, these inhomogeneities in the films may be avoided by annealing in ultra-high vacuum for long times prior to annealing in O_2.

Analysis of even thinner Al_2O_3 films, namely 2.4 nm Al_2O_3/SiO_2 (1 nm)/Si, subjected to RTA in ultra-high vacuum at temperatures ranging from 900 to 1000°C were investigated. MEIS analysis showed that samples annealed at 900°C in vacuum differed insignificantly from the as-deposited sample. However, higher-temperature anneal for shorter time (980°C, 20 s) suggested O loss as shown in figure 3.1.5. Besides, annealing at 1000°C for 20 s caused O and Al loss, as well as Si appearance at the Al_2O_3 film surface due to the formation of pinholes, which were confirmed by atomic force microscope images. The depths of the pits, approximately 100 Å, indicated that substantial silicon etching has taken place, probably through volatile SiO formation.

Furthermore, annealing in O_2 at different pressures was performed. MEIS showed a broadening of the oxygen peak and an increase in the Si portion of the combined Si/Al peak as displayed in figure 3.1.6. This is consistent with an increase of an underlying SiO_2 layer. Unfortunately, the presence of Al in this region or of Si atoms in the initially aluminium oxide film could not be determined by MEIS. Assuming an $Al_2O_3/SiO_2/Si$ structure, the authors estimated the grown SiO_2 layer thickness. They found a roughly logarithmic increase with pressure, as shown in figure 3.1.6, where the result for bare silicon oxidation for 30 min at 600°C at 10^{-2} mbar is also presented, for sake of comparison. The authors concluded that the silicon oxide layer growth is not significantly different with or without the thin aluminium oxide film.

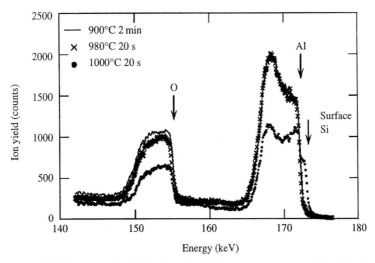

Figure 3.1.5. MEIS spectra for Al_2O_3 films after high-temperature annealing. The decrease in oxygen intensity at high temperature is caused by the growth of voids in the film. Regions of the films that have not desorbed still contribute to oxygen intensity, retaining linewidth. From [14].

Reaction–diffusion in extreme ultrathin (less than 1 nm) Al_2O_3 films on Si

Kundu *et al* [15–17] investigated ultrathin (0.6 nm thick) aluminium oxide films on silicon. The films were produced by depositing aluminium on a silicon wafer and oxidizing at 2×10^{-6} mbar of O_2 at 400°C. An advantage of this route is to build Al_2O_3 films free from carbon, hydrogen and OH groups.

The samples were analysed by complementary techniques after being subjected to *in situ* reoxidation in O_2 pressures of 2×10^{-6}, 5×10^{-6}, 2×10^{-5}, and 5×10^{-5} mbar, for 10, 20, 30, 40, and 60 min, and at 400, 550, 700, and 750°C. The first issue addressed was the characterization of the interface. Images of scanning reflection electron microscopy showed a clean Si surface for the as-deposited sample and for samples oxidized for 20 min at 700°C at the lower oxidation pressures, indicating that the interface is still intact and that it is abrupt. However, oxidation at 700°C for 20 min at 5×10^{-5} mbar causes an intermediate SiO_2 layer to grow. Absence of the intermediate silicon oxide layer has been also observed for the other temperatures when considering the three lower O_2 pressures. At higher O_2 pressure (5×10^{-5} mbar), an SiO_2 layer is observed to form, increasing in thickness as temperature increases. At this last pressure, after 20 min of annealing at 750°C, the aluminium oxide layer is observed to crystallize, in agreement with previous observations.

The growth rate of the silicon oxide layer is found to be much less than the growth rate of SiO_2 on Si in the absence of an Al_2O_3 film. This is not

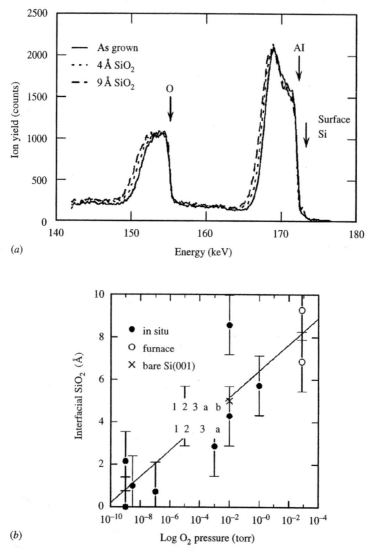

Figure 3.1.6. Effect of oxidation of Al_2O_3 deposited by ALCVD directly on Si(001). (*a*) MEIS spectra show broadening of the oxygen peak as well as the Si component of the combined Si/Al peak, indicating an increase in the underlying SiO_2 thickness. (*b*) Measured thickness of underlying SiO_2 layer versus oxidation pressure. Oxidations were at the indicated O_2 pressure for 30 min at 600°C. The line is drawn to guide the eye. From [14].

in agreement with the findings by Copel *et al* [14] discussed above. It remains a point to be clarified whether this slowing down of the oxygen diffusion through the Al_2O_3 film is a consequence of the different deposition routes. Indeed, ALCVD deposition using TMA/H_2O cycles leaves H and OH groups in the aluminium oxide layer, which could serve as sources for oxygen or as alternative mechanisms for oxygen transport.

Besides oxidation of the substrate, the thermal treatment of samples for long times at high temperatures and pressures also affects the aluminium oxide film.

Some points remain open, when comparing the results obtained by different authors:

(a) How crystallization temperature of the Al_2O_3 film on Si depends on deposition route and on thermal treatments parameters.
(b) Oxidation of substrate-Si and the consequent mobility of either interstitial Si or volatile SiO seem to play an important role in the Al loss from the films. However, some authors observed measurable quantities of SiO_2 forming at the interface without any significant Al loss.
(c) Where and which phases Si may form when it migrates into the Al_2O_3 films. Separate SiO_2 and Al_2O_3 phases were observed by certain authors with some SiO_2 at the surface, while some other authors found evidence of Si spread out through the whole region of the aluminium oxide film and not only at the surface, either as Al silicates or as oxidized Si.
(d) Finally, oxygen diffusion through Al_2O_3 films on Si presents several controversial aspects that will be discussed in the following.

Oxygen diffusion

Diffusion of oxygen was investigated in 35 nm thick Al_2O_3 films on Si deposited by ALCVD [18]. Thick films were used aiming at keeping the Al_2O_3/Si interface well apart from the surface, isolating the oxygen transport across the Al_2O_3 films from the effects induced by substrate oxidation.

^{18}O profiles as determined by NRP in samples annealed for different times and temperatures in $^{18}O_2$ are shown in figure 3.1.7. The profiles clearly showed that, differently from the case of thermal growth of SiO_2 films on Si in O_2, oxygen from the gas phase diffuses in the films strongly interacting with the Al_2O_3 film network, being thus incorporated in near-surface and bulk regions, to amounts and ranges that increased with annealing time, temperature, and pressure. The main reaction channel for the propagating ^{18}O front in the Al_2O_3 film network is $^{16}O-^{18}O$ exchange. The second channel of ^{18}O incorporation, namely oxidation of the silicon substrate, is not present here, since in almost all cases the ^{18}O front has not yet attained the interface.

There are different channels through which molecules coming from a gas phase may diffuse in Al_2O_3 films. Alternative diffusion channels happen when besides diffusive O_2, there is atomic oxygen diffusing both interstitially or through vacancies and other defects. Atoms coming from the gas phase may also be exchanged for oxygen in the network.

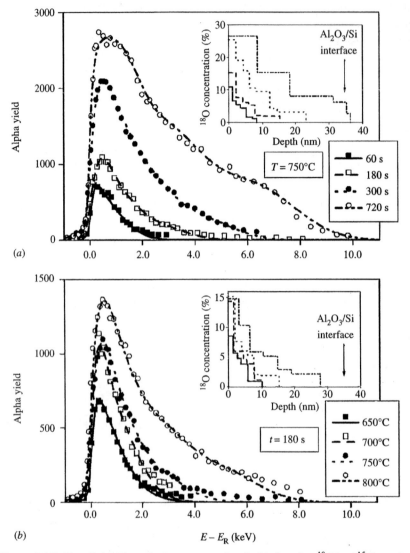

Figure 3.1.7. Experimental excitation curves (symbols) for the $^{18}O(p,\alpha)^{15}N$ nuclear reaction around the 151 keV resonance and the corresponding simulations (lines) for Al_2O_3 films on Si after annealing in 6 mbar of $^{18}O_2$: at 750°C for different times (top) and for 180 s for different temperatures (bottom). ^{18}O profiles assumed in the simulations are shown in the insets. From [18].

As a result of reaction (including exchange) of oxygen in the Al_2O_3 network, a trace is left in the films of the presence of a diffusing front of ^{18}O. It also implies that the macroscopic diffusion coefficient of oxygen in the Al_2O_3 is smaller than that in SiO_2. One concludes then that Al_2O_3 does not prevent oxygen diffusion, but it must considerably slow it down in comparison to oxygen diffusion in silicon oxide. In fact, diffusion coefficients were estimated based on the growth rate of interfacial SiO_2 in Al_2O_3/Si structures at different temperatures and, using the Arrhenius rate equation, they found extremely small values of E_a (0.09 eV) and D_0 (3×10^{-2} nm^2 min^{-1}). In any case, it is very useful to consider reaction–diffusion modelling of these systems.

Oxygen diffusion in zirconium oxide-based films on Si

Zirconium oxide

Different authors investigated atomic transport in ultrathin ZrO_2 films deposited on Si using MEIS. Ultrathin intermediate layers of either SiO_2, SiO_xN_y, or $ZrSi_xO_y$ were present in all cases. Such layers were either thermally grown at high temperature previous to ZrO_2 deposition (usually SiO_2 or SiO_xN_y) or spontaneously grown during the first stages of deposition (usually $ZrSi_xO_y$). The MEIS spectrum of 3.5 nm thick ZrO_2 film grown on SiO_2/Si has a Zr signal that is nearly trapezoidal in shape, with a sharp leading edge. The film was modelled as a continuous 3.8 ± 0.3 nm layer of ZrO_2 (having a little excess of oxygen) on 1.3 ± 0.3 nm of SiO_2. The ZrO_2/SiO_2 interface was seen to be abrupt within experimental depth resolution (approximately 0.5 nm). On the other hand, the ZrO_2 films grown on HF etched Si exhibit a much narrower Zr peak than the oxidized sample, with a more gradual slope on the leading and trailing edges. Significantly less Zr is determined from the Zr peak in this case. AFM results [19] (not shown) indicate an increase in rms roughness from 0.15 nm in the ZrO_2/SiO_2 sample to 0.57 nm in the HF etched sample. Furthermore, the Si peak is greatly altered in the HF etched sample, with a significant component visible at the surface. After a 900°C anneal for 2 min the $ZrO_2/SiO_2/Si$ structure undergoes almost no change, but a 1000°C flash for 30 s reduces the ZrO_2 film to a silicide. The consequence of this last thermal treatment is seen in the MEIS spectrum as a broadening of the Zr peak and its intermixing with the underlying Si, and by the absence of the O peak, as shown in figure 3.1.8.

Upon annealing in oxygen atmosphere (oxygen-pressure range 0.4–10 mbar, temperature range 300–930°C) ZrO_2 films were stable against silicate formation. The Zr and Si MEIS signals did not intermix. Significant oxygen diffusion was observed [20, 21] promoting growth of the SiO_2 layer between the ZrO_2 film and the Si substrate. This interfacial SiO_2 growth saturated quickly in time (no change after 5 min at 500°C) and in pressure (no change seen from 0.4 to 10 mbar). The thickness of the grown SiO_2 layer increased with annealing temperature as shown in figure 3.1.9 (top). MEIS showed that Zr and Si were essentially immobile up to 800°C, such that interfacial layer formation could be only attributed to oxygen from the gas

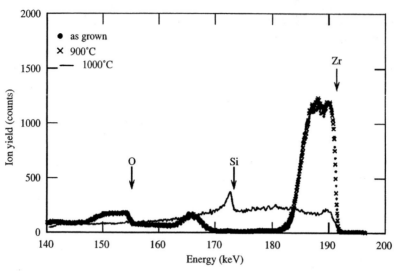

Figure 3.1.8. MEIS spectra from as-deposited and annealed ZrO_2 films on Si. From [19].

phase diffusing across the film to react with substrate-Si at the ZrO_2/Si interface.

The atomic transport in ZrO_2 films on Si during thermal treatment in O_2-containing atmospheres was investigated using MEIS in samples annealed in $^{18}O_2$-enriched oxygen. The results [22] indicated a substantial amount of isotopic substitution of ^{18}O from the gas phase for ^{16}O previously existing in the ZrO_2 films ($^{16}O-^{18}O$ exchange), as shown in figure 3.1.9 (bottom). According to figure 3.1.10, ^{18}O is uniformly distributed in depth in the ZrO_2 films, without major changes of the film stoichiometry. The method of isotopic substitution allowed characterization of a fast oxygen transport through ZrO_2, which was attributed to atomic oxygen (O) diffusion either in a vacancy sublattice or grain boundary. This interpretation comes from the fact that ZrO_2 films can partially recrystallize under thermal annealing, while stabilized cubic and tetragonal zirconia are well known as fast ionic conductors having doping-induced oxygen vacancies [23], and the oxygen transport readily proceeds via a very mobile O sublattice. The propagating ^{18}O front from the gas/solid interface reacts with the ZrO_2 network, the main reaction channels being most probably $^{16}O-^{18}O$ exchange and completion of the ZrO_2 stoichiometry, which resulted in ^{18}O fixation in the films. Oxygen has a much lower diffusivity in the growing SiO_2 film at the interface than in the metal oxide layer, and this becomes the rate-limiting step.

Instabilities were observed by annealing at 930°C in 0.1 mbar of O_2. A small amount of Si appears at the surface as shown in the MEIS spectra

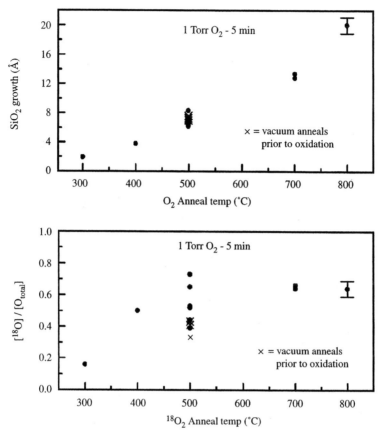

Figure 3.1.9. (Top) Observed amount of interfacial SiO_2 growth versus annealing temperature for re-oxidation in O_2 of ZrO_2 films on Si. (Bottom) The fraction of incorporated ^{18}O observed versus annealing temperature for re-oxidation in $^{18}O_2$. From [20].

of figure 3.1.11. These Si atoms at the surface are oxidized, forming approximately 0.3 nm of SiO_2. An instability concerning migration of substrate-Si atoms to the surface can be explained within the reaction–diffusion framework used in the 'Oxygen diffusion in aluminium oxide films on Si' section to model the thermal behaviour of Al_2O_3 films on Si. Si interstitials are injected into the ZrO_2/SiO_2 film (and into the Si substrate as well) following the formation of SiO_2 by reaction between substrate-Si and diffusive oxygen arriving at the interface. The Si interstitials are then transported to the surface as part of the reaction–diffusion dynamics established in the ZrO_2 films. In the equations of the previous section, the transport of injected Si interstitials is represented by the mobile species Si^i.

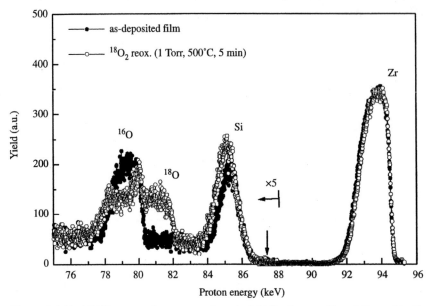

Figure 3.1.10. MEIS spectra from an as-deposited ZrO$_2$ film on Si and from this film re-oxidized in ^{18}O$_2$. The arrow indicates the energy where scattering from surface Si would be detected. From [21].

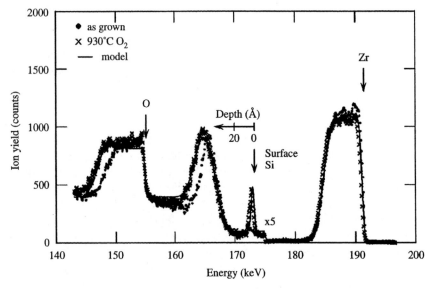

Figure 3.1.11. Effect of oxidation in O$_2$ of ZrO$_2$ thin films on Si as observed by MEIS. From [19].

When arriving at the surface, mobile Si^i is fixed as SiO_2 by reaction with oxygen from the gas phase.

The very high ^{16}O–^{18}O exchange observed in isotopic substitution experiments and the uniform ^{18}O incorporation throughout the whole ZrO_2 film evidence a striking difference from the same kind of experiment performed in SiO_2 films on Si. In the latter, isotopic exchange and ^{18}O incorporation from the gas phase occur only in the near-surface regions of SiO_2 films, owing to peroxyl defects existing therein, and in the near-interface region, but not in the bulk of the films where the diffusing species (O_2) does not react with the silica network. In the present ZrO_2 films, on the other hand, the fast diffusing species ($^{18}O_2$ and ^{18}O) from the gas phase do react with the ZrO_2 network, a small proportion being eventually consumed for oxide stoichiometry completion, since ZrO_2 is an O-deficient oxide. However, most of the ^{18}O atoms replace ^{16}O atoms from the network that are then released and diffuse. ^{16}O atoms that migrate toward the surface may desorb whereas those migrating toward the interface can again be fixed by reaction with substrate-Si forming SiO_2.

Zirconium silicate and aluminates

Zr and Si concentrations in silicate films were determined [24] before and after annealing, by Rutherford backscattering (RBS) in random, highly tilted geometry as well as in channelling (c-RBS) geometry. The spectra shown in figure 3.1.12 reveal that these concentrations are homogeneous across the films and constant within the sensitivity of RBS. However, surface selective, far more sensitive analysis with low-energy ion-scattering (LEIS) evidenced that after annealing in vacuum or in $^{18}O_2$ at 600°C the concentration of Si at the film surface is higher than in bulk. This is confirmed by NRP of Si, as shown in figure 3.1.13. Thus, analogous to what happens in ZrO_2 films, Si atoms accumulate at the film surface where they are prone to react with any deposited metal electrode. The electronic consequence of that would be modification of electrode resistivity and dielectric capacitance due to this reaction layer. Annealing in $^{18}O_2$ atmosphere revealed intensive oxygen exchange (over 60%) between the gas and solid phases as seen in the channelled-RBS spectra of figure 3.1.12(*b*) and in the LEIS spectra of figure 3.1.13.

Another atomic scale transport process was revealed [25] by the removal of the silicate film after annealing at 1000°C by HF-etching, and subsequent analysis by time-of-flight secondary ion mass spectrometry (TOF-SIMS) and RBS, which revealed the presence of the order of 10^{11} (RTA) and 10^{13} (furnace annealing) Zr per cm^2 in the Si substrate. This transport of Zr into the Si substrate could be responsible for degradation of the mobility of charge carriers in the transistor channel.

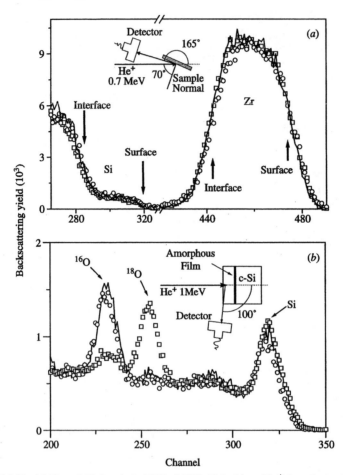

Figure 3.1.12. (*a*) Zr and Si signals in RBS of 700 keV incident He$^+$ ions from zirconium silicate films on Si; (*b*) channelled-RBS and grazing-angle detection spectra of 1 MeV incident He$^+$ ions. Solid lines represent the as-deposited sample; empty circles and squares represent the vacuum- and $^{18}O_2$-annealed samples, respectively. From [37].

Reaction–diffusion processes involving transport of Si and Zr atoms are responsible for the observations described above. Besides, a very intensive oxygen transport is in force, explaining the fact that O atoms from all depths in the silicate films are exchanged.

Zirconium aluminates ($ZrAl_xO_y$) and ZrO_2/Al_2O_3 stacks are other serious candidates for gate dielectric. They present some unique advantages arising from the fact that Al_2O_3 films on Si form an Si/SiO$_2$-type interface after annealing in O$_2$ at 800°C, besides the fact that the presence of Al increases the overall crystallization temperature.

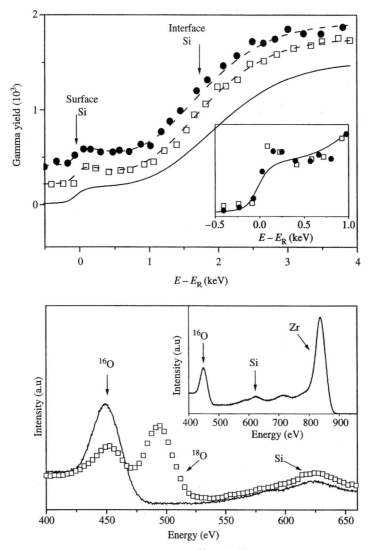

Figure 3.1.13. (Top) Excitation curves of the $^{29}Si(p,\gamma)^{30}P$ around 414 keV for the same samples and symbols as figure 3.1.12. (Bottom) Si and O signals in LEIS spectra and complete spectrum in the inset for the same samples and symbols figure 3.1.12. From [37].

Stability studies were undertaken in zirconium aluminates ($ZrAl_xO_y$) and in ZrO_2/Al_2O_3 stacks. Zr_4AlO_9 films deposited on Si by sputtering in O_2 atmosphere were rather stable against annealing in vacuum and in O_2 [26, 27]. Nevertheless, some atomic scale instabilities were also revealed by NRP and LEIS analyses following furnace annealing at 600°C, such as oxygen

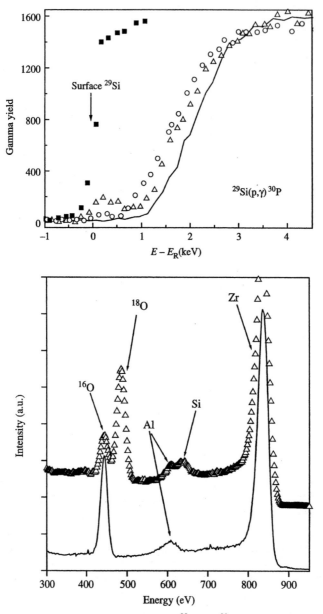

Figure 3.1.14. (Top) Excitation curves of the ^{29}Si(p,γ)^{30}P reaction around 414 keV for Zr$_4$AlO$_9$ films on Si as-deposited (solid line), vacuum annealed (circles), and ^{18}O$_2$-annealed (triangles). Also shown is an excitation curve from a bare Si wafer, which establishes the energy position corresponding to surface Si. (Bottom) LEIS spectra for as-deposited and ^{18}O$_2$-annealed samples. From [30].

exchange and Si migration from the substrate to the film surface as shown in figure 3.1.14.

Excellent thermal stability was obtained [28] in Al_2O_3 (0.5 nm)/ZrO_2 (4 nm)/Al_2O_3 (1.5 nm)/Si stacks deposited by ALCVD on Si(001) wafers. The interfacial layer was composed by Zr, Al, Si, and O having a 0.6 nm thick Zr-deficient layer in contact with the Si substrate. In contrast, the top layer was stoichiometric Al_2O_3. After UHV annealing at 1000°C for 20 s, which are temperature and time figures sufficient for most source and drain dopant annealing processes, almost no change was observed in the MEIS spectra as shown in figure 3.1.15 (top). This is in contrast to what was shown above, where annealing of ZrO_2/SiO_2 film structures on Si at 1000°C completely decomposed the metal oxide layer. However, figure 3.1.15 (top) shows that annealing in UHV at 1100°C for 10 s completely decomposed the nanolaminate layer. A temperature as high as 1100°C is not of real interest for practical purposes, but nevertheless it is important to have the maximum possible information on the thermodynamic stability of the films. Annealing in N_2 and O_2 led to growth of a silicon oxide (silicate) layer at the interface. In figure 3.1.15 (centre), complementary MEIS and angle-resolved XPS show that annealing in N_2 and O_2 not only led to growth of an interfacial oxide layer but also to the transport of Si atoms to the surface, as observed previously for zirconium oxide, silicates, and aluminates. The growth of the interfacial oxide was minimized and rendered self-saturating by remote plasma nitridation of nanolaminate structures, as shown in figure 3.1.15 (bottom).

Oxygen diffusion in gadolinium oxide-based films on Si

Gadolinium oxide films were successful deposited on Si [29] by electron-beam evaporation with a very thin (0.5 nm) SiO_2 interlayer. Combined HRTEM and x-ray reflectivity measurements established a three-layer structure for the as-deposited sample. The top layer is polycrystalline Gd_2O_3 and the intermediate layer is amorphous, consisting of a mixture of Gd_2O_3 and SiO_2, although the interface between the top and intermediate layers is not well defined. The layer next to the Si substrate is SiO_2.

Annealing in vacuum of this structure at 700°C produced a densification (reduction in thickness, as observed by TEM) of the Gd_2O_3 and (Gd_2O_3, SiO_2) layers. Annealing in 10^{-4} mbar of O_2 at 700°C also reduced the thickness of these layers but the thickness of the SiO_2 layer next to the Si substrate increased substantially.

Sputter–Auger depth profiling was used to determine the O/Gd ratios versus depth in the films as shown in figure 3.1.16. For the as-deposited film a uniform 1.5 ratio was obtained as expected, whereas for the oxygen-annealed sample the formation of a thick SiO_2 layer beneath the Gd_2O_3–(Gd_2O_3/SiO_2)

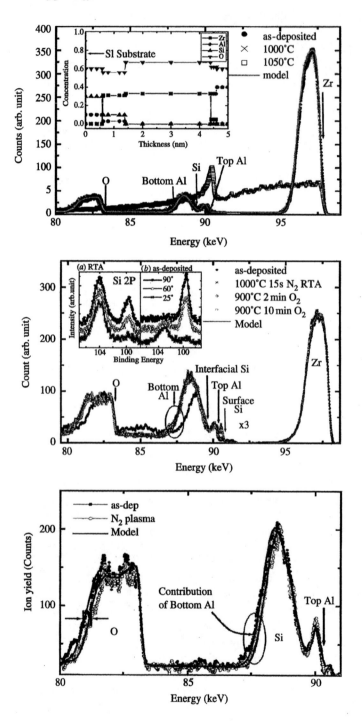

structure was observed. The O/Gd ratio in the Gd_2O_3 film remained essentially the same after oxygen annealing, which means that within the sensitivity of Auger electron spectroscopy and excepting oxygen exchange between gas and solid phases, oxygen diffused in the Gd_2O_3–(Gd_2O_3/SiO_2) structure without interacting with it to react with substrate Si forming SiO_2. Si was observed to be mobile, being transported into the top-most Gd_2O_3 layer.

Silicon migration into the films was prevented by directly depositing pseudobinary Gd–Si oxide (here termed gadolinium silicate) films on Si with average composition $GdSi_{1.4}O_{4.1}$ [30]. These films were annealed in 70 mbar of $^{18}O_2$ for 60 s at temperatures between 600 and 800°C. Another set was prepared by pre-annealing the as-deposited samples in vacuum at 700 or 800°C for 2 min and then performing the $^{18}O_2$ annealing just described.

Profiling of Si was performed by NRP using the narrow resonance at 414 keV in the nuclear reaction $^{29}Si(p,\gamma)^{30}P$ ($\Gamma_R \sim 100$ eV). Excitation curves before and after annealing in the different conditions exposed above were identical, as shown in figure 3.1.17, meaning that there was neither transport of Si from the substrate into the gadolinium silicate film nor across the film toward the surface or interface regions. RBS analysis revealed that the areal densities of Si and Gd in the films remained constant after annealing. ^{18}O from the annealing atmosphere, on the other hand, did diffuse in the gadolinium silicate films as shown by the excitation curves and profiles of figure 3.1.18. Oxygen diffusion is lower in the films pre-annealed in vacuum than in those directly annealed in $^{18}O_2$ as a result of the densification of the films during pre-annealing. The ^{18}O profiles showed that an oxygen front propagates from the gas phase into the gadolinium silicate films, the range of the propagating oxygen front increasing with the increase of annealing temperature. This allows flexible engineering and control of the electrical properties of the interface by growing the adequate SiO_2 interlayer thickness.

The apparent tendency of gadolinium silicate films to have only a very slight oxygen deficiency after annealing in oxygen, rather than having a pronounced oxygen excess or deficiency, may be an important factor determining the observed stability of the gadolinium silicate films against annealing in oxygen.

Figure 3.1.15. MEIS spectra from Al_2O_3 (0.5 nm)/ZrO_2 (4 nm)/Al_2O_3 (1.5 nm)/Si stacks deposited by ALCVD on Si(001) as-deposited and annealed in different atmospheres. The insets represent profiles (top) and Si2p XPS spectra (centre). From [28].

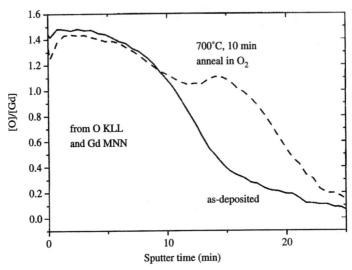

Figure 3.1.16. O/Gd concentration ratios for GdO$_3$ films as-deposited on Si(001) and annealed in O$_2$ at 700°C for 10 min. From [29].

Reaction–diffusion in lanthanum and yttrium oxide films on Si

Lanthanum oxide (La$_2$O$_3$) films grown on Si(001) in O$_2$ atmosphere in a modified molecular beam epitaxy chamber exhibited [31] significant substrate-Si penetration into the films even for depositions at 200°C.

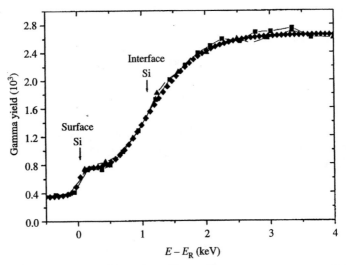

Figure 3.1.17. Excitation curves of the ^{29}Si(p,γ)^{30}P reaction around 414 keV from gadolinium silicate films annealed in vacuum and in ^{18}O$_2$ at 800°C. From [30].

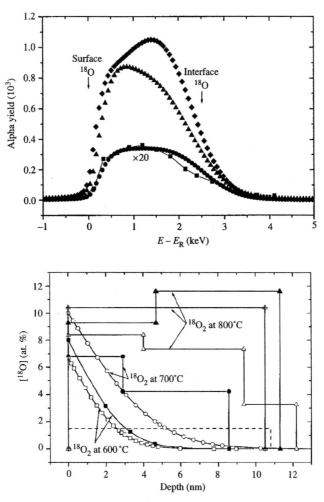

Figure 3.1.18. (Top) Alpha yield of the $^{18}O(p,\alpha)^{15}N$ nuclear reaction around 151 keV from gadolinium silicate films as-deposited (squares) and annealed at 800°C in vacuum (circles), in $^{18}O_2$ (diamond), and in vacuum $+^{18}O_2$ (triangles). (Bottom) ^{18}O profiles for as-deposited and annealed gadolinium silicate films on Si. From [30].

In another deposition route [32] consisting of a high-temperature La effusion cell in the presence of molecular oxygen, amorphous La_2O_3 was deposited on oxidized (2 nm) Si(001). Figure 3.1.19 shows EELS spectra taken at different distances from the SiO_2/Si interface in the annealed sample indicating atomic transport of La into SiO_2. Although no information on Si in the La_2O_3 layer could be obtained by EELS, crystallization of the sample annealed at 800°C suggested that the uppermost layer, next to the Pt electrode, contained less than 40% SiO_2.

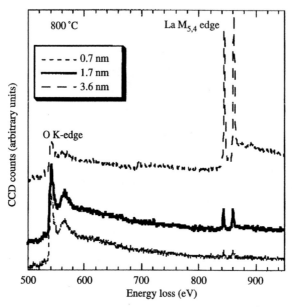

Figure 3.1.19. Oxygen K and L edges in EELS spectra for La_2O_3 films annealed in N_2 at 800°C. The approximate distance from the Si interface is indicated. From [32].

Yttrium oxides have also been deposited on Si or prepared by thermally oxidizing deposited yttrium films in O_2 [33].

The same problem of Si penetration into the films observed in La_2O_3 was observed for Y_2O_3 as shown in the MEIS spectra of figure 3.1.20. In the case of Y deposition on Si followed by thermal oxidation of the metal film, various Si surface pretreatments were explored to modify and control metal/Si interaction and oxidation behaviour. Pre-treatments included plasma oxidation, nitridation, and oxynitridation of Si, forming interlayers of thickness around 1 nm. When Y was deposited on these pretreated surfaces and oxidized, no evidence of SiO_2 was seen on the samples by XPS analysis, indicating conversion of SiO_2 to silicate during oxidation. Nitridation or oxynitridation of the samples reduced markedly the Si penetration into the films as shown in figure 3.1.20, as well as silicate formation. The resulting film after Y deposition and oxidation has a composition close to Y_2O_3.

Reaction–diffusion in hafnium oxide-based films on Si

Hafnium oxide and oxynitrides

The stability of hafnium oxide films deposited by chemical vapour deposition (CVD) on thermally oxynitrided Si wafers (SiO_xN_y/Si) in a laboratory-scale

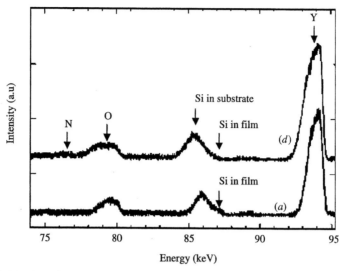

Figure 3.1.20. MEIS energy spectra for yttrium oxide films formed by oxidation at 600°C in O$_2$ of deposited Y films on (*a*) Si and (*d*) nitrided Si. The shoulder representing Si in the film is clearly larger for the film formed on Si (*a*) than for nitrided Si (*d*). From [33].

chemical vapour deposition/atomic layer deposition (CVD/ALD) growth chamber was investigated [35]. Compositional depth profiles as determined by MEIS for as-deposited and vacuum-annealed at 800°C samples are shown in figure 3.1.21. The as-deposited sample showed little or no substrate-Si diffusion toward the outer surface. The Si concentration in the film was found to be less than 2% of the Hf concentration. A silica/Hf silicate layer was identified at the interface to the Si substrate with a thickness of about 1.4 nm. Thus, the MEIS data could be modelled in a three-layer model (step function) as 5.9 nm HfO$_2$, 0.9 nm HfSiO$_4$ (probably with a very low nitrogen content), and little SiO$_2$ or SiO$_x$N$_y$.

After vacuum annealing at 800°C, the structure remains essentially stable and there is still no Si diffusion into the film. A 0.5 nm thick interfacial layer, most probably SiO$_2$, appears between the HfO$_2$ and the Si substrate, which can be seen by HRTEM. The resulting structure was better modelled by 6.2 nm HfO$_2$ film and a pure 0.5 nm SiO$_2$ with rather sharp interfaces.

The effect of vacuum annealing at 1000°C for 60 s on 2.5 nm thick HfO$_2$ films grown by CVD was followed by MEIS, as shown in figure 3.1.22. In the as-deposited sample, no intensity is observed at the energy corresponding to scattering events on Si atoms at the sample surface, whereas Si is observed at the surface in the annealed sample. Other significant changes are observed, namely the reduction of Hf and O intensities, which can be explained by SiO desorption and conversion of the HFO$_2$ film into hafnium silicide.

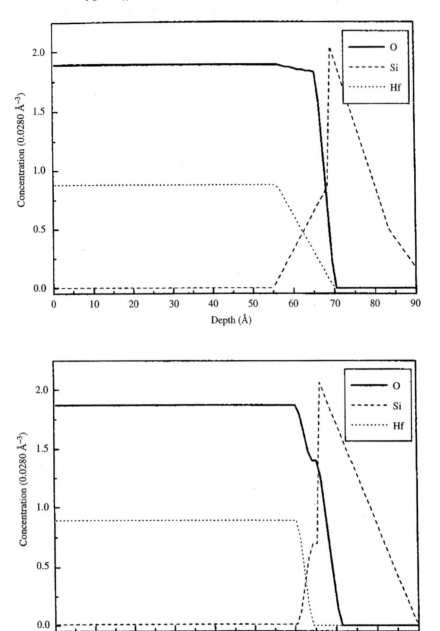

Figure 3.1.21. MEIS profiles for $HfO_2/SiO_xN_y/Si$ samples as-deposited (top) and annealed at 800°C in vacuum (bottom). From [35].

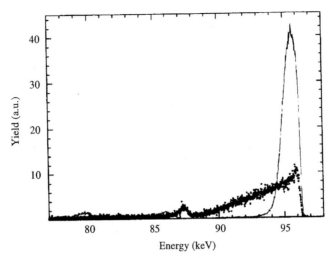

Figure 3.1.22. MEIS energy spectra obtained from a 2.5 nm HfO_2 film on Si. The solid line corresponds to the as-deposited film. The dots represent the spectrum obtained after annealing in vacuum at 1000°C for 60 s. From [21].

Better stability against annealing at 1000°C was obtained [36] for 5 nm HfO_2/SiO_xN_y films on Si(001) substrates. 5.0 nm (ellipsometric determination) HfO_2 films were deposited on Si(001) substrates, which were thermally oxynitrided in NO prior to metal oxide deposition. The starting structure was prepared in the following sequence: HF cleaning of silicon wafers, followed by annealing in NO, and followed by MOCVD of HfO_2 at 550°C using the precursor Hf-*t*-butoxide. The wafers were then subjected to post-deposition RTA in Ar:N_2 at 1000°C for 10 s (Ar-annealing), in order to simulate a typical dopant-annealing step. Finally, the wafers were subjected to RTA in O_2 at 800°C for 10 or 60 s (O_2-annealing), simulating some of the several usual thermal processing steps, including those intentionally performed in O_2. This last annealing step was also alternatively performed in 70 mbar of O_2 97% enriched in the ^{18}O isotope ($^{18}O_2$-annealing), allowing us to distinguish between oxygen incorporated from the gas phase and that previously existing in the as-prepared films. Furthermore, these $^{18}O_2$-annealings were performed with and without Ar pre-annealing, aiming at investigating the effects of the two thermal steps separately. The areal densities and concentration versus depth distributions of the different atomic species were determined by RBS of He^+ ions in a channelling geometry with grazing angle detection of the scattered ions (channelled-RBS), by nuclear reaction analysis (NRA), and by narrow nuclear resonant reaction profiling (NRP) of O and Si. Chemical composition and reaction were accessed by XPS using a Mg $K\alpha$ x-ray source and detection angle between the sample normal and the axis of the electron energy analyser of 25°.

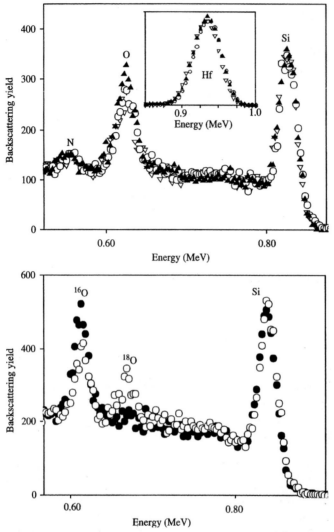

Figure 3.1.23. (Top) Channelled-RBS spectra of 1 MeV incident He$^+$, with detection of the scattered ions at 100° with the direction of incidence, from an as-prepared HfO$_2$/SiO$_x$N$_y$/Si sample (solid dots), Ar-annealing at 1000°C for 10 s followed by O$_2$-annealing at 800°C for 10 s (open dots), and 60 s (solid triangles). The corresponding Hf signals are shown in the inset. (Bottom) Ar-annealing at 1000°C for 10 s followed by ^{18}O$_2$-annealing at 800°C for 10 s (solid dots) and ^{18}O$_2$-annealing only at 800°C for 10 s (open dots). From [36].

Figure 3.1.23 (top) shows the Si, O, and N signals in channelled-RBS spectra of 1 MeV incident He$^+$ ions from as-prepared and O$_2$-annealed samples, with the Hf signals in the inset. The areal densities of Hf and Si (as determined by channelled-RBS) and of ^{16}O and ^{14}N (as determined by NRA)

Table 3.1.2. Areal densities (in units of $10^{15}\,\mathrm{cm}^{-2}$) of different atomic species and isotopes. Ar-annealings at 1000°C, O_2- and $^{18}O_2$-annealings at 800°C. Errors in the areal densities are 10% for Si and ^{14}N, 5% for Hf and ^{16}O, and 3% for ^{18}O. From [36].

	Hf	Si	^{16}O	^{18}O	^{14}N
1. As-prepared	9.5	1.4	22.2	0.02	1.6
2. Ar anneal + O_2 anneal (10 s)	9.6	1.4	24.3	0.02	1.5
3. Ar anneal + O_2 anneal (60 s)	9.8	1.8	28.5	0.02	1.2
4. Ar anneal	9.6	1.5	23.2	0.02	1.6
5. Ar anneal + $^{18}O_2$ anneal (10 s)	9.6	1.5	22.4	0.51	1.5
6. $^{18}O_2$ anneal (10 s)	9.8	1.7	19.1	6.54	1.4
7. Ar anneal + $^{18}O_2$ anneal (60 s)	9.5	1.7	22.3	4.76	1.2
8. $^{18}O_2$ anneal (60 s)	9.6	2.2	15.4	11.13	1.1

are given in table 3.1.2. O_2-annealing for 60 s (sample 3 in table 3.1.2) produces an increase in the O areal density of about 30% with respect to the as-prepared sample (sample 1) and a comparable percentage decrease in the N areal density. Figure 3.1.23 (bottom) shows Si, ^{16}O, and ^{18}O signals in channelled-RBS spectra of $^{18}O_2$-annealed samples for 10 s, with and without Ar pre-annealing (samples 5 and 6 in table 3.1.2). One notices the presence of ^{16}O–^{18}O exchange, which increased with time of annealing in $^{18}O_2$ and was much larger for samples that were annealed in $^{18}O_2$ only than for those that were pre-annealed in Ar. N–^{18}O exchange probably also occurs as a parallel process to ^{16}O–^{18}O exchange.

Excitation curves of the $^{18}O(p,\alpha)^{15}N$ nuclear reaction around the resonance at 151 keV ($\Gamma_R = 100$ eV) and ^{18}O profiles of as-prepared and $^{18}O_2$-annealed samples are shown in figure 3.1.24. The ^{18}O profiles in samples 5 and 7 (pre-annealing in Ar + 10 or 60 s $^{18}O_2$-annealing) indicate a propagating ^{18}O front from the surface and reaction (eventually ^{16}O–^{18}O and N–^{18}O exchange reactions only) with the HfO_2 network. Furthermore, ^{18}O profiles in samples 6 and 8 ($^{18}O_2$-annealing only for 10 and 60 s) are deeper and higher. When compared to similar studies in aluminium, zirconium and gadolinium oxides and silicates, this HfO_2/SiO_xN_y structure displays higher resistance to oxygen migration from the gas into the solid phase and incorporation therein, as well as smaller isotopic exchanges.

Figure 3.1.25 shows excitation curves of the $^{29}Si(p,\gamma)^{30}P$ nuclear reaction around the resonance at 414 keV and the corresponding ^{29}Si profiles, revealing that Si remains immobile during annealing, in contrast to several of the above-mentioned materials where substrate-Si is seen to migrate into the oxide film.

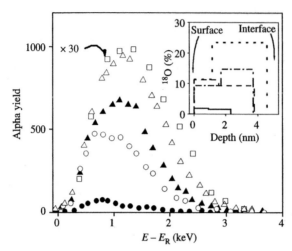

Figure 3.1.24. Excitation curves of the $^{18}O(p,\alpha)^{15}N$ nuclear reaction around the resonance at 151 keV with the corresponding ^{18}O profiles in the inset for $HfO_2/SiO_xN_y/Si$; as-prepared sample (open squares), Ar-annealing at 1000°C for 10 s followed by $^{18}O_2$-annealing at 800°C for 10 s (solid dots, solid curve) and 60 s (solid triangles, dash–dot curve), and $^{18}O_2$-annealing only at 800°C for 10 s (open dots, dash curve) and 60 s (open triangles, dot curve). From [36].

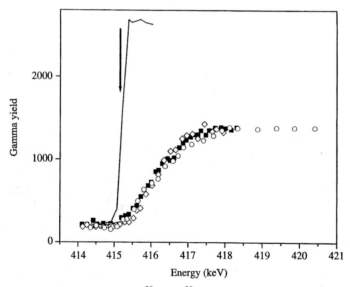

Figure 3.1.25. Excitation curves of $^{29}Si(p,\gamma)^{30}P$ around 414 keV for $HfO_2/SiO_xN_y/Si$ as-prepared sample (solid squares), Ar-annealing at 1000°C for 10 s followed by $^{18}O_2$-annealing at 800°C for 60 s (diamond) and $^{18}O_2$-annealing only at 800°C for 60 s (circles). The solid line is the excitation curve for a bare Si wafer and the arrow indicates the energy position for surface Si. From [36].

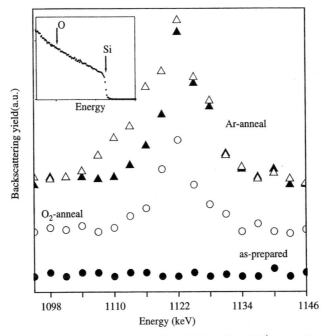

Figure 3.1.26. Hf signal in RBS spectra of 1.2 MeV incident He$^+$ ions of samples after oxide and oxynitride layers removal by HF-etching: as-prepared sample (solid circles), $^{18}O_2$-annealed only at 800°C for 60 s (open circles), and Ar-annealed only at 1000°C for 10 s at normal incidence (solid triangles) and 85° tilt (open triangles). The O and Si signals for the Ar-annealed sample are shown in the inset. From [36].

Figure 3.1.26 shows typical Hf-signal regions in RBS spectra from samples whose oxide and oxynitride layers were tentatively removed by means of concentrated HF-etching. Indeed, the absence of O signal in the inset indicates that, within the sensitivity of RBS, the removal process was effective. Thus, except for a possible incomplete etching of the oxide and oxynitride layers, thermally activated Hf migration into Si seems to occur, consistent with previous observation in Zr and Hf silicate films on Si. An Hf profile could be inferred in the Ar-annealed sample using a tilted geometry, which gives a depth resolution of approximately 2 nm. These RBS analyses indicate that Hf would penetrate into Si to areal densities of the order of 10^{12} cm^{-2} and a maximum range of approximately 4 nm.

This structure is essentially stable against Ar pre-annealing at 1000°C and O$_2$-annealing at 800°C. Oxygen migration proceeds by means of a propagating front from the surface that reacts with the HfO$_2$/SiO$_x$N$_y$ network as it advances, the main reaction channels being O–O and O–N exchanges.

For these annealing temperatures and times, which are characteristic of semiconductor device fabrication processing steps, the HfO$_2$/SiO$_x$N$_y$

structure proved to be more resistant to O and Si migration and incorporation than others studied previously, with the Ar pre-annealed samples exhibiting a higher resistance than those directly annealed in O_2. The intermediate layer is also essentially stable against annealing, apart from moderate N loss due to N–O exchange reaction and SiO_2 formation due to oxidation of the substrate. This stability can be attributed to a synergism between the properties of HfO_2 films on Si and the reaction–diffusion barrier constituted by both the SiO_xN_y interlayer and N eventually incorporated into the HfO_2 films. The only possible deleterious effect of annealing observed was thermally activated migration of Hf into the Si substrate. This has still to be confirmed and a much more careful investigation of this last aspect is currently being undertaken using the much higher sensitivity of LEIS.

Hafnium silicates

Hafnium silicates and oxynitrosilicates seem to be even better replacements for SiO_2 or SiO_xN_y as gate dielectric in advanced, Si-based complementary metal–oxide–semiconductor (CMOS) transistors. In order to be effectively incorporated into ultra-large-scale integration (ULSI) fabrication technology, the gate dielectric material must maintain its integrity during further processing steps. In particular, RTA of source and drain dopants, usually performed at and above 1000°C, has been indicated as the most aggressive step.

The deleterious consequences of post-gate dielectric deposition annealing that have been reported so far include thickening of the SiO_2 interface layer and chemical reactions at both gate electrode/dielectric and dielectric/substrate interfaces, with the consequent lowering of the equivalent oxide thickness.

Furthermore, there is a need for controlling the effects of annealing in intentionally or unintentionally O_2-containing atmospheres (even at very low O_2 partial pressures), which renders post-deposition annealing even more aggressive to gate dielectric integrity, especially in the region near the dielectric–semiconductor interface.

The effects of RTA at 1000°C on the structure, composition, atomic transport, and oxygen incorporation kinetics in hafnium silicate films on Si have been recently investigated [37]. The hafnium silicate films, which are actually $(HfO_2)_{1-x}(SiO_2)_x$ pseudobinary alloys or compounds, were deposited by reactive sputtering in O_2 on HF-cleaned, 200 mm Si(100) p-type substrates. RTA was performed in N_2 (N_2-annealing) and O_2 (O_2-annealing) at atmospheric pressure. RTA in O_2 was also performed in 7×10^3 Pa of O_2 97% enriched in the ^{18}O isotope ($^{18}O_2$-annealing). This allows differentiation between oxygen incorporated from the gas phase and that previously existing in the films.

Elemental composition of the as-deposited samples was determined by RBS and NRA to be approximately $HfSi_2O_5$, which remained essentially the same after RTA annealing in N_2 and O_2. Channelled-RBS spectra of 1 MeV incident He^+ ions from 9 and 3 nm as-deposited and N_2- and O_2-annealed samples are shown in figure 3.1.27, evidencing that while the areal density of oxygen does not increase significantly after annealing of the 9 nm samples, annealing of the 3 nm samples in N_2 leads to a slight increase in the oxygen areal density and annealing in O_2 leads to a substantial increase in the oxygen areal density. This is another consequence of the reaction–diffusion nature of oxygen transport in high-κ films, because although oxygen diffuses in the films reacting with the hafnium silicate network, the ratio between 'diffusion length' of oxygen in the films and film thickness determines the amount of oxygen that arrives at the interface being fixed by reaction with substrate Si.

$^{18}O_2$-annealing of the samples shown in figure 3.1.27 leads to the channelled-RBS spectra of figure 3.1.28. One notices the immobility of Hf and the incorporation of ^{18}O from the gas phase into the $(HfO_2)_{1-x}(SiO_2)_x/SiO_2$ film structure. Previous work revealed that Hf from hafnium silicate is much less likely to penetrate into Si than other studied cases like, for example, Zr from zirconium silicate. Isotopic exchange (^{16}O–^{18}O exchange) is observed, being far more pronounced in 3 nm films than in 9 nm films.

^{18}O profiles in these 9 and 3 nm hafnium silicate film samples were determined by NRP using the $^{18}O(p,\alpha)^{15}N$ nuclear reaction around the resonance at 151 keV ($\Gamma_R = 100$ eV). Excitation curves and extracted ^{18}O profiles are shown in figure 3.1.29. One can see that an oxygen front from the gas phase diffuses from the surface reacting with the hafnium silicate network. According to figure 3.1.28, ^{16}O–^{18}O exchange is probably the main reaction channel. However, since the total amount of oxygen ($^{16}O + {}^{18}O$) was found to increase during anneals, other reaction channels must be active, such as completion of the silicate stoichiometry and oxidation of the Si-substrate. Samples that were pre-annealed in either O_2 or N_2 are more resistant to the propagation of the ^{18}O front than the as-deposited one. Therefore, less ^{18}O will reach and oxidize the Si substrate in the pre-annealed samples than in the as-deposited one. Furthermore, in the 9 nm films, the propagating ^{18}O front does not trespass across the $HfSi_xO_y/Si$ interface whereas in the 3 nm films the ^{18}O front goes far beyond this interface, consistently with the larger thickening of the SiO_2 interlayer.

^{29}Si profiles in 9 and 3 nm films were determined for the $^{18}O_2$-annealed samples using the $^{29}Si(p,\gamma)^{30}P$ nuclear reaction around the resonance at 414 keV ($\Gamma_R \cong 100$ eV). Excitation curves and extracted profiles are shown in figure 3.1.30, where one notices an accumulation of Si at the film surfaces of the 9 nm samples and not in the 3 nm samples. In the bulk of the as-deposited samples a roughly constant Si concentration is observed, which remained

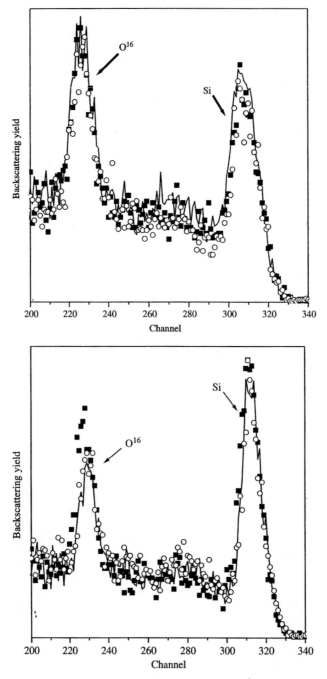

Figure 3.1.27. Channelled-RBS spectra of 1 MeV incident He$^+$ ions from 9 nm (top) and 3 nm (bottom) as-deposited samples and after annealing in N$_2$ and O$_2$ at 1000°C for 60 s. From [37].

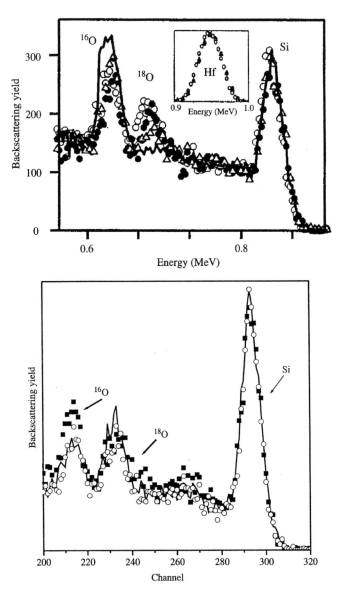

Figure 3.1.28. Channelled-RBS spectra of 1 MeV incident He$^+$ ions from 9 nm (top) and 3 nm (bottom) samples after RTA in $^{18}O_2$ at 1000°C for 60 s. From [37].

essentially immobile during 1000°C annealing. There is no observable transport of Si from the substrate into the HfSi$_x$O$_y$ films after annealing, revealing high stability of the bulk and interface regions in contrast with several materials described above such as Al$_2$O$_3$, ZrO$_2$, Zr–Si–O and Zr–Al–O.

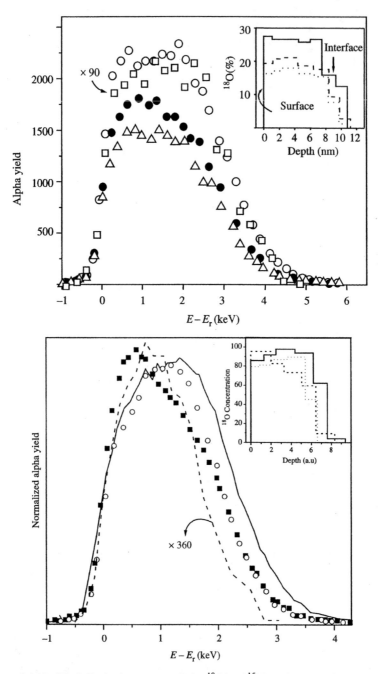

Figure 3.1.29. (Top) Excitation curves of the $^{18}O(p,\alpha)^{15}N$ nuclear reaction around the resonance at 151 keV with the corresponding ^{18}O profiles in the inset for a 9 nm hafnium silicate film on Si: as-deposited sample (open squares), $^{18}O_2$-annealed at 1000°C for 60 s

Thus, oxygen reaction–diffusion is a significant atomic scale concern regarding instability of the $(HfO_2)_{1-x}(SiO_2)_x/SiO_2/Si$ system. It has to be controlled before hafnium silicate films can be seriously considered as replacements for SiO_2 or SiO_xN_y as the gate dielectric. In order to further investigate this aspect, the kinetics of ^{18}O incorporation during RTA of as-deposited 9 nm film at 1000°C in $^{18}O_2$ was determined by NRA. The result is shown in figure 3.1.31 (top). For annealing times up to 30 s, a time interval that would be sufficient for most dopant annealing processing steps, the ^{18}O incorporation is fairly moderate.

Furthermore, excitation curves of the $^{18}O(p,\alpha)^{15}N$ nuclear reaction around the resonance at 151 keV shown in figure 3.1.31 (bottom) indicate that, for 30 s annealing, the ^{18}O front barely reaches the dielectric/substrate interface and thus only minor oxidation of the Si substrate is expected. For 60 s annealing time there is an abrupt increase in ^{18}O incorporation which is accompanied by a substantial part of the ^{18}O propagating front reaching the dielectric/substrate interface, leading to a large thickening of the SiO_2 interlayer. The ^{18}O concentration in the $(HfO_2)_{1-x}(SiO_2)_x$ film reaches a saturation at approximately 30%, after which only the propagation of the ^{18}O front in depth is observed. For the longest RTA time, namely 480 s, the excitation curve exhibits a distinct zone towards higher values of $E - E_R$ with a much higher concentration of ^{18}O. The extracted ^{18}O profile reveals that this is an $Si^{18}O_2$ layer, approximately 5 nm thick, formed beneath the $(HfO_2)_{1-x}(SiO_2)_x/SiO_2$ thin-film structure by reaction of diffusing $^{18}O_2$ (or ^{18}O) with the Si substrate. This aggressive annealing in oxygen causes dramatic changes in thickness and abruptness of the interfaces.

The above-discussed results for hafnium silicate films on Si having two different, characteristic thicknesses (9 and 3 nm) constitute clear evidence of the reaction–diffusion nature of the processes taking place during oxygen annealing; on one hand, oxygen is transported across the films by reacting with the $HfSi_xO_y$ network, which renders the transport driven by reaction; on the other hand, the amount of oxygen that arrives at the near-interface region to react therein forming additional hafnium silicate depends on the film thickness, establishing a net competition between diffusion length and film thickness, characteristic of diffusion-driven transport processes.

(open circles, solid line), O_2-annealed at 1000°C for 60 s followed by $^{18}O_2$-annealing at 1000°C for 60 s (solid circles, dashed line), N_2-annealed at 1000°C for 60 s followed by $^{18}O_2$-annealing at 1000 °C for 60 s (open triangles, dotted line). (Bottom) Excitation curves of the $^{18}O(p,\alpha)^{15}N$ nuclear reaction around the resonance at 151 keV with the corresponding ^{18}O profiles in the inset for a 3 nm hafnium silicate film on Si. From [37].

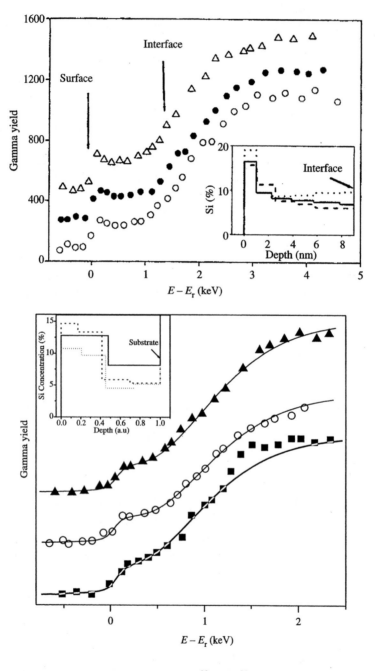

Figure 3.1.30. (Top) Excitation curves of the $^{29}Si(p,\gamma)^{31}P$ nuclear reaction around the resonance at 414 keV with the corresponding ^{29}Si profiles in the inset for 9 nm hafnium silicate film on Si: as-deposited sample (open squares, ×90), $^{18}O_2$-annealed at 1000°C for

In the particular case of $HfSi_xO_y$ films on Si described above, good compositional and atomic transport integrity are observed when they are subjected to RTA at 1000°C, especially if we consider that 30 s of RTA processing is more than enough for most dopant annealing processing steps. Pre-annealing in a non-reactive atmosphere like N_2 increases the stability and resistance to oxidation of the films. Annealing in an oxygen-containing atmosphere reveals a rapidly propagating front of oxygen from the film surface, which reacts with the $(HfO_2)_{1-x}(SiO_2)_x$ network as it advances.

Oxygen exchange is the main reaction channel, although oxygen incorporation to complete the pseudobinary alloy stoichiometry and oxygen incorporation to oxidize the Si substrate are alternative, active reaction channels. All those instabilities are moderate for up to 30 s of RTA. For 60 s RTA times and above there is a sudden increase in the concentration and penetration depth of the oxygen propagating front, leading to a substantial increase in the thickness of the SiO_2 interlayer. For 480 s RTA a complete collapse of the structure and interface abruptness of the film are observed. When compared to previous materials considered for SiO_2 replacements as gate dielectric, such as Al_2O_3, ZrO_2, Zr–Si–O, Zr–Al–O, Gd_2O_3, Gd–Si–O and others, the present hafnium silicate films present higher stability under annealing at 1000°C.

Interfacial layers

Silicon oxide, nitride, and oxynitride interfacial layers play a beneficial role in improving the thermodynamical stability of high-κ dielectrics on Si. The main reasons for this improvement are:

(a) The action of N in forming more efficient barriers to oxygen diffusion as described in the 'Reaction–diffusion in high-κ dielectrics on Si' section. In the case of high-κ dielectrics, this has strong implication on the limitation of the thickness of the intermediate layer, which in turn determines the overall dielectric constant and therefore the overall capacitance.

(b) The action of N in reducing the concentration of reactive sites for the chemically active species, such as O_2, O, Si, and metals.

---

60 s (open circles, solid line), O_2-annealed at 1000°C for 60 s followed by $^{18}O_2$-annealing at 1000°C for 60 s (solid circles, dashed line), N_2-annealed at 1000°C for 60 s followed by $^{18}O_2$-annealing at 1000°C for 60 s (open triangles, dotted line). (Bottom) Excitation curves of the $^{29}Si(p,\gamma)^{31}P$ nuclear reaction around the resonance at 414 keV with the corresponding ^{29}Si profiles in the inset for 3 nm hafnium silicate film on Si.

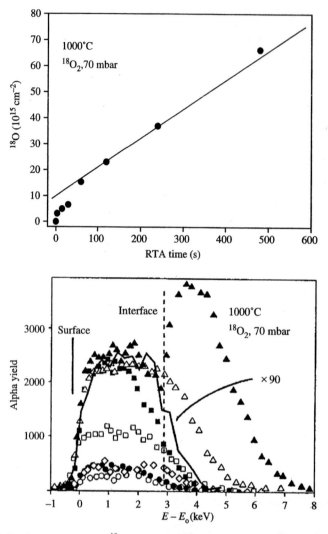

Figure 3.1.31. (Top) Kinetics of ^{18}O incorporation during RTA in $^{18}O_2$ at 1000°C of a hafnium silicate film on Si as determined by NRA. (Bottom) Excitation curves of the $^{18}O(p,\alpha)^{15}N$ nuclear reaction around the resonance at 151 keV for: 3 s (open circles), 15 s (solid circles), 30 s (open diamond), 60 s (open squares), 120 s (solid squares), 240 s (open triangle), 480 s (solid triangle). The solid line represents the excitation curve for the as-deposited sample (×90). From [37].

Furthermore, owing to the previously discussed reaction–diffusion nature of the phenomena taking place during thermal processing of high-κ dielectrics on Si, an efficient diffusion barrier to oxygen constituted by the presence of N will also limit the mobility of other species such as substrate-Si and metallic species from the high-κ film itself. Thus, nitridation of the intermediate layer seems to be essential for the electrical performance of high-κ dielectrics, as well as to provide the required stability in further thermal processing steps as illustrated in the following.

However, although it has been clear for quite some time that an extremely thin (typically 0.5 nm or less) interfacial layer of silicon oxide, or even better of silicon oxynitride, grown or deposited on the surface of the active region of the silicon semiconductor substrate before or after high-κ dielectric deposition, would be a very convenient way of engineering the best replacement for silicon oxide, the recognition of the key role of intermediate N-containing layers was only very recently made clear, such that the details on atomic transport and stability are only starting to appear in the literature. Thus, one cannot at this stage discuss atomic scale mechanisms, as this is not yet clarified.

The rest of this section will be limited to a presentation of some experimental observations of the effects of the different routes followed to create a nitrided intermediate layer on reaction–diffusion in high-κ dielectrics on Si.

Thermal nitridation in ammonia (NH_3) is a commonly used route to deposit an ultrathin, non-stoichiometric silicon nitride film (SiN_x) previously to high-κ dielectric deposition [38–40]. Rapid thermal nitridation of Si in NH_3 for 30 s at 700°C inhibited the formation of SiO_2 near the interface during deposition of 1.5 nm HfO_2 films, as well as during RTA of the structure, favouring the formation of hafnium silicates, which eventually contain N. RTN in NH_3 at 700°C for 30 s should grow a silicon nitride (oxynitride) layer with thickness of approximately 0.5 nm. Furthermore, nitridation in NH_3 was also effective in inhibiting the re-crystallization during post-deposition annealing.

Another common route to produce nitrided intermediate layers is remote plasma nitridation in ionized N_2O^*, NO^*, and N_2^* atmospheres. This route provides very fine control in thickness, allowing monolayer control of the intermediate layer. For example, figure 3.1.20 shows how Si migration into Y_2O_3 films is inhibited by remote plasma deposition of approximately one monolayer of N previous to metallic Y film oxidation to grow Y_2O_3 films on Si.

Thermal oxynitridation of the Si-substrate surface in NO is the most well-established method in current CMOSFET technology, offering a self-limited oxynitrided growth which is in current use for the production of gate dielectrics based on SiO_xN_y, constituting an interim solution in industrial practice. This route has also proven itself effective in providing an intermediate oxynitride layer that improves stability and control of reaction–diffusion in high-κ dielectrics. Figure 3.1.26 shows again that Si migration into HfO_2 films during annealing in Ar at 1000°C and in O_2

at 800°C is inhibited by an intermediate layer of SiO_xN_y thermally grown in NO previous to HfO_2 deposition.

A final route that deserves mention here is deposition of high-κ dielectrics in either NO, N_2O, or N* atmospheres, leading to metal oxynitride or metal oxynitrosilicate films, in which case not only the interface is rendered more stable by the incorporation of N, but also the high-κ film itself. This was in reality the case in the illustration given in figure 3.1.26, since NO was also one of the precursors for MOCVD deposition of HfO_2 following thermal oxynitridation of the Si substrate in NO. HRTEM images indicate that there is no visible intermediate layer formed during deposition and that recrystallization following annealing at 1000°C is significantly reduced.

In summary, there are different routes leading to nitrided intermediate layers providing significant improvements in the stability of high-κ dielectrics on Si. Further atomic scale investigation of the stability of the intermediate layer itself, as well as investigation of the electrical properties of the resulting high-κ/intermediate layer/Si structures, are required to establish valid criteria for choosing the optimum route in each case.

Reaction–diffusion models

Diffusion and reaction terms

When diffusing through high-κ material films deposited on Si, oxygen may react with either the high-κ material or with the substrate, modifying the diffusion medium. A realistic model should then take into account not only the medium non-homogeneity but also its time evolution, caused by the existing reactions. In this section, we consider specifically the use of reaction–diffusion equations to model oxygen diffusion in high-κ materials on Si. Usually, the output of these models is measurable quantities, such as density profiles of the involved elements and kinetics curves as well as diffusion coefficients and reaction rates.

The phenomenological definition of the diffusion coefficient D is given by

$$D = \lim_{t \to \infty} \frac{|\Delta \vec{r}|^2}{t}$$

where $|\Delta \vec{r}|^2 = \langle |\vec{r} - \langle \vec{r} \rangle|^2 \rangle$ is the variance of the spatial distribution of the diffusing quantity. In this definition, D is a macroscopic quantity and, in the case of isotropic and homogeneous systems, it coincides with the diffusion coefficient appearing in the usual diffusion equation.

Diffusion is rooted in displacements of some microscopic quantity in contact with a heat bath or other random movement source. A general method to obtain macroscopic diffusion from microscopic random displacements is to consider random walkers on discrete lattices and time, where we associate the diffusion coefficient with the walker jump probability.

Consider a random walk in one dimension, here taken as the horizontal axis labelled by $x \in (-\infty, +\infty)$. At each time interval Δt the walker has defined probabilities to jump a distance Δx to the right or to the left of its current position. Define $\rho(x, t)$ as the probability distribution of finding the walker at site x at time t. To consider any diffusive medium, that may in general be non-homogeneous, one can consider jump probabilities that depend, for example, on the path linking the starting and the final sites. The probability distribution evolves then as a Markov chain and this process may be characterized by a transition matrix $W(x \rightarrow x', t)$ giving the probability that the walker jumps from x to x' during the time interval between t and $t + \Delta t$, provided that it was at site x at time t. Considering a symmetric transition matrix, we can write that $W(x, t) = W(x \rightarrow x - \Delta x, t) = W(x - \Delta x \rightarrow x, t)$, for $\Delta x > 0$, and the process is described by

$$\rho(x, t + \Delta t) - \rho(x, t) = W(x + \Delta x)\rho(x + \Delta x, t) + W(x, t)\rho(x - \Delta x, t)$$

$$- W(x, t)\rho(x, t) - W(x + \Delta x, t)\rho(x, t) \qquad (3.1.1)$$

By defining $D(x, t) = (\Delta x^2 / \Delta t) W(x, t)$ in the limit that $\Delta x \rightarrow 0$ and $\Delta t \rightarrow 0$, with $(\Delta x^2 / \Delta t)$ finite, the above equation can be written as a differential equation:

$$\frac{\partial \rho}{\partial t} = \frac{\partial}{\partial x} \left[D(x, t) \frac{\partial \rho}{\partial x} \right] \qquad (3.1.2)$$

which reduces to the usual diffusion equation when D is constant in time and space. Now the functions $\rho(x, t)$ are probability density functions and from a non-dimensional hop probability W we arrive at a diffusion coefficient with the expected dimension of length^2/time. Different diffusion equations may be obtained when hopping probabilities are taken to depend differently on the starting site, on the arriving site, and/or on the hop direction. For example, considering that the hop probability depends on the hop final point only, that is, $W(x \rightarrow x \pm \Delta x, t) = W(x \pm \Delta x, t)$, equation (3.1.1) may be rewritten as

$$\rho(x, t + \Delta t) - \rho(x, t) = W(x, t)\,\rho(x + \Delta x, t) + W(x, t)\,\rho(x - \Delta x, t)$$

$$- W(x + \Delta x, t)\,\rho(x, t) - W(x - \Delta x, t)\,\rho(x, t) \qquad (3.1.3)$$

In the continuous time and space limit this yields the following differential equation:

$$\frac{\partial \rho}{\partial t} = D(x,t)\frac{\partial^2 \rho(x,t)}{\partial x^2} - \rho(x,t)\frac{\partial^2 D(x,t)}{\partial x^2} \qquad (3.1.4)$$

Equations (3.1.2) and (3.1.4) describe different microscopic phenomena and yield different solutions, and their generalization to two or three dimensions is straightforward.

Here we are interested in modelling oxygen diffusion and reaction in high-κ material films on Si. When atoms are diffusing through a network, the path followed by the particles is extremely diverse and complex, even when treated as classical particles. There are many different forms to cross a given small path and each of these ways may have a different occurring probability. Here, we assume that $W(x,t)$ describes a sum of many independent events. This will be the case when the typical time interval for the actual events to occur is much smaller than the smallest observational time and/or Δx describes a 'mesoscopic' length—large enough to justify the application of the Central Limit Theorem but still small enough to be taken as a differential quantity.

The first step to model reaction–diffusion systems is to specify the relevant species, namely those that are involved in changing measurable quantities. Examples of quantities that may vary are the amounts of incorporated nitrogen or oxygen to a silicon substrate exposed to nitridant or oxidant atmosphere. Different chemical reactions are modelled by different terms added to the diffusion equations, based on the chemical reaction kinetics. They act as sink or source terms added to the conservation equation represented by the generalized diffusion equation. In what follows, we discuss some examples of reaction–diffusion models intended to model thermal treatment of thin films of metal oxides grown or deposited on silicon.

Thermal growth of silicon oxide films on Si

Thermal growth of silicon oxide films on Si in dry O_2 has been extensively studied by isotopic substitution methods. Sequential oxidation of bare Si in $^{16}O_2$ followed by $^{18}O_2$ indicates that oxygen incorporation takes place mostly where there are oxygen deficiencies when compared to stoichiometric oxide. For thick films (20 nm or more), where there exists a layer of stoichiometric oxide, oxygen diffuses through this film without reacting, until it reaches the interface. For very thin films, the near-interface region covers the near-surface region and we expect oxygen incorporation also near the surface.

To meet these experimental findings, it is explicitly assumed in the theoretical model that reaction is only possible between diffusing oxygen and silicon that is not fully oxidized (Si^{nfo}), that is, when crossing a stoichiometric SiO_2 film, oxygen diffuses without reacting. With this assumption, thermal growth of silicon oxide films in O_2 was successfully modelled [41] in the whole thickness range as a dynamical system. The only diffusing species was taken to be O_2. Si was considered to be immobile according to the Si isotopic substitution experiment results. Growth is promoted by reaction of diffusing O_2 with Si^{nfo}, wherever the two species meet. The reaction–diffusion equations for the time evolution of the relative concentrations of the involved species are then

$$\frac{\partial \rho_o}{\partial t} = D\frac{\partial^2 \rho_o}{\partial x^2} - k\rho_o(x,t)\rho_{Si}(x,t)$$

$$\frac{\partial \rho_{Si}}{\partial t} = k\rho_o(x,t)\rho_{Si}(x,t)$$

(3.1.5)

where ρ_{Si} and ρ_o are the relative concentrations of non-oxidized bulk silicon and diffusive oxygen, respectively. Silicon oxide concentration, represented as ρ_{SiO}, is obtained from bulk silicon relative concentration as $\rho_{SiO} = 1 - \rho_{Si}$. Here, we consider a gauge concentration, namely the silicon bulk concentration ($C_G = C_{Si}^{bulk}$), relative to which all other concentrations are taken, such that $0 \leq \rho_{Si} \leq 1$. The diffusion coefficient D is assumed constant in time and space and we considered a one-dimensional problem. Equations (3.1.5) are intended to model the diffusion of O_2 molecules through a network of silicon oxide, reaching bulk silicon and reacting with it. The actual local concentrations are not continuous functions of space at the microscopic level and the problem is certainly not one dimensional, meaning that the above differential equations imply approximations. Since in the problem of thermal growth of a thin film the directions perpendicular to the surface should in principle behave symmetrically, a mean field approximation over these two directions, is implied in the above equations, with both oxygen and silicon homogeneously distributed over the directions perpendicular to growth. This approximation should remain valid whenever diffusion coefficients are constant in space and time.

We take the displacement dx as a length interval small enough to be treated as an infinitesimal in the differential equations, but still large enough that $\rho_i(x)$, $i = $ Si or O, may stand for the average relative number per unit length of atoms or molecules of i-species in the slice between x and $x + $ dx. In a real system when an oxygen molecule is diffusing in silicon oxide, for example, there are many different paths and hence different hopping probabilities, that in general depend on space and eventually on time. As we are considering an average over the two directions perpendicular to the surface, we are taking into account a macroscopic number of such hoppings and the diffusion coefficient is associated with a weighted mean over all possible paths for oxygen movements.

The numerical solution of reaction–diffusion models yields time-dependent profiles for each species and other measurable quantities, such as total amounts of incorporated species, are obtained from these profiles.

Comparing these data with experimental results, we can produce quantitative estimates for the model parameters.

Parameter estimation depends then on the number and quality of available experimental data that must yield enough information to discern among different possible situations. However, it is always possible to define 'natural units' in reaction–diffusion equations, which especially helps numeric solutions. In the differential equations for the relative concentrations ρ, as equations (3.1.5), the only dimensional variables are space and time, since the relative concentrations were made non-dimensional. We can hence use one diffusion coefficient and one reaction rate to define length and time units. Define non-dimensional position u and time τ, respectively, as $u = x/\sqrt{D/k}$ and $\tau = kt$, and relative concentration functions as $\phi(u, \tau) = \rho(u\sqrt{D/k}, \tau/k)$. With these units of length and time the model equations for $\phi(u, \tau)$ do not have any parameter. This is only possible because in the original model equations there are only two parameters. We must stress, however, that further parameters may appear from initial and boundary conditions, for example, the parameter associated with the adsorbed gaseous species concentration on the film surface.

To solve these equations initial and boundary conditions must be specified, which we take as $\phi_{Si}(x, 0)$ for all $x > 0$ and $\phi_O(0, t) = p_0$, with p_0 being the relative concentration of adsorbed O_2 at the surface, which is directly proportional to the O_2 gas pressure [42, 43].

Numerical solution of the reaction–diffusion equations by finite difference methods gives the depth profiles of Si and diffusive O_2 showing that: (a) the oxide/Si interface is graded, in strict agreement with observations, and (b) the thickness of the growing silicon oxide film is given by the distance from the surface to the leading edge of the Si concentration profile. More quantitatively, the oxide thickness can be calculated by integration of $\phi_{oxide}(x, t) = 1 - \phi_{Si}(x, t)$ giving, for example, the oxide thickness as a function of oxidation time or oxidation kinetics. Comparison of calculated and experimental kinetics in the temperature, time and oxide thickness intervals of 800–1000°C, 0–300 min and 1–20 nm, respectively, is shown in figure 3.1.32 where a very good agreement is evidenced. The extension of the reaction–diffusion model to larger thickness (parabolic or Deal and Grove regime) is trivial.

Figure 3.1.32. Experimental growth kinetics from different experiments and the corresponding model fittings. The oxygen partial pressures in the experiments were (a) 1 atm, (b) 0.1 atm, and (c) 0.01 atm. From [41].

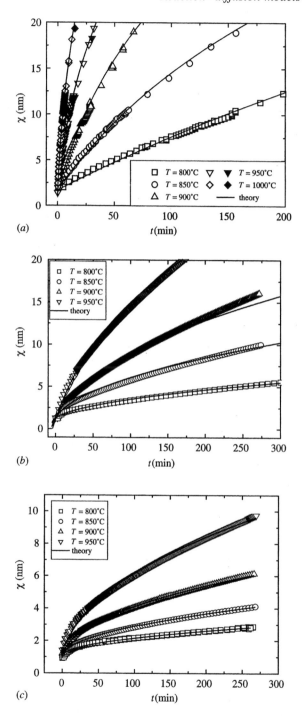

Non-homogeneous diffusion: thermal growth of silicon nitride films on Si

A markedly different case from that of SiO_2 growth is the thermal growth of silicon nitride films on Si in ammonia (NH_3), owing to the diffusion barrier action of stoichiometric or non-stoichiometric silicon nitride. The growth kinetics is very fast at initial stages, rapidly changing into a very slow, maybe self-limited growth regime that lasts even for long processing times and high nitridation temperatures.

Experimental evidences indicate that: (a) nitrogenous species are reactive and mobile during thermal growth of silicon nitride films on Si in NH_3, whereas Si is reactive and immobile, and (b) thermal nitridation of Si in NH_3 begins with the dissociation of the gas molecule into smaller fragments, like NH_2 and H that react with silicon generating silicon nitride. Film growth proceeds by the migration of NH_2 towards the nitride/Si interface to react with Si, either as an interstitial or a substitutional defect. Since the silicon nitride/Si transition region, containing non-fully nitrided Si or sub-nitrides, corresponds usually to more than half of the film thickness, the mobile nitrogenous species react substantially within the sub-nitride region, promoting growth and releasing H. Reaction of ammonia fragments within the sub-stoichiometric transition region leads to the formation of a stoichiometric Si_3N_4 layer.

Reaction–diffusion equations were used [44] to model the observed self-limited growth kinetics attributed to diffusion barriers, assuming that the nitridant species diffuses through sub-stoichiometric (sub-nitride) layers as well as through Si, having zero diffusivity in stoichiometric Si_3N_4. Since Si_3N_4 does not react with NH_3, after establishing a stoichiometric silicon nitride layer in near-surface regions, nitridation may still occur by reaction within the subjacent sub-nitride layer, but film growth will eventually stop when the previously adsorbed nitridant species have been consumed.

Model equations are built up to describe the time evolution of relative concentrations of silicon and nitridant species,

$$\frac{\partial \rho_N}{\partial t} = D^* \frac{\partial^2 \rho_N}{\partial x^2} - \rho_N \frac{\partial^2 D^*}{\partial x^2} - k\rho_N \rho_{Si}$$
$$\frac{\partial \rho_{Si}}{\partial t} = -k\rho_N \rho_{Si}$$

(3.1.6)

where ρ_{Si} and ρ_N stand for concentrations of, respectively, Si (non-fully reacted or substrate) and diffusing nitridant species. These concentrations are non-dimensional since they are taken relative to a gauge value that is equal to the concentration of bulk silicon for ρ_{Si}, and equal to the necessary amount of nitridant species to form stoichiometric silicon nitride for ρ_N, such that $0 \le \rho \le 1$ for all species. Concentration of silicon nitride is taken as $\rho_{SiN}(x,t) = 1 - \rho_{Si}(x,t)$ and the diffusion coefficient D^* in equations (3.1.6) varies with position and time. This intends to model the variation in diffusion coefficient due to the change in the composition of the film. Assuming that it varies linearly from D to zero with the relative amount of silicon nitride we can write that $D*(x,t) = D\rho_{Si}(x,t)$. A physical interpretation of such a relation is that the probability that a site at position x accepts a diffusing nitridant species is directly proportional to the amount of nitride at that position at that time, representing an example of the non-homogeneous diffusion term described by equation (3.1.4).

The reaction diffusion equations for this model are:

$$\frac{\partial \rho_N}{\partial t} = D\rho_{Si}\frac{\partial^2 \rho_N}{\partial x^2} - D\rho_N\frac{\partial^2 \rho_{Si}}{\partial x^2} - k\rho_N\rho_{Si}$$

$$\frac{\partial \rho_{Si}}{\partial t} = -k\rho_N\rho_{Si}$$

(3.1.7)

with

$$D = \lim_{\Delta t \to 0}\lim_{\Delta x \to 0}\frac{W_0\Delta x^2}{\Delta t}$$

being the diffusion coefficient of the nitridant species in bulk silicon. This restriction upon nitridant species diffusion through stoichiometric silicon nitride, together with the dependence of the diffusion coefficient on the concentration of nitrogen, constitute the origin of the differences between kinetics of silicon nitridation and silicon oxidation.

Numerical solutions of equation (3.1.7) give the density profiles shown in figure 3.1.33 where one can see that surface Si is initially consumed, being transformed into silicon nitride and, as time proceeds, the reaction front advances into the c-Si matrix. The amount of non-fully nitrided silicon in the near-surface region decreases fast, progressively preventing incorporation of nitridant species and consequently slowing down film growth. For the two longest nitridation times the relative widths of the near-surface, stoichiometric Si₃N₄ layer and the near-interface layer presenting gradual transition from stoichiometric Si₃N₄ to substrate c-Si are comparable, modelling well experimental observations.

Growth kinetics can be obtained by integrating $\rho_{SiN}(x,t) = 1 - \rho_{Si}(x,t)$ in all space. The experimental kinetics, measured from early stages of very fast growth up to saturation into a self-limited regime, are well fitted by the calculated kinetics, and the NH₃ pressure dependence of the silicon nitride film thickness in the saturation region can also be calculated from the solution of the model equations, in very good agreement with experiments.

Modelling reaction–diffusion in Al₂O₃ films on Si

^{18}O Incorporation into the Al₂^{16}O₃ films during annealing in ^{18}O₂ is mainly due to exchange reaction (^{16}O–^{18}O exchange), constituting a

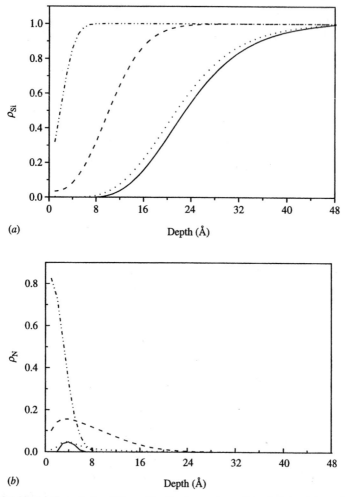

(a)

(b)

Figure 3.1.33. (*a*) Snapshots of Si profiles for increasing—from left to right—nitridation times; (*b*) the corresponding nitridant species profiles for each nitridation time considered in (*a*).

reaction–diffusion situation that can be modelled considering a thick, amorphous Al_2O_3 layer exposed to $^{18}O_2$ that starts diffusing in the Al_2O_3 network reacting with it. ^{16}O and ^{18}O are modelled in two different states: diffusive (d) and fixed (f). The diffusive state represents mobile, non-incorporated oxygen, which exists only when the film is under annealing conditions ($^{18}O_2$-pressure and high temperature). Diffusive ^{18}O may displace fixed ^{16}O, with probability k, fixing ^{18}O in the network and generating mobile ^{16}O. The reversal reaction is also possible. The time evolution of the density functions is described by the following differential equations.

$$\frac{\partial \rho_{16}^{d}}{\partial t} = D \frac{\partial^2 \rho_{16}^{d}}{\partial x^2} + k[\rho_{18}^{d}\rho_{16}^{f} - \rho_{16}^{d}\rho_{18}^{f}]$$

$$\frac{\partial \rho_{16}^{f}}{\partial t} = -k[\rho_{18}^{d}\rho_{16}^{f} - \rho_{16}^{d}\rho_{18}^{f}]$$

$$\frac{\partial \rho_{18}^{d}}{\partial t} = D \frac{\partial^2 \rho_{18}^{d}}{\partial x^2} - k[\rho_{18}^{d}\rho_{16}^{f} - \rho_{16}^{d}\rho_{18}^{f}]$$ (3.1.8)

$$\frac{\partial \rho_{18}^{f}}{\partial t} = k[\rho_{18}^{d}\rho_{16}^{f} - \rho_{16}^{d}\rho_{18}^{f}]$$

where $\rho_{16}^{d}, \rho_{16}^{f}, \rho_{18}^{d},$ and ρ_{18}^{f} are, respectively, the relative density functions of diffusive or fixed ^{16}O and ^{18}O, where the gauge concentration is taken as the oxygen concentration in stoichiometric aluminium oxide. The reaction terms consider that the probability that a reaction effectively occurs is proportional to the concentration of both reactants at the same time, at the same place, modulated by the reaction rate k. The requirement that the film remains stoichiometric, that is, with the same amount of fixed oxygen, is translated by the equation below,

$$\rho_{16}^{f} = 1 - \rho_{18}^{f}.$$ (3.1.9)

As initial conditions, we assumed that $\rho_{16}^{d}(x, t = 0) = 0, \rho_{16}^{f}(x, t = 0) = 1,$ ... $\rho_{18}^{d}(x, t = 0) = 0,$ and $\rho_{18}^{f}(x, t = 0) = 0 \; \forall x > 0.$

The boundary condition is $\rho_{18}^{d}(x = 0, t) = p_0, \; \forall t > 0,$ stating that ^{18}O pressure is kept at the relative value p_0 at the film surface.

Typical numerically calculated density function profiles are shown in figure 3.1.34, reproducing the trend of experimental results, namely a decreasing incorporated ^{18}O profile from the surface into the bulk of the Al₂O₃ films.

Figure 3.1.35 shows the results of the time and pressure dependences of ^{18}O incorporation determined experimentally, which could be modelled in the framework of equations (3.1.9).

Indeed, for all those combinations of annealing parameters leading to amounts of incorporated ^{18}O much smaller than those of remaining ^{16}O, that is, $\rho_{18}^{f} \ll \rho_{16}^{f}$, the reaction terms in equation (3.1.9) can be linearized:

$$k \left[\rho_{18}^{d} \rho_{16}^{f} - \rho_{16}^{d} \rho_{18}^{f} \right] \cong k\rho_{18}^{d}.$$ (3.1.10)

This approximation simply states that the incorporation of ^{18}O depends on its concentration and reaction rate only, since it is diffusing in a 'sea' of ^{16}O. The model becomes then analytically soluble and the amount of ^{18}O incorporated in the network, calculated as

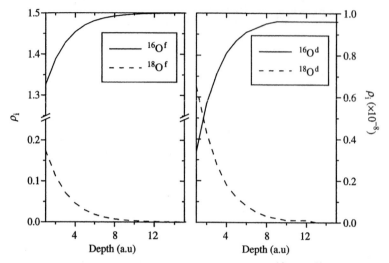

Figure 3.1.34. Typical calculated profiles of fixed and diffusive ^{16}O and ^{18}O in Al_2O_3 films. The profile of $^{16}O^d$, after reaching a maximum, decreases towards greater depths (tail not shown). The amount of $^{16}O^d$ (obtained by integrating the area under the curve) is of the order of the amount of $^{18}O^f$. From [18].

$$\chi_{18}(t) = \int dx\, \rho^f_{18}(x, t) \tag{3.1.11}$$

is given by

$$\chi_{18}(t) = \chi_{18}(t = 0)$$
$$+ 2p_0 \sqrt{\frac{D}{1.5k}} \left[\frac{\sqrt{1.5kt}\, e^{-1.5kt}}{2\sqrt{\pi}} + \frac{3kt - 1}{4} \operatorname{erf}\left(\sqrt{1.5kt}\right) \right] \tag{3.1.12}$$

reproducing the incorporation kinetics as shown by the solid lines in figure 3.1.35. In this figure it is also shown that the amount of incorporated ^{18}O depends linearly on the oxygen pressure at the surface, supporting the linear dependence given above.

The above results demonstrate that oxygen diffuses through Al_2O_3 films, interacting with the network, but conserving the stoichiometry of the film, or in other words conserving diffusive oxygen amounts. One expects then a larger difference in energy between diffusive and fixed states for oxygen atoms in Al_2O_3 than for oxygen diffusion in SiO_2 films on Si. Thus, incorporation of ^{18}O in the bulk of SiO_2 films on Si is not observed, whereas it is always present in the bulk of Al_2O_3 films on Si.

Modelling of reaction–diffusion in thinner Al_2O_3 films on SiO_2/Si annealed in O_2, where other reactions involving Si are triggered by the arrival of the propagating oxygen front to the interfaces, was also accomplished.

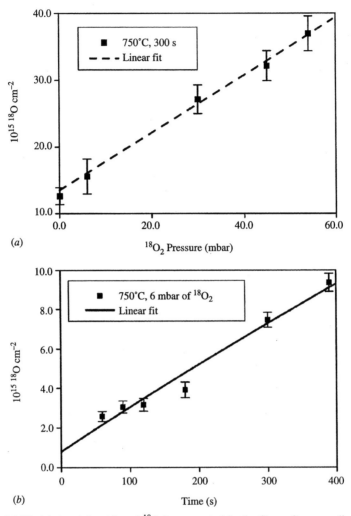

Figure 3.1.35. (*a*) Areal densities of ^{18}O incorporated in the films after annealing under pressures between 0.06 and 54 mbar. The dashed line is from a linear fit to the data; (*b*) kinetics of the oxygen incorporated in the films. The solid curve represents the theoretical result. Error bars account for measurement uncertainties. From [18].

The experimental findings concerning this situation have been reported above, and may be summarized as follows:

(a) Sharpness of the initial interfaces—the as-deposited samples have very sharp interfaces (within one to two atomic layers) both on Si and SiO₂/Si substrates.

(b) Oxygen exchange—annealing of $Al_2{}^{16}O_3$ films in $^{18}O_2$ showed ^{16}O–^{18}O exchange taking place along the whole aluminium oxide films.

(c) Mechanism of oxygen transport—oxygen is transported differently in Al_2O_3 films on Si and in SiO_2 films on Si. Indeed, when 20 nm thick SiO_2 films are annealed in $^{18}O_2$, a minor amount of ^{18}O is fixed in the near-surface region mainly by exchange for ^{16}O atoms from the SiO_2 network, while most of the ^{18}O is incorporated near the SiO_2/Si interface where reaction with substrate-Si and most of the growth takes place, with no reaction and consequently no incorporation of ^{18}O in the bulk of the growing films. In the case of Al_2O_3 films, in contrast to SiO_2, oxygen from the gas phase diffuses in the films strongly interacting with the Al_2O_3 network, being thus incorporated in near-surface and bulk regions.

(d) Oxidation of the substrate—for high enough temperatures, O_2 pressures, and times, or thin enough films, the diffusing oxygen front may reach the interface and a film of SiO_2 may be grown by oxidation of substrate-Si.

(e) Substrate-Si transport and incorporation in Al_2O_3 films—when the substrate is oxidized by reaction at the interface, there is evidence of silicon diffusing into the Al_2O_3 film. The origin of mobile Si seems to be the injection of Si interstitials formed during oxidation due to volume differences of Si in SiO_2 as compared to Si in bulk Si. The precise location of mobile Si in the Al_2O_3 films can vary, Si having been found either spread through the film or located only in near-surface regions. This may be markedly dependent on the initial thickness of the Al_2O_3 film and is still under investigation. Furthermore, annealing of the films in vacuum or neutral atmospheres previous to oxygen annealing may strongly reduce O diffusion as well as Si migration.

(f) Aluminium loss—when substrate-Si is oxidized at the interface, there is evidence for aluminium loss. As aluminium has very low solubility in Si, this loss must have occurred through the film surface. One notes however that Kundu *et al* also report results where there is substrate oxidization without Al loss, indicating that this phenomenon may depend also on the thermal treatment and sample deposition routes.

(g) Crystallization—annealing at high temperatures may cause the aluminium oxide film to crystallize. The crystallization temperature, however, depends on the sample preparation routes and/or thicknesses.

Reaction–diffusion equations were proposed assuming OH-free films with sharp initial interfaces with SiO_2/Si. The involved species, namely O, Al, and Si are modelled in two different states: mobile and fixed. Concentrations of each species were described relative to a gauge concentration, here taken as the atomic concentration of bulk silicon, C_{Si}^{bulk}. That is, the dynamical variables of the model consist of six relative local, adimensional concentrations, defined as

$$\rho^f_{Si}(x,t) = \frac{C^f_{Si}(x,t)}{C^{bulk}_{Si}}, \quad \rho^i_{Si}(x,t) = \frac{C^i_{Si}(x,t)}{C^{bulk}_{Si}}$$

$$\rho^f_O(x,t) = \frac{C^f_O(x,t)}{C^{bulk}_{Si}}, \quad \rho^d_O(x,t) = \frac{C^d_O(x,t)}{C^{bulk}_{Si}} \qquad (3.1.13)$$

$$\rho^f_{Al}(x,t) = \frac{C^f_{Al}(x,t)}{C^{bulk}_{Si}}, \quad \rho^i_{Al}(x,t) = \frac{C^i_{Al}(x,t)}{C^{bulk}_{Si}}$$

where the subindices indicate the species and the superindices f, d, i, are, respectively, associated with the fixed, diffusive, and interstitial states.

Initially the samples consist of a thin film of stoichiometric Al$_2$O$_3$, of thickness Δx_1, on a ultrathin layer of SiO$_2$, of thickness Δx_2, on an infinite Si substrate. The initial conditions, representing the system before annealing in O$_2$, are given by

$$\rho^f_{Si}(x,t=0) = \begin{cases} 0 \text{ for } 0 \leq x < \Delta x_1 \\ 0.5 \text{ for } \Delta x_1 \leq x < \Delta x_1 + \Delta x_2 \quad \rho^i_{Si}(x,t=0) = 0.0 \; \forall x > 0, \\ 1.0 \text{ for } x \geq \Delta x_1 + \Delta x_2 \end{cases}$$

$$\rho^f_O(x,t=0) = \begin{cases} 1.5 \text{ for } 0 \leq x < \Delta x_1 \\ 1.0 \text{ for } \Delta x_1 \leq x < \Delta x_1 + \Delta x_2 \quad \rho^d_O(x,t=0) = 0.0 \; \forall x > 0, \\ 0.0 \text{ for } x \geq \Delta x_1 + \Delta x_2 \end{cases}$$

$$(3.1.14)$$

$$\rho^f_{Al}(x,t=0) = \begin{cases} 1.0 \text{ for } 0 \leq x < \Delta x_1 \\ 0.0 \text{ for } \Delta x_1 \leq x < \Delta x_1 + \Delta x_2 \quad \rho^i_{Al}(x,t=0) = 0.0 \; \forall x > 0. \\ 0.0 \text{ for } x \geq \Delta x_1 + \Delta x_2 \end{cases}$$

In the above equations, Si concentration in SiO$_2$ is taken as half the value of Si concentration in bulk silicon or Al concentration in Al$_2$O$_3$, according to approximate densities of these materials. Fixed O concentrations ρ^f_O are taken assuming that both oxide layers are initially stoichiometric, fully oxidized. As in the 'Reaction–diffusion in high-κ dielectrics on Si' section, a one-dimensional problem is considered, where x is the direction perpendicular to the sample surface, implying an average over the other two directions. Also, allowing the possibility that in a given slice of width dx there is a mixture of Al$_2$O$_3$, SiO$_2$, and non-oxidized silicon, a normalization condition is imposed, reflecting volume conservation in each layer, that is

$$\rho_{Al}^f + 2\rho_{Si}^f - \rho_{Si}^{nfo} = 1 \qquad (3.1.15)$$

where $\rho_{Si}^{nfo} = \rho_{Si}^f + \frac{3}{4}\rho_{Al}^f - \frac{1}{2}\rho_O^f$ is the non-oxidized fraction of fixed Si. The amount of fixed oxygen can be calculated as

$$\rho_O^f = 2 - \frac{1}{2}\rho_{Al}^f - 2\rho_{Si}^f \qquad (3.1.16)$$

This normalization considers that (a) for bulk silicon, $\rho_{Si}^f = 1.0$ and $\rho_{Al}^f = \rho_{Si}^{nfo} = \rho_O^f = 0$; (b) for pure Al_2O_3, $\rho_{Al}^f = 1.0$, $\rho_O^f = 1.5$, and $\rho_{Si}^f = \rho_{Si}^{nfo} = 0$, and for (c) pure SiO_2, $\rho_{Si}^f = 0.5$, $\rho_O^f = 1.0$, and $\rho_{Al}^f = \rho_{Si}^{nfo} = 0$.

One must also specify a boundary condition, which models the exposure of the sample to the oxygen atmosphere. It is

$$\rho_O^d(x = 0, t) = P_0 \ \forall t > 0 \qquad (3.1.17)$$

where P_0 is a relative (non-dimensional) concentration of O_2 available at the surface. It is reasonable to assume that P_0 depends monotonically on the O_2 pressure p_0. However, the correct functional form depends on the details of the surface. In the reaction–diffusion model for thermal growth of silicon oxide, for example, there has been proposed a linear relation, such that the available concentration of O_2 is proportional to the product of the concentration of O_2 in the gas phase and the fractional free volume in the sample surface. Many factors can modify this assumption, for example, roughness or pores in the sample surface, electric fields generated by surface defects, etc. In the present case, we take P_0 as a parameter of the model, which should increase linearly with O_2 pressure.

The following scenario thus emerges: the initial structure $Al_2O_3/SiO_2/Si$ is exposed to O_2 at a given (non-dimensional) constant concentration P_0, available at the sample surface to diffuse into the sample. Oxygen diffuses through the initial oxide, being eventually exchanged for O already existent in these films. As, in the present formulation of the model, there is no discrimination between diffusing O and the oxygen already in the films, this exchange does not change the relative concentration densities. Upon reaching bulk silicon, oxygen reacts, forming silicon oxide, and since oxidized Si occupies a larger volume, it generates interstitial Si that is prone to move, as suggested by molecular dynamics simulations. Hence, silicon oxidation has a twofold effect: it transforms oxygen from mobile to fixed and some silicon from fixed to mobile (interstitial) species. Mobile silicon spreads through the sample, towards both bulk silicon and Al_2O_3 regions. Interstitial Si cannot be trapped in SiO_2. However, when in the Al_2O_3 region, mobile Si can displace Al, since silicon oxide formation is thermodynamically favoured over that of aluminium oxide. This reaction implies fixing Si in the original Al_2O_3 network and transferring fixed Al and O from the Al_2O_3 network to mobile states. Mobile Al and O atoms that reach the surface may escape, reducing their total amounts in the sample. Since interstitial Si is generated by oxidation of Si, whereas interstitial Al is created by reaction involving interstitial Si, in the absence of Si oxidation there is neither Si nor Al transport or Al loss. Therefore, the experimental observation that annealing in vacuum does not change the initial profiles, in the absence of OH contamination, is consistent with the depicted scenario.

To write the differential equations for the density functions for this system we consider one equation for each density function as follows

$$
\begin{aligned}
\frac{\partial \rho_O^d}{\partial t} &= D_1 \frac{\partial^2 \rho_O^d}{\partial x^2} - 2k_1 \rho_{Si}^{nfo} \rho_O^d + k_2 \rho_{Si}^i \rho_{Al}^f, & \frac{\partial \rho_O^f}{\partial t} &= 2k_1 \rho_{Si}^{nfo} \rho_O^d - k_2 \rho_{Si}^i \rho_{Al}^f \\
\frac{\partial \rho_{Si}^i}{\partial t} &= D_2 \frac{\partial^2 \rho_{Si}^i}{\partial x^2} + k_1 \rho_{Si}^{nfo} \rho_O^d - k_2 \rho_{Si}^i \rho_{Al}^f, & \frac{\partial \rho_{Si}^f}{\partial t} &= -k_1 \rho_{Si}^{nfo} \rho_O^d + k_2 \rho_{Si}^i \rho_{Al}^f \\
\frac{\partial \rho_{Al}^i}{\partial t} &= D_3 \frac{\partial^2 \rho_{Al}^i}{\partial x^2} - D_3 \frac{\partial^2 \rho_{Al}^f}{\partial x^2} + 2k_2 \rho_{Si}^i \rho_{Al}^f, & \frac{\partial \rho_{Al}^f}{\partial t} &= -2k_2 \rho_{Si}^i \rho_{Al}^f
\end{aligned}
$$

$$(3.1.18)$$

where d and i stand respectively for diffusive and interstitial species. D_1 and D_2 are the diffusion coefficients for mobile O and Si, taken as constant functions of depth. The diffusion coefficient $D_{Al}(x,t)$ for interstitial aluminium, however, is assumed to depend on the composition of the medium, to account for the very low solubility of aluminium in Si, that is

$$
D_{Al}(x,t) = D_3 \rho_{Al}^f(x,t) \tag{3.1.19}
$$

where D_3 is the diffusion coefficient of interstitial aluminium in Al₂O₃. This assumption implies that interstitial aluminium is non-soluble in pure SiO₂ as well, and decreases from D_3 to zero with the concentration of fixed aluminium in the oxide. The diffusion equation for interstitial aluminium is written taking the diffusion non-homogeneity as described by equation (3.1.12).

Equation (3.1.18) considers two reactions. The first one, related to the reaction constant k_1, is proportional to the product $\rho_{Si}^{nfo} \rho_O^d$ and describes the oxidation of fixed silicon not yet fully oxidized. It obviously changes the concentration of fixed and diffusive oxygen, and changes also the concentration of fixed and interstitial silicon, since Si in silicon oxide occupies roughly double the volume as in bulk Si. The term $k_1 \rho_{Si}^{nfo} \rho_O^d$ appears then in four of the six equations for the density functions (3.1.48). The second reaction term, associated with the reaction constant k_2, is proportional to the product $\rho_{Si}^i \rho_{Al}^f$ and describes the displacement of fixed Al by interstitial silicon. As silicon and aluminium have different valences, when interstitial silicon is exchanged by fixed aluminium, some oxygen is liberated, increasing the amount of diffusive oxygen and correspondingly decreasing the local concentration of fixed oxygen. Consequently, the reaction term $k_2 \rho_{Si}^i \rho_{Al}^f$

appears in all six equations (3.1.18). The numeric coefficients in some of the reaction terms guarantee the correct stoichiometry and normalization.

Equations (3.1.18) together with the initial and boundary conditions are solved by finite difference methods yielding density function profiles. Figure 3.1.36 shows density profiles for all six species at initial time and after 8000 (*a*) and 15 000 (*b*) iteration steps. The values of the parameters were conveniently chosen. The profiles reproduce the experimental results with the following features:

(a) Aluminium being lost and its profile $(Al^f + Al^i)$ presenting a maximum between surface and interface. As aluminium oxide is amorphous, it is reasonable to consider that at room temperature interstitial

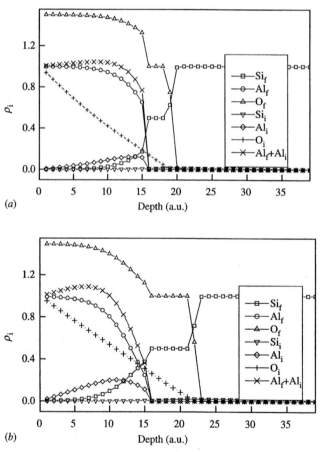

Figure 3.1.36. Calculated density function profiles for all species and for the sum of both aluminium species.

aluminium becomes trapped. However, it remains a point to be experimentally verified whether there is an excess of aluminium in the film after annealing.

(b) Silicon migrating into the aluminium oxide region, forming a compound containing Si, Al, and O.

(c) For initial times, before further growth of the initial silicon oxide layer, species such as Al^i and Si^i are absent; there is no loss of aluminium or transport of silicon into the aluminium oxide region.

(d) When the oxygen front reaches the substrate, interstitial silicon and aluminium are created, aluminium is lost and there is transport of silicon into the aluminium oxide region.

(e) Due to the exchange of fixed aluminium for fixed silicon in the aluminium oxide region, there is a partial loss of fixed oxygen.

These results show that the model is capable of describing the mobility and reaction of Al, O, and Si during RTA of Al$_2$O$_3$/SiO$_2$/Si nanostructures in O$_2$ and their immobility during annealing in vacuum, since the mobility of Al and Si is possible, following this model, only when there is oxidation of the substrate. However, this description is semi-quantitative at the present stage. The estimation of more precise values for the parameters such as the diffusion coefficients and reaction constants requires far more experimental data than those currently available.

The modelling of RTA of Al$_2$O$_3$/SiO$_2$/Si nanostructures in O$_2$ demonstrates also the combinatorial explosion in the number of parameters of reaction diffusion models. The number of parameters increases due to the many different possible situations, depending on the interactions among the different species.

A different path in modelling diffusion and reaction in solid materials is to consider molecular dynamics techniques, where more fundamental physical principles and details may be taken into account. These efforts point in the direction of describing the possible paths and estimating their probabilities for the transport of particles through a complicated network. Besides calculating at each time step the relative contribution of all relevant events in a careful account, many time steps must be considered to obtain an estimate of some quantity that can be compared to observable quantities, for example, the phenomenological diffusion coefficient (that requires the long- time limit) or a kinetics curve for oxide film growth. These calculations demand an enormous amount of memory and computational time, and to make things finite some cut-offs and approximations are necessary. Nevertheless, they do provide extremely relevant information as regards details of the physical interactions subjacent to diffusive processes as relevant reactions, different diffusion mechanisms, etc. This information is essential to propose terms and estimate parameters in reaction–diffusion equations. Rather than *alternative* theoretical modelling techniques, molecular dynamics (first-principles) techniques and reaction–diffusion equations are *complementary* theoretical tools.

References

[1] Baumvol I J R 1999 *Surf. Sci. Rep.* **36** 1 and references therein
[2] Trimaille I and Rigo S 1989 *Appl. Surf. Sci.* **39** 65
[3] Baumvol I J R, Krug C, Stedile F C, Gorris F and Schulte W H 1999 *Phys. Rev. B* **60** 1492
[4] Chaneliere C, Autran J L, Devine R A B and Balland B 1998 *Mater. Sci. Eng.* **R22** 269
[5] Hubbard K J and Schlom D G 1996 *J. Mater. Res.* **11** 2757
[6] Yamamoto K, Hayashi S, Kubota M and Niwa M 2002 *Appl. Phys. Lett.* **81** 2053
[7] Gosset L G *et al* 2002 Alternatives to SiO$_2$ as gate dielectrics for future Si-based microelectronics *Mater. Res. Soc., Workshop Series, Warrendale* eds J Morais and I J R Baumvol p 5
[8] Gosset L G, Damlecourt F, Renault O, Rouchon D, Holliger Ph, Ermolieff A, Trimaille I, Ganem J-J, Holliger Ph, Martin F and Semeria M-N 2002 *J. Non-Cryst. Solids* **303** 17
[9] Krug C, da Rosa E B O, de Almeida R M C, Morais J, Baumvol I J R, Salgado T D M and Stedile F C 2000 *Phys. Rev. Lett.* **85** 4120
[10] Krug C, da Rosa E B O, de Almeida R M C, Morais J, Baumvol I J R, Salgado T D M and Stedile F C 2001 *Phys. Rev. Lett.* **86** 4714
[11] Copel M 2001 *Phys. Rev. Lett.* **86** 4713
[12] Gusev E, Copel M, Cartier E, Baumvol I J R, Krug C and Gribelyuk M A 2000 *Appl. Phys. Lett.* **76** 176
[13] Klein T, Niu D, Li W, Maher D M, Hobbs C C, Hedge R I, Baumvol I J R and Parsons G N 1999 *Appl. Phys. Lett.* **75** 4001
[14] Copel M, Cartier E, Gusev E P, Guha S, Bojarczuk N N and Poppeller M 2001 *Appl. Phys. Lett.* **78** 2670
[15] Kundu M, Miyata N and Ichikawa M 2001 *Appl. Phys. Lett.* **78** 1517
[16] Kundu M, Ichikawa M and Miyata N 2002 *J. Appl. Phys.* **91** 492
[17] Kundu M, Ichikawa M and Miyata N 2002 *J. Appl. Phys.* **92** 1914
[18] da Rosa E B O, Baumvol I J R, Morais J, de Almeida R M C, Papaleo R M and Stedile F C 2002 *Phys. Rev. B* **86** 121303
[19] Copel M, Gribelyuk M and Gusev E 2000 *Appl. Phys. Lett.* **76** 436
[20] Busch B W, Schulte W H, Garfunkel E, Gustafsson T, Qi W, Nieh R and Lee J 2000 *Phys. Rev. B* **62** 13290
[21] Starodub D *et al* 2002 *Mater. Res. Soc. Workshop Series, Warrendale* eds J Morais and I J R Baumvol pp 5–99
[22] Gustafsson T, Lu H C, Busch B W, Schulte W H and Garfunkel E 2001 *Nucl. Instrum. Meth. Phys. Res. B* **183** 146
[23] Ramanatham S, Muller D A, Wilk G D, Park C M and McIntyre P C 2001 *Appl. Phys. Lett.* **79** 3311

[24] Morais J, da Rosa E B O, Miotti L, Pezzi R P, Baumvol I J R, Rotondaro A L P, Bevan M J and Colombo L 2001 *Appl. Phys. Lett.* **78** 2446

[25] Quevedo-Lopez M, El-Bouanani M, Addepalli S, Duggan J L, Gnade B E, Visokay M R, Douglas M, Bevan M J and Colombo L 2001 *Appl. Phys. Lett.* **79** 2958

[26] Morais J, da Rosa E B O, Pezzi R P, Miotti L and Baumvol I J R 2001 *Appl. Phys. Lett.* **79** 1998

[27] da Rosa E B O, Morais J, Pezzi R P, Miotti L and Baumvol I J R 2001 *J. Electrochem. Soc.* **148** G44

[28] Chang H S, Jeon S, Hwang H and Moon D W 2002 *Appl. Phys. Lett.* **80** 3385

[29] Landheer D, Gupta J A, Sproule G I, McCaffrey J P, Graham M J, Yang K-C, Lu Z-H and Lennard W N 2001 *J. Electrochem. Soc.* **148** G29

[30] Landheer D, Wu X, Morais J, Baumvol I J R, Pezzi R P, Miotti L, Lennard W N and Kim J K 2001 *Appl. Phys. Lett.* **79** 2618

[31] Guha S, Cartier E, Gribelyuk M A, Bojarczuk N A and Copel M 2000 *Appl. Phys. Lett.* **77** 2710

[32] Stemmer S, Maria J-P and Kingon A I 2001 *Appl. Phys. Lett.* **79** 102

[33] Chambers J J and Parsons G N 2000 *Appl. Phys. Lett.* **77** 2385

[34] Parsons G N, Chambers J J, Niu D, Ashcraft R and Klein T M 2002 *Mater. Res. Soc., Workshop Series, Warrendale* eds J Morais and I J R Baumvol p 13

[35] Sayan S, Aravamudhan S, Busch B W, Schulte W H, Cosandey F, Wilk G D, Gustafsson T and Garfunkel E 2002 *J. Vac. Sci. Technol.* **A20** 507

[36] Bastos K P, Morais J, Miotti L, Pezzi R P, Soares G V, Baumvol I J R, Tseng H-H, Hedge R I and Tobin P J 2002 *Appl. Phys. Lett.* **81** 1669

[37] Morais J, Miotti L, Soares G V, Pezzi R P, Teixeira S R, Bastos K P, Baumvol I J R, Chambers J J, Rotondaro A L P, Visokay M and Colombo L 2002 *Appl. Phys. Lett.* **81** 2995

[38] Kirsch P D, Kang C S, Lozano J, Lee J C and Ekerdt J G 2002 *J. Appl. Phys.* **91** 4353

[39] Jeon S, Choi C-J, Seong T-Y and Hwang H 2001 *Appl. Phys. Lett.* **79** 245

[40] Baumvol I J R, Stedile F C, Ganem J-J, Rigo S and Trimaille I 1995 *J. Electrochem. Soc.* **142** 1205

[41] de Almeida R M C, Gonçalves S, Baumvol I J R and Stedile F C 2000 *Phys. Rev. B* **61** 12992 and references therein

[42] Bouchaud J-P and Georges A 1990 *Phys. Rep.* **195** 127

[43] Havlin S and Ben-Avraham D 1987 *Adv. Phys.* **36** 695–798

[44] de Almeida R M C and Baumvol I J R 2000 *Phys. Rev. B* **62** 16255 and references therein

Chapter 3.2

Defects in stacks of Si with nanometre thick high-κ dielectric layers: characterization and identification by electron spin resonance

A Stesmans and V V Afanas'ev

Introduction

It is not an overstatement that the realization of the semiconductor era, better known as, the 'silicon age', owes much to the simple fact that silicon has a native insulator, SiO_2, of superb quality. In this epic, relentless scaling, much more than diversification, was, and still appears as, a natural law, propelling dazzling progress and leading to fascinating technological achievements. The benefits include the fact that the more an integrated circuit (IC) is scaled, the higher becomes its packing density and circuit speed, and the lower its power consumption. Progress in Si IC technology projects scaling from the current $0.18-0.25\,\mu m$ minimum lithographic line width devices towards the next $0.1\,\mu m$ generation, implying that the current SiO_2 gate dielectric must scale below 2 nm thickness [1].

This relentless miniaturization continuously re-emphasizes the vital role of the gate dielectric in complementary metal–oxide–silicon (MOS) devicing. For one, this drastic lateral scaling mandates to maintain efficient signal transmission, compensation for the lost areal gate capacitance, which may be realized by reducing the gate oxide thickness and/or enhancing the dielectric permittivity $\varepsilon = \kappa\varepsilon_0$, where ε_0 is the vacuum permittivity and κ the dielectric constant. So far, conventional thermal SiO_2, with such unique and superb properties, has served so extraordinarily well as an unexcelled gate dielectric that stretching its usage as far as possible appears natural. Yet, with the projected SiO_2 thicknesses in the sub-2 nm range for the near future [1], the application of this foremost insulator in MOS devices faces fundamental limits such as excessive direct (tunnelling) leakage current, boron penetration, electron mobility degradation, and reliability problems [2–6]. Because of the allowable gate leakage current, the need for an alternative higher-κ gate dielectric has become increasingly more acute. This has stimulated enhanced research in replacing SiO_2 with an alternative layer of high dielectric constant (κ) for future MOS generations. See [5] for a recent detailed overview. The application of higher-κ insulators should enable usage of thicker films of equivalent SiO_2 electrical thickness (EOT, defined as $d_{EOT} = d_{high-\kappa}\varepsilon_{SiO_2}/\varepsilon_{high-\kappa}$) with an expected reduction in leakage current and improved gate reliability.

Obviously, like the superb thermal SiO_2 dielectric on Si, the aimed for high-κ materials should also exhibit high thermodynamic stability, large barriers for Schottky emission (wide bandgaps), low impurity diffusion coefficients, low leakage currents and inertness. And, very crucially, when serving as gate insulators, they should form high-quality smooth interfaces with low interface trap (defect) density. This has been dealt with at length in many works (see [5] and references therein), and the many aspects are detailed in this book.

As the IC world is rapidly approaching the projected turning point in its still young history, research efforts are intensified, both experimentally [1, 7, 8] and theoretically [9–13]. In view of the immense value at stake, it is natural that an immense arsenal of complementary top-sensitivity analysing techniques is being applied in conjunction to explore microstructural, compositional, and bonding chemistry aspects of ultrathin (sub-5 nm) high-κ films on Si [13–18]. Applied techniques include medium-energy ion scattering (MEIS), high-resolution transmission electron microscopy (HRTEM), nuclear resonance profiling (NRP), x-ray photoelectron and Auger spectroscopy (some overviews on these techniques as applied to high-κ dielectrics may be found in [17, 18]). Evidently, as the next step, the electrical performance (transport, charge trapping, presence of detrimental electrically active interface traps) is thoroughly investigated using electrical techniques, such as capacitance–voltage (C–V) and current–voltage (I–V) [5, 17, 19–23]. Investigated as potential alternative gate dielectrics are high-κ metal oxides, including Al_2O_3, Y_2O_3, ZrO_2, $SrTiO_2$, TiO_2, Ta_2O_5, HfO_2, and La_2O_3 [24], with the latter two currently receiving much attention. Some have already been disfavoured [5, 17, 18], among other things, because of stability reasons; see the various contributions in this book for up-to-date details.

Inherently, however, the above methods reveal little about the atomic nature of crucial point defects related to or at the origin of observed electrical deficiencies. Yet, as amply documented for the Si/SiO_2 structure, point defects were demonstrated or hinted at as the origin of a realm of detrimental aspects, such as adverse interface traps [25], oxide fixed charge, irradiation-induced degradation [26], stress-induced leakage current, and thermally induced oxide breakdown. A most vital issue is the quality of the c-Si/insulator interface, particularly as regards the presence of electrically active inherent defects [25]. So, investigating the likely altered interface situation in c-Si/alternative dielectric structures *vis-à-vis* standard c-Si/SiO$_2$ will be crucial for advancing semiconductor device technology. The Si/dielectric interface control is a most critical issue for high-κ material to move to integration in IC manufacturing.

Here, in a complementary approach, electron spin resonance (ESR)— the technique of choice for atomic identification of defects—is applied to newly formed Si/insulator stacks, with the aim to provide atomic identification and quantification of defects. The known characteristics of the amply studied Si/SiO$_2$ structure are used as backdrop. In a first part, ESR study is applied to stacks of ultrathin SiO$_x$ and Al$_2$O$_3$ films sandwiched between (100) Si and thin ZrO$_2$ layers; the latter is an intensively investigated [5] alternative dielectric (κ ~ 14–25). Besides aiming at information on Si/metal oxide interfaces and dielectric properties, some of the aspects of interlayers are also addressed. A second case study concerns the Si/HfO$_2$ entity, where a comparative analysis is carried out for three types of state-of-the-art deposition method of HfO$_2$ on (100) Si. Nowadays, high-κ-on-Si research appears dominated by HfO$_2$, either in pure form or as Hf silicates, silicides, and aluminates [27–31]. From the vast amount of knowledge acquired over the various decades from meticulous intense research on the Si/SiO$_2$ system, it may be expected that with regard to the grown interfaces, different fabrication (deposition) methods will ensure an impact on interfacial particularities, such as atomic coordination, kinds of occurring defect and densities, and their passivation behaviour in, e.g., H$_2$. It may generally be expected that a comparative in-depth analysis may be very revealing.

The interest in inserting interlayers (Al$_2$O$_3$, SiO$_x$) may derive from various sides. On the basic physics side, e.g., as to ZrO$_2$, one may try to improve on the much-reduced conduction band offset with respect to Si of ZrO$_2$ (2.0 ± 0.1 eV) [32] as compared to SiO$_2$ (3.15 ± 0.05 eV), in order to reduce Schottky emission leakage currents. On the chemical-manufacturing side, there is the influence of the growth base on the film deposition characteristics. Regarding ALD of ZrO$_2$ layers, it has been found that the nucleation on H-terminated last-HF-dipped (100) Si surfaces is inhibited [14], resulting in nonuniform and discontinuous films. However, after pregrowth of a thin thermal oxide, uniform ZrO$_2$ film growth is observed, with an abrupt interface between SiO$_2$ and ZrO$_2$. Finally, in a more general context, investigation of the impact of interlayers may be desirable in the light of the concern that fabrication of whatever Si/metal oxide structure may always be attendant, to a greater or lesser extent, with the occurrence, however thin, of some chemically intermingled transitional layer threatening the aimed for 'ideal' abrupt transition.

Currently, HfO$_2$ on Si is the subject of intensified research activity for application as high-κ gate dielectric material, for which there may be several reasons. First, the oxide possesses interesting properties such as high κ (15–26) value, large Gibbs energy of formation and as, e.g., compared to ZrO$_2$, a better thermodynamic stability in contact with Si, which would enable one to limit interfacial intermixing upon post-deposition heat treatments. Hence, the HfO$_2$-on-Si case is addressed from a broad range of technological/chemical angles of view; it varies from single layers of simple HfO$_2$ [31] over single layers of more complex HfO$_2$-based composites (such as Hf silicate, HfSiO$_x$N$_y$) [27, 28, 30] on Si to combinations of HfO$_2$ layers in stacks (nanolaminates) with other dielectric layers [17, 29, 33], such as TiO$_x$N$_y$, Al$_2$O$_3$, and SiO$_x$.

As a main result from the ESR study, it is revealed that the ruling paramagnetic inherent interface defects in Si/SiO$_2$, i.e., P_b-type centres [25], also appear as major players at the interface of (100) Si with the studied ultrathin high-κ metal oxide layers. Probing their properties provides pertinent information about the nature of the interfaces concerned, indicating, among others, these to be under enhanced strain in the as-grown state. The properties of these defects serve as a criterion for interface quality. The influence of post-deposition heat treatment is addressed to assess interface and layer stability.

As stated, the principal defects observed by ESR are the P_b-type defects, known from the Si/SiO$_2$ interface. Hence, after the 'Introduction', the work starts with a short overview of the properties of these defects. A brief synopsis of ESR work performed on Si/high-κ structures, followed by a description of ESR modalities, are dealt with in the subsequent sections. We then proceed to addressing two particular cases investigated: we overview our ESR results obtained on the ZrO$_2$ and Al$_2$O$_3$ layers deposited on (100) Si, with involvement of interlayers; results of the Si/HfO$_2$ system, manufactured by various methods, are addressed next. Finally, the last section contains the main conclusions and final remarks.

P_b-type defects at the Si/SiO$_2$ interface: salient properties

At least some of the electrically revealed interface traps in Si/SiO$_2$ structures were found to be paramagnetic, as evidenced by their ESR response [34]. As detected by ESR, they are generally termed P_b-type centres, the class of paramagnetic trivalent Si defects, intensely studied before (see [25, 34–38]). Various works have overviewed their properties [25, 37, 38]. They are mismatch-induced inherent point defects at the Si/SiO$_2$ interface, correlating with interface orientation. Three types are common [25, 34]: at the (111) Si/SiO$_2$ interface, only one type is generally observed, specifically termed P_b, identified as [35] prototype interfacial trivalent Si, denoted

$Si_3 \equiv Si^{\bullet}$. It was shown to be an amphoteric trap of effective correlation energy $U_e = 0.5\,eV$ with the $+/0$ and $0/-$ levels deep in the Si bandgap at 0.3 and 0.8 eV above the valence band, respectively [39]. The technologically foremost (100) Si/SiO_2 interface usually exhibits two prominent types, called P_{b0} and P_{b1} [34]. All three variants were shown to be trivalent Si centres, with P_{b0} chemically identical to P_b, but now residing at (imperfections of, e.g., facets) a macroscopic (100)-oriented Si/SiO_2 interface. Both were established as adverse electrical interface traps [38, 40]. The electrical activity of the P_{b1} centres as interface traps, initially accepted, is still in dispute [40]. The P_{b1} is concluded to be a distorted defected interfacial dimer (an approximately $\langle 211 \rangle$ oriented $Si_3 \equiv Si^{\bullet}$ [41, 42]). For standard oxidation temperatures T (800–960°C), areal densities of physical defect sites (including both ESR active and inactive ones) of $[P_b] \sim 5 \times 10^{12}\,cm^{-2}$ and $[P_{b0}], [P_{b1}] \sim 1 \times 10^{12}\,cm^{-2}$ occur naturally [36, 43, 44].

Their thermochemical properties appear dominated by interaction with hydrogen [45–50]. When heating in H_2 ambient, defect inactivation, i.e., passivation, is pictured as chemical saturation of the failing Si bond by hydrogen, symbolized as SiH formation. Their passivation is part of the goal of the technologically routinely applied post-oxidation anneal in forming gas (FG). Studied most intensively, ESR analysis has inferred a transparent fully reversible H–P_b interaction scheme. Passivation in molecular H_2 and dissociation in vacuum (examined in the key temperature ranges ~ 230–260 and 500–600°C, respectively, are found to be well modelled by the chemical reactions

$$P_b + H_2 \rightarrow P_bH + H \qquad (3.2.1)$$

$$P_bH \rightarrow P_b + H \qquad (3.2.2)$$

proceeding with activation energies $E_f = 1.51 \pm 0.04$ and $E_d = 2.83 \pm 0.03\,eV$, respectively. Similar values were obtained for P_{b1}. More details can be found elsewhere [49].

ESR work performed on Si/high-κ film entities

The scaling engine has shrunk devices down to such dimensions that the workhorse gate oxide SiO_2, and related nitrided Si oxide, is running out of steam. As soon as it had outlined and, next, clearly projected that to keep the device scaling train running, the replacement of the conventional SiO_2 gate insulator by a 'new', higher-κ dielectric would be a necessity in the near future, it originated a boom in the research on high-κ insulators, particularly in combination with Si.

The search for an alternative high-κ gate insulator is propelled by industrial demands. Quite naturally, when addressing new Si/insulator combinations, the first mainstream research activity was to investigate the microstructural, compositional, and bonding chemistry aspects of the newly constructed Si/insulator combinations using the available high-sensitivity physical analysing techniques [14–18]. As the enhanced scientific activity concerned a search for a new gate dielectric, obviously, this was shortly followed or accompanied by studying electrical properties [20–23, 51]. This implied intrinsic properties of the grown layers as well as their behaviour in combination with Si, in an, e.g., MIS capacitor or even device arrangement [52] using standard techniques such as $C-V$ and conductance–voltage ($G-V$) measurements. One crucial aspect of the electrical quality of the entity concerns occurring detrimental charge traps and recombination centres. The control of these, and reduction to acceptable levels, is an absolute requirement for successful introduction in devices of any alternative dielectric. And, as compiled in a recent overview on high-κ gate dielectrics [5], several works have already appeared, addressing electrical properties and characterization of traps. Then, in a next step, with defects and traps present, one may want to deepen insight, with an ultimate goal identification of occurring defects (charge traps, recombination centres, interface defects). Here, experimentally, the number of atomic probes available is very limited[1], ESR being the technique of choice when it comes to atomic identification of point defects. Coming in a later stadium, it may then perhaps not come as a surprise that the ESR application as an investigation tool of Si/insulator entities in the rush for an alternative dielectric for SiO_2 is still in its infancy. Only a few works have so far appeared.

For clarity, however, before proceeding, we should add that over the decades since the ESR spectroscopy was initiated in the late 1940s, there has of course appeared a vast amount of ESR research on insulators, with numerous publications, mostly, however, in their singular bulk form. The results provide a rich database for reference. These are not the subjects here. Instead, the current work wants to focus only on the application of ESR study on alternative gate oxides in combination with Si.

First K-band ESR works reported on an investigation of defects in entities of (100) Si with ultrathin ALD layers of ZrO_2 and Al_2O_3 [53, 54].

---

[1]Although on a somewhat different level, there are some other methods that may provide information on defect structure and chemical composition. Notably, we refer to high-resolution infrared spectroscopy and UV–visible absorption and pholuminescence spectroscopy, probing the vibronic modes and electron levels of defects, respectively. These optical methods can yield the symmetry of a defect, information on involved impurities, its electronic energy levels and the vibrational modes. However, provided it can be resolved, conclusive information is generally inferred from ESR observations from analysis of the hyperfine interaction structure. Unlike ESR though, the above spectroscopies do not require unpaired spins, which would render them of much wider potential use.

Simultaneously, it also addressed the role of SiO_x and Al_2O_3 interlayers. The defects observed were the trivalent Si P_b-type interface defects P_{b0} and P_{b1}—the archetypical defects for the interface of (100) Si with SiO_2 and Si oxynitride, located right at the interface. Over the years of intense research on the Si/SiO_2 system, the P_b-type centres have emerged as the workhorse ESR-active atomic probes of the nature of the interface they are residing in, injected there by nature as *horses of Troy* revealing their environment. Hence the work mainly provided information on the nature of occurring interfaces. Main findings included that in the as-manufactured state, the interface of Si with the dielectric layers is basically Si/SiO_2-like. They are under enlarged strain as reflected, among other things, by the significantly enhanced P_{b0}, P_{b1} defect densities, with P_{b0} being very dominant (six- to sevenfold as compared to standard Si/SiO_2). More technological grade interface quality, in terms of P_b-type interface defects, could be attained by 'mild' post-growth heating (~650°C). Some of these results will be the subject of the present compilation.

In a next X-band ESR (4–20 K) work, Cantin and von Bardeleben [55] studied (100) Si/a-Al_2O_3 (40 Å) entities manufactured using atomic layer deposition (300°C) on HF-dipped-last (100) Si. For reason of comparison, also studied were structures with a thin (20 Å) thermal oxide interlayer grown by rapid thermal oxidation (RTO) at 800°C before Al_2O_2 deposition, as well as Si/SiO_2(RTO) control samples. Relying on computer-assisted decomposition of entangled spectra, the observation of the P_{b0}, P_{b1} defects was reported with concentrations in the range 10^{12}–10^{13} cm^{-2}, with P_{b0} being dominant—similar to previous work [53]. Additionally, an intense third centre was observed at $g = 2.0056$–2.0059, characteristic of unpaired Si bonds in disordered/a-Si [56, 57]. It is commonly referred to as the D centre. Though it was realized that a main fraction could have originated from sample cleavage (damage in cracks at edge cleavage planes [58]), the authors suggested some fraction of the centres may reside at the Si/Al_2O_3 interface or in the dielectric. They also studied passivation of defects in forming gas (FG; 5% H_2 in N_2) at 450°C, generally applied to passivate interface defects to below 10^{10} $cm^{-2}eV^{-1}$. Yet, the passivation was found to be highly inefficient—only ~50% reduction in $[P_{b0}]$. They concluded the effect not to be due to hindrance by the top Al_2O_3 layer of H_2 diffusion on its way to the Si/SiO_2 interface, thus concluding a problematic situation.

A most recent work [59] using X-band (300 K) ESR reported the observation of the P_b defect at the interface of nominally (111) Si/HfO_2 (14.5 nm) entities manufactured by ALD using $Hf(NO_3)_4$ as the precursor. For the (100) Si face, simplest for ESR spectroscopy, the observation of the archetypical P_b defect was reported referring to an Si/SiO_2-type interface. The defect was reported to be somewhat different from the P_b defect in standard thermal (111) Si/SiO_2, herewith apparently disregarding the well-known fact that the P_b modalities, including g tensor, may somewhat vary (quite naturally) depending on the exact $Si/SiO_{2(x)}$ interface type and oxide (interface) relaxation state [36, 60, 61]. In this regard, new naming, as proposed, would be merely superfluous and confusing. Though measured by the conventional $C-V$ method on a separate (100) Si/HfO_2 sample, the authors found post-formation heating in FG (60 s) highly efficient in interface trap passivation.

Clearly, as an initial common result, the described ESR works almost exclusively deal with the observation of interface defects at the Si/dielectric interface. Apparently, with the kind of limited ESR sensitivity achievable, they appear as most readily observed. This must be considered as the first step. There is no doubt that with increasing ESR research efforts in the field, other defects in various densities will be atomically unveiled in the large number of newly explored Si/high-κ insulator structures.

ESR technique

Application to high-κ dielectrics: general modalities

Generally, as to the potentiality of spectroscopy, application of ESR to Si/high-κ stacks will in principle not much differ from the investigation on the basic Si/SiO_2 entity. When focusing on interface and dielectric film properties, it will face similar hurdles, such as the limited number of defects one can physically pack into one ESR sample fitting into the microwave cavity, inherent to studying interfaces and thin layers.

Next, as concerns Si substrates, extreme care has to be exercised about left sample cutting damage, i.e., cracks at lateral edges, unavoidably giving rise to an intense, often prohibitively interfering, D signal [56] on top of investigated weaker signals. Also, when studying one-side-deposited Si/high-κ structures, as is often the case, care should be taken to exclude interference from unwanted possible signals originating from the less controlled back surface/interface.

However, Si/high-κ entities might also provide some benefit for ESR, e.g., when measuring at reduced temperature T. For some types of defect, reduced T may result in higher sensitivity through enhancement of paramagnetism by the Boltzmann factor. Moreover, this may advantageously be accompanied by a higher loaded Q due to freezing out of any substrate carriers. True, microwave saturability of signals generally increases fast with decreasing T because of increasing spin–lattice relaxation time T_1 [62]; but, with the currently envisioned Si/high-κ entities, with necessarily ultrathin dielectric layers, saturability of P_b-type interface defects appears naturally reduced, similarly as observed for Si/SiO_2 with ultrathin SiO_2 layers [45, 60, 63]. This may be related to significant changes in the density of phonon states in the dielectric layer, or more generally, in the local environment of the P_b-type defects. Then higher microwave power P_μ can be applied, with attendant gain in sensitivity. Current top quality spectrometers may detect $\sim 1 \times 10^9$ centres ($S = 1/2$) of 1 G linewidth at low T within acceptable signal averaging times.

Defect activation

As well outlined for the case of P_b-type interface centres, paramagnetic point defects are prone to ESR inactivation (at moderate temperatures) through binding with H [45, 47]. In this respect, see, e.g., also E'-type defects in Si/SiO_2 physics [64, 65]. Hence, with the application of some particular fabrication techniques (H-rich ambient) for high-κ film deposition, such as, e.g., ALD, it may be anticipated that in such cases, most point defects will be left effectively passivated (invisible for ESR). In the case of P_b-type defects, the usual method [40] to efficiently detach H is thermal treatment at elevated T (typically ~600°C; ~1 h). Clearly, however, such steps might irreversibly alter the initial physicochemical structure of interfaces and dielectric layers, with embedded defects, thus obfuscating the study of the initial state. Instead, as preferred method, prior to ESR observations, samples were generally subjected to 8.48 eV VUV-irradiation obtained from a Xe-resonant discharge lamp (flux 5×10^{14} photons cm^{-2} s^{-1}; ~5 min; 300 K) in air to photo-dissociate H-terminated dangling bonds [63, 66] and/or possibly unveil non-ideal (strained, weak) bonding. The method has been demonstrated to be most efficient for both oxide and interface defects in Si/SiO_2. The front side interfaces and dielectric layers are found not to be affected by the VUV treatment, except for the ESR defect activation through H photodesorption.

ESR spectroscopy

The ESR results discussed in this work have been obtained by conventional CW absorption-derivative K-band (~20.5 GHz) ESR measurements, performed at 4.3 K under conditions of adiabatic slow passage, as described elsewhere [36]. The applied magnetic field B was rotated in the $(0\bar{1}1)$ Si substrate plane over a range of 0–90° (relative accuracy ~0.1°) with respect to the [100] interface normal n. Particular care was exercised to avoid any (noticeable) signal distortion due to overmodulation of B or incident microwave power (P_μ) induced saturation effects; P_μ was generally restricted to the sub-0.8 nW range. Defect ($S = 1/2$) densities were determined relative to the signal of a co-mounted intensity marker through comparison of the signal intensities (I) obtained by orthodox double numerical integration of the detected derivative-absorption spectra (dP_μ/dB), recorded in one trace. If not indicated explicitly, generally, the attained absolute and relative accuracy is estimated at ~10 and 5%, respectively, for the largest spin densities. The backside of the samples was HF-etched immediately before initiating ESR observations. Typically, an ESR sample bundle comprises ~10 slices.

Nanometre-thin layers of ZrO$_2$ and Al$_2$O$_3$ on (100) Si

Samples

Samples studied were Si/SiO$_x$/ZrO$_2$ and Si/Al$_2$O$_3$/ZrO$_2$ stacks, prepared on low-doped one-side-polished (100) Si wafers (p-type; $n_a \sim 10^{15}$ cm^{-2}; 0.8 mm thick) last-cleaned in aqueous HF (5% by weight) prior to dielectric layer deposition. Uniform stoichiometric ultrathin layers of Al$_2$O$_3$ (~ 0.5–3 nm) and ZrO$_2$ (5–20 nm), as well as ultrathin SiO$_x$ layers, were deposited at 300°C in an ALD reactor [67] using Al(CH$_3$)$_3$ and H$_2$O, and ZrCl$_4$ and H$_2$O as precursors, respectively, at a pressure of ~ 1 torr. The SiO$_x$ layers (~ 5 Å) were grown in H$_2$O at 300°C for 20 min. Details can be found elsewhere [20]. From these wafers, 2×9 mm^2 area ESR samples were cut with their 9 mm edge along a [0 $\bar{1}$ 1] direction. To analyse the thermal stability of deposited layers and interfaces, after initial ESR tests, some samples were subjected to additional annealings at ~ 650 and 800°C in vacuum ($< 4 \times 10^{-6}$ torr) or dry O$_2$ (99.9995%; ~ 1.1 atm) using a conventional resistively heated laboratory facility. A fresh sample was used for each additional thermal step.

Experimental results and analysis

As illustrated in figure 3.2.1, only weak ESR signals could be observed in the as-deposited state, which was somewhat anticipated given the H-rich ambient in the employed ALD apparatus during dielectric film deposition. However, as shown in figure 3.2.1 by the spectra observed at 4.3 K on Si/Al$_2$O$_3$ (0.5)/ZrO$_2$ (5 nm), distinct signals do appear upon VUV irradiation. For the applied magnetic field $B \parallel$ [100] interface normal n (i.e. magnet angle $\phi_B =$ 0°), two clear signals are observed: a prominent one at $g = 2.0059 \pm 0.0001$, with peak-to-peak width $\Delta B_{pp} = 9.7 \pm 0.4$ G, and a weaker one characterized by $g = 2.003\,75 \pm 0.0001$ and $\Delta B_{pp} = 3.75 \pm 0.2$ G, with spin densities ($S = 1/2$) determined as $(5.9 \pm 0.3) \times 10^{12}$ cm^{-2} and $(1.5 \pm 0.2) \times 10^{12}$ cm^{-2}, respectively, corresponding to a [P$_{b0}$]/[P$_{b1}$] intensity ratio $R \sim 3.9$.

The signals exhibit distinct anisotropy, as illustrated by the spectra (*b*) and (*c*) in figure 3.2.1—the property reminiscent of the well-known P$_{b0}$, P$_{b1}$ centres at the thermal Si/SiO$_2$ interface. Conclusive identification of the signals as originating from P$_{b0}$, P$_{b1}$ defects followed from g mapping, is shown for example in figure 3.2.2 for B rotating in the (0 $\bar{1}$ 1) plane. The inferred principal g matrix values for P$_{b0}$ ($g_\parallel = 2.001\,85$, $g_\perp = 2.0081$) and P$_{b1}$ ($g_1 = 2.005\,77$, $g_2 = 2.007\,35$, $g_3 = 2.0022$) match those of previous work [60] on conventional (100) Si/SiO$_2$. As is the case for Si/SiO$_2$, this strict crystallographic correlation with the Si substrate implies that the defects pertain to the nominal (100) Si/Al$_2$O$_3$ interface. This conclusion is confirmed by the study of simple (100) Si/Al$_2$O$_3$ (3 nm) ALCVD structures, where an identical P$_{b0}$/P$_{b1}$ pattern is observed.

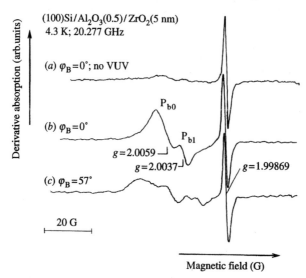

Figure 3.2.1. Derivative-absorption ESR spectra observed at 4.3 K for two directions of B in the $(0\bar{1}1)$ plane on an (100) Si/Al_2O_3 (0.5)/ZrO_2 (5 nm) stack (ALD; 300°C): (*a*) as grown; (*b*), (*c*) after RT VUV irradiation (8.48 eV; ~5 min) to photodissociate H from passivated defects, clearly revealing the presence of P_{b0}, P_{b1} defects at the Si/Al_2O_3 interface. The signal at $g = 1.99869$ stems from a co-mounted Si:P marker sample. Applied modulation field amplitude (100 kHz) was 0.4 G, and incident P_μ ~2 nW.

As shown by the ESR traces in figure 3.2.3, similar observations were made on the (100) Si/SiO_x (0.5)/ZrO_2 (5 nm) structure as for Si/Al_2O_3 (0.5)/ZrO_2 (5 nm), i.e., observation of a prominent P_{b0} signal and a weaker P_{b1} feature. Yet, there are at least two noticeable differences: first, as compared to the latter case, the signals are narrower, i.e., for $B \parallel n$, $\Delta B_{pp} = 6.8 \pm 0.4$ and 3.5 ± 0.2 G, for P_{b0} and P_{b1}, respectively. Second, relative to the P_{b0} signal (density $= 4.3 \times 10^{12}$ cm^{-2}), the P_{b1} signal is weaker ($R \gtrsim 9$). The observed linewidths may be compared with those observed in thermal Si/SiO_2 grown at $T > 900$°C, given as $\Delta B_{pp} = 6.0 \pm 0.2$ and 3.3 ± 0.2 G, for P_{b0} and P_{b1}, respectively [60]. It thus appears that as compared to standard Si/SiO_2, the linewidths observed in the structures here are broadened, the more so for the Si/Al_2O_3 (0.5)/ZrO_2 (5 nm) stack. These data bear out two noteworthy results: (a) P_{b0}, P_{b1} are observed as key defects at the Si/dielectric interface in both types of ALD-grown stack studied; (b) as the P_{b0}, P_{b1} fingerprint is generally unique for the thermal (100) Si/SiO_2 interface, the results reveal that while reassuring for the (100) Si/SiO_x (0.5)/ZrO_2 (5 nm) stack, the initial nominal (100) Si/Al_2O_3 interface is basically also (100) Si/SiO_2 type.

The P_{b0} origin of the principal signal in the ESR spectrum for $B \parallel n$ [curve (*a*) in figure 3.2.3] is also clearly evidenced by spectra (*b*) and (*c*) in

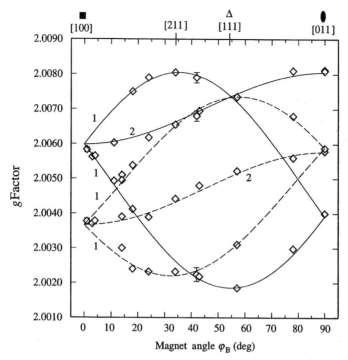

Figure 3.2.2. Angular *g* map of ESR signals observed in ALD-grown (100) Si/Al₂O₃ (0.5)/ZrO₂ (5 nm) for *B* rotating in the (0 $\bar{1}$ 1) plane. The curves represent the various *g* branches previously inferred for the P_{b0} and P_{b1} defects in standard thermal (100) Si/SiO₂, with principal values $g_{\parallel}$ = 2.00185, $g_{\perp}$ = 2.0081 for P_{b0} (solid curves) and g_1 = 2.00577, g_2 = 2.00735, g_3 = 2.0022 for P_{b1} (dashed curves). The close agreement identifies the observed signals as originating from P_{b0}, P_{b1} defects residing at the Si/Al₂O₃ interface. The added numbers indicate expected relative branch intensities.

figure 3.2.3 measured for different *B* directions on (100) Si/SiO$_x$ (0.5)/ZrO₂ (5 nm). For ϕ_B ~25°, the P_{b0} resonance is seen to split into three components, while two components are observed for ϕ_B ~55° (*B* ∥ a ⟨111⟩ direction): a broader one at *g* ~2.0074 (*B* at ~71° with the unpaired Si orbital of three crystallographically equivalent P_{b0} orientations) [60] and a narrow one at *g* ~ 2.0019 (*B* along the corresponding ⟨111⟩ unpaired Si sp₃ hybrid). Again, their respective linewidths, i.e., 8.5 ± 0.5 and 4.1 ± 0.2 G, are broadened as compared to standard thermal Si/SiO₂, typically exhibiting the values 7.5 ± 0.5 and 2.2 ± 0.2 G, respectively.

In addition to excessive broadening, the P_{b0}, P_{b1} line widths also exhibit field angle dependence (see figures 3.2.1 and 3.2.3), a well-known characteristic first clearly quantified [68, 69] for P_b in (111) Si/SiO₂, and later also for P_{b0}, P_{b1} [60]. It originates from mismatch- (stress-) induced local variations in the defect morphology over the various P_{b0} sites, resulting in a

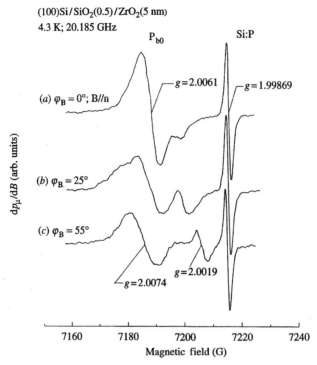

Figure 3.2.3. ESR spectra observed on an (100) Si/SiO$_x$ (0.5)/ZrO$_2$ (5 nm) structure after 300 s VUV irradiation for three directions of **B** in the (0 $\bar{1}$ 1) plane. The angular behaviour of the principal signal in spectrum (*a*), that is, splitting into three and two components (curves (*b*), (*c*), respectively) is consistent with known properties of P$_{b0}$ in (100) Si/SiO$_2$. Applied modulation field amplitude (100 kHz) was 0.4 G; P_μ = 2 nW.

strain-induced (supposedly Gaussian) spread σ_g in g, predominantly in $g_\perp$. With respect to the P$_b$ defect, though not expected in first order, it has been shown [36] that there also exists a spread $\sigma g_\parallel$, albeit an order of magnitude smaller than in $\sigma g_\perp$. Thus, this enables quantification of excessive broadening of the P$_{b0}$ response in terms of variations in g spread. Analysing along the lines outlined elsewhere (neglecting, to a good approximation, any $\sigma g_\parallel$ contribution) [60], the values $\sigma g_\perp$ (P$_{b0}$) = 0.0011 ± 0.0001 and 0.001 75 ± 0.000 15 are obtained for the (100) Si/SiO$_x$ (0.5)/ZrO$_2$ (5 nm) and (100) Si/Al$_2$O$_3$/ZrO$_2$ structures, respectively. As compared to the value [60] $\sigma_{g\perp}$ = 0.000 95 ± 0.000 05 for thermal (100) Si/SiO$_2$, we thus find enhanced stress, in particular for the Si/Al$_2$O$_3$ interface. Physically, this constitutes an example of how probing the salient spectroscopic ESR properties of the P$_b$-type defects may reveal pertinent and insightful information regarding the interface they reside in, which might otherwise appear hard to access by another approach.

Pertinently, it is interesting to note here that in agreement with previous reports [63], the observed P_{b0}, P_{b1} signature, i.e., large P_{b0} system, strongly reduced P_{b1} signal—the large P_{b0}/weak P_{b1} attribute—and broadened lines as compared to standard thermal Si/SiO₂, is the one characteristic of thermal Si/SiO₂ interfaces grown at reduced *T*. This is illustrated in figure 3.2.4, showing an ESR spectrum ($\boldsymbol{B} \parallel \boldsymbol{n}$) measured on a (100) Si/SiO₂ entity grown at 260°C (1 atm O₂; ~160 min), with inferred densities of $[P_{b0}] \sim 5 \times 10^{12}$ cm⁻² and $[P_{b1}] \sim 4 \times 10^{11}$ cm⁻² (i.e. $R \sim 8$) (together with broadened lines), close to the values reported for the (100) Si/SiO$_x$/ZrO₂ stack. For comparison, also shown in figure 3.2.4 is a spectrum (*a*) from a more conventional Si/SiO₂ entity grown at ~800°C.

Influence of post-deposition heating

More interesting information is obtained after post-deposition annealing (PDA). The influence of various PDA treatments is diagrammed in figures 3.2.5 and 3.2.6 in terms of P_{b0}, P_{b1} densities—the latter quantity being chosen as representative for probing the influence of POA treatments. We first address the Si/SiO$_x$ (0.5)/ZrO₂ (5 nm) case (figure 3.2.5). A first noteworthy

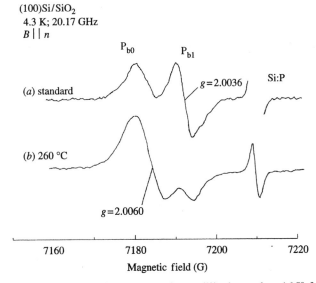

Figure 3.2.4. K-band P_b, P_{b1} ESR spectrum (curve (*b*)) observed at 4.3 K for $\boldsymbol{B} \parallel$ [100] surface normal on thermal (100) Si/SiO₂ (260°C; 1 atm O₂; ~160 min), with $[P_{b0}] \sim 5 \times 10^{12}$ cm⁻². Note the very weak P_{b1} signal intensity as compared to that of P_{b0}. Top spectrum: thermal (100) Si/SiO₂ grown at ~800°C, with P_{b0} and P_{b1} spin densities inferred as ~1.6 and 1.9 × 10¹² cm⁻². Applied magnetic field modulation amplitude was 0.3 G; $P_\mu \sim 0.25$ nW.

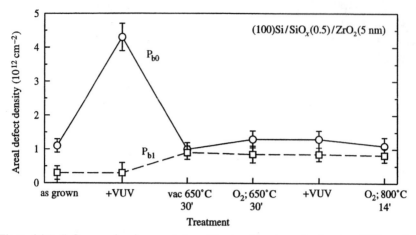

Figure 3.2.5. Influence of various post-deposition treatments on the density of ESR-active P_{b0}, P_{b1} defects at the (100) Si/SiO_x interface in ALD grown (100) Si/SiO_x (0.5)/ZrO_2 (5 nm) stacks.

observation is that upon annealing in vacuum at 650°C, the P_{b0}, P_{b1} densities attain values more typical for standard (relaxed) Si/SiO_2: [P_{b0}] decreases drastically to $\sim 1 \times 10^{12}\,cm^{-2}$, while [$P_{b1}$] grows to $\sim 1 \times 10^{12}\,cm^{-2}$. This is observed to be attendant with narrowing of the line width to standard values, that is 6.0 ± 0.2 and 3.3 ± 0.2 for P_{b0} and P_{b1}, respectively, for **B ∥ n** bearing out reduced strain broadening. Interestingly, in line with the findings for the as-deposited state, these characteristics also comply with those of thermal Si/SiO_2 structures grown at low *T*. Here, it was also observed that upon simple PDA in vacuum ($\sim$750°C), more standard-like narrowed P_b and P_{b1} signals [with increased [P_{b1}] of $\sim(1.5-3) \times 10^{12}\,cm^{-2}$] are retrieved [47, 63].

In a next step, the Si/SiO_x/ZrO_2 structures were subjected to PDA in oxygen at 650 and 800°C. However, this did not result in any noticeable further change in the ESR spectrum, the respective P_{b0}, P_{b1} densities remaining unchanged within experimental error, as shown in figure 3.2.5. This may not be surprising, once a (close to) standard type Si/SiO_2 interface has been realized; as additional oxide is grown, the continuously inwardly re-grown interface remains the one characteristic for higher-*T* oxidation, with attendant standard P_b-type defect pattern.

Interestingly, the influence of PDA on the (100) Si/Al_2O_3 (0.5)/ZrO_2 (5 nm) stack is different from that on (100) Si/SiO_x/ZrO_2. As shown in figure 3.2.6, in terms of P_{b0}, P_{b1} densities, there is now only a modest impact of PDA in vacuum at 650°C (30 min); [P_{b0}] decreases (only) about 30%, while [P_{b1}] remains unchanged at a density of $(1.45 \pm 0.3) \times 10^{12}\,cm^{-2}$. This contrasts with the (100) Si/SiO_x/ZrO_2 case, where the P_{b0}, P_{b1} densities evolve to those of standard thermal Si/SiO_2. In broad terms, this finding for (100) Si/Al_2O_3/ZrO_2 is corroborated by results from previous microstructural

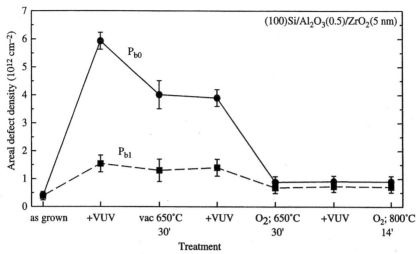

Figure 3.2.6. Evolution of the density of P$_{b0}$, P$_{b1}$ defects at the (100) Si/Al$_2$O$_3$ interface in ALD grown (100) Si/Al$_2$O$_3$ (0.5)/ZrO$_2$ (5 nm) stacks under various post-deposition treatments.

and compositional analysis [16] using nuclear reaction analysis and x-ray photoelectron spectroscopy, reporting all species to remain immobile under PDA in vacuum (~30 s) for T up to 800°C. This was later confirmed in other work [70]. However, while generally in agreement, it is important to note that the present ESR data reveal a limited change does occur during PDA in vacuum. It then likely concerns only a spatially very limited (perhaps on the scale of one atomic layer) modification (e.g. structural rearrangement/relaxation, some additional interfacial SiO$_x$ growth), which has inherently escaped detection in the previous structural/compositional probing. However, as shown in figure 3.2.6, a drastic difference is observed when changing the annealing ambient to oxygen—standard Si/SiO$_2$ [P$_{b0}$], [P$_{b1}$] values are attained after PDA at 650°C in O$_2$ (30 min), with no further change for T increasing to 800°C.

For reasons of completeness, also included in figure 3.2.6 are data observed on the Si/Al$_2$O$_3$/ZrO$_2$ entity annealed at 650°C after additional VUV irradiation. Unchanged defect densities are observed, reassuringly confirming that the applied VUV treatment does not create additional P$_b$-type defects. At the same time, it demonstrates that no P$_b$-type centres are left passivated by H after PDA in vacuum at 650°C.

Discussion

It is thus found that the Si/dielectric interface in Si/SiO$_x$/ZrO$_2$ may be upgraded (closely) to that of standard thermal Si/SiO$_2$ by moderate annealing at $T \gtrsim 650°C$ in vacuum, i.e., no additional O_2 supply needed—an interesting result when put in broader context. Indeed, previous work reported the ZrO_2 layer, under annealing at T up to $\sim 900°C$ (in vacuum or O_2) to be very stable against silicate formation and intermixing [14, 20, 71, 72], hinting that no O is likely extracted from the ZrO_2 network to grow additional SiO_2 interlayer during such thermal treatment. Thus, when combined with the present ESR results, it may be inferred that with an initial SiO_x interlayer ~ 0.5 nm thick, annealing at $T \gtrsim 650°C$ suffices to transform the Si/SiO_x interface in Si/SiO_x (0.5)/ZrO_2 (5 nm) to standard-like without (or with marginal) additional SiO_x interlayer growth.

In the case of the $Si/Al_2O_3/ZrO_2$ stack, by contrast, additional supply of O_2 is required to upgrade the interface, i.e., there is likely a need for additional growth of the SiO_x interlayer. This may indicate that in the as-grown state, the Si/Al_2O_3 interface, although clearly Si/SiO_2 type, is very abrupt indeed, in agreement with previous work [15], and that essentially no thermal reshaping can occur without attendant growth of additional SiO_x layer(s). It suggests that to enable thermally induced upgrading of the interface, a minimal SiO_x thickness $d_{SiO_x}^{min}$ is required; the $Si/SiO_x/ZrO_2$ data suggest $d_{SiO_x}^{min} \gtrsim 0.5$ nm. This may be compared with the value (~ 0.6 nm) inferred for the Si/Si_3N_4 case on the basis of electrical ($C-V$, $I-V$) data [7]. Interestingly, $d_{SiO_x} \sim 0.7$ nm was also concluded to be the minimum layer thickness to obtain bulk (e.g. regarding bandgap) SiO_2 properties [5, 73]. Pertinently, it may also be added that model calculations based on constraint theory bear out the necessity for interposing an ultrathin SiO_2 layer between the Si substrate and high-κ oxides (Al_2O_3, Ta_2O_5, TiO_2) to attain device grade interface properties. [7]

Importantly, one may remark that the very specific ESR-signature of the P_b-type interface defects is invoked here as the criterion for technological interface quality. As amply evidenced before [25, 36, 40], technological interface quality is inextricably connected with a standard P_{b0}, P_{b1} fingerprint, both regarding inherent defect densities and salient ESR features—whatever the manufacturing facility. This criterion finds support from other disciplines. Considering inherent defect densities, from the theoretical side, the P_{b0}, P_{b1} criterion, i.e., existence of an inherent interface defect density (dangling bonds) for standard thermal Si/SiO_2, is also corroborated by the extension of constraint theory [previously developed to describe (ideal) glass formation networks] to Si/dielectric interfaces [74]. On the experimental side, it is consistent with previous notions that with oxide layers deposited at reduced T on Si, low interface trap density can only be obtained by deposition on thin oxides pregrown by standard high-T processing [75]—effectively stating that only thermally grown Si/SiO_2 meets microelectronics requirements—or by providing a sufficient (higher-T) thermal budget.

In a nutshell, the principal ESR results include: (a) the P_{b0}/P_{b1} signature, while reassuring for the (100) $Si/SiO_x/ZrO_2$ case, reveals the initial Si/Al_2O_3 interface also to be basically Si/SiO_2 type. This apparently natural fact may constitute a fundamental bonus as to successful application of high-κ metal oxides in future Si-based devices. (b) Yet, the pertinent interfaces are found to be under enhanced stress, inherent to low-T fabrication. (c) Fortunately, upgrading (close) to that of more standard thermal (100) Si/SiO_2 may be realized through providing appropriate ($T \gtrsim 650°C$) thermal budget, where for the $Si/Al_2O_3/(ZrO_2)$ entity, an O_2 ambient appears mandatory; a minimum SiO_x interlayer thickness ($\gtrsim 0.5$ nm) appears necessary. It heralds the feasibility of carrying over the unexcelled properties of (100) Si/SiO_2 towards the Si/metal oxide entities. But obviously, the insertion of an SiO_2 interlayer will weigh upon the ultimate EOT optimization.

The (100) Si/HfO$_2$ entity

Interest

At present, the search for the most suitable alternative high-κ gate insulator appears dominated by HfO_2. The interest in HfO_2, or, more generally, Hf-based multicomponent dielectrics (such as $HfSi_xO_y$, Hf–Si–O–N films) derives from various factors. It combines a high κ (15–26) with a bandgap of 5.6 eV, with favourable conduction and valence band offsets with respect to Si (2.0 $\pm$ 0.1 and 2.5 $\pm$ 0.1 eV, respectively) [76]. Compared, for example, to a previous high-κ contender, ZrO_2 ($\kappa = 14$–25), HfO_2 exhibits better thermodynamic stability, as would follow from the more negative Gibbs energy of formation for HfO_2 than for ZrO_2 (cf -260 kcal mol^{-1} versus -248 kcal mol^{-1}). The projected stability when in contact with Si [77, 78] would enable one to prevent or minimize interfacial intermixing (SiO_x, silicate, silicide formation) upon post-formation thermal treatments (including S/D/poly-Si activation). Generally, considering processing conditions needed for device fabrication, stability against crystallization upon post-formation heat treatments of deposited amorphous dielectrics is also of concern. Here, an interesting observation is that crystallization of $HfSi_xO_yN_z$ is suppressed for annealing up to 1100°C. [28, 30]. Additionally, HfO_2 has been shown to be compatible with an n$^+$poly-Si gate in a high-mobility MOS field effect transistor [79] and HfO_2 has been scaled down to EOT ≤ 10 Å [80]. The formation of interfacial layers, however, still appears an issue (see, e.g. [31]), and little is known about the formed technologically relevant interface with (100) Si. So, an obvious focus of research concerns the electrical quality of the aimed 'technological' Si/HfO_2 interface and the oxide layer itself. Here ESR research may contribute with detailed atomic information, wherein lies the interest in such study.

From the vast amount of knowledge acquired over the various decades from meticulous intense research on the Si/SiO_2 system [81], it may be expected that, with regard to the grown interfaces, different fabrication (deposition) methods will ensure an impact on interfacial particularities, such as atomic coordination, kinds of occurring defect and densities, and their passivation behaviour in, e.g., H_2. It may generally be expected that a comparative in-depth analysis may be much revealing. Such study is carried out in the present work using ESR, an atomic scale point defect probe, in combination with electrical $C-V$ and $G-V$ measurements, for three types of state-of-the-art deposition method of HfO_2 on (100) Si.

Even if HfO_2-based oxides will not make it ultimately as a workable future alternative gate dielectric, it still constitutes an interesting case study.

Samples

The study concerns three types of sample, labelled A, B, and C, obtained by depositing 5–7 nm thick HfO_2 films on n- and p-type (100) Si substrates. The IMEC pre-deposition surface cleaning was applied in the case of samples A and B. Samples of type A were grown by ALD at 300°C using $HfCl_4$ and H_2O precursors. Samples B were produced by using metallo-organic (MO) CVD at 485°C from tetrakis-diethylaminohafnium and O_2 precursors. For the third type, C, films were deposited on HF-dipped-last Si surfaces in H and C-free conditions by CVD using the nitrato precursor $Hf(NO_3)_4$ (N-CVD). Heating effects were studied by subjecting some samples to 10 min PDA in O_2 (99.9995%; 1.1 atm) at 650°C. Defect passivation was examined by subjecting samples to H_2 (99.9999%; 1.1 atm; 30 min) at 400°C. Then, the samples were split in two batches, one for ESR, the other for electrical analysis. For the latter, MOS structures were fabricated by thermoresistive evaporation of circular 0.4 mm^2 area Au electrodes in high vacuum from a resistively heated W boat. The electrical study implied $C-V$ and ac $G-V$ measurements. Prior to some ESR observations, samples were subjected to the photodissociation treatment (8.48 eV photons).

Experimental results and analysis

Examples of ESR spectra for all three samples A–C observed after the hydrogen photodissociation treatment are shown in figure 3.2.7 for $B \parallel n$. No ESR signals could be discerned in the as-deposited samples A and B, suggesting efficient passivation by H. By contrast, for sample C, the ESR signals observed prior to and after photodissociation treatment appear identical, indicative of negligible presence of H during N-CVD. Apart from the Si:P marker signal at $g = 1.998\,69$, two signals are clearly observed, a larger one at $g = 2.0060 \pm 0.0001$ and one at 2.0036 ± 0.0001. For the

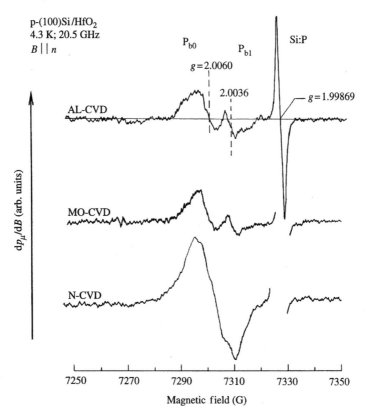

p-(100)Si/HfO₂
4.3 K; 20.5 GHz
$B \parallel n$

P_{b0}   P_{b1}   Si:P

$g = 2.0060$

2.0036

$g = 1.99869$

AL-CVD

MO-CVD

N-CVD

dp_μ/dB (arb. units)

7250   7270   7290   7310   7330   7350

Magnetic field (G)

Figure 3.2.7. Derivative-absorption K-band ESR spectra observed at 4.3 K on (100) Si/HfO₂ entities of types A, B, and C, subjected to RT VUV irradiation (hydrogen photodissociation) prior to observation. The signals observed at $g = 2.0060$ and 2.0036 stem from P_{b0} and P_{b1} interface centres, respectively, while the signal at $g = 1.998\,69$ stems from a co-mounted marker sample. The applied modulation field amplitude was 0.4 G, and incident $P_\mu \sim 0.8$ nW.

applied B orientation ($B \parallel n$), these values are characteristic, respectively, for the P_{b0} and P_{b1} defects in standard (100) Si/SiO₂ and were observed, similarly, at the previously studied interfaces of (100) Si with ALD grown Al₂O₃ and ZrO₂ layers (see the previous section 'Experimental results and analysis'). The identification is confirmed by field angular dependent ESR measurements for B rotating in the (0 $\bar{1}$ 1) plane.

It is interesting to compare the inferred P_{b0}, P_{b1} defect densities, listed in table 3.2.1. Several aspects are worth mentioning. First, the P_{b0} defect density appears to be substantially larger than typical for standard thermal (100) Si/SiO₂ ([P_{b0}] $\sim 1.0 \times 10^{12}$ cm⁻²). Two, the observation of the same defect densities, within experimental accuracy, in samples A and B suggests that neither the variation in deposition temperature in the range 300–485°C

Table 3.2.1. Sample properties and density of ESR-active interface defects observed after VUV hydrogen photodissociation treatment in differently prepared (100) Si/HfO$_2$ entities.

Sample no.	HfO$_2$ deposition method	Deposition temperature (°C)	[P$_{b0}$] (10^{12} cm^{-2})	[P$_{b1}$] (10^{12} cm^{-2})
A	ALCVD	300	2.3 ± 0.4	1.6 ± 0.2
B	MOCVD	485	2.4 ± 0.4	1.2 ± 0.2
C	N-CVD	350	8.1 ± 0.3	0.9 ± 0.2

nor the presence of particular impurities (e.g. Cl in sample A and carbon in sample B) have a significant effect on the incorporated defect density. Given the H- and C-poor conditions of the N-CVD process, the observed considerably enhanced P$_{b0}$ density in sample C may then likely be related to the lack of H during deposition and/or incorporation of N in the near-interfacial oxide layers. Third, the P$_{b0}$ density in sample A (ALD) is found to be about 50–60% lower than in ALD samples with Al$_2$O$_3$ and ZrO$_2$ layers (cf figures 3.2.5 and 3.2.6). In one interpretation, this may indicate that the initial IMEC cleaning of the Si surface may yield a better Si/oxide interface than starting from an oxidation in H$_2$O at 300°C of HF-last Si, used in the latter case. Fourth, the P$_{b1}$ density appears less sensitive to the pre-deposition surface processing than of P$_{b0}$. The currently observed P$_{b1}$ density is close to that ([P$_{b1}$] = (1.5 ± 0.2) × 10^{12} cm^{-2}) observed in as-grown (100) Si/Al$_2$O$_3$ after VUV treatment (cf. figure 3.2.6), i.e., somewhat enhanced as compared to that of standard (100) Si/SiO$_2$, indicating enhanced strain. The P$_{b1}$ defect, altogether, seems to play a less important role.

Influence of post-deposition thermal treatment

The effect of PDA on the structures is also interesting. ESR measurements show that PDA in O$_2$ (1.1 atm; 650°C; 10 min) results in a significant reduction (spectra not shown) of the P$_b$-type defect densities, now approaching those of standard thermal (100) Si/SiO$_2$, i.e., [P$_{b0}$], [P$_{b1}$] ∼ 1 × 10^{12} cm^{-2}. In this regard, the (100) Si/HfO$_2$ system behaves much the same as the previously studied (100) Si/Al$_2$O$_3$ and (100) Si/ZrO$_2$ entities. The impact of PDA in O$_2$ at 650°C (10 min) on electrical properties has been studied in conjunction with ESR by *C–V* and *G–V* observations. Results are shown in figure 3.2.8 for sample C, before and after subjection to PDA. The peaked features in the data curves are much less pronounced after PDA. A significant reduction is observed in the electrically active interface traps, in strong correlation with the large decrease in P$_{b0}$ density, suggesting, as expected [39, 40], the latter to be the dominant source of fast interface states in (100) Si/HfO$_2$.

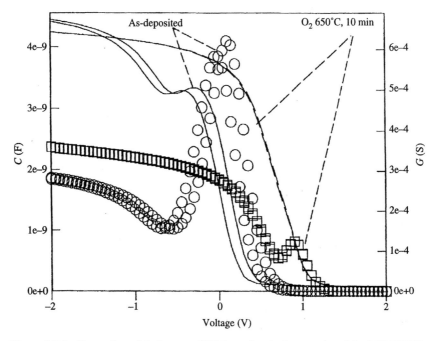

Figure 3.2.8. Illustration of the impact of PDA on electrical properties of the (100) Si/HfO₂ entity: observed 100 kHz $C-V$ (lines) and $G-V$ (symbols) curves on as-deposited p-type (100) Si/HfO₂/Au samples (MOS) of type C (solid lines, circles) and after 10 min PDA in O_2 (1.1 atm) at 650°C (dashed line, squares).

This conclusion is further corroborated by ESR results from hydrogen passivation experiments; no measurable P_{b0}, P_{b1} density is left, indicating efficient passivation. In agreement, the electrical $C-V$ and $G-V$ measurements after heating in H_2 also show the disappearance of all features (peaks, humps) associated with the presence of interface states. Thus, the passivation of interface defects in (100) Si/HfO₂ appears to proceed similarly as in standard Si/SiO₂, from where it may be concluded that neither the HfO₂ layer nor the N-doped interfacial oxide interlayer between Si and HfO₂ present in sample C do observably hinder the passivation of interface traps by hydrogen. A similar conclusion was obtained previously for a 14.7 nm HfO₂ top layer on (111) Si [59].

The information on passivation may appear useful. Indeed, one may perhaps not *a priori* expect the passivation in H_2 ambient to be equally efficient *per se* over the various types of Si/insulator stack studied, even in the case of nm thin oxide layers. There may be several reasons. As outlined recently [50], one is that due to the unavoidable simultaneous competition between passivation and the reverse dissocation reactions, the attained passivation efficiency depends on the type of interface established, i.e., the 'strain' (relaxation) state of the interface, through affecting the spreads in pertinent activation energies for both chemical reactions (cf equations (3.2.1) and (3.2.2)). With the limited amount of work done, we know little about the nature of the grown new interfaces, and occurring strain. However, from initial ESR work, we do know they are generally under enhanced strain (hence also enlarged spreads in activation energies), ascribed to the generally low deposition temperatures (strongly reduced thermal budget) used in fabricating the Si/high-κ entities. For completeness, we should add that the passivation efficiency in the present work—carried out in 1.1 atm H_2—may be expected naturally to be somewhat higher than attained with the conventional anneal in FG, viz., 100% H_2 *vis-à-vis* 5–30% H_2 in N_2.

There are several conclusions from this study on (100) Si/HfO_2. First, combined with previous results on the (100) Si/Al_2O_3 and (100) Si/ZrO_2 entities, the present ESR data on (100) Si/HfO_2 show that for all three types of metal oxide studied, the basic defects occurring at the (100) Si/dielectric interface are identical to those found in conventional thermally grown (100) Si/SiO_2—the established interfaces are effectively Si/SiO_2-like. Two, upon applying PDA in O_2 with sufficiently high thermal budget, the initial Si/SiO_2 interface character between Si and HfO_2, or, more generally, high-κ metal oxides, is enhanced—the differences in interface defect densities observed in the as-deposited Si/high-κ insulator with thermal Si/SiO_2 disappear, the interface now more approaching device grade quality in terms of interface traps.

Conclusions and final remarks

Results have been discussed of ESR probing of paramagnetic defects in (100) Si/dielectric entities with nm thin layers of ZrO_2, Al_2O_3, and HfO_2 deposited at 300–485°C by state-of-the-art CVD-based methods. As a main finding, the principal defects observed, almost exclusively, in all investigated stacks include a prominent P_{b0} and a weaker P_{b1} signal. This almost inseparable P_b-type defect duo generally constitutes the unique ESR fingerprint of a (100) Si/SiO_2 interface—on the basis of this hallmark, investigated Si/high-κ interfaces are also revealed to be basically Si/SiO_2 type. The defects are likely to be fully exposed after appropriate VUV-induced hydrogen photo-desorption.

The results indicate the phenomenon of the formation of some kind of Si/SiO_x to be generally 'endemic' to interfaces between Si and a metal oxide. But in view of the well-known electrical quality of the Si/SiO_2 interface, this apparently natural fact might constitute a fundamental bonus for successful application of high-κ metal oxides in future Si-based devices in terms of realizing interfaces of technologic quality. But of course, insertion of low-κ SiO$_2$, either inadvertently or deliberately, is at the cost of the ultimate EOT. So, the potential merit of this phenomenon will much depend on whether the required minimum SiO$_{2(x)}$ interlayer thickness can be reconciled with the overall low EOT requirement.

Also, as in the present case we are dealing with more complex (multilayer) structures, with more chemical elements involved than 'simple' Si/SiO$_2$, the thermodynamic stability of such initially optimally construed high-κ structures remains to be found out.

As exposed by the spectroscopic properties of the P$_{b0}$, P$_{b1}$ defects, the pertinent interfaces in the as grown state are under enhanced stress, inherent to low-T fabrication. Upgrading to more standard (100) Si/SiO$_2$ properties can be achieved through additional annealing in appropriate ambient with sufficient thermal budget.

Generally, as to the probing of paramagnetic defects, ESR spectroscopy of the Si/high-κ metal oxide structures turns out to wind up in another P$_b$-type interface defect saga, where we can beneficially draw upon the immense amount of knowledge accumulated for standard thermal Si/SiO$_2$. This may be very advantageous with respect to getting insight into the newly fabricated Si/dielectric structures.

So far, almost only P$_b$-type defects, pertaining to the Si/dielectric interface, have been observed and reported on, with little information on other defects possibly present. Obviously, a next step in the future investigation will concern detection of such defects. For clarity, the fact that ESR fails to trace other defects does not necessarily mean none are present; they may, e.g., just reside in the wrong (diamagnetic) spin state or their responses may be broadened beyond detection. Given the basically altered initial interface nature (e.g. different chemical elements involved), there may occur additional defects at the key Si/dielectric interface. In addition, dealing with Si/multilayer structures, there may occur defects related to other layer-to-layer interfaces. But definitely, it will also imply defects in the bulk of the deposited new high-κ and other layers. In this regard, there is no reason to expect a different behaviour from the conventional SiO$_2$ or SiO$_x$N$_y$ dielectrics, where several pertinent defects, some of crucial electrical impact, have been revealed and identified—much depending on the applied fabrication method, the dielectric formation is never ideal. Here, we should add that the properties of the deposited high-κ layers, being necessarily very thin, may deviate from the bulk ones, as already evidenced. This should be put in the light of the growing intensification and diversification of the high-κ gate research, where to promote chemical and phase (amorphicity) stability, alternative multicomponent dielectrics, such as silicates, silicites, nitrided silicates, etc) are investigated. It appears we have just entered an exciting, yet exacting activity.

As may appear from the present work, the field of applying ESR spectroscopy in the expanding Si/high-κ insulator world is just emerging, with only a few works so far carried out. Without doubt, as was the case with the exploration of the epic Si/SiO_2 entity, it will soon expand, with application of improved and extended ESR spectroscopy; conventional ESR observations will be expanded to multiple observational frequencies and complemented by related techniques of higher sensitivity and/or different specificity, such as spin-dependent electrically detected ESR. Quite naturally, upon manufacturing of new Si/high-κ insulator entities, physical micro-structural analysis is carried out first, followed by, most importantly, electrical characterization; in turn, this is complemented by fundamental atomic probing. Much is to be expected also from ESR analysis in close conjunction with electrical study.

Acknowledgments

The authors are indebted to M. M. Heyns and S. De Gendt (IMEC, Belgium) and S. A. Campbell (University of Minnesota, USA) for providing samples. Stimulating discussions with M Houssa (University of Provence, France, and IMEC, Belgium) are gratefully acknowledged. The Flanders Fund for Scientific Research is acknowledged for financial support.

References

[1] See *The International Technology Roadmap for Semiconductors* 2001 edn (San Jose, CA: Semiconductor Industry Association)
[2] Buchanan D A and Lo S H 1997 *Microelectron. Eng.* **36** 13
[3] Lo S H, Buchanan D A, Taur Y and Wang W 1997 *IEEE Electron Device Lett.* **18** 209
[4] Green M, Sorsch T, Timp G, Muller D, Weir B, Silverman P, Moccio S and Kim Y 1999 *Microelectron. Eng.* **48** 25
[5] Wilk G D, Wallace R M and Anthony J M 2001 *J. Appl. Phys.* **89** 5243
[6] Green M L, Gusev E P, Degraeve R and Garfunkel E 2001 *J. Appl. Phys.* **90** 2057
[7] Lucovsky G, Wu Y, Niimi H, Misra V and Philips J C 1999 *J. Vac. Sci. Technol.* **B17** 1806
[8] Lu H C, Yasuda N E, Garfunkel E, Chang J P, Opila R L and Alers G 1999 *Microelectron. Eng.* **48** 287
[9] Robertson J 2000 *Vac. Sci. Technol.* **B18** 1785
[10] Peacock P W and Robertson J 2002 *J. Appl. Phys.* **92** 4712
[11] Lucovsky G 2001 *International Workshop on Gate Insulator* eds S Ohmi, F Fujita and H S Momose (Tokyo: Jap. Soc. of Appl. Phys.) p 14
[12] Foster A S, Lopez-Gijo F, Shluger A L and Nieminen R M 2002 *Phys. Rev.* **B65** 174117
[13] Riganese G M, Detraux F, Gonze X and Pasquarello A 2002 *Phys. Rev. Lett.* **89** 117601
[14] Copel M, Gribelyuk M and Gusev E P 2000 *Appl. Phys. Lett.* **76** 436
[15] Gusev E P, Copel M, Cartier E, Baumvol I J R, Krug C and Gribelyuk M A 2000 *Appl. Phys. Lett.* **76** 176

[16] Krug C R, da Rosa E B O, de Almeida R M C, Morais J, Baumvol I J, Salgado T D M and Stedile F C 2000 *Phys. Rev. Lett.* **85** 4120

[17] Gusev E G, Cartier E, Buchanan D A, Gribelyuk M, Copel M, Okorn-Schmidt H and D'Emic C 2001 *Microelectron. Eng.* **59** 341

[18] Gusev E P 2000 *Defects in SiO$_2$ and Related Dielectrics; Science and Technology* NATO Science Series eds G Pacchioni, L Skuja and D L Griscom (Dordrecht: Kluwer) p 557

[19] Fukumoto H, Morita M and Osaka Y 1989 *J. Appl. Phys.* **65** 5210

[20] Houssa M, Naili M, Zhao C, Bender H, Heyns M M and Stesmans A 2001 *Semicond. Sci. Technol.* **16** 31

[21] Houssa M, Tuominen M, Naili N, Afanas'ev V, Stesmans A, Haukka S and Heyns M M 2000 *J. Appl. Phys.* **87** 8615

[22] Xu Z, Houssa M, De Gendt S and Heyns M 2002 *Appl. Phys. Lett.* **80** 1975

[23] Carter R J *et al* 2001 *International Workshop on Gate Insulator 2001* eds S Ohmi, F Fujita and H S Momose (Tokyo: Japan. Soc. Appl. Phys.) p 94

[24] Yamada H, Shimizu T and Suzuki E 2002 *Japan. J. Appl. Phys.* **41** L368

[25] See, for example, Helms R and Poindexter E H 1998 *Rep. Prog. Phys.* **83** 2449

[26] Lenahan P M and Dressendorfer P V 1983 *J. Appl. Phys.* **54** 1457

[27] Kirsch P D, Kang C S, Lozano J, Lee J C and Ekerdt J G 2002 *J. Appl. Phys.* **91** 4353

[28] Visokay M R, Chambers J J, Rotondaro A L P, Shanware A and Colombo L 2002 *Appl. Phys. Lett.* **80** 3183

[29] Cho M-H, Roh Y S, Whang N, Jeong K, Choi H J, Nam S W, Ko D-H, Lee J H, Lee N I and Fujihara K 2002 *Appl. Phys. Lett.* **81** 1071

[30] Quevedo-Lopez M A, El-Bouanani M, Kim M J, Gnade B E, Wallace R M, Visokay M R, LiFatou A, Bevan M J and Colombo L 2002 *Appl. Phys. Lett.* **81** 1074

[31] Lee J H and Ichikawa M 2002 *J. Appl. Phys.* **92** 1929

[32] Afanas'ev V, Houssa M, Stesmans A and Heyns M M 2001 *Appl. Phys. Lett.* **78** 3073

[33] Ahn Y S, Ban S H, Kim K J, Kang H, Yang S, Roh Y and Lee N-E 2002 *Japan. J. Appl. Phys.* **41** 7282

[34] Poindexter E H, Caplan P, Deal B and Razouk R 1981 *J. Appl. Phys.* **52** 879

[35] Brower K L 1983 *Appl. Phys. Lett.* **43** 1111

[36] Stesmans A 1993 *Phys. Rev.* **B48** 2418

[37] Poindexter E and Caplan P 1983 *Prog. Surf. Sci.* **14** 211

[38] Poindexter E H 1989 *Semicond. Sci. Technol.* **4** 961

[39] Gerardi G J, Poindexter E H, Caplan P J and Johnson N M 1986 *Appl. Phys. Lett.* **49** 348

[40] Stesmans A and Afanas'ev V V 1998 *Phys. Rev.* **B57** 10030

[41] Stesmans A, Nouwen B and Afanas'ev V 1998 *Phys. Rev.* **B58** 15801

[42] Stirling A, Pasquarello A, Charlier J-C and Car R 2000 *Phys. Rev. Lett.* **85** 2773

[43] Futako W, Nishikawa M, Yasusa T, Isoya J and Yamasaki S 2001 *International Workshop on Gate Insulator 2001* eds S Ohmi, F Fujita and H S Momose (Tokyo: Japan. Soc. Appl. Phys.) p 130

[44] Stesmans A and Afanas'ev V V 1998 *J. Vac. Sci. Technol.* **B16** 3108

[45] Brower K L 1988 *Phys. Rev.* **B38** 9757

[46] Brower K L 1990 *Phys. Rev.* **B42** 3444

[47] Stesmans A 1996 *Appl. Phys. Lett.* **68** 2076; Stesmans A 1996 *Appl. Phys. Lett.* **68** 2723

[48] Stesmans A 2000 *Phys. Rev.* **B61** 8393

[49] Stesmans A 2000 *J. Appl. Phys.* **88** 489

[50] Stesmans A 2002 *J. Appl. Phys.* **92** 1317
[51] Michaelashvili V, Betzer Y, Prudnikov I, Orenstein M, Ritter D and Eisenstein G 1998 *J. Appl. Phys.* **84** 6747
[52] See, for example, Eisenbeiser K *et al* 2000 *Appl. Phys. Lett.* **76** 1324
[53] Stesmans A and Afanas'ev V V 2001 *J. Phys. Condens. Matter.* **13** L1
[54] Stesmans A and Afanas'ev V V 2002 *Appl. Phys. Lett.* **80** 1957
[55] Cantin J L and von Bardeleben H J 2002 *J. Non-Cryst. Solids* **303** 175
[56] See, for example, Taylor P C 1984 *Semiconductors and Semimetals* vol 21C ed J I Pankove (New York: Academic) p 99
[57] Poindexter E H and Caplan P J 1987 *J. Vac. Sci. Technol.* **A6** 1352
[58] Lemke B P and Haneman D 1987 *Phys. Rev.* **B17** 1893
[59] Kang A Y, Lenahan P M, Conley J F Jr and Solanski R 2002 *Appl. Phys. Lett.* **81** 1128
[60] See, for example, Stesmans A and Afanas'ev V V 1998 *J. Appl. Phys.* **83** 2449
[61] Yount J T, Lenahan P M and Wyatt P W 1995 *J. Appl. Phys.* **77** 699
[62] See, for example, Orbach R and Stapleton M J 1972 Electron spin–lattice relaxation *Electron Paramagnetic Resonance* ed S Geschwind (New York: Plenum)
[63] Stesmans A and Afanas'ev V V 2000 *Appl. Phys. Lett.* **77** 1469
[64] See, for example, Griscom D 1985 *J. Appl. Phys.* **58** 2524 and references therein
[65] Edwards A H 1995 *J. Non-Cryst. Solids* **187** 232
[66] Pusel A, Wetterauer U and Hess P 1998 *Phys. Rev. Lett.* **81** 645
[67] Hot wall flow type F-450 reactor, ASM Microchemistry Ltd., Finland
[68] Brower K L 1986 *Phys. Rev.* **B33** 4471
[69] Stesmans A and Braet J 1986 *Insulating Films on Semiconductors* eds J J Simone and J Buxo (Amsterdam: North-Holland) p 25
[70] Copel M, Cartier E, Gusev E P, Guha S, Bojarczuk N and Poppeller M 2001 *Appl. Phys. Lett.* **78** 2670
[71] Jeon T S, White J M and Kwong D L 2001 *Appl. Phys. Lett.* **78** 368
[72] Watanabe H 2001 *Appl. Phys. Lett.* **78** 3803
[73] Tang S, Wallace R M, Seabaugh A and King-Smith D 1998 *Appl. Surf. Sci.* **135** 137
[74] Lucovsky G, Wu Y, Niimi H, Misra V and Phillips J C 1999 *Appl. Phys. Lett.* **74** 2005
[75] Nguyen S V, Dobuzinsky D, Dopp D, Gleason R, Gibson M and Fridmann S 1990 *Thin Solid Films* **193/194** 595
[76] Afanas'ev V V, Stesmans A, Chen F, Shi X and Campbell S A 2002 *Appl. Phys. Lett.* **81** 1053
[77] Lee B H, Kang L, Nieh R, Qi W J and Lee J C 2000 *Appl. Phys. Lett.* **80** 2135
[78] Wilk G D and Wallace R M 2000 *Appl. Phys. Lett.* **76** 11
[79] Campbell S A, Ma T Z, Smith R, Gladfelter W L and Chen F 2001 *Microelectron. Eng.* **59** 361
[80] Lee B H, Choi R, Kang L, Gopalan S, Nieh R, Onishi K, Jeon Y, Qi W-J, Kang C S and Lee J C 2000 *IEDM Tech. Dig.* **2000** 39
[81] Edwards A H and Fowler W B 1999 *Microelectron. Reliab.* **39** 3

Chapter 3.3

Band alignment at the interfaces of Si and metals with high-permittivity insulating oxides

V V Afanas'ev and A Stesmans

Introduction

Thin insulating layers of metal oxides with dielectric constant higher than that of SiO_2 ($\kappa = 3.9$) are currently considered as candidate gate dielectric materials for metal–oxide–silicon (MOS) devices capable of reducing the direct tunnelling of electrons [1–5]. To suppress the tunnelling effect between metal and silicon, the energy barriers for electrons (Φ_e) and holes (Φ_h) at the metal/oxide and Si/oxide interfaces must be sufficiently high. Consequently, these barrier heights, together with the dielectric constant of an oxide, represent the parameters of primary importance for selection of an insulator for device application. So far, due to the lack of experimental data, the interfacial barriers were evaluated theoretically using the *bulk* parameters of solids involved [6, 7]. However, they may not necessarily reflect the properties of *thin films*. Aluminium oxide (Al_2O_3) provides a clear example of such differences: the reduction of the oxide thickness from 5.5 to 3.5 nm was reported to result in a 0.5 eV Al/Al_2O_3 barrier lowering [8]. Moreover, fabrication of laterally uniform ultra-thin insulators requires a low process temperature to avoid any reaction and/or intermixing with the Si substrate which may affect the barrier height. For instance, the Al/Al_2O_3 interface barrier height increases from 1.8–2.2 eV for the oxides formed at room temperature [9–13] to 2.45 eV for thermal oxidation in air at 300°C [14] and 2.9 eV for thermal oxidation in O_2 at 425°C [15], to reach 3.05 and 3.2 eV for the Al_2O_3 layers deposited at 850 [16] and 900°C [17], respectively. Worth mentioning here is also the fact that the amorphous low-temperature alumina has the bandgap width (E_g) of only 5.1 eV [13] as compared to

217

$E_g = 8.8 \, \mathrm{eV}$ for bulk Al_2O_3 crystal. Such an uncertainty in Φ_e and E_g leads not only to large inaccuracy in the evaluation of the electron tunnelling probability, but also to a largely undefined metal–silicon work-function difference (ϕ_{ms}) thus hampering determination of the electric field strength and the fixed charge density in the oxide from the electrical measurements.

The knowledge of barrier heights is also needed to analyse the electron transport mechanism(s) across the MOS insulator. While the band-to-band tunnelling is highly sensitive to the energy offset, it is also possible that the traps in the oxide will mediate the electron transport between the Si and metal, as happens in the Si MOS structures subjected to electrical stress or irradiation [18–20]. In the latter case, the most important parameter might be the trap energy depth which determines the trap-to-trap transition and trap-to-band emission probabilities [21, 22]. Clarification of the current transport mechanism is of vital importance for optimization of the gate dielectric technology: were the trap-assisted injection to be dominant, *bulk* properties of the oxide must be improved; otherwise, *interface* engineering will be needed to reduce the gate leakage current.

Barrier determination by internal photoemission spectroscopy

The internal photoemission (IPE) phenomenon represents an optically stimulated emission of a charge carrier (electron or hole) from one solid (emitter) into another solid over the interface energy barrier, as illustrated in figure 3.3.1 for a MOS structure biased by voltage of different polarity. Similarly to the case of electron photoemission into vacuum (the external photoemission), the IPE can be described as a multi-step process that includes the optical excitation of a charge carrier in the emitter, its transport to the surface of the emitter, surmounting the potential barrier of a height Φ, and transport in the second solid [23]. General analysis of the IPE quantum yield (Y) dependence on the exciting photon energy $h\nu$ indicates that it represents a power function of difference between the photon energy and the barrier height, $Y \sim (h\nu - \Phi)^p$, where power p is uniquely determined by the shape of energy distribution of excited carriers at the surface of the emitter [24]. Then Φ can be found by linear extrapolation of the $Y^{1/p} - h\nu$ dependence to $Y = 0$ providing the most straightforward way of interface barrier height determination.

The inaccuracy of the IPE method is primarily determined by the spectral resolution of the illumination system because it gives the largest uncertainty in the extrapolation procedures. For the typical spectral width of a monochromator slit of $0.02 \, \mathrm{eV}$ the threshold determination error appears to be about $0.05 \, \mathrm{eV}$ if the IPE signal is detected starting from 0.1–$0.15 \, \mathrm{eV}$

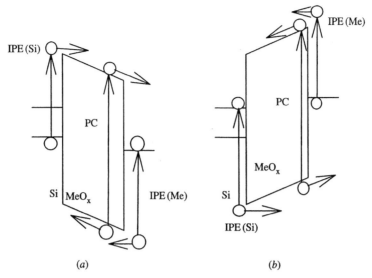

Figure 3.3.1. Optically excited electron transitions in MOS structure biased with positive (*a*) and negative (*b*) voltage applied to the metal electrode. The transitions related to IPE and PC are indicated.

above the spectral threshold. The additional contribution to the error may come from the uncertainty in the p-value arising from deviation of the excited carrier energy distribution from ideally rectangular ($p = 2$) or triangular ($p = 3$) shape [24]. This might happen, for instance, when studying IPE from transition metals with electron orbitals of different types, e.g. s, p and d states, contributing to the density of states (DOS) near the Fermi level. In this case the uncertainty in p-value may increase the Φ determination error to 0.1–0.15 eV [17]. However, one can improve the accuracy of the procedure by considering p as an independent experimental parameter describing the energy distribution of charge carriers and determine its value, not necessary an integer, from the experimental $Y–h\nu$ curve.

Experimental realization of IPE into an insulator requires illumination of the interface with monochromatic light with simultaneous application of electric field to extract the injected electrons/holes. From the point of view of yield determination, the most direct way of counting the injected charge carriers consists in measuring photocurrent they generate when moving to the opposite collecting electrode (cf figure 3.3.1). In the case of thin insulators on Si this naturally leads to the configuration of the test sample as a structure with optical input through semitransparent (thin metal layer) or transparent (electrolyte) conducting top electrode. Though the latter provides lower light absorption in the top electrode [13], the metal gates give more experimental flexibility including possible extension of the optical measurement range to the deep UV region and choice of the metal workfunction. Therefore, we have used MOS capacitors defined by thermal evaporation of semitransparent (15 nm thick) metal electrodes onto the oxides. A set of metals (Mg, Al, Cu, Ni, Au) with substantial difference in electronegativity was used to allow reliable separation between the photocurrents related to IPE from metal from that caused by IPE of charge carriers of opposite sign from the Si electrode of MOS structure.

The photocurrent measurements in MOS structures allow us, in addition to IPE, to analyse the intrinsic and defect-related photoconductivity (PC) of the insulating material, which is also shown in figure 3.3.1. In contrast to the IPE, the intrinsic PC is insensitive to the direction of electric field in the insulator or to the material of the metal electrode. In its turn, for the same strength of electric field, the IPE current is insensitive to the insulator thickness if the carrier trapping probability is low [23], while the PC current increases with the insulator thickness (d) in the range $d < 1/\alpha$, where α is the optical absorption coefficient in the analysed spectral range. Therefore, analysis of the photocurrent dependence on insulator thickness can also be of help in attributing it to the IPE or the PC process.

To complete the Si/oxide or metal/oxide interface band diagram determination, the insulator bandgap width E_g can be determined from spectral dependences of intrinsic PC [25]. Assuming the mean free path of photogenerated charge carriers to be larger than the thickness of the insulating layer, the PC current will be proportional to the joint density of states (J-DOS) for optical transitions. In amorphous insulators this dependence is experimentally observed to be close to $I_{ph} \sim (h\nu - E_g)^2$ [23, 25–27] which allows determination of E_g by extrapolation of the current normalized to the incident photo-flux n_{ph} to zero using $(I_{ph}/n_{ph})^{1/2}$–$h\nu$ plot. Again, as in the case of IPE from metals, the power of the PC spectral dependence can be considered as an independent parameter corresponding to the energy dependence of J-DOS (intrinsic PC) or to the energy dependence of photoionization cross-section (defect-related PC). Worth noticing here is that for the intrinsic PC, i.e., for the excitation of a valence electron to the oxide conduction band, the J-DOS represents convolution of electron state energy distributions in the valence and conduction bands. Therefore, the use by several authors [28–30] of the same functional (linear) dependence for both the valence band DOS and the J-DOS involved in the electron energy losses is only permitted if the conduction band DOS is the δ-function $\delta(E - E_c)$, which means physically *the absence of a conduction band*. Obviously, the bandgap width and band offset evaluation using such an unrealistic assumption are unreliable. Indeed, the elaborate analysis of the Al_2O_3 electron energy loss spectra indicates that the loss function substantially differs from the linear dependence on the loss energy [31].

Deposition of high-permittivity metal oxides on (100) Si

To ensure the outmost lateral uniformity of the insulating layers, one can use the atomic-layer deposition (ALD) providing an atomically controlled growth mode through saturation of each adsorbed monolayer of the metal oxide atoms [32, 33] (see also the chapter of M. Ritala in the present book). The Al_2O_3 (0.5–100 nm), ZrO_2 (3–100 nm) and HfO_2 (5–100 nm) layers were deposited on clean (HF dip prior to ALD) or SiO_2-covered low-doped (100) Si substrates (n- or p-type, n_d, $n_a \approx 1 \times 10^{15}$ cm^{-3}) at 300°C using H_2O, $Al(CH_3)_3$, $ZrCl_4$ and $HfCl_4$ precursors, respectively, as described elsewhere [32, 34]. ALD-grown Al_2O_3/ZrO_2 stacks and Al–Hf mixed oxides were also analysed. After deposition, some samples were subjected to post-deposition anneal (PDA) in dry O_2 or in a mixture of $N_2 + 5\%$ O_2 at temperatures in the range 500–1000°C to grow additional oxide at the interface between Si and the Al_2O_3 or ZrO_2 layers.

For the sake of comparison, thin layers of HfO_2 (5–120 nm) deposited on (100) Si using two other CVD processes were also studied—the CVD from metallo-organic precursor tetrakis-diethylaminohafnium and O_2 at 485°C (MO-CVD) and the deposition from the nitrato precursor $Hf(NO_3)_4$ at 350°C (N-CVD) [35] (see also the chapter of S. Campbell in this book). The latter process involves no carbon- or hydrogen-containing species, which allows us to evaluate the possible impact of these impurities on electronic properties of the deposited oxide. For electrical measurements, MOS capacitors of 0.5 mm^2 area were defined by thermal evaporation of semitransparent (15 nm thick) metal (Mg, Al, Cu, Ni, Au) electrodes onto the oxide in high ($\approx 10^{-6}$ torr) vacuum from a resistively heated W boat. The same metal electrodes were also evaporated onto 4–5 nm thick thermal SiO_2 layers grown on (100) Si at 800°C. No post-metallization anneal was performed to ensure a minimal metal/oxide chemical interaction. MOS structures were characterized at room temperature by dc current–voltage ($I–V$), capacitance–voltage ($C–V$) (10^2–10^6 Hz) and by IPE. The $I–V$ curves of the MOS structures were measured not only in darkness but also under excitation with photons from an Ar$^+$ ion laser ($h\nu = 2.41–2.73$ eV) to reveal the photon-stimulated electron transitions [36]. From the accumulation capacitance of MOS structures with as-deposited uniform Al_2O_3, ZrO_2 and HfO_2 layers, the low-frequency relative dielectric constants were found to be 8, 15 and 17–20 (depending on deposition method), respectively. Upon oxidation, the dielectric constants of both the interfacial silicon oxide and the high-permittivity oxide were found to increase [34]. The IPE spectral curves were measured in the photon energy range $h\nu = 2–5$ eV (the photon energy was always kept well below the insulator bandgap width—5.4 eV for ALD ZrO_2 [26], 5.6 eV for N-CVD HfO_2 [37]) with spectral resolution of 2 nm. IPE yield Y was defined as photocurrent normalized to the incident photon flux [23, 38, 39]. The density of electrons injected during the IPE experiments was kept below $10^{12}\,\mathrm{cm}^{-2}$ to avoid substantial trapping in oxide. The $C-V$ measurements after the IPE experiments were used to monitor the charging.

Bandgap width determination from PC measurements

The PC spectra of samples with six different high-permittivity oxide insulators of 6–20 nm thickness are compared in figure 3.3.2. In addition to the above-mentioned Al_2O_3, ZrO_2 and HfO_2 layers obtained using ALD, the results are also shown for ALD Ta_2O_5 and TiO_2, and for Nb_2O_5 grown using the liquid source mist-CVD [40]. Being plotted in the $(I_{ph}/n_{ph})^{1/2}-h\nu$ co-ordinates, the spectra show well-defined linear portions, which allows determination of the oxide bandgap as the intrinsic PC threshold. The obtained results are summarized in table 3.3.1 for the as-deposited oxides on (100) Si. They importantly indicate that, at least for the oxides with $E_g > 5\,\mathrm{eV}$, the bandgap width of the ALD films is substantially lower than in the bulk crystals of the same oxides. For instance, the obtained 5.4 eV bandgap of the ALD ZrO_2 can be compared to 6.1, 5.78 and 5.83 eV direct

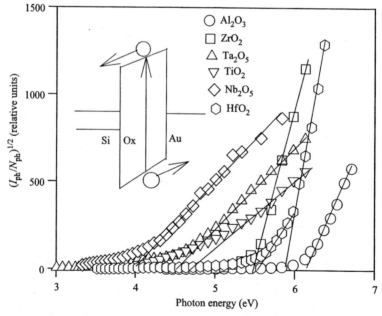

Figure 3.3.2. PC spectra of (100) Si/oxide/Au samples with various high-permittivity oxide insulators: ALD Al_2O_3, ZrO_2, HfO_2, Ta_2O_5 and TiO_2, and for Nb_2O_5 grown using liquid-source mist deposition measured at a strength of electric field in the oxide of $1\,\mathrm{MV\,cm}^{-1}$ with metal biased negatively. Lines illustrate the procedure of PC threshold determination. The inset shows electronic transitions in the oxide responsible for PC.

Table 3.3.1. Oxide bandgap width E_g determined from PC spectra of ALD oxides.

	TiO_2	Nb_2O_5	Ta_2O_5	ZrO_2	HfO_2	Al_2O_3
E_g (± 0.1 eV)	4.4	4.0	4.4	5.5	5.6; 5.9	6.2

bandgap energies in the cubic, tetragonal and monoclinic zirconia, respectively [41]. Even larger bandgap narrowing is observed in ALD alumina, in which the bandgap width appears to be approximately 30% smaller than in the Al_2O_3 crystal (8.7 eV [42]). These results indicate that the fundamental electronic structure of the deposited oxides is not necessary identical to that of the bulk crystal of the same oxide.

Examination of electrical properties of the MOS structures prepared using insulating oxides indicated in figure 3.3.2 revealed that, for the as-deposited films, an acceptably low leakage current density can be obtained only in the cases of ZrO_2, HfO_2 and Al_2O_3 insulators. For this reason, most of the attention will be paid to them below. In figure 3.3.3 are shown the PC

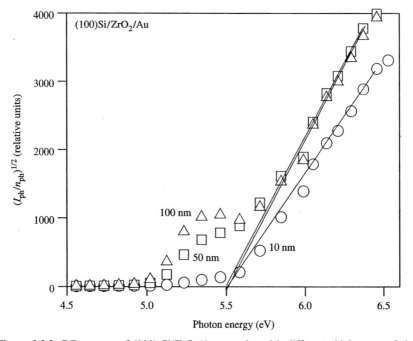

Figure 3.3.3. PC spectra of (100) $Si/ZrO_2/Au$ samples with different thicknesses of the high-permittivity oxide (in nm): 10 (O), 50 (□) and 100 (△) measured at a strength of electric field in the oxide of $1\,MV\,cm^{-1}$ with metal biased negatively.

spectra of ALD ZrO_2 layers of different thicknesses, all exhibiting the same spectral threshold at 5.5 eV. With increasing oxide thickness, progressively intense sub-threshold PC is observed, suggesting excitation of the oxide bulk states. As the annealing at 800°C has no effect on the relative intensity of the sub-threshold PC, one can exclude possible presence of different crystalline phases in the zirconia film as the origin of the complex PC spectrum because the phase composition of the film is strongly affected by high-temperature treatment [34]. This would rather suggest the relationship of the sub-threshold PC to some defect or impurity left in the film.

To investigate the possible impact of the deposition process on the bandgap width of the high-permittivity oxide, we compared the PC spectra of HfO_2 layers of different thicknesses deposited using three different techniques, as shown in figure 3.3.4. Reaffirming the fundamental character of the PC thresholds observed, all the HfO_2 layers exhibit two thresholds [37] which correspond to two bandgap values, 5.6 and 5.9 eV as indicated in table 3.3.1. These are likely related to the co-presence of different phases in the polycrystalline HfO_2 films [43]. However, the relative intensity of the two PC bands appears to depend strongly on the deposition chemistry: the lowest band is much more pronounced in the ALD and MOCVD films than in the N-CVD ones. The latter suggests that the phase composition of the film is a strong function of the deposition chemistry, with hydrogen, available both during ALD and MOCVD, the most likely candidate for an 'active' impurity.

The impact of the annealing and the type of interlayer between Si and the metal oxide on PC spectra is illustrated in figure 3.3.5 for N-CVD HfO_2 [38]. In the samples with a SiON interlayer grown during N-CVD on a clean Si surface [35], 30 min treatment in O_2 at 650°C results in an increase of the volume fraction of the phase with $E_g = 5.6$ eV (figure 3.3.5(a)). This effect is even more pronounced if an O-free Si_3N_4 layer with ≈ 1.5 nm thickness is grown on the (100) Si surface prior to the HfO_2 deposition (figure 3.3.5(b)). It is likely then that the annealing-induced crystallization of the oxide film is largely determined by its interface with the substrate.

The ALD Al_2O_3 provides another example of the annealing-induced phase transformation [44]. This is illustrated by the PC spectra shown in figure 3.3.6 for 100 nm thick alumina subjected to 10 min PDA in $N_2 + 5\%$ O_2 at various temperatures: it is seen that for $T > 800$°C the PC with threshold at 6.1 eV rapidly decreases indicating 'opening' of the alumina bandgap. The x-ray diffraction data reveal formation of cubic γ-alumina phase after high-temperature PDA [44], which has the bandgap width of 8.7 eV [42]. The volume fraction of the narrow-bandgap amorphous alumina can be determined from the slope of the PC spectra, and its evolution with the PDA temperature is shown in the inset in figure 3.3.6. Such dramatic impact of the PDA on the electronic properties of Al_2O_3 allows one to explain the discrepancy between the results obtained in samples with

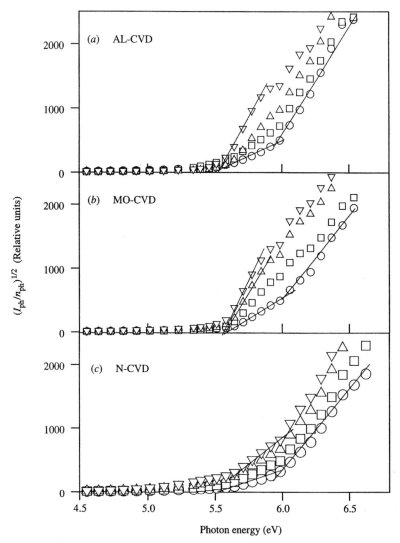

Figure 3.3.4. PC spectra of (100)Si/HfO$_2$/Au samples with different thicknesses of the high-permittivity oxide (in nm): 5 (O), 20 (□), 50 (△), and 100 (▽). Panels (*a*), (*b*) and (*c*) correspond to the HfO$_2$ grown on Si by ALD, MO-CVD and N-CVD, respectively.

low-temperature and high-temperature grown Al$_2$O$_3$ layers indicated in the 'Introduction'.

One might consider using not only the pure oxides, but also their solid solutions in MOS devices. This makes it necessary to analyse the bandgap evolution as a function of the oxide composition [45]. The PC spectra are shown in figure 3.3.7 for as-deposited 5 nm layers of pure Al$_2$O$_3$, HfO$_2$ and

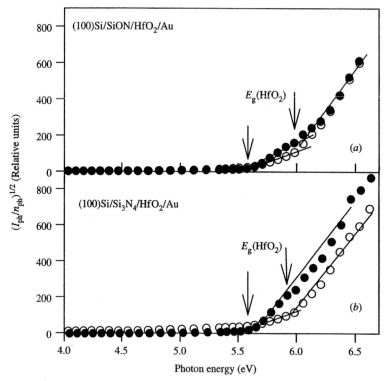

Figure 3.3.5. The HfO_2 PC spectra measured in the p-Si/SiON/HfO_2/Au (*a*) and p-Si/Si_3N_4/HfO_2/Au (*b*) samples, both with 8 nm thick HfO_2 layers at an average strength of electric field in the oxide of 1 MV cm^{-1} with the metal biased negatively. The open symbols correspond to the as-deposited oxide layers, the filled ones to those oxidized for 30 min in O_2 at 650°C.

two mixed oxides. The latter were grown by ALD using (1:1) and (3:1) cycle ratio between Al(CH$_3$)$_3$ and HfCl$_4$, which yield the molar fraction of alumina in the film of 40 and 70%, respectively. The addition of Al is seen to immediately increase the bandgap width from 5.6 to the $\approx 6\,$eV value typical for pure alumina. What is remarkable in the spectra shown in figure 3.3.7 is the gradual decrease in the slope of the $(I_{ph}/n_{ph})^{1/2}-h\nu$ dependences which suggests decrease in the J-DOS with increasing Al concentration. Apparently, when Al is introduced into the oxide in substantial concentration, the conduction band states derived from the empty 5d states of Hf [46] shift upwards and appear to be energetically located well above the lowest conduction band originating from the 3p states of Al [45]. The latter remains at approximately the same energy as in pure Al$_2$O$_3$, but their density even appears to be reduced in the mixed oxide because of the lower mole fraction of alumina in the film (cf $\triangle$ and ∇ in figure 3.3.7).

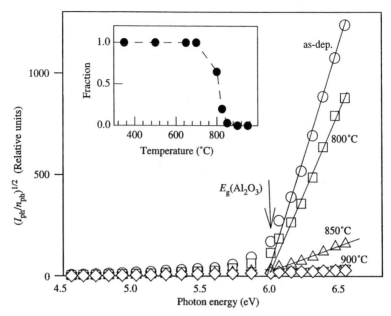

Figure 3.3.6. PC spectra of (100) Si/Al$_2$O$_3$ (50 nm)/Au capacitors measured under -5 V bias on the Au for the as-deposited (O) Al$_2$O$_3$ layers and after annealing for 10 min in N$_2$ + 5% O$_2$ at 800°C (□), 850°C (△), 900°C (▽) and 950°C (◇). The arrow E_g(Al$_2$O$_3$) indicates the spectral threshold of intrinsic PC in Al$_2$O$_3$. The inset shows the relative content of the low-bandgap (amorphous) alumina phase at various annealing temperatures.

IPE at (100) Si/insulator interfaces

Electron IPE spectra and thresholds

The electron IPE spectra for n-Si/Al$_2$O$_3$/Au and n-Si/ZrO$_2$/Au MOS structures (metal biased positively) are shown in figure 3.3.8(*a*) and (*b*), respectively. Because replacement of the Au gate electrode with Al has no effect on the IPE curves, the latter correspond to emission of electrons from the Si valence band to the conduction band of the insulator, as shown in the inset in figure 3.3.8(*b*). This is affirmed independently by observation of the IPE spectrum modulation by direct optical transitions in the Si crystal [23, 47] as indicated by arrows E_1 and E_2 in figure 3.3.8(*a*). These transitions cause deviation of spectral curves from the $Y \approx (h\nu - \Phi_e)^3$ dependence thus limiting the spectral interval of linear fit and causing a substantial increase of the IPE threshold determination error. For the as-deposited Al$_2$O$_3$ the IPE spectral threshold appears to be slightly above 3 eV, while for ZrO$_2$ it is ≈ 0.15 eV lower.

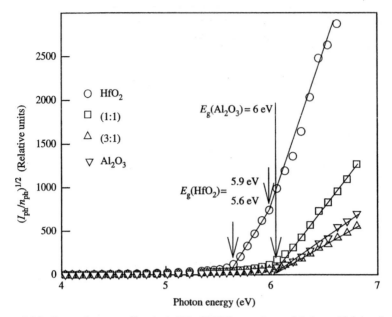

Figure 3.3.7. Spectral curves of intrinsic PC of MOS capacitors with 5 nm thick insulating layers of different composition: HfO_2 in (○), mixed Al–Hf oxides with 40% (□) and 70% (△) mole fraction of alumina, and pure Al_2O_3 (▽). Solid circles show the effect of PDA on HfO_2 spectra. The curves are measured under –1.5 V bias applied to the Au electrode. Lines guide the eye; arrows indicate the PC thresholds of pure oxides.

An IPE spectral threshold close to that of ZrO_2 is observed in HfO_2 layers, as illustrated from the IPE spectra shown in figure 3.3.9 for HfO_2 deposited using three different methods: ALD (□), MO-CVD (△), and N-CVD (○). The reproducibility of the threshold, which was demonstrated above for the HfO_2 bandgap width, suggests its fundamental character, i.e. independence of the band structure of the Si/HfO_2 interface from the particular HfO_2 deposition technique. However, the data shown in figure 3.3.9 indicate substantial variation of the electron IPE yield when changing the oxide deposition method, suggesting different probability of emission for electrons photoexcited in Si [48]. This is likely related to differences in thickness and/or composition of the interfacial Si oxide layer grown during HfO_2 deposition, which determines the electron scattering probability: It is seen that higher deposition temperature (300°C for ALD vs. 485°C for MO-CVD) leads to a reduced electron emission which is consistent with a thicker Si oxide layer, while the N-CVD suppresses the IPE yield further because of more aggressive Si oxidation by the by-products of $Hf(NO_3)_4$ decomposition [35].

The spectra shown in figure 3.3.10 correspond to electron IPE from Si into pure ALD HfO_2 and Al_2O_3 and two Al–Hf mixed oxides of ≈ 20 nm thickness grown using the procedure described above. Open symbols

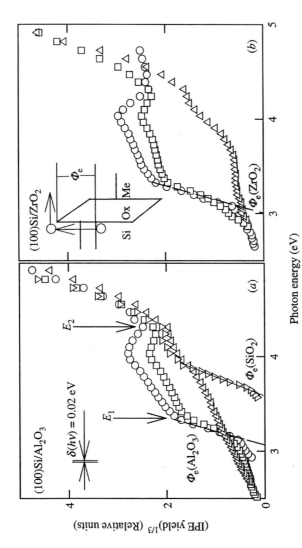

Figure 3.3.8. Cube root of the IPE yield as a function of photon energy for MOS structures with different dielectrics: (*a*) as-deposited 5 nm thick Al_2O_3 (○) and oxidized at 650°C for 30 min (□) or at 800°C for 10 min (△) as compared to a 4.1 nm thick thermal SiO_2 (▽); (*b*) as-deposited 7.4 nm ZrO_2 (○) and oxidized at 650°C for 30 min (□) or at 800°C for 10 min (△). All curves are measured under an applied electric field of 2 MV cm^{-1} in the insulating layer closest to Si. The arrows E_1 and E_2 indicate onsets of direct optical transition in the Si crystal. The spectral thresholds Φ_e are indicated for different oxides. The energy width of the monochromator slit is indicated in figure 3.3.1(*a*). The error in the IPE yield determination is smaller than the symbol size.

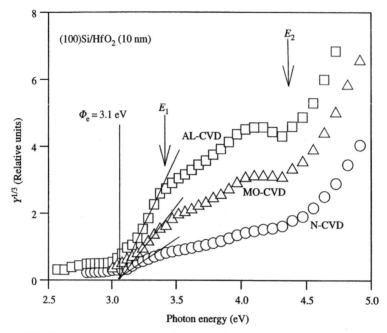

Figure 3.3.9. Spectral dependences of electron IPE from the (100) Si valence band into 10 nm thick as-deposited HfO_2 layers grown by ALD at 300°C (□), CVD from metallo-organic precursors at 485°C (△) and nitrato precursor at 350°C (○). The arrows mark the energies of transitions E_1 and E_2.

correspond to the as-deposited layers; filled ones show the results for the samples subjected to 10 min PDA at 900°C in $N_2 + 5\%$ O_2. The spectral threshold of IPE from the Si valence band into the oxide conduction band determined using a $Y^{1/3}-h\nu$ plot is found to be at 3.1 eV for both pure Al_2O_3 and mixed oxides of (1:1) and (3:1) composition. Within the experimental error of 0.1 eV, this value coincides with those of pure ALD alumina and HfO_2 determined at the same strength of electric field in the oxide (3.0–3.1 eV, cf [26, 27, 37]). Despite marginal influence of the oxide composition on spectral thresholds, IPE yield is seen to be reduced in the oxides of mixed composition. This suggests a reduced DOS in the oxide conduction band which is consistent with the reduced J-DOS revealed by the PC experiments. The PDA shifts the spectral onset of the IPE to a higher photon energy, approaching the threshold of 4.25 eV for IPE from (100) Si into SiO_2. This observation suggests growth of an interfacial SiO_2-like layer, probably a kind of silicate as indicated by a 'tail' in the sub-threshold IPE. The latter is particularly pronounced in the case of pure Al_2O_3 insulator suggesting a much greater extent of intermixing between silicon and aluminium oxides than in structures with Hf–Al mixed oxides.

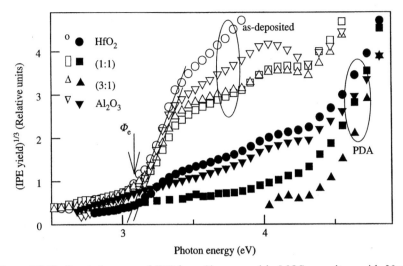

Figure 3.3.10. Spectral curves of IPE from Si measured in MOS capacitors with 20 nm thick pure HfO_2 (O), mixed Al–Hf oxides with 40% (□) and 70% (△) mole fraction of alumina, and pure Al_2O_3 (▽). Open symbols correspond to the samples in the as-deposited state, closed to those after 10 min PDA at 900°C in $N_2 + 5\%$ O_2. The curves are measured under +4 V bias applied to the Au electrode. Lines guide the eye.

Barrier height determination

To obtain the exact barrier energies, the spectral thresholds were measured at different strengths of electric field and then extrapolated to zero field in the Schottky coordinates as shown in figure 3.3.11. The resulting barriers Φ_{ev} measured between the top of the Si valence states and the bottom of the oxide conduction states are 3.25 ± 0.08 eV for Al_2O_3 and 3.1 ± 0.1 eV for both ZrO_2 and HfO_2 as compared to $\Phi_{ev} = 4.25 \pm 0.05$ eV at the Si/SiO_2 interface which is also shown for comparison. The Schottky plot allows us to determine the effective image-force dielectric constant ε_i related to the refraction index n of the insulator near the Si surface ($\varepsilon_i \approx n^2$). From the slopes of the lines in figure 3.3.11, ε_i values are obtained as 2.1 ± 0.2, 3.5 ± 0.5, 5 ± 1 and 6 ± 2 for SiO_2, Al_2O_3, ZrO_2 and HfO_2, respectively. Within experimental error, these values coincide with the square of the refractive index of bulk SiO_2 ($n = 1.456$), Al_2O_3 ($n = 1.77$) and ZrO_2 ($n = 2.2$).

Comparison of the above results with earlier data on IPE from Si into Al_2O_3 layers deposited at high temperature or subjected to high-temperature anneal [16, 17, 49] indicates that in the as-deposited state the barrier height is much reduced (3.25 eV vs. 3.9–4.1 eV). The 2.1 eV conduction band offset resulting from our value also appears to be significantly lower than the theoretically predicted 2.8 eV value [7] thus affirming the importance of interface effects. Upon oxidation, the major IPE emission shifts to the spectral range of $h\nu > 4$ eV, suggesting as explanation of earlier experimental

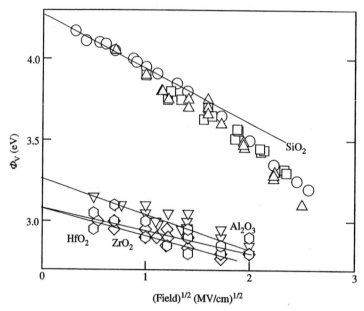

Figure 3.3.11. Schottky plot of the spectral thresholds for IPE from the Si valence band into the conduction band of SiO_2 layers of different thicknesses (in nm): 55 (○), 5.8 (□), 4.1 (△), as compared to 5 nm thick Al_2O_3 (▽), 7.4 nm thick ZrO_2 (□) and 10 nm thick HfO_2 (○) layers. The symbol size corresponds to the error of the threshold determination.

data [16, 17, 49] both formation of a silicate interlayer and alumina crystallization. In the Si/ZrO_2 and Si/HfO_2 systems the barrier increases to a similar value of ≈ 4 eV starting from 3.1 eV for the as-deposited sample. Interestingly, crystallization of ZrO_2 observed upon oxidation at $T > 500°C$ [34] seems to have no substantial effect neither on the energy of the ZrO_2 conduction band edge (with respect to energy bands of the Si substrate) nor on the bandgap width of the oxide itself.

Post-deposition oxidation reduces the IPE yield in the spectral range of 3 eV $< h\nu < 4.5$ eV, which indicates a decrease in the DOS corresponding to the conduction band of the high-permittivity metal oxide. Instead, the spectral features characteristic for SiO_2 become pronounced as the thickness of the thermally grown interlayer exceeds 1 nm [48]. Worth noticing here is the fact that the IPE curves of the oxidized gate stacks do not reach the shape corresponding to the IPE into pure thermal SiO_2 (▽ in figure 3.3.8(*a*)), not even at the highest oxidation thermal budgets. A significant 'tail' is observed in the sub-threshold region, which becomes particularly pronounced in the case of Al_2O_3. More details for ZrO_2 are revealed by the spectral curves measured in the sub-barrier spectral range using Ar^+ laser excitation, shown in figure 3.3.12. This photocurrent caused by excitation of electrons via the oxide conduction band tail states decreases with increasing oxidation

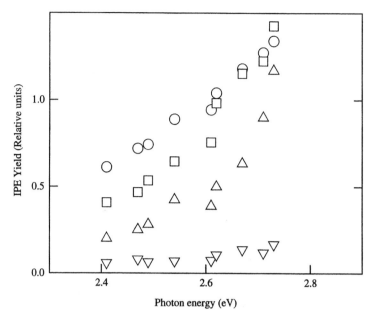

Figure 3.3.12. IPE yield as a function of photon energy for n-Si/7.4 nm ZrO$_2$/Au MOS structures in as-deposited state (O) and after oxidation at 500°C for 60 min (□), 650°C for 30 min (△) and at 800°C for 10 min (▽). The measurements are made with applied field strength in ZrO$_2$ layer of 3 MV cm^{-1}, metal biased positively. The error in the IPE yield determination is smaller than the symbol size.

temperature, but does not disappear entirely. Apparently, the electron exchange between Si and the band tail states in ZrO$_2$ gets constrained upon growth of a SiO$_2$-like interlayer. However, the latter is likely to be a silicate containing a considerable density of gap states, which provide under-barrier pathways for electrons.

Hole IPE at (100) Si/HfO$_2$ interface

In contrast to the external photoemission case, IPE of holes can also be observed [13, 23]. This is evidenced by the negative metal bias IPE spectra [38] shown in figure 3.3.13 which pertain to the p-Si/SiON/HfO$_2$ (9 nm) (*a*) and p-Si/Si$_3$N$_4$/HfO$_2$ (10 nm) (*b*) N-CVD samples with Au and Al metallization biased at −2 and −2.5 V, respectively. Open symbols correspond to the as-deposited insulators, the filled ones to those oxidized for 30 min in pure O$_2$ at 500°C (▲) and 650°C (▼). The spectra change significantly when the type of metal and the kind of interlayer are changed, indicating more than one contribution to the photocurrent (cf figure 3.3.1). In the samples with a SiON interlayer (figure 3.3.13(*a*)), the large difference in spectral thresholds between Au and Al suggests, as the dominant

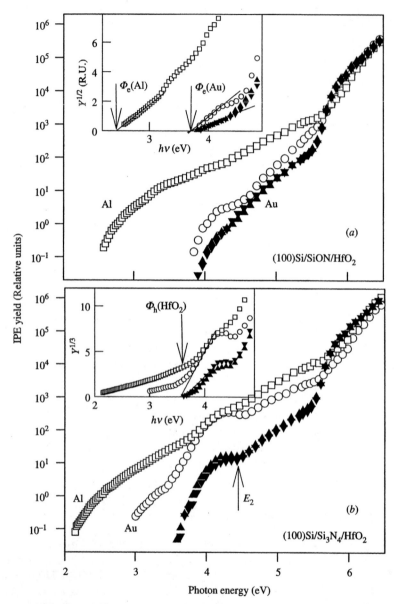

Figure 3.3.13. IPE yield as a function of photon energy in the as-deposited p-Si/SiON/HfO$_2$ (9 nm) (*a*) and p-Si/Si$_3$N$_4$/HfO$_2$ (10 nm) (*b*) MOS capacitors with Au (O) and Al (□) electrodes measured under negative voltages of −2.0 and −2.5 V, respectively. Filled symbols correspond to the samples additionally oxidized for 30 min in O$_2$ at 500°C (▲) and 650°C (▼), measured under −2.0 V bias on Au electrodes. The arrow E_2 indicates the onset of direct optical transitions in the Si crystal. The insets illustrate determination of spectral thresholds using $Y^{1/2}$–$h\nu$ and $Y^{1/3}$–$h\nu$ plots.

contribution to photocurrent, the emission of electrons from the electron states near the Fermi levels of these metals. At the same time, the post-deposition oxidation weakly influences the IPE: only for Au is some yield reduction detectable, while for Al the spectral curve remains unaffected (not shown). These observations allowed us to associate the observed photocurrent with IPE of electrons from Au and Al into HfO_2. The corresponding spectral thresholds were determined by extrapolating the $Y^{1/2}-h\nu$ plot to zero yield, as shown in the inset in figure 3.3.13(a). These thresholds show a weak dependence on applied electric field similarly to the case of IPE from metal into Al_2O_3 and ZrO_2 [26, 27], and yield the barrier heights $\Phi_e(Al) = 2.5 \pm 0.1\,eV$ and $\Phi_e(Au) = 3.7 \pm 0.1\,eV$, close to the spectral thresholds.

In the samples with an Si_3N_4 interlayer, an additional emission is observed at $h\nu > 3.6\,eV$ for both Al and Au electrodes (cf figure 3.3.13(b)), suggesting its relationship to some optical excitation in Si or in the insulator itself. The absence of this intense IPE band in the spectra measured with positive metal bias (cf figure 3.3.9) indicates that it is not related to the excitation of electron transitions in Si_3N_4 or HfO_2. Rather, the yield reduction at $h\nu = 4.3\,eV$ (E_2 reflection peak of Si in figure 3.3.13(b)) suggests this IPE to be due to the light absorption in the Si crystal. The same conclusion is suggested by the yield reduction observed after additional oxidation, shown in figure 3.3.13(b): because the barriers for both electrons and holes are higher at the Si/SiO_2 interface (IPE gives $\Phi_{eV} = 4.25\,eV$ and $\Phi_h = 5.7\,eV$ [23]) than those at the Si/Si_3N_4 interface ($\Phi_{eV} = 3.17\,eV$ and $\Phi_h = 3.06\,eV$ [50]), incorporation of oxygen at the interface increases the barrier thickness and height for a charge carrier emitted from Si. This would also explain why this IPE process is highly efficient in the $Si/Si_3N_4/HfO_2$ stack but not in the $Si/SiON/HfO_2$ one (cf panels (b) and (a) in figure 3.3.13). As a result, we associate this IPE band with photoemission of holes from deep levels in the Si valence band into the HfO_2 valence band (cf figure 3.3.1(b)). The injection of holes is evidenced by the observation of positive oxide charging upon prolonged illumination of the samples with an Si_3N_4 interlayer with white light through a 4 eV low-pass filter. The spectral threshold of this excitation found from the $Y^{1/3}-h\nu$ plot shown in the inset of figure 3.3.13(b) appears to be field independent (within the accuracy limit of 0.1 eV), which is consistent with the hole IPE behaviour from Si and 4H-SiC into SiO_2 [23, 51], yielding the barrier height for holes $\Phi_h(HfO_2) = 3.6 \pm 0.1\,eV$. This barrier corresponds to the energy difference between the bottom of the Si conduction band and the top of the HfO_2 valence band. Worth noticing here is the high quantum efficiency of hole IPE: even after oxidation at 650°C it remains one order of magnitude more intense than the IPE from Au (cf figure 3.3.13(a) and (b)). Finally, one can evaluate the HfO_2 bandgap width by combining the independently measured barriers for electrons and holes and the known Si bandgap width as

$E_g(HfO_2) = \Phi_e(HfO_2) + \Phi_h(HfO_2) - E_g(Si) = 5.6 \pm 0.15\,eV$, i.e., the same value as the lowest PC threshold of HfO_2 (cf figure 3.3.4). This result importantly indicates that the bulk bandgap value of HfO_2 is preserved across the entire oxide layer.

IPE at the metal/insulator interfaces

Examples of IPE spectral curves from a metal into 5 nm thick Al_2O_3 and ZrO_2 layers are presented in figure 3.3.14. For Al_2O_3 only data for Al and Au are shown in figure 3.3.14(a) which are similar to the data reported by DiMaria for IPE into high-temperature deposited aluminium oxide [17]. For ZrO_2 the spectra of IPE from Mg (O), Al (□), Ni (△), Cu (▽) and Au (◇) are presented in figure 3.3.14(b). As the IPE from the same metals into thermal SiO_2 of the same thickness (curves not shown) obeys the Fowler law [51] $Y \sim (h\nu - \Phi)^2$ (in agreement with literature results for thicker oxides [52–54]), Fowler co-ordinates ($Y^{1/2}$–$h\nu$) were used to determine the IPE spectral thresholds. A cube fit $Y^{1/3}$–$h\nu$ was found to give less good fit; it results in a systematic 0.1 eV red shift of the IPE threshold which is comparable to the error arising from the extrapolation inaccuracy. It needs to be recalled here that the power of the yield dependence on photon energy is determined exclusively by the shape of the energy distribution of excited electrons in the emitter and by the energy dependence of the barrier surmounting probability [24]. For electrons emitted from the same metal, their energy distribution must be the same and the variation of the spectral curves will reflect mostly the properties of the potential barrier. Therefore, deviation of the IPE spectra from the Fowler law in the near-threshold region (figure 3.3.14) indicates a lateral non-uniformity of the interfacial barrier. Two more features are revealed by the data shown. First, the thermal oxidation at a temperature as high as 800°C does not change the metal/insulator barrier significantly as compared to the as-deposited state (cf data for Al in figure 3.3.14(a)). Second, the electric field has only a weak effect on spectral thresholds (within the indicated accuracy limit of 0.1 eV) suggesting a smaller Schottky barrier reduction as compared to the IPE from Si into Al_2O_3 and ZrO_2 (cf figure 3.3.2). The barrier heights Φ are listed in table 3.3.2 for the studied insulator/metal pairs, and Φ_V values are given for (100)Si/insulator interfaces.

To enable a meaningful comparison between different oxides, the measured barrier heights are plotted in figure 3.3.15 together with literature data for SiO_2 and Al_2O_3 as a function of the Pauling metal electronegativity X_M [55]. For thermal SiO_2, panel (a), the barriers measured on 5 nm thick oxides (□) perfectly follow the trend of barrier increase with the metal electronegativity reported earlier for thicker oxides (symbol O from [52], [53] for W). In the case of Al_2O_3, panel (b), the literature data

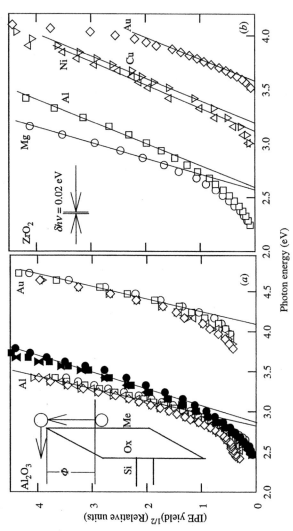

Figure 3.3.14. (*a*) Square root of the IPE yield from the metal as a function of photon energy for MOS structures with Al and Au electrodes deposited on a 5 nm thick as-deposited (open symbols) and oxidized at 800°C for 10 min (filled symbols) Al_2O_3 insulator. The strength of the electric field in the oxide is (in $MV\,cm^{-1}$) as follows: Al on as-deposited oxide, 0.2 (○), 0.5 (□), 1.0 (△), 1.5 (▽), 3.0 (◇); Al on oxidized Al_2O_3, 0.5 (●), 1.0 (■), 2.0 (▲), 4.0 (▼); Au on as-deposited oxide, 0.5 (○), 1.0 (□), 2.0 (△), 3.0 (▽), 4.0 (◇), metal bias is negative. (*b*) Square root of the IPE yield from the metal as a function of photon energy for MOS structures with different metals deposited on a 5 nm thick as-deposited ZrO_2 layer. The electric field strength in the oxide is 2 $MV\,cm^{-1}$, metal biased negatively. The lines illustrate the determination of the IPE spectral thresholds. The energy width of the monochromator slit is indicated in figure 3.3.4(*b*). Inaccuracy of the IPE yield determination is smaller than the symbol size.

Table 3.3.2. Barrier heights (Φ_V, Φ) at the interfaces between (100) Si, metals, and SiO$_2$, Al$_2$O$_3$ and ZrO$_2$ insulators.

Insulator	(100) Si	Mg	Al	Ni	Cu	Au
SiO$_2$ ($\pm$0.05 eV)	4.25	2.50	3.15	3.70	3.85	4.10
Al$_2$O$_3$ ($\pm$0.1 eV)	3.25	2.6	2.9	3.5	3.6	4.1
ZrO$_2$ ($\pm$0.1 eV)	3.1	2.6	2.7	3.2	3.25	3.5
HfO$_2$ ($\pm$0.1 eV)	3.1		2.5	3.4		3.7

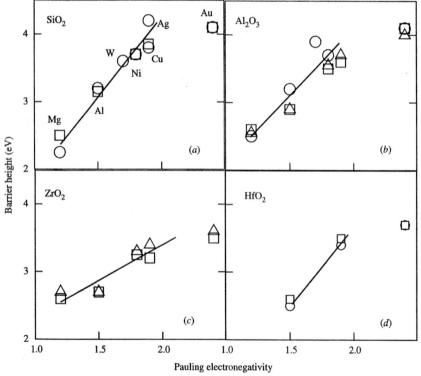

Figure 3.3.15. Barrier height at the metal/insulator interface as a function of the gate metal electronegativity for 5 nm thick thermal SiO$_2$ (*a*), 5 nm thick ALD Al$_2$O$_3$ (*b*), ZrO$_2$ (*c*) and HfO$_2$ (*d*). Circles indicate the literature values for thicker oxides, squares refer to the experimental results on as-fabricated oxide layers, triangles refer to the ALD layers oxidized in O$_2$ at 650°C for 30 min. Lines guide the eye. The symbol size corresponds to 0.1 eV inaccuracy of the barrier height determination.

(symbol ○ from [17], [49] for W), on average, lie slightly above the values determined for both as-deposited (□) and thermally oxidized (△) AL-CVD layers, though for Mg and Au the agreement is perfect. Taking into account the spread of the data, the increase of the interfacial barrier height with the metal electronegativity appears to be the same as in the case of SiO_2 and thus close to the ideal case $d\Phi/dX_M = 1$ [56]. For both as-deposited (□) and oxidized (△) ZrO_2 layers, panel (*c*), the increase of the barrier height with metal electronegativity is found to be considerably smaller (ideality factor $d\Phi/dX_M \approx 0.75$) than for SiO_2 and Al_2O_3. The behaviour of HfO_2, panel (*d*), seems to resemble that of SiO_2 and Al_2O_3, but an insufficient number of experimental data points corresponding to different metals prevents us from making a solid conclusion.

The above data indicate that in the temperature range up to 800°C, one can use the barrier heights listed in table 3.3.2. However, as demonstrated earlier (cf the 'Bandgap width determination from PC measurements' section) ALD alumina undergoes crystallization at around 800°C accompanied by significant widening of its bandgap [44]. The IPE spectra measured with −5 V bias applied to the Au electrode are shown in figure 3.3.16 for MOS capacitors with 50 nm thick as-deposited Al_2O_3 (○) or

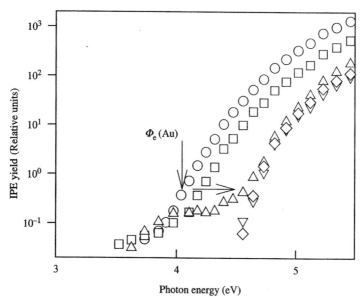

Figure 3.3.16. Logarithmic plot of IPE spectra measured in (100) Si/Al_2O_3 (50 nm)/Au capacitors under −5 V bias on the Au for the as-deposited (○) Al_2O_3 layers and after annealing for 10 min in N_2 + 5% O_2 at 800°C (□), 850°C (△), 900°C (▽) and 950°C (◇). The arrow Φ_e(Au) indicates the spectral threshold of electron IPE from Au into the oxide, the horizontal one shows the direction of the annealing-induced threshold shift.

subjected to 10 min anneal in an $N_2 + 5\%$ O_2 mixture at 800°C ($\square$), 825°C ($\triangle$), 850°C ($\triangledown$) and 950°C ($\lozenge$). With anneal temperature exceeding 800°C, the spectral threshold of electron IPE from Au into Al_2O_3 shifts by $\approx 0.5\,eV$ towards higher energy, indicating formation of a higher barrier. A similar trend is observed in the samples with Al electrodes (curves not shown) suggesting that the barrier increase is related to an upward shift of the Al_2O_3 conduction band edge. This shift suggests that crystallization affects the atomic surroundings of the Al^{3+} ions in the oxide which mostly contribute to the DOS near the bottom of the conduction band [46], eliminating configurations yielding the alumina phase with narrow bandgap.

Workfunction differences

On the basis of the barrier values listed in table 3.3.2, the metal/silicon workfunction difference ϕ_{ms} was calculated for both the n- and p-type Si samples [26, 27]. The results are shown by lines in figure 3.3.17 and are compared with the flatband voltage (V_{FB}) values determined from the high-frequency C–V curves in the n- (filled symbols) and p-type (open symbols) MOS structures with Al_2O_3 (*a*), Al_2O_3/ZrO_2 (*b*) and ZrO_2 (*c*) dielectrics, both as deposited (circles) and oxidized at 650°C for 30 min (squares) or at 800°C for 10 min (triangles). The dotted lines on the panels indicate the ϕ_{ms} behaviour in n-Si MOS capacitors with thermally grown SiO_2 dielectric. The comparison between ϕ_{ms} and V_{FB} values reveals three features. First, for the same metal, ϕ_{ms} values measured in the MOS structures with thermal SiO_2 are considerably smaller than those in MOS structures with an ALD-deposited insulator. Second, the systematically observed trend $V_{FB} > \phi_{ms}$ for the as-deposited ALD layers indicates the presence of a substantial density of negative charge in metal oxide, particularly in ZrO_2-based insulators. The density of this charge is effectively reduced by oxidation of the deposited dielectric. Third, and by contrast, the Al-gated samples exhibit a systematic V_{FB} shift to below ϕ_{ms} indicating a positive charge build-up in the insulator upon Al evaporation. The latter may be related, for instance, to a chemical interaction of Al with the H-containing fragments at the oxide surface, leading to the formation of protons.

The important result of this ϕ_{ms} study consists in revealing a large difference in the values measured in MOS structures with thermal SiO_2 and with an ALD-deposited insulator, which indicates the presence of a significant interface dipole. Growth of a thin silicon oxide at the interface between the Si substrate and the ALD layer has no impact neither on the Si/ALD oxide band alignment nor on this dipole (cf figure 3.3.17), suggesting that the latter is associated with the metal/oxide interface. The latter suggestion finds support from the observed non-idealities in the IPE spectra of the metal/oxide interfaces. First, the tails of the Y–$h\nu$ curves refer to

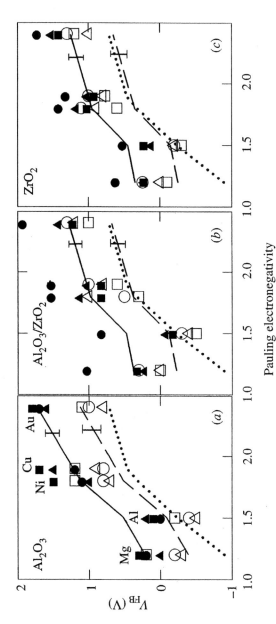

Figure 3.3.17. Flatband voltages of n- (filled symbols) and p-type (open symbols) MOS capacitors with 5 nm thick Al_2O_3 (*a*), stacked 1.5 nm Al_2O_3/5 nm ZrO_2 (*b*) and 5 nm thick ZrO_2 (*c*) insulators as a function of gate metal electronegativity. Solid and dashed lines show correspondingly the behaviour of the metal–semiconductor workfunction difference ϕ_{ms} for n- and p-type Si, dotted lines indicate ϕ_{ms} for an n-type MOS structure with thermal SiO_2 insulator. The error of VFB determination is smaller than the symbol size.

non-uniformity of the barrier. As may be seen from figure 3.3.14, they show a clear trend to increase for metals with a lower barrier (Al, Mg) as compared to Au. The latter suggests that, in addition to the conduction band tail states, the barrier may also be affected by a non-uniform polarization layer possibly related to negative charge arising from electrons trapped in the oxide at levels below the metal Fermi level [57].

Next, a weak dependence of the metal/oxide barrier height on the strength of electric field in the oxide suggests the presence of a strong built-in electric field which is also consistent with the formation of an interfacial polarization layer. Finally, the weaker Φ-electronegativity dependence for ZrO_2 (cf figure 3.3.17(c)) may also be related to the relative enhancement of the negative dipole for the metals with smallest barriers (Mg, Al). The latter effect may be of concern for practical applications of the ALD gate insulators: the threshold voltage control by using n^+/p^+-polycrystalline silicon gates or metal electrodes with different workfunction may appear not so efficient as in the MOS devices with a SiO_2 gate dielectric.

Electron tunnelling vs. trap-assisted transport

The determined interface barrier heights and the contact potential values allowed us to analyse in detail the *I–V* curves of MOS structures with high-κ dielectrics. Examples of such curves for n-Si/ZrO_2/Au structures (positive metal bias) are shown in figure 3.3.18(a) for the as-deposited samples (○) and oxidized at various thermal budgets (500°C for 60 min (□); 650°C for 30 min (△); 800°C for 10 min (∇)) samples as well as one annealed in high vacuum (650°C for 30 min (▲)). Similar curves were observed when Au was replaced with another metal indicating that the measured current is due to electron injection from Si substrate. Oxidation of the Si/ZrO_2 structure results in significant reduction of the leakage current probably related to the growth of a silicon oxide at the interface. The importance of oxygen is clearly indicated by lack of current reduction upon thermal treatment in high vacuum. To reveal the injection mechanism, the Fowler–Nordheim (FN) plots [58] of the dark (filled symbols) current and the photocurrent measured under excitation with 2.71 eV photons at incident power of 100 mW (open symbols) are shown in figure 3.3.18(b) for both as-deposited and oxidized samples. It is clearly seen that with increasing oxidation temperature both currents decrease dramatically. In the low-field range the currents do not exhibit the FN behaviour typical for electron tunnelling, suggesting a trap-assisted electron transport. The slope of the FN-type *I–V* curves observed at high field values increases with oxidation temperature approaching that for the case of FN tunnelling from Si into SiO_2. The latter indicates the formation of an SiO_2-like interlayer as the major reason for the reduction of the leakage current in the oxidized Si/ZrO_2 structures.

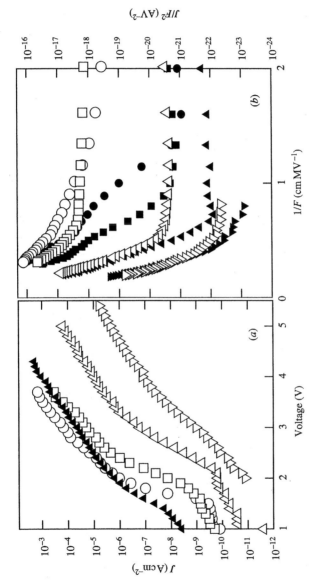

Figure 3.3.18. (*a*) *I–V* curves (positive metal bias) of n-Si MOS structures with 7.4 nm ZrO₂ insulator in as-deposited state (○), after oxidation at 500°C for 60 min (□), at 650°C for 30 min (△), or at 800°C for 10 min (▽), compared to the sample annealed in high vacuum at 650°C for 30 min (▲). (*b*) The Fowler–Nordheim plots of dark (filled symbols) and photocurrent (2.71 eV, 100 mW, open symbols) measured in n-Si MOS structures with 7.4 nm ZrO₂ insulator in as-deposited state (circles), after oxidation at 500°C for 60 min (squares), at 650°C for 30 min (triangles), or at 800°C for 10 min (inverted triangles) under positive metal bias.

The influence of an interlayer between the high-permittivity insulator and silicon is also seen from the FN plots of the dark and photocurrent ($h\nu = 2.71$ eV, 100 mW) measured in n-type Si MOS structures with as-deposited and oxidized stacks of 1.3 nm thermal SiO_2/5 nm ZrO_2 and 1.5 nm Al_2O_3/5 nm ZrO_2, shown in the figure 3.3.19(a) and (b), respectively. The presence of a pre-grown SiO_2 layer (figure 3.3.19(a)) significantly reduces the leakage current already in the as-deposited (not oxidized) sample. After oxidation, a further drop of the trap-related currents in the low-field range is observed, but it has only a marginal influence on the high-field electron tunnelling. The reduction of both the dark and photocurrent by the Al_2O_3 interlayer (figure 3.3.19(b)) is not as strong as by SiO_2 suggesting a lower interfacial barrier for electrons at the Si/Al_2O_3 than at the Si/SiO_2 interface, in accordance with the IPE results. Only after oxidation at 800°C are the currents typical for the SiO_2/ZrO_2 stack reached.

A behaviour very different from the one discussed above with respect to oxidation is revealed by I–V curves of electron injection from Au into ZrO_2 measured in p-type Si MOS structures (metal biased negatively) and shown in figure 3.3.20(a) as an FN plot. The dark current (filled symbols) is barely affected by oxidation. The laser-induced current (open symbols) decreases in the low-field range with increasing temperature of oxidation suggesting a reduction in the density of traps available for photoionization; yet, there is little change at high field. The FN plots of the dark current measured in similar MOS structures with Al (open symbols) and Au (filled symbols) electrodes are compared in figure 3.3.20(b) (three curves are shown for each sample). Despite significant difference in the barrier height for electrons at the interfaces of ZrO_2 with Al and Au (cf table 3.3.2), hardly any variation in the slope of the FN plots is observed in figure 3.3.20(b). The latter indicates that the emission of an electron from the metal into the conduction band of ZrO_2 does not constitute the rate-limiting step. Rather, the insensitivity to the metal type suggests that the current is determined by emission of an electron from some deep, stable against oxidation, gap state in ZrO_2. This may account for a lower current from Al than from Au measured at nominally the same electric field strength at the metal/ZrO_2 interface (cf figure 3.3.20(b)) despite the 0.8 eV lower barrier height for Al as compared to Au (table 3.3.2). Apparently, a negatively charged polarization layer is formed at the Al/ZrO_2 interface effectively reducing the electron emission probability from oxide traps. This picture is consistent with the non-ideal behaviour of the IPE characteristics discussed earlier and points towards the trap-assisted tunnelling as a dominant conduction mechanism in ALD Al_2O_3 and ZrO_2 layers. Worth noticing here that the trap-assisted tunnelling mechanism is independently affirmed by observation of significant temperature dependence of the leakage current in the ZrO_2-based insulating stacks [22].

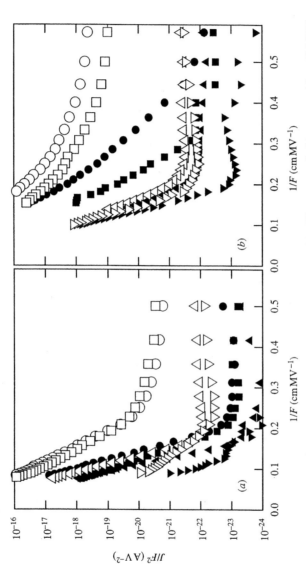

Figure 3.3.19. The FN plots of dark (filled symbols) and photocurrent (2.71 eV, 100 mW, open symbols) measured in n-Si MOS structures with stacks of 1.3 nm SiO_2/5 nm ZrO_2 (*a*) and 1.5 nm Al_2O_3/5 nm ZrO_2 (*b*) insulators in as-deposited state (circles), after oxidation at 500°C for 60 min (squares), at 650°C for 30 min (triangles), or at 800°C for 10 min (inverted triangles) under positive metal bias.

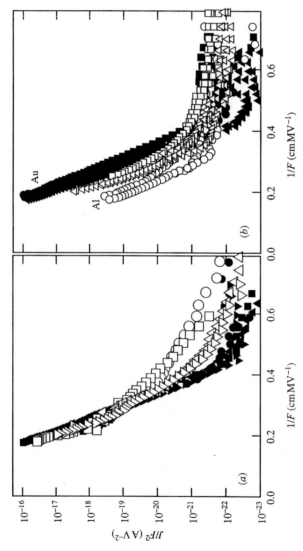

Figure 3.3.20. (*a*) FN plots of dark (filled symbols) and photocurrent (2.71 eV, 100 mW, open symbols) measured in p-Si MOS structures with 5 nm ZrO₂ insulators in as-deposited state (circles), after oxidation at 500°C for 60 min (squares), at 650°C for 30 min (triangles), or at 800°C for 10 min (inverted triangles) under negative metal bias. (*b*) FN plots of dark current measured in p-Si MOS structures with 5 nm ZrO₂ insulators and Al (open symbols) and Au (filled symbols) electrodes in as-deposited state (circles), after oxidation at 650°C for 30 min (squares), or at 800°C for 10 min (triangles) under negative metal bias. Three curves are shown for each sample.

The impact of traps on the current transport across the high-permittivity insulator is also suggested by I–V curves of p-Si-MOS capacitors with Au electrodes and 5 nm thick Al–Hf mixed oxides of different composition shown in figure 3.3.21. To suppress the possible hole injection from Si, the samples were prepared by ALD on IMEC-cleaned (100) Si which leaves a thin (<1 nm) Si oxide interlayer between the semiconductor and the metal oxide. As indicated earlier, the (1:1) and (3:1) cycle ratio between Al and Hf precursor gases yields the mixed oxide with the mole fraction of alumina of 30 and 70%, respectively. Despite the IPE measurements indicating that the barrier height for electrons at the Au/oxide interface is little sensitive to the oxide composition in the studied Al concentration range, the incorporation of Al leads to significant reduction of the leakage current. As the decrease of the conduction band DOS in the mixed oxides as compared to the pure HfO$_2$ by a factor of 2–3 (cf figure 3.3.10) is unlikely to account for the leakage current decrease by several orders in magnitude, it might be due to the impact of oxide mixing on the defect/impurity density in the oxide films.

Conclusions

IPE analysis of interfaces of (100) Si and various metals with thin thermally grown SiO$_2$ and ALD Al$_2$O$_3$, ZrO$_2$ and HfO$_2$ insulators provided directly the

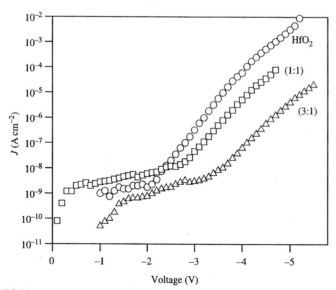

Figure 3.3.21. I–V curves measured at negative metal bias in p-type Si MOS structures with 5 nm thick ALD mixed Al–Hf oxide insulators of different compositions. The indicated (1:1) and (3:1) cycle ratio between Al and Hf precursor gases yields the mixed oxide with the mole fraction of alumina of 30 and 70%, respectively.

electron barrier energies between the filled states of the metals and Si and the conduction band of the respective insulators. The as-grown ALD oxides show a barrier with Si by ≈ 1 eV lower than SiO_2. In metal/Al_2O_3 contacts, the barrier height increases with electronegativity of the metal in the nearly ideal way $d\Phi/dX_M \approx 1$, while for as-deposited ZrO_2 insulators a less ideal behaviour is observed with $d\Phi/dX_M \approx 0.75$. The obtained barrier values remain unchanged down to the oxide layer thickness of ≈ 3 nm for Al_2O_3, ZrO_2 and HfO_2 and ≈ 1 nm for SiO_2 [30]. The metal/silicon workfunction differences in the MOS structures determined on the basis of the measured barrier heights appear to be different from those for the SiO_2 insulator. This is related to the formation of a negative dipole layer at the interfaces of metals with CVD oxides indicating the presence of a substantial density of electron traps. The traps present in CVD insulators determine the electron transport and their influence can be efficiently reduced by introduction of a SiO_2 interlayer. The latter suggests that the observed insulating properties of the high-permittivity dielectric stacks have not yet reached their fundamental limit determined by band offsets at the interfaces and can thus be improved further.

Acknowledgments

The authors would like to thank M.M. Heyns, S. DeGendt, M. Houssa (IMEC, Belgium), C. Fan, N. Hollien, X. Shi and S. A. Campbell (University of Minnesota, USA) and W. Tsai (SEMATECH/IMEC, Belgium) for providing samples used in the present work.

References

[1] Devine R A B, Vallier L, Autran J-L, Paillet P and Leray J-L 1996 *Appl. Phys. Lett.* **68** 1775
[2] Nakagawara O, Toyota Y, Kobayashi M, Yoshino Y, Katayama Y, Tabata H and Kawai T 1996 *J. Appl. Phys.* **80** 388
[3] Alers G B, Fleming R M, Wang Y H, Dennis B, Pinczuk A, Redinbo G, Urdahl R, Ong E and Hasan Z 1998 *Appl. Phys. Lett.* **72** 1308
[4] Wilk G D, Wallace R M and Anthony J M 2000 *J. Appl. Phys.* **87** 484
[5] Copel M, Gribelyuk M and Gusev E 2000 *Appl. Phys. Lett.* **76** 436
[6] Robertson J and Chen C W 1999 *Appl. Phys. Lett.* **74** 1168
[7] Robertson J 2000 *J. Vac. Sci. Technol.* B **18** 1785; Peacock P W and Robertson J 2002 *J. Appl. Phys.* **92** 4712
[8] Kadlec J and Gundlach K H 1975 *Solid State Commun.* **16** 621
[9] Braunstein A I, Braunstein M, Picus G S and Mead C A 1965 *Phys. Rev. Lett.* **14** 219
[10] Braunstein A I, Braunstein M and Picus G S 1965 *Phys. Rev. Lett.* **15** 956
[11] Schuermeyer F L and Crawford J A 1966 *Appl. Phys. Lett.* **9** 317
[12] Nelson O L and Anderson D E 1966 *J. Appl. Phys.* **37** 77
[13] Goodman A M 1970 *J. Appl. Phys.* **41** 2176

[14] Ludwig W and Korneffel B 1967 *Phys. Status Solidi* **24** K 137

[15] Shepard K W 1965 *J. Appl. Phys.* **36** 796

[16] Szydlo N and Poirier P 1971 *J. Appl. Phys.* **42** 4880

[17] DiMaria D J 1974 *J. Appl. Phys.* **45** 5454

[18] DiMaria D J and Cartier E 1995 *J. Appl. Phys.* **78** 3883

[19] Scarpa A, Paccagnella A, Montera F, Gibaudo G, Pananakakis G, Ghidini G and Fuochi P G 1997 *IEEE. Trans. Nucl. Sci.* **NS-44** 1818

[20] Afanas'ev V V and Stesmans A 1999 *J. Electrochem. Soc.* **146** 3409

[21] Svensson C and Lundsrom I 1973 *J. Appl. Phys.* **44** 4657

[22] Houssa M, Tuominen M, Naili M, Afanas'ev V V, Stesmans A, Haukka S and Heyns M M 2000 *J. Appl. Phys.* **87** 8615

[23] See, e.g., Adamchuk V K and Afanas'ev V V 1992 *Prog. Surf. Sci.* **41** 111

[24] Powell R J 1970 *J. Appl. Phys.* **41** 2424

[25] DiStefano T H and Eastman D E 1971 *Solid State Commun.* **9** 2259

[26] Afanas'ev V V, Houssa M, Stesmans A and Heyns M M 2002 *J. Appl. Phys.* **91** 3079

[27] Afanas'ev V V, Houssa M, Stesmans A, Adriaenssens G J and Heyns M M 2002 *J. Non-Cryst. Solids* **303** 69

[28] Miyazaki S, Narasaki M, Ogasawara O and Hirose M 2001 *Microelectron. Eng.* **59** 373

[29] Miyazaki S 2001 *J. Vac. Sci. Technol.* B **19** 2212

[30] Yu H Y, Li M F, Cho B J, Yeo C C, Joo M S, Kwong D L, Pan J S, Ang C H, Zheng J Z and Ramanathan S 2002 *Appl. Phys. Lett.* **81** 376

[31] French R H, Müllejans H and Jones D J 1998 *J. Am. Ceram. Soc.* **81** 2549

[32] Ritala M and Leskela M 1994 *Appl. Surf. Sci.* **75** 333

[33] Ritala M, Kukli K, Rahtu A, Raisanen P I, Leskela M, Sajavaara T and Keininen J 2000 *Science* **288** 319

[34] Houssa M, Naili M, Zhao C, Bender H, Heyns M M and Stesmans A 2001 *Semicond. Sci. Technol.* **16** 31

[35] Campbell S A, Ma T Z, Smith R, Gladfelter G L and Chen F 2001 *Microelectron. Eng.* **59** 361

[36] Afanas'ev V V and Stesmans A 1997 *Phys. Rev. Lett.* **78** 2437

[37] Afanas'ev V V, Stesmans A, Chen F, Shi X and Campbell S A 2002 *Appl. Phys. Lett.* **81** 1053

[38] Afanas'ev V V, Bassler M, Pensl G, Schulz M J and Stein von Kamienski E 1996 *J. Appl. Phys.* **79** 3108

[39] Afanas'ev V V, Stesmans A and Andersson M O 1996 *Phys. Rev.* B **54** 10820

[40] Lee D O, Roman P, Wu C T, Mahoney W, Horn M, Mumbauer P, Brubaker M, Grant R and Ruzyllo J 2001 *Microelectron. Eng.* **59** 405

[41] French R H, Glass S J, Ohuchi F S, Xu Y-N and Ching W Y 1994 *Phys. Rev.* B **49** 5133

[42] Ealet B, Elyakhloufi M H, Gillet E and Ricci M 1994 *Thin Solid Films* **250** 92

[43] Nam S-W, Yoo J-H, Nam S, Choi H-J, Lee D, Ko D-H, Moon J-H, Ku J-H and Choi S 2002 *J. Non-Cryst. Solids* **303** 139

[44] Afanas'ev V V, Stesmans A, Mrstik B J and Zhao C 2002 *Appl. Phys. Lett.* **81** 1678

[45] Afanas'ev V V, Stesmans A and Tsai W 2003 *Appl. Phys. Lett.* **83** 245

[46] Lucovsky G 2002 *J. Non-Cryst. Solids* **303** 40

[47] Di Stefano T H and Lewis J E 1974 *J. Vac. Sci. Technol.* **11** 1020

[48] Afanas'ev V V, Houssa M, Stesmans A and Heyns M M 2001 *Appl. Phys. Lett.* **78** 3073

[49] Ludeke R, Cuberes M T and Cartier E 2000 *Appl. Phys. Lett.* **76** 2886; *J. Vac. Sci. Technol.* B **18** 2153

[50] Goodman A M 1968 *Appl. Phys. Lett.* **13** 275
[51] Fowler R H 1931 *Phys. Rev.* **38** 45
[52] Deal B E, Snow E H and Mead C A 1966 *J. Phys. Chem. Solids* **27** 1873
[53] Powell R J and Beairsto R C 1973 *Solid State Electron.* **16** 265
[54] Solomon P M and DiMaria D J 1981 *J. Appl. Phys.* **52** 5867
[55] Pauling L 1960 *The Nature of The Chemical Bond* 3rd edn (Ithaca, NY: Cornell University Press)
[56] Kurtin S, McGill T C and Mead C A 1969 *Phys. Rev. Lett.* **22** 1433
[57] Wang C J and DiStefano T H 1975 *CRC Crit. Rev. Solid State Sci.* **5** 327
[58] Fowler R H and Nordheim L 1928 *Proc. R. Soc. A* **119** 173

Chapter 3.4

Electrical characterization, modelling and simulation of MOS structures with high-κ gate stacks

Jean-Luc Autran, Daniela Munteanu and Michel Houssa

Introduction

High-κ materials have been intensively studied for gate dielectric application in advanced metal–oxide–semiconductor (MOS) devices and integrated circuits. A great challenge for these new material replacement solutions is the control of the interface region between the deposited dielectric and the silicon substrate. Since an SiO_x or a silicate film is possibly formed, the presence of this interfacial layer can significantly impact the electrical characteristics of the global MOS system, leading to significant modifications of: (a) the equivalent oxide thickness (EOT) of the gate dielectric stack; (b) the interface properties in terms of interface trap density and/or near-interface oxide trap/charge distributions; (c) the gate leakage conduction mechanisms. As a direct consequence, the development of dedicated experimental and simulation methods to address these effects is of primary importance for the accurate characterization of MOS devices with alternative gate dielectrics. In this framework, the purpose of this chapter is to give an overview of our modelling methodology and some of our recent results: calculation and analysis of the MOS capacitance–voltage (C–V) response and current–voltage (I–V) characteristics, taking into account various parasitic phenomena like electrical instabilities or physical limitations due to the measurement set-up, electrically active defects, and multi-layered nature of the gate stack. This chapter is organized as follows: in the section 'Mercury-probe technique', some frequency-related effects are emphasized from experimental and modelling viewpoints when using a mercury probe as the measurement set-up or when considering C–V measurements in the presence of interface states. Other performance limitations of ultra-thin oxide or high-κ MOS capacitors are described in the sections 'Performance limitation of ultra-thin oxide or high-κ MOS capacitors' to 'Investigation of the stretch-out effect on C–V curves', such as the polysilicon depletion phenomenon, the carrier quantum confinement effects and the possible stretch-out of the C–V characteristics due to a lateral non-uniform oxide charge distribution in the gate dielectric stacks. Finally, in the section 'Current–voltage (I–V) characteristics', different charge carrier conduction mechanisms through high-κ gate stacks are examined and some recent results concerning the current–voltage response of poly-Si gated structures and the effect of temperature on I–V characteristics are discussed in the sections 'Effect of temperature on the current–voltage characteristics' and 'Current–voltage characteristics of poly-Si gated structures', respectively.

Capacitance–voltage (C–V) characteristics

This section is dedicated to the electrical characterization and modelling of high-κ layers deposited on silicon in terms of capacitance–voltage (C–V) response. In the section 'Mercury-probe technique', a simple analytical model for the C–V technique using a mercury probe is presented. A time-resolved analysis of the C–V and an inverse modelling approach is detailed in the section 'Performance limitation of ultra-thin oxide or high-κ MOS capacitors', to determine the energy distribution and the capture cross-section of interface traps. Finally, in the section 'Frequency modelling of C–V characteristics', we present a tentative modelling approach based on the electrostatic effect of a lateral non-uniform oxide charge distribution to explain the stretch-out effect often observed on high-κ related C–V characteristics.

Mercury-probe technique

The integration of high-κ dielectrics in MOS technologies possibly requires some preliminary 'material studies' to optimize parameters related to the deposition step or post-deposition annealing treatments. To perform such electrical measurements at the front-end level, the use of a mercury probe [1, 2] is a possible alternative solution to the elaboration of patterned MOS capacitors. This offers the possibility to investigate material properties just after the dielectric deposition. Nevertheless, the resulting C–V characteristics cannot be directly analysed in so far as the capacitance measured in the accumulation regime appears to be strongly dependent on the measurement frequency.

To illustrate this point, typical uncorrected C–V measurements obtained on a Hg-MOS structure with HfO_2 gate dielectrics are shown in figure 3.4.1.

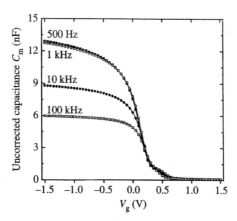

Figure 3.4.1. Uncorrected $C-V$ characteristics obtained on a p-type Hg-MOS structure (boron-doped silicon, $5-10\,\Omega\,\mathrm{cm}$) with HfO_2 gate dielectric stack using a mercury probe and an HP4284A *LCR* meter at different measurement frequencies. The 3.5 nm hafnium oxide was deposited by ALCVD on 0.7 nm native SiO_2 oxide. The total EOT is 1.2 nm. The area of the mercury drop is $0.59\,\mathrm{mm}^2$ (after Garros *et al* [5]).

For this curve, classical corrections of series resistance [3] are not valid here because the large frequency dispersion of the capacitance is not only due to series resistance effects but also to the electrical response of the $\mathrm{Hg/HfO_2}$ interfacial layer of unknown composition (moisture, organic compounds or mercury oxide) [2, 4]. This section presents the modelling of the mercury-dielectric–substrate capacitance and the complete procedure, recently proposed by Garros *et al* [5], to correct measurements from parasitic impedances in the frequency measurement range.

Equivalent circuit model of the Hg-MOS structure
Figure 3.4.2 shows a schematic representation of the Hg-MOS structure formed by the silicon substrate, the dielectric stack and the mercury drop. The key-point of the analysis developed by Garros *et al* [5, 6] is to consider the parasitic effect induced by the interfacial layer between the dielectric stack and the mercury drop contact on the frequency response of the structure. In the following, both the $C-V$ and the conductance–voltage $(G-V)$ characteristics of the structure are assumed to be measured for different measurement frequencies. The different equivalent electrical models of the Hg-MOS structure in accumulation regime are reported in figure 3.4.3. The complete impedance Z takes into account the top interfacial layer, modelled by a capacitance C_p in parallel with a conductance G_p, both in series with the dielectric impedance $(G_{\text{high-}\kappa}, C_{\text{high-}\kappa})$ and with the circuit series resistance R_s. The corresponding impedance measured by the capacitance-metre in parallel mode is Z_m. For convenience, an equivalent impedance of the circuit, Z_{eq}, corresponding to the limit case $R_s = 0$, is also introduced.

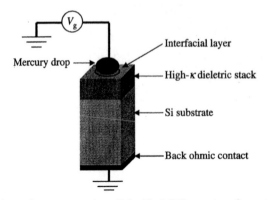

Figure 3.4.2. Schematic representation of the Hg-MOS structure formed by the mercury drop contact, the mercury/dielectrics interfacial layer, the high-κ dielectric layer and the silicon substrate.

Extraction procedure for the model parameters

Figure 3.4.4 shows the experimental variations of G_m and C_m in the accumulation regime when varying the measurement pulsation ω. Over a wide frequency range, both the capacitance and the conductance show large variations which can be perfectly fitted using the complete equivalent model

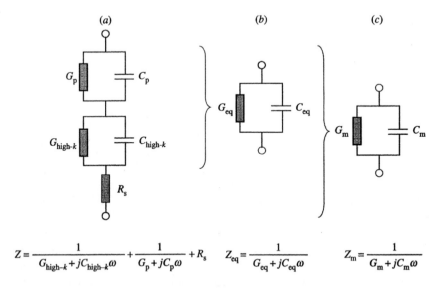

Figure 3.4.3. Small-signal equivalent models of the Hg-MOS structure in the accumulation regime ($V_g \ll 0$). (a) complete impedance Z with $R_s \neq 0$; (b) equivalent impedance Z_{eq} with $R_s = 0$; (c) equivalent parallel impedance Z_m measured by the capacitance meter (after Garros *et al* [5]).

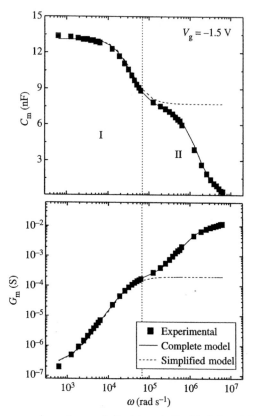

Figure 3.4.4. Frequency dependence of C_m (*a*) and G_m (*b*) deduced from experimental measurements at $V_g = -1.5\,\mathrm{V}$ (same sample as in figure 3.4.1) and calculated using the complete ($R_s \neq 0$) and the simplified models ($R_s = 0$) (after Garros *et al* [5]).

$Z(\omega)$ with a single set of parameters (R_p, C_p, G_p, $C_{\text{high-}\kappa}$, $G_{\text{high-}\kappa}$) given below. The procedure to correctly extract these parameters can be summarized as follows:

(a) At high frequencies (domain II in figure 3.4.4) and as Z_{eq} remains constant, Z variations are only due to series resistance effects. R_s is thus directly determined from experimental data at high frequencies using the following equation:

$$R_s = \frac{G_m}{G_m^2 + C_m^2 \omega^2} \tag{3.4.1}$$

(b) If one remarks that the simplified model $Z_{\text{eq}}(\omega)$ successfully fits experimental data at low and medium frequencies (domain I in figure 3.4.4), the following set of equations can be deduced from the equation $Z = Z_{eq}$ with $R_s = 0$, i.e.:

$$C_{eq} = C_r \frac{1 + \omega^2/\omega_1^2}{1 + \omega^2/\omega_2^2} \qquad (3.4.2)$$

$$G_{eq} = G_r \frac{1 + \omega^2/\omega_3^2}{1 + \omega^2/\omega_4^2} \qquad (3.4.3)$$

$$C_r = \frac{G_p^2 C_{high-\kappa} + G_{high-\kappa}^2 C_p}{(G_p + G_{high-\kappa})^2} \qquad (3.4.4)$$

$$G_r = \frac{G_p G_{high-\kappa}}{G_p + G_{high-\kappa}} \qquad (3.4.5)$$

where ω_1, ω_2, ω_3 and ω_4 are four cut-off pulsations.

Considering that the present approach is limited to the reasonable case $G_{high-\kappa} \ll G_p$, i.e., the conductance of the interfacial layer is much more important than that of the gate dielectrics, the constant and maximum value of C_m obtained at low frequency can be identified with $C_{high-\kappa}$ (equations (3.4.2) and (3.4.4)), and the minimum value of G_m with $G_{high-\kappa}$ (equations (3.4.3) and (3.4.5)). In this latter case, if $G_m(\omega)$ still does not exhibit saturation as shown in figure 3.4.4, $G_{high-\kappa}$ can be more accurately measured through an *I–V* characteristic measured on the same structure.

(c) Once R_s, $C_{high-\kappa}$ and $G_{high-\kappa}$ are determined, the complex impedance of the interfacial layer $(G_p + jC_p\omega)^{-1}$ can be obtained by minimizing the difference ΔZ between modelled and experimental impedances, respectively, Z and Z_m, over the whole frequency interval. One has to numerically solve the implicit equation $\Delta Z = 0$ with:

$$\Delta Z = Z_m \frac{1}{G_{high-\kappa} + jC_{high-\kappa}} - R_s - \frac{1}{G_p + jC_p\omega} \qquad (3.4.6)$$

(d) Using the extracted set of parameters (R_s, C_p, G_p, $C_{high-\kappa}$, $G_{high-\kappa}$) and assuming that these values do not depend on the gate voltage, it is then possible to correct experimental *C–V* data from the effects of interfacial layer and series resistance for all the different measurement frequencies. The corrected capacitance $C_C(V_g)$ is then given by:

$$C_C(V_g) = \frac{1}{\omega} \text{Im} \left\{ \left[Z_m(V_g) - R_s - \frac{1}{G_p + jC_p\omega} \right]^{-1} \right\} \qquad (3.4.7)$$

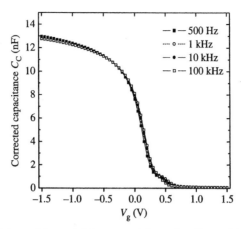

Figure 3.4.5. *C–V* data of figure 3.4.1 corrected from the effects of interfacial layer and series resistance using the proposed model.

Experimental verification

To illustrate this last point, figure 3.4.5 shows $C_C(V_g)$ characteristics calculated from data of figure 3.4.1 with the above procedure. R_s, C_p, G_p, $C_{\text{high-}\kappa}$ and $G_{\text{high-}\kappa}$ are found to be, respectively, equal to 83 Ω, 18.4 nF, 1.18 mS, 13.1 nF and 0.17 μS. It turns out that the interfacial layer has an EOT = 1.1 nm and behaves as a very leaky oxide according to the conductance value. The gate dielectric stack is found to have an EOT = 1.2 nm after quantum mechanical correction (see the section 'Performance limitation of ultra-thin oxide or high-κ MOS capacitors'). Experimentally, this example also shows that the technique is suitable for EOTs down to 1.2 nm for high-κ materials whereas it is limited to ~3 nm for SiO_2 layers since leakage currents for such films induce large errors on low-frequency measurements using the capacitance-meter.

Performance limitation of ultra-thin oxide or high-κ MOS capacitors

The EOT is one of the most fundamental parameters that define a gate dielectric stack. As shown in the section 'Mercury-probe technique', the extraction of this EOT can be dramatically impacted, if it is not corrected, by parasitic phenomena due to the measurement set-up and/or the experimental procedure. But other physical phenomena, intrinsic to the MOS system, can also induce significant deviations with respect to the response of an ideal structure. In this section, such physical limitations are reported: they concern the depletion phenomenon that occurs in polysilicon gates and the carrier quantum confinement effects present in silicon regions (substrate and semiconductor gate) where energy band-bending can become significant.

Polysilicon depletion phenomenon

It is well known that, from an electrical viewpoint, a MOS capacitor with a metallic gate can be regarded as two capacitors (the gate dielectric stack capacitance C_{ox} and the semiconductor capacitance C_{SC}) mounted in series [2]. The total gate-to-substrate capacitance C can therefore be written as

$$\frac{1}{C} = \frac{1}{C_{\text{ox}}} + \frac{1}{C_{\text{SC}}} \tag{3.4.8}$$

where C_{ox} depends on the different permittivity constants ε_i and physical thicknesses t_i of the insulating layers entering the composition of the multi-layer gate stack:

$$(C_{\text{ox}})^{-1} = \sum_i \frac{t_i}{\varepsilon_i} \tag{3.4.9}$$

and C_{SC} varies as:

$$C_{\text{SC}} = -\frac{\mathrm{d}Q_{\text{SC}}}{\mathrm{d}\Psi_{\text{S}}} \tag{3.4.10}$$

where Ψ_{S} is the potential drop in the semiconductor and Q_{SC} is the total semiconductor charge induced by the applied gate voltage V_{G}.

For strongly inverted or accumulated semiconductor surface and in absence of quantum effects, C_{SC} becomes very high and can generally be neglected in front of C_{ox}, ensuring $C = C_{\text{ox}}$. However, for an MOS structure with highly doped polysilicon gate materials, the potential drop in the gate electrode is no longer negligible and has to be accounted for. This phenomenon, known as the polysilicon depletion effect (as manifested in the inversion regime in standard CMOS technologies), significantly reduces the gate-to-substrate MOS capacitance. The total capacitance now reads:

$$\frac{1}{C} = \frac{1}{C_{\text{ox}}} + \frac{1}{C_{\text{SC}}} + \frac{1}{C_{\text{G}}} \tag{3.4.11}$$

where $C_{\text{G}} = -\mathrm{d}Q_{\text{G}}/\mathrm{d}\Psi_{\text{G}}$ is the polysilicon capacitance, Q_{G} is the poly-silicon charge and Ψ_{G} is the polysilicon potential drop.

Calculated characteristics demonstrating the impact of the polysilicon doping level on the theoretical C–V response of an MOS capacitor are shown in figure 3.4.6 [7]. As can be seen from this figure, the polydepletion effect can only be reduced at high doping levels ($5 \times 10^{20}\,\text{cm}^{-3}$) but not entirely suppressed due to the incompressible Debye capacitance in the polysilicon electrode.

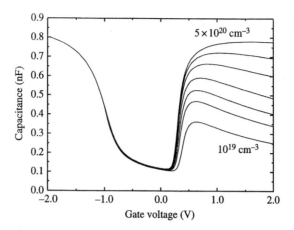

Figure 3.4.6. Theoretical C–V curves calculated for a p-type $HfO_2/SiO_x/Si$ structure with a N^+ polysilicon gate. The polysilicon doping levels are: 10^{19}, 2×10^{19}, 3×10^{19}, 5×10^{19}, 1×10^{20}, 2×10^{20} and 5×10^{20} cm^{-3}. The other simulation parameters are: $V_{FB} = -1.0$ V, EOT $= 1.5 \times 10^{-9}$ m, NA $= 1 \times 10^{18}$ cm^{-3}, $A_G = 160 \times 250$ μm^2 (after Autran [7]).

Carrier quantum confinement effects

As the dimensions of MOS transistors further decrease, scaling rules involve higher doping levels and thinner dielectric layers. This leads to higher electric fields at the dielectric–silicon interface and to a narrowing of the inversion and accumulation layers. Therefore, quantified energy levels appear in these potential wells. The carrier density is thus modified with regard to the classical model and this modification has a serious effect on the C–V characteristics of MOS structures, as explained in the following.

First, let us consider the band diagram of a p-type MOS structure shown in figure 3.4.7. In such a system, the electrostatic potential ϕ verifies the one-dimension Poisson equation:

$$\frac{d}{dx}\left(\varepsilon(x)\frac{d}{dx}\right)\phi(x) = -\frac{q}{\varepsilon_0}\{p(x) - n(x) - N_{A^-}\} \qquad (3.4.12)$$

where ε is the permittivity, p and n are the free hole and electron concentrations, respectively, and N_{A^-} is the acceptor impurity concentration.

In the presence of quantum effects, electrons and holes are confined in the quantum well formed by the band bending in the semiconductor and the interface band offset (figure 3.4.7). Their eigenenergies (E_{ij}) and envelope wave functions (Ψ_{ij}) are solutions of the time-independent, effective-mass and one-dimensional Schrödinger equation:

$$-\frac{\hbar^2}{2}\frac{d}{dx}\left(\frac{1}{m_j^*(x)}\frac{d}{dx}\right)\Psi_{ij}(x) + V(x)\Psi_{ij}(x) = E_{ij}\Psi_{ij}(x) \qquad (3.4.13)$$

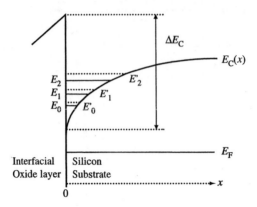

Figure 3.4.7. Band diagram of a p-type MOS structure in the inversion regime showing the quantization of the conduction band into discrete energy levels. Two series of levels are considered in this case, for electrons with a longitudinal mass and for electrons with a transverse mass respectively. ΔE_C refers to the conduction band offset at the interfacial oxide/Si interface.

where $\hbar$ is the reduced Planck constant, m_j^* is the effective mass of carriers and V is the potential energy given by $V(x) = -q\phi(x) + \Delta E_C(x)$, where $\Delta E_C(x)$ is the band offset at the oxide–substrate interface.

In equation (3.4.13), subscript i denotes the energy sub-band ($i = 1, 2, 3, \ldots$) and subscript j denotes the semiconductor valley. As a result, the electron and hole concentrations in the structure can be expressed as follows:

$$n(x) = \frac{kT}{\pi\hbar^2}\sum_j\sum_i g_j m_{d,j}^* \ln\left[1 + \exp\left(\frac{E_F - E_{ij}}{kT}\right)\right]|\Psi_{ij}(x)|^2 \qquad (3.4.14)$$

$$p(x) = \frac{kT}{\pi\hbar^2}\sum_j\sum_i g_j m_{d,j}^* \ln\left[1 + \exp\left(\frac{E_{ij} - E_F}{kT}\right)\right]|\Psi_{ij}(x)|^2 \qquad (3.4.15)$$

where k is the Boltzmann constant, T is the absolute temperature, $m_{d,j}^*$ is the mass of the 2D density of states, g_j is the valley degeneracy and E_F is the Fermi level.

Equations (3.4.12) and (3.4.13) form a system of coupled equations that must be self-consistently solved [8]. It must be also noticed that, in these two equations, both the dielectric constant and the carrier effective mass are position dependent.

Due to the particular band structure of silicon, equation (3.4.13) must be separately solved twice for electrons, considering electrons with a longitudinal mass ($m_\ell^*, g_\ell = 2, m_{d,\ell}^* = m_t^*$) and electrons with a transverse mass $\left(m_t^*, g_t = 4, m_{d,t}^* = \sqrt{m_\ell^* m_t^*}\right)$. Recent experimental works and theoretical considerations give: $0.9163 \times m_0 < m_\ell^* < 0.98 \times m_0$ with a relatively weak temperature dependence and $m_t^* = 0.191 \times m_0 \times E_g(0)/E_g(T)$ where m_0 is the electron mass and $E_g(T)$ is the silicon bandgap at temperature T [9, 10]. The uncertainty on these values should be $\sim 5\%$.

For holes, the silicon band structure is more complex than for electrons and three different sub-bands must be considered to solve equation (3.4.13), respectively for heavy (m_{hh}^*), light (m_{lh}^*) and splitting (m_{sh}^*) holes [11]. Unfortunately, the coupling as well as the nonparabolicity of these sub-bands is important [12]: the use of a constant effective mass per energy sub-band should be considered as a poor approximation which may introduce an uncertainty on hole-related quantities. It could also explain why the values reported in the literature are quite dispersed and form an indeterminate set of data [10]. Moreover, experimental hole mass values are obtained from cyclotron resonance experiments at low temperature (1–4 K), and are used in a lot of works for much higher temperatures up to 300 K. Table 3.4.1 summarizes different data collected in the literature for heavy, light and splitting holes and corresponding density-of-states masses (when available). Data show differences up to 40% especially for heavy-hole effective masses. Note that the uncertainty may be higher for effective masses related to dielectric materials for which only a few data are currently available in the literature.

When computing the carrier distributions with equations (3.4.14) and (3.4.15), quantum confinement effects are found to induce: (a) a drastic reduction of the charge density with respect to the one obtained in the classical case (i.e. the Fermi–Dirac distribution for a three-dimensional electronic gas) and (b) a displacement of the charge centroid of a few Angstroms in the direction of the substrate bulk. Because this increase of the charge sheet in accumulation or in inversion is gate voltage dependent, it results in a non-constant capacitance value in these regimes, as clearly shown in figure 3.4.8. This figure illustrates the quantum calculation of a complete $C–V$ curve and its comparison with experimental data obtained on a p-type W/TiN/HfO$_2$/Si structure. Figure 3.4.8 also shows the very good matching between experimental data and simulated characteristics that can be used for

Table 3.4.1. Effective masses ($m*$) and density-of-state masses (m_d^*) for holes in silicon reported in the literature. Subscripts lh, hh and s refer to light, heavy and splitting holes, respectively.

	[10]	[13]	[14]	[11]	[15]
m_{lh}^*	0.537	0.49	0.52	0.29	0.49
m_{hh}^*	0.153	0.16	0.16	0.20	0.16
m_s^*	0.234	0.29	0.24	0.29	0.24
$m_{d,lh}^*$				0.251	
$m_{d,hh}^*$				0.645	

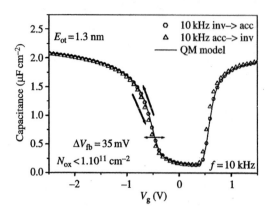

Figure 3.4.8. Experimental and quantum-mechanical simulated $C-V$ curves related to a W/TiN/HfO$_2$/Si structure with EOT = 1.5 nm (after Guillemot *et al* [16]).

the extraction of the oxide thickness. With such an extraction procedure based on the complete fit of the $C-V$ characteristic in accumulation, values of extracted thickness have been found to coincide within ~1.5 Å with values measured by spectroscopic ellipsometry and high-resolution transmission electron microscopy (HRTEM) [8].

Frequency modelling of $C-V$ characteristics

For high-κ materials used as gate insulators in MOS devices, great efforts are being focused on improving electrical properties of the interfacial region, more particularly in terms of interface state density or surface roughness which can directly impact carrier mobility [16–18]. Interface trap density in such high-κ-based devices can be basically extracted from $C-V$ measurements in quasi-static mode or at low frequencies (typically a few Hertz, depending on the value of the capture cross-sections of the traps). For higher frequencies, the trapping kinetic of the traps must be correctly modelled in order to extract quantitative information about their energy distribution and/or their capture cross-section. The first modelling of such trapping transient effects on $C-V$ curves was originally developed by Heiman *et al* [19]. In this section, we present an alternative modelling, recently developed by Masson *et al* [20]. This method is based on the evaluation of the filling probability of a trap versus its energy position in the silicon bandgap. It presents the main advantage to treat any interface trap density or capture cross-section distributions (continuum or no-continuum) at any tunnelling distance in the dielectric from the interface. It can also be used to compute the dynamic response of the traps under transient bias, such as encountered in several MOS characterization techniques: charge pumping, $C-V$, current transient spectroscopy or conductance. In the following, this method is applied, through an inverse modelling approach, to the extraction of interface trap parameters related to the $Si/HfSi_xO_y/HfO_2$ system.

Transient modelling of trap activity and capacitance response

In the approach proposed by Masson *et al*, the calculation of the *C–V* curve at a given measurement frequency (F) is based on the evaluation of the differential capacitance associated with the interface traps (C_{it}) when a small modulation signal ($\delta V_G(t) = A\sin(2\pi F \times t)$) is superimposed on the static gate voltage (V_G).

Considering an interface trap associated with the energy level E_t in the silicon bandgap, the time evolution of its filling probability (f_t) verifies the first-order differential equation [21]:

$$\frac{df_t(t,E_t)}{dt} = c_n + e_p - (e_n + c_n + e_p + c_p)f_t \qquad (3.4.16)$$

where c_n, c_p, e_n and e_p are the capture and emission rates for electrons and holes.

In the framework of the Schockley–Read–Hall (SRH) theory [22], these parameters are dependent on the capture cross-sections for electrons and holes (respectively σ_n and σ_p), the carrier thermal velocity (v_{th}) and the trap energy level in the silicon bandgap (E_t). The relationship between the surface potential, Ψ_S, and V_G is given by:

$$V_G = \Phi_{MS} - \frac{Q_{ox}}{C_{ox}} + \Psi_S - \frac{Q_{SC}}{C_{ox}} - \frac{Q_{it}}{C_{ox}} \qquad (3.4.17)$$

where Φ_{MS} is the gate-semiconductor work function difference, Q_{ox} is the oxide fixed charge, Q_{SC} is the total semiconductor charge, Q_{it} is the charge trapped in interface traps and C_{ox} is the oxide capacitance (per unit area).

The quantity Q_{it} is calculated (at any time) from the following equation

$$Q_{it}(t) = q\int_{E_{VS}}^{E_{iS}} D_{it}(E_t)(1 - f_t(t,E_t))dE_t - q\int_{E_{iS}}^{E_{VC}} D_{it}(E_t)f_t(t,E_t)dE_t \qquad (3.4.18)$$

where $D_{it}(E_t)$ is the energy-distribution of the traps present at the interface (we suppose here an amphoteric behaviour), E_{VS} and E_{CS} are the energy levels of the semiconductor valence band and conduction band at the interface, respectively, and E_{iS} is the intrinsic level at the interface.

The differential capacitance of the traps C_{it} can thus be estimated as follows:

$$C_{it} = \left| \frac{Max(Q_{it}(t)) - Min(Q_{it}(t))}{\delta\psi_S} \right| \qquad (3.4.19)$$

where $\delta \Psi_{\text{S}}$ corresponds to the small surface potential variation imposed by δV_{G} around the particular value of the surface potential $\Psi_{\text{S}}(V_{\text{G}})$ (determined by the static gate voltage V_{G}).

Finally, the total capacitance C of the structure is given by:

$$\frac{1}{C} = \frac{1}{C_{\text{ox}}} + \frac{1}{C_{\text{SC}} + C_{\text{it}}} \qquad (3.4.20)$$

where C_{ox} and C_{SC} are the oxide and the semiconductor capacitances, respectively.

Trap characterization of the HfSi$_x$O$_y$/Si system

Equations (3.4.16)–(3.4.20) can be used to extract, via an 'inverse modelling approach', interface trap parameters with a good sensitivity and accuracy. We illustrate this for n-type MOS capacitors with a gate dielectric stack of 7.7 nm HfO$_2$ oxide deposited by atomic layer chemical vapour deposition (ALCVD) on a 0.7 nm native Si oxide film. Gate electrodes were formed by low pressure chemical vapour deposition (LPCVD) of 3500 Å N^{2+} poly-silicon at 580°C. Finally, doping activation was obtained by a 750°C rapid thermal processing (RTP) anneal under nitrogen ambient for 1 min. The value of the dielectric constant of the interfacial layer is 4.0 before gate deposition, which seems to indicate, in the absence of any chemical data, that this layer is close to stochiometric SiO$_2$. After gate deposition, this value becomes approximately three times higher, demonstrating a partial transformation into a hafnium silicate HfSi$_x$O$_y$ layer [20, 23].

Figure 3.4.9 shows typical C–V characteristics obtained on an MOS structure that was not subjected to any post-gate-deposition annealing

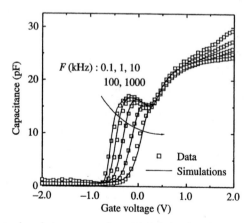

Figure 3.4.9. Comparison between experimental and simulated C–V characteristics for an n-type MOS structure with unannealed HfO$_2$ gate dielectric. The measurement frequency F ranges from 100 Hz to 1 MHz.

treatment. The frequency dependence of the hump, evidenced on the C–V curves near $V_G \sim -0.25\,\mathrm{V}$, suggests the presence of a large number of acceptor interface traps located in the upper part of the Si bandgap. This trap density distribution was determined by such an inverse modelling approach from the shape of the D_{it} hump and the electron capture cross-section of the traps from the 'extinction' of the hump versus the measurement frequency. Using an iterative optimization scheme and starting with a Gaussian distribution of traps, the following parameters are deduced from experimental data: $D_{it-max} = 13.2 \times 10^{12}\,\mathrm{eV}^{-1}\,\mathrm{cm}^{-2}$, $\Delta E = 0.32\,\mathrm{eV}$, $E_{max} = E_i + 0.26\,\mathrm{eV}$, $\sigma_n = 1.5 \times 10^{-17}\,\mathrm{cm}^2$. Figure 3.4.9 also shows the corresponding theoretical curves calculated at the end of the optimization process. A very good agreement between the two sets of data is observed which confirms *a posteriori* the initial choice of a Gaussian distribution for the interface traps. Figure 3.4.10 shows the resulting D_{it} spectrum and the different distribution parameters obtained from the comparison between the experimental data and the simulations. Concerning the extracted capture cross-section $(1.5 \times 10^{-17}\,\mathrm{cm}^2)$, this value appears to be approximately 10 times lower than typical values relative to the dangling bonds (P_b centres) at the Si/SiO$_2$ interface [24]. This suggests a good agreement with a Coulombic centre model that predicts a capture cross-section inversely proportional to the square of the dielectric permittivity [25], i.e., $\sigma \sim \varepsilon^{-2}$. Indeed, the dielectric constant of the HfSi$_x$O$_y$ layer is expected to be around $\varepsilon \sim 10$–12, which is about three times the value of $\varepsilon_{SiO_2} \sim 3.9$. Consequently, the capture cross-section of P_b-like centres at the HfSi$_x$O$_y$/Si interface should be about $(3)^2 \sim 9$ times lower than the cross-section related to the SiO$_2$/Si interface. In future works, the present approach should be easily generalized to other

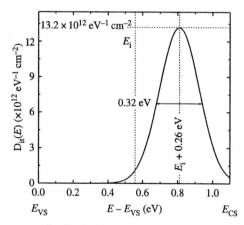

Figure 3.4.10. Energy trap distribution in the silicon band-gap extracted with the present model from C–V data of figure 3.4.7.

dielectric–semiconductor systems to provide rapid and accurate characterization of interface traps or near-interfacial oxide traps.

Investigation of the stretch-out effect on $C-V$ curves

It is well known that capacitance–voltage ($C-V$) characteristics of non-ideal MOS structures can be stretched out along the voltage axis. This voltage stretch-out is classically attributed to an increase in the interface trap level density whereas a pure shift of the $C-V$ curve towards negative (respectively, positive) voltages is due to the presence of positive (respectively, negative) charge in the oxide [2, 26]. Other sources of fluctuations have been also identified, for example the random nature of the doping distribution [27] or local oxide thickness fluctuations in ultra-thin oxide devices [28]. In several studies, it has been experimentally shown that high-κ gate dielectric stacks can present a rather good interface quality with the underlying substrate (due to the presence of a thin SiO$_x$ interfacial layer) but, at the same time, exhibit highly stretched $C-V$ curves [29–34]. Because interface states cannot be invoked in this case, the non-homogenous nature of such deposited gate stacks, and especially in terms of fixed oxide charges, could be at the origin of the particular $C-V$ behaviour. In this section, we report recent results [35, 36] which clearly show that a lateral non-uniform oxide charge distribution in MOS devices can also induce a stretch-out of the capacitance–voltage characteristics. The origin of this effect is due to fluctuations in the surface potential, which varies laterally over the area of the device.

Simulated structure and numerical details

The present analysis is based on the two-dimensional solution of the Poisson equation using a finite difference scheme on a uniform mesh. The schematic description of the simulated structure is presented in figure 3.4.11. The 2D domain simulation (2D grid of NX × NY nodes shown in figure 3.4.11) includes the high-permittivity dielectric layer (NX × NHK points), the interfacial oxide layer (NX × NOX points) and the silicon substrate. The high-κ layer permittivity considered in this work is $\varepsilon_R = 20$ and the p-type substrate doping is $N_A = 1 \times 10^{18}\,\mathrm{cm}^{-3}$. The gate is biased from accumulation to inversion regimes (gate voltage V_G), while the substrate is grounded. Floating boundary conditions have been assumed for the two vertical boundaries of the structure.

Effect of a discrete oxide charge

The first step of this work was to highlight and quantify the effect of a local density of fixed oxide charge on the 2D potential distribution. This effect is illustrated in figure 3.4.12 for a point charge located in the middle of the

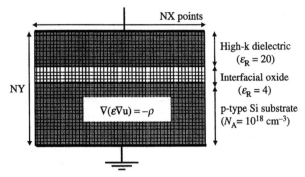

Figure 3.4.11. Schematic representation of a MOS structure with a two-layer gate dielectric stack. The regular grid (NX × NY points) corresponds to the simulation domain on which is solved the bi-dimensional Poisson equation. *x* and *y* directions, respectively, refer to horizontal and vertical co-ordinate axis.

structure $(x = NX/2)$, in the oxide just above the oxide/semiconductor interface $(y = NHK + NOX)$. The presence of such a local charge density induces important 2D potential fluctuations in the underlying semiconductor region and especially at the semiconductor surface. The direct consequence is a visible stretching of the C–V curve, as compared to the reference curve for which the gate stack is totally free of charge. This very simple example immediately shows that this stretching effect is different from the one obtained when considering interface traps. Indeed, the total capacitance of the structure appears to be greater or smaller than the reference capacitance, which is fundamentally different than in the case where interface traps are present. In this case, the capacitance contribution of

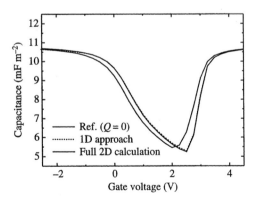

Figure 3.4.12. Theoretical C–V curve resulting from the presence of a point density of fixed oxide charge $(\rho = 10^{27}\,\mathrm{cm}^{-3})$ on a single node of the 2D grid $(x = NX/2,\ y = NHK + NOX)$ located on the adjacent line above the oxide/silicon interface. The reference curve (no charge in the gate stack) and the one obtained with the one-dimensional approach are also plotted.

the interface traps (in parallel with the semiconductor capacitance) always contribute to increase the total capacitance of the structure, in all semiconductor regimes.

Another point highlighted by this elementary example is that a bi-dimensional approach is mandatory for the study of the stretching effects, even if only a point oxide charge is present in the gate stack. Considering the same point charge repartition as previously, we verified this statement by comparing the $C-V$ calculated using a uni-dimensional (1D) approach with the one obtained in full 2D calculation. The 1D approach considers NX independent capacitors in parallel: NX $-$ 1 elements with zero charge and a single one with the charge density previously defined. The 1D calculation of the total capacitance reduces to the evaluation of the elementary $C-V$ response of the NX capacitors connected in parallel. In other words, the Poisson equation is solved NX times in one dimension along the [1,NY] lines (NX independent systems). Figure 3.4.12 shows that the 1D curve fits the reference curve (i.e. without any oxide charge), since the 2D curve presents an important stretching effect. As a result, the 1D approach fails to take into account local charge in the stack structure, since this method does not take into account the effect of the charge on the adjacent nodes of the grid, as in the case of the 2D approach. In the next sub-sections, the influence of several oxide charge patterns on $C-V$ is investigated using full 2D calculation.

Linear distribution in the interfacial oxide
One of the most elementary charge patterns to be investigated is a linear gradient of oxide charge in the interfacial oxide layer. We thus consider a charge density in the interfacial layer which varies linearly from 0 for $x = 0$ to Q_{max} for $x = $ NX. Figure 3.4.13 shows the corresponding $C-V$ curves simulated for different Q_{max}. We observe an important deformation of the curves with respect to the reference one, mainly characterized by: (a) a voltage

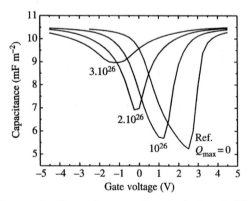

Figure 3.4.13. Influence of a linear charge gradient in the interfacial layer on the $C-V$ response of the MOS structure defined in figure 3.4.10. Q_{max} is expressed in $q\,\mathrm{m}^{-3}$.

shift towards negative voltages, (b) an increase in the minimum capacitance and (c) an enlargement of the $C–V$ curve. The first deviation is a direct consequence of the positive charge in the oxide (a negative oxide charge would induce a positive shift), the second one is due to the fact that the presence of this charge reduces the voltage domain corresponding to the depletion regime and, finally, the third one is a direct consequence of the x-coordinate dependence of the surface potential.

Random distribution in the stack
We now examine the effect of a random charge distribution in the gate stack. The charge pattern is generated by writing N times a charge increment, ΔQ, in the two-dimensional domain corresponding to the dielectric stack. At each iteration, the (x,y) coordinates of the written node are calculated using a random number generator. Figure 3.4.14 shows an example of a $C–V$ curve corresponding to a random pattern defined by $N = 20$ and $\Delta Q = 10^{27}\, q\, \mathrm{m}^{-3}$. The potential perturbations induced in this case primarily affect the semiconductor region in contact with the interfacial layer. As a result, the corresponding $C–V$ curve presents a visible stretching effect. Note that this stretch-out disappears when increasing the parameter N, simply because the charge pattern tends to become uniform.

Charged grain boundaries in the high-permittivity layer
The last oxide charge pattern examined in this section is inspired from the situation encountered in MOS devices with crystallized high-permittivity materials. This tentative modelling approach is summarized in figure 3.4.15 for a typical high-permittivity/interfacial oxide bi-layer system. From this HRTEM image, we model the high-κ layer as a juxtaposition of crystalline

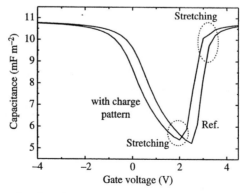

Figure 3.4.14. Example of random pattern defined by the parameters $N = 20$ and $\Delta Q = 10^{27}\, q\, \mathrm{m}^{-3}$ (see text) and its impact on the $C–V$ response of the MOS structure of figure 3.4.10. The reference curve (no charge in the gate stack) is also plotted.

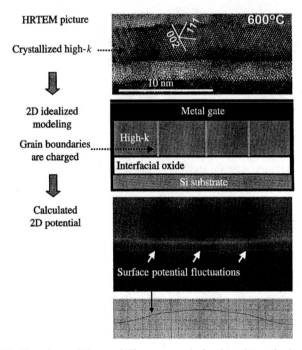

Figure 3.4.15. Flowchart of the modelling approach developed to take into account the fixed charge present at grain boundaries in the crystalline high permittivity dielectric of a bi-layer MOS structure. The two-dimensional stack has been idealized from an HRTEM cross-section image. Vertical grain boundaries have been considered with a charge density of $10^{27} q\,m^{-3}$. The two-dimensional potential chart and the corresponding cross-section at the semiconductor surface are represented for $V_G = 2\,V$.

grains with vertical and charged grain boundaries. Introducing this particular charge pattern into the simulator, we evaluate its effect on the 2D potential repartition and we quantify the induced surface potential fluctuations, as illustrated in figure 3.4.15. We see that the charge localized at grain boundaries significantly affects the potential repartition not only in the high-permittivity layer, but also in the underlying oxide and semiconductor regions. The surface potential profile, also shown in figure 3.4.15, quantifies these modifications induced at the semiconductor surface. The C–V response of the complete MOS system described in figure 3.4.15 is shown in figure 3.4.16. This C–V curve is shifted towards negative voltages always for the same reason (the charge present in the gate stack is positive), but if we translate this curve to compare its shape with the reference C–V, we note evidence of the stretching effect induced by the crystalline state of the high-permittivity layer, as previously modelled. This result is important because it shows that the microscopic structure of the layer could affect the electrical

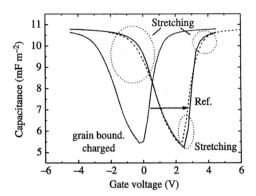

Figure 3.4.16. *C–V* response related to the MOS system of figure 3.4.13. The dotted line corresponds to a pure translation of the same curve to highlight stretching effects by comparison with the reference curve (no charge in the gate stack).

properties of the semiconductor surface, of course if we make the assumption that grain boundaries are electrically charged. This last hypothesis is only speculative for the moment because, to our knowledge, there are no experimental data available for such high-permittivity materials. But, we can reasonably suppose that charge trapping should increase at grain boundaries and lead to fixed charge densities, as in the well-known case of polysilicon, for example. The estimation of the grain boundary charge density in crystalline high-permittivity materials, for example, by using an electrostatic force microscope, should provide an interesting research direction for future experimental or theoretical works.

Current–voltage (*I–V*) characteristics

In this section, we detail theoretical and experimental aspects of charge transport in high-κ materials. Experimental details about device preparation and measurement procedures are indicated in the section 'Experimental details'. The tunnelling response of high-κ gate stacks is carefully examined in the section 'Tunnelling current through the gate stack' and some recent results concerning the *I–V* response of poly-Si gated structures and the effect of temperature on *I–V* characteristics are discussed in the two last sections.

Experimental details

Current–voltage data reported in this section are related to MOS capacitors fabricated with the following technological parameters. (100) n- and p-type 200 mm wafers were first cleaned using a standard wet cleaning process.

An ultra-thin thermal silicon dioxide (SiO_2) or silicon oxynitride (SiON) layer was grown on the substrate. The SiO_2 and SiON layer thicknesses were estimated to be 1.1 and 1.5 nm from spectroscopic ellipsometry and high-resolution cross-sectional transmission electron microscopy measurements (HRTEM). The nitrogen content in the SiON layer was estimated to be about 10 at.% from secondary ion mass spectrometry measurements. A ZrO_2 or HfO_2 layer was next deposited on the Si/SiON substrate, using atomic layer deposition (ALD), with $ZrCl_4$ or $HfCl_4$ and H_2O sources (see the chapter by M. Ritala in the present book). The thickness of this layer was estimated from HRTEM measurements to be either 3 or 4 nm.

MOS capacitors were next defined on these stacks ($100 \times 100 \, \mu m^2$ gate area), with either metal (Au, Al or TiN) or n^+ poly-Si electrodes. Al was sputtered on the poly-Si gate in order to improve the electrical contacts, and the wafer backside also was Al metallized. The samples were finally annealed in N_2/H_2 at 400°C for 30 min. A schematic illustration of the devices analysed in this work is shown in figure 3.4.17, together with an HRTEM picture of an SiO_2/ZrO_2 gate stack.

The capacitance–voltage (C–V) characteristics of the structures were measured using an Agilent 4275A multi-frequency *LCR* meter, with a frequency in the range 10^4–10^6 Hz. The measurements of the current–voltage (I–V) characteristics of the structures were performed with Agilent 4156C or Keithley 4200 semiconductor parameter analysers. I–V measurements were performed in accumulation, i.e., with a positive (negative) gate bias for n-type (p-type) capacitors.

Tunnelling current through the gate stack

The current–voltage characteristics of MOS capacitors with 1.5 nm SiON/3 nm ZrO_2 gate stacks are presented in figure 3.4.18(a), for structures with Al or Au gates, respectively, under substrate (filled symbols) and gate

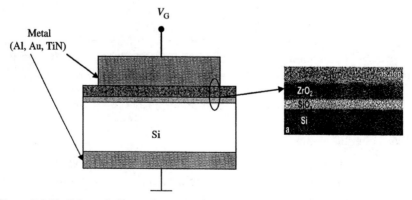

Figure 3.4.17. Schematic illustration of the MOS capacitors with SiO_x/ZrO_2 gate stacks.

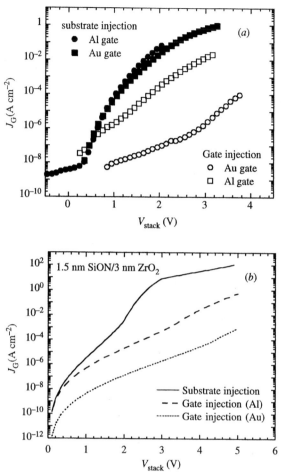

Figure 3.4.18. (*a*) Current–voltage characteristics of MOS capacitors with 1.5 nm/3 nm ZrO$_2$ gate stacks, under substrate (bold symbols) and gate (open symbols) injection. Circles correspond to Al gates and squares to Au gates. (*b*) Simulations of the current–voltage characteristics of the capacitors, considering the tunnelling effect through the SiON/ZrO$_2$ stack.

(open symbols) injection [37]. These data are presented as a function of the voltage across the gate stack, $V_{stack} = V_{ox} + V_{hk}$, obtained by subtracting the flat-band voltage of the structure (determined by *C–V* measurements) and the Si surface potential ψ_s from the gate voltage V_G. A polarity effect is clearly observed in these figures; the current is higher under substrate injection, as compared to gate injection. Besides, the current depends much on the type of gate under gate injection, by contrast to the results observed under substrate injection.

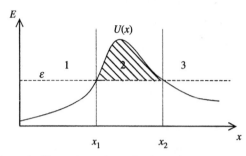

Figure 3.4.19. Schematic illustration of an electron of energy ε tunnelling through a potential barrier $U(x)$ between the classical turning points x_1 and x_2. The hatched zone corresponds to the area defined by the tunnelling distance $(x_1 - x_2)$ and the energy barrier $U(x) - \varepsilon$.

These data can be explained, at least qualitatively, by assuming that the major conduction mechanism, at room temperature, is the tunnelling of electrons through the gate stack [38, 39]. Let us consider an electron of energy ε, tunnelling through a potential barrier $U(x)$, as illustrated in figure 3.4.19. Within the Wentzel–Kramers–Brillouin (WKB) approximation, which is often used to calculate the tunnelling current flowing through MOS devices [40, 41], the electron transmission probability $T(\varepsilon)$ through the potential barrier is given by [42]

$$T(\varepsilon) = \exp\left[\frac{-2i}{\hbar} \int_{x_1}^{x_2} \sqrt{2m(\varepsilon - U(x))}dx\right] \qquad (3.4.21)$$

where x_1 and x_2 are the classical turning points, as illustrated in figure 3.4.19. According to this expression, the transmission probability increases exponentially when the area defined by the tunnelling distance $(x_2 - x_1)$ and the potential barrier $U(x)$ decreases.

The energy band diagram of an Si/1.5 nm SiON/3 nm ZrO$_2$/Al, as determined from internal photoemission measurements [43] (see also the chapter of V.V. Afanas'ev and A. Stesmans in the present book), is presented in figure 3.4.20, under substrate and gate injection, for a stack voltage $V_{stack} = V_{ox} + V_{hk} = 2.5$ V. The barrier height $q\phi_1$ defined by the conduction band offset of the Si substrate and the SiON layer is fixed at 3.2 eV, the conduction band offset $q\phi_2$ between the SiON and ZrO$_2$ layer is fixed at 1.2 eV, and the barrier height between the Al Fermi level and the ZrO$_2$ conduction band is fixed at 1.6 eV. The band gaps of SiON and ZrO$_2$ are fixed at 8.9 and 5.4 eV, respectively. The potential drops in the SiON layer (V_{ox}) and the ZrO$_2$ layer (V_{hk}) are calculated using the Gauss law, i.e.

$$V_{ox} = V_{stack}\left(\frac{\kappa_{ox}t_{hk}}{k_{hk}t_{ox}} + 1\right)^{-1} \qquad (3.4.22)$$

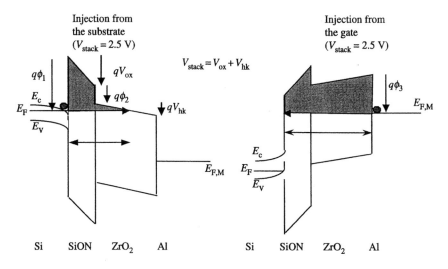

Figure 3.4.20. Schematic energy band diagram of an $Si/SiON/ZrO_2/Al$ structure under substrate and gate injection of electrons.

$$V_{hk} = V_{stack} \left(\frac{\kappa_{hk} t_{ox}}{\kappa_{ox} t_{hk}} + 1 \right)^{-1} \tag{3.4.23}$$

where κ_{ox} (3.9) and κ_{hk} (20) are the relative dielectric constants of SiON and ZrO_2, respectively; t_{ox} and t_{hk} are the thicknesses of these layers. These potential drops are presented in figure 3.4.21 as a function of V_{stack}.

Due to the asymmetry of the band structure of the $SiON/ZrO_2$ stack, and the potential distribution across these layers, it appears clearly from figure 3.4.20 that the area defined by the tunnelling distance and the

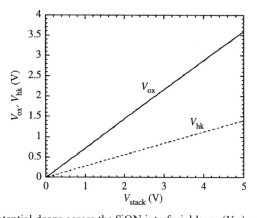

Figure 3.4.21. Potential drops across the SiON interfacial layer (V_{ox}) and the ZrO_2 layer (V_{hk}) as a function of the potential drop across the gate stack (V_{stack}).

potential barrier (shaded areas) is much smaller under substrate injection, compared to gate injection. Consequently, according to equation (3.4.21), the tunnelling current flowing through the structure should to be much higher under substrate injection, in agreement with the data presented in figure 3.4.18(*a*).

The current–voltage characteristics of the devices are first modelled by calculating the tunnelling current J_{tunnel} flowing through the high-κ gate stack, using the expression [38–40]

$$
\begin{aligned}
&J_{tunnel}(V) \\
&= \frac{(m^*_{ox}m^*_{hk})^{1/2}k_BT}{2\pi^2\hbar^3} \int_{E_F}^{\infty} T(E,V)\ln\left[\frac{1+\exp[(E_F-E)/k_BT]}{1+\exp[(E_F-E-qV_G)/k_BT]}\right]dE
\end{aligned}
$$

(3.4.24)

where m^*_{ox} and m^*_{hk} are the electron tunnelling effective masses in the SiON and ZrO_2 layers, respectively, E_F the cathode Fermi level and V_G the applied bias. Under substrate injection, the electron tunnelling probability $T(E,V)$ reads, within the WKB approximation,

$$
\begin{aligned}
T(E,V) &= T_{ox}(E,V)T_{hk}(E,V) \\
&= \exp\left[\frac{-2}{\hbar}\int_0^{x_1}\sqrt{2m^*_{ox}(E-E_{ox}(x))}dx\right] \\
&\quad \times \exp\left[\frac{-2}{\hbar}\int_{x_1}^{x_2}\sqrt{2m^*_{hk}(E-E_{hk}(x))}dx\right]
\end{aligned}
$$

(3.4.25)

where T_{ox} and T_{hk} are the tunnelling probabilities through the SiON and ZrO_2 layers, respectively, x_1 and x_2 are the classical turning points in the SiON and ZrO_2 layers, respectively, $E_{ox}(x)$ and $E_{hk}(x)$ are the conduction bands in the SiON and ZrO_2 layers, respectively,

$$
E_{ox}(x) = q\phi_1 + E_F - E - q\frac{V_{ox}}{t_{ox}}x
$$

(3.4.26)

$$
E_{hk}(x) = q\phi_1 + E_F - E - qV_{ox} - q\phi_2 - q\frac{V_{hk}}{t_{hk}}(x - t_{ox})
$$

(3.4.27)

The expressions of the classical turning points x_1 and x_2 depend on the form of the potential barrier, which can be trapezoidal (direct tunnelling), or triangular (Fowler–Nordheim tunnelling); see, e.g. [38]. Under gate injection, the electron tunnelling probability reads:

$$T(E,V) = T_{hk}(E,V)T_{ox}(E,V)$$

$$= \exp\left[\frac{-2}{\hbar}\int_0^{x_1}\sqrt{2m_{hk}^*(E - E_{hk}(x))}dx\right]$$

$$\times \exp\left[\frac{-2}{\hbar}\int_{x_1}^{x_2}\sqrt{2m_{ox}^*(E - E_{ox}(x))}dx\right] \qquad (3.4.28)$$

where

$$E_{hk}(x) = q\phi_3 + E_F - E - q\frac{V_{hk}}{t_{hkx}}x \qquad (3.4.29)$$

$$E_{ox}(x) = q\phi_3 + E_F - E - qV_{hkx} + q\phi_2 - q\frac{V_{ox}}{t_{ox}}(x - t_{hk}). \qquad (3.4.30)$$

The tunnelling current, calculated according to equations (3.4.24)–(3.4.30), is shown in figure 3.4.18(*b*) as a function of the potential drop across the gate stack. The solid lines represent the current calculated under substrate injection (with Al and Au gates), and the dashed and dotted lines correspond to the case of gate injection, with Al and Au gates, respectively. These simulated current–voltage characteristics were obtained by fixing the following values of the physical parameters: $T = 300\,\text{K}$, $\kappa_{ox} = 3.9$, $\kappa_{hk} = 20$, $t_{ox} = 1.5\,\text{nm}$, $t_{hk} = 3\,\text{nm}$, $m_{ox} = 0.5\,m_0$ [41], $m_{hk} = 0.3\,m_0$ [38, 39], $q\phi_1 = 3.2\,\text{eV}$ [43], $q\phi_2 = 1.2\,\text{eV}$ [43], $q\phi_3 = 1.6\,\text{eV}$ (Al) [44] and $q\phi_3 = 2.4\,\text{eV}$ (Au) [44]. Comparison between the simulations and the experimental results leads to the following findings:

(a) The model explains the polarity effect observed on the current–voltage characteristics of the Si/SiON/ZrO$_2$/metal structures, which can be attributed to the asymmetry of their energy band diagram.

(b) Under gate injection, the simulated tunnelling current is lower for the Au gate, as compared to the Al gate, in agreement with the experimental data. This result can be explained by the different work function of Al (4.2 eV) [26] and Au (5.4 eV) [26], which leads to a higher energy barrier at the Au/ZrO$_2$ interface, compared to the Al/ZrO$_2$ interface.

(c) Under substrate injection, the simulated tunnelling current does not depend on the type of metal gate, consistently with the experimental data. This result arises from the fact that the barrier heights $q\phi_1$ and $q\phi_2$ are fixed by the conduction band offsets at the Si/SiON and SiON/ZrO$_2$ interfaces, respectively, and are thus independent of the gate.

However, the current–voltage characteristics measured under substrate and gate injection cannot be reproduced simultaneously by the tunnelling model described by equations (3.4.24)–(3.4.30), with a consistent set of physical parameters; i.e., other charge transport mechanisms could be involved. This has been checked by investigating the effect of temperature on the current–voltage characteristics of the structures.

Effect of temperature on the current–voltage characteristics

The current–voltage characteristics of an $Si/1\,nm\ SiO_2/4\,nm\ HfO_2/TiN$ structure, measured at 25 and 125°C, under substrate and gate injection, are shown in figures 3.4.22(*a*) and (*b*), respectively [45]. Under substrate injection, one observes that the current depends weakly on temperature for

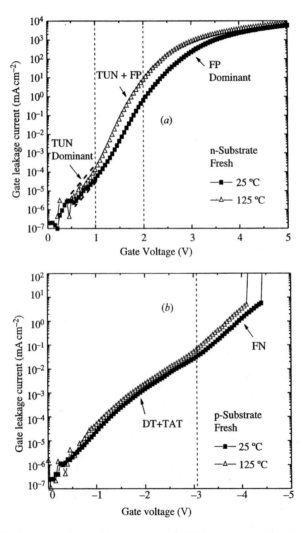

Figure 3.4.22. Current–voltage characteristics of MOS capacitors with $1\,nm\ SiO_2/4\,nm$ HfO_2 gate stacks, at 25 and 125°C, measured under substrate (*a*) and gate (*b*) injection.

$V_G < 1$ V. For $1.0 < V_G < 3.0$ V, the current increases with temperature by about one order of magnitude, from 25 to 125°C. For $V_G > 3$V, the current is again much less dependent on temperature. On the other hand, under gate injection, the current depends weakly on temperature, over most of the gate voltage range. A polarity effect is thus also observed on the temperature dependence of the current–voltage characteristics of the structures, a finding than can be explained as follows.

Under substrate injection, electrons are tunnelling through the SiO_2 interfacial layer, and arrive at the SiO_2/HfO_2 interface. Since the HfO_2 layer is deposited at relatively low temperature (300°C), defects are most probably present in this layer. Defects that occupy energetic levels close (within 1 eV) to the bottom of the conduction band of HfO_2 can participate in the conduction mechanism, via a field- and temperature-assisted electron trapping and detrapping process, the so-called Poole–Frenkel mechanism [26], as illustrated in figure 3.4.23. For a coulombic attraction potential between the electrons and the trapping centres, the current flowing through the material due to the Poole–Frenkel effect reads [26]

$$J_{PF} = CF_{ins} \exp\left[\frac{-q\left(\phi_t - \sqrt{qF_{ins}/\pi\varepsilon_0 n^2}\right)}{k_B T}\right] \qquad (3.4.31)$$

where C is a constant depending on the density of trapping centres, F_{ins} is the electric field across the dielectric layer, $q\phi_t$ is the energetic level of defects with respect to the bottom of the conduction band of the material, ε_0 is the permittivity of free space and n is the refractive index of the material. Let us recall that this mechanism has been suggested to occur in gate dielectrics like Si_3N_4 [46], Ta_2O_5 [47, 48], and ZrO_2 [49].

When $qV_{stack} < q(\phi_1 - \phi_2 - \phi_t)$, the Fermi level in the Si substrate lies below the energy level $q\phi_t$ of the trapping centres in HfO_2, as illustrated in figure 3.4.24(a), i.e., electrons cannot be trapped by these centres. In this case, electrons are tunnelling through the SiO_2/HfO_2 stack and the current depends weakly on the temperature. From the data shown in figure 3.4.22(a), this situation occurs when $V_G < 1$ V. For $q\phi_1 = 3.2$ eV and $q\phi_2 = 1.2$ eV, this corresponds to $q\phi_t = 1$ eV.

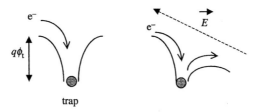

Figure 3.4.23. Schematic illustration of the Poole–Frenkel effect in an insulator subjected to an electric field E.

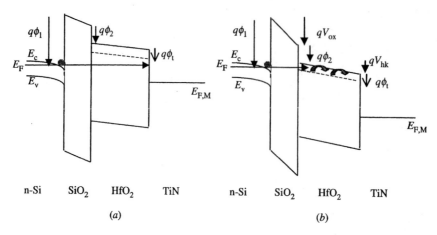

Figure 3.4.24. Schematic energy band diagram of an $Si/SiO_2/HfO_2/TiN$ structure under substrate injection, illustrating the tunnelling effect through the gate stack (*a*) and the tunnelling effect in the SiO_2 layer and the Poole–Frenkel effect in the HfO_2 layer (*b*).

When $qV_{stack} > q(\phi_1 - \phi_2 - \phi_t)$, the Fermi level in the substrate lies above $q\phi_t$, and electrons injected into the HfO_2 layer can be trapped and detrapped by the defects, as illustrated in figure 3.4.24(*b*). In this case, the Poole–Frenkel mechanism becomes the dominant charge transport mechanism (the more so at high temperature), and the current flowing through the structure is strongly dependent on temperature. The Poole–Frenkel current, calculated according to equation (3.4.31), is shown in figure 3.4.25(*a*) as a function of the potential drop across the HfO_2 layer (calculated from equation (3.4.23)). The following values of the physical parameters were fixed for the computation: $t_{ox} = 1\,nm$, $t_{HfO_2} = 4\,nm$, $\kappa_{ox} = 3.9$, $\kappa_{HfO_2} = 20$, $q\phi_t = 1\,eV$, $n = 2$. The experimental results, measured under substrate injection at 25 and 125°C, are shown in figure 3.4.25(*b*). Reasonable agreement between the data and simulations are obtained with this simple model. Consequently, the strong temperature dependence of the current observed in the gate voltage range [1 V, 3 V] can be attributed to the Poole–Frenkel mechanism in the HfO_2 layer.

Under gate injection, and due to the asymmetry of the energy band diagram of the structure, the Fermi level in the TiN gate lies below the energy levels of the trapping centres when $qV_{HfO_2} < q(\phi_3 - \phi_t)$, where $q\phi_3 = 2.5\,eV$ is the potential barrier at the TiN/HfO_2 interface, as illustrated in figure 3.4.26. For $q\phi_t = 1\,eV$, this situation is observed when $V_{HfO_2} < 1.5\,V$, corresponding to $V_G < 3.8\,V$. Consequently, the tunnelling effect is the dominant charge transport mechanism in the structure for $V_G < 3.8\,V$, and the current then depends weakly on temperature, consistently with the experimental results shown in figure 3.4.22(*b*). For $V_G > 3.8\,V$, the Poole–Frenkel mechanism

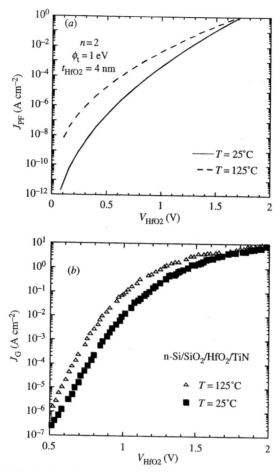

Figure 3.4.25. (*a*) Simulation of the Poole–Frenkel conduction mechanism through a 4 nm HfO$_2$ layer, at 25 and 125°C. (*b*) Current density through an Si/SiO$_2$/HfO$_2$/TiN structure as a function of the voltage drop across the HfO$_2$ layer, at 25 and 125°C.

should occur in the HfO$_2$ layer and the current should be much more sensitive to temperature. However, dielectric breakdown of the gate stack occurs between 4 and 4.5 V, and this mechanism is practically not observed under gate injection. This reasoning allows us to explain the polarity effect observed on the temperature dependence of the current in the Si/SiO$_2$/HfO$_2$/TiN structures.

Current–voltage characteristics of poly-Si gated structures

The current–voltage characteristics of MOS structures with high-κ gate stacks and poly-Si gates are very different from those observed with metal gates. Indeed, as shown in figure 3.4.27, the current measured under gate

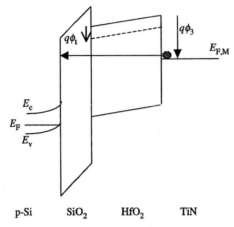

Figure 3.4.26. Schematic energy band diagram of an Si/SiO$_2$/HfO$_2$/TiN structure under gate injection, illustrating the tunnelling effect through the gate stack.

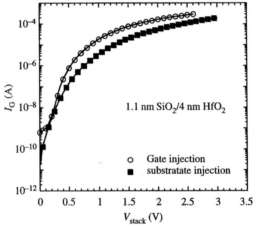

Figure 3.4.27. Current–voltage characteristics of an Si/1.1 nm SiO$_2$/4 nm HfO$_2$/polySi structure under substrate and gate injection, respectively.

injection is (slightly) higher than under substrate injection in Si/1.1 nm SiO$_2$/4 nm HfO$_2$/n$^+$ poly-Si structures, unlike the results observed with metal gates, cf. figure 3.4.18(*a*). In addition, the temperature dependence of the current is very similar under substrate and gate injection for capacitors

***

Figure 3.4.28. Current–voltage characteristics of an Si/1.1 nm SiO$_2$/4 nm HfO$_2$/polySi structure under substrate (*a*) and gate (*b*) injection, at different temperatures. (*c*) Current through the 1.1 nm SiO$_2$/4 nm HfO$_2$ gate stack as a function of inverse temperature, measured under substrate and gate injection, respectively.

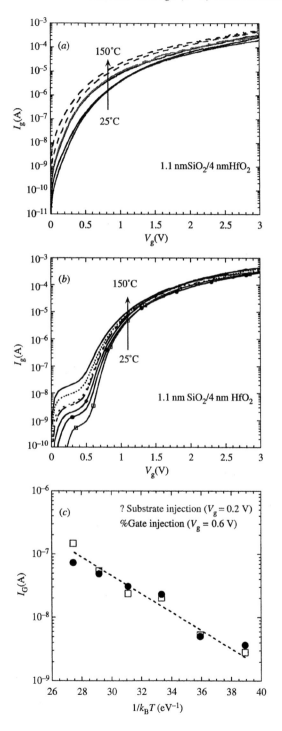

with poly-Si gates, as shown in figure 3.4.28(*a*) and (*b*). Both injection polarities are compared in figure 3.4.28(*c*). From this figure, it appears that the current is thermally activated, as expected for the Poole–Frenkel mechanism, and that the temperature dependence of the current does not depend much on the gate bias polarity. These results are, thus, strikingly different from those observed with metal gates, and can be explained by

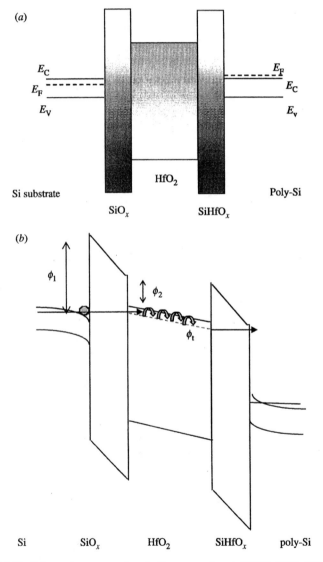

Figure 3.4.29. Schematic energy band diagram of an Si/SiO$_x$/HfO$_2$/SiHfO$_x$/polySi structure, at flat band voltage (*a*) and under substrate injection (*b*).

assuming that a chemical reaction occurs between the HfO$_2$ and poly-Si layer, either during the CVD deposition of poly-Si, or during the subsequent activation anneals (see also the chapter by E. Young and V. Kaushik in the present book); this reaction most probably leads to the formation of an interfacial layer at the poly-Si/HfO$_2$ interface, either SiO$_x$ or SiHfO$_x$. As illustrated in figure 3.4.29(*a*), the presence of this layer gives a symmetric shape of the energy band diagram of the structures, hence leading to the 'symmetrization' of their current–voltage characteristics. It should be also

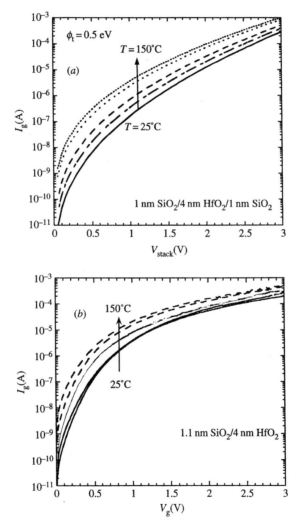

Figure 3.4.30. Simulated (*a*) and measured (*b*) current–voltage characteristics of an Si/SiO$_x$/HfO$_2$/SiHfO$_x$/polySi structure at different temperatures.

pointed out that the extracted EOT of the structures with poly-Si gates (extracted from high-frequency C–V characteristics, not shown here) is systematically higher (by about 0.5 nm) than those extracted in structures with metal gates. These results are also consistent with the presence of a low-κ interfacial layer at the poly-Si/HfO$_2$ interface.

Consistent with the discussion about the charge carrier conduction mechanism through the SiO$_2$/HfO$_2$ layer presented here above, we assume that the major conduction mechanism through the interfacial layers is the tunnelling effect, while the Poole–Frenkel effect occurs in the HfO$_2$ layer, as illustrated in figure 3.4.29(b). The simulated current–voltage characteristics of a 1 nm SiO$_2$/4 nm HfO$_2$/1 nm SiO$_2$ gate stack are presented in figure 3.4.30(a) at different temperatures. A good qualitative agreement is observed between the theoretical curves and the experimental data measured under substrate injection, which are reproduced in figure 3.4.30(b) for comparison. Let us notice that the energy of the trapping centres with respect to the conduction band of HfO$_2$ is found here to be about 0.5 eV, as compared to 1 eV in the case of metal gates. Since the type of gate has an influence on this energy level, these trapping centres could be induced by the interaction between the gates and the high-κ material.

Conclusions

We have presented in this chapter recent results about the characterization, modelling and simulation of the electrical properties of MOS structures with high-κ gate stacks. The capacitance–voltage characteristics of these structures were first examined in detail. The mercury-probe technique was shown to be a useful tool for the fast characterization of these structures, provided that the contribution of the Hg/high-κ interfacial layer to the impedance of the system is properly taken into account. The impact of the poly-Si depletion and carrier quantum confinement on the CV response of the structures was next discussed. An original model for the frequency response of interface traps on the CV characteristics was next presented, considering an energetic distribution of traps within the silicon band-gap. From the comparison between simulations and experimental results, the energetic distribution of interface traps was found to be Gaussian-like, as expected for Pb centres at the Si/SiO$_2$ interface. The trap cross-section was found to be about 10^{-17} cm^2, suggesting that these defects are Pb-like centres (Si trivalent dangling bonds) at an Si/HfSi$_x$O$_y$ interface. Finally, the impact of a non-uniform distribution of fixed charges on the CV characteristics of high-κ gate stacks was discussed. It was found that this distribution could lead to a stretch-out of the CV curves of the capacitors. From a material point of view, this non-uniform distribution of charges could be related to the presence of charged grain boundaries in poly-crystalline high-κ layers.

The second part of this chapter was devoted to the characterization of the current–voltage characteristics of MOS capacitors with high-κ gate stacks. The *IV* curves of such devices are polarity-dependent, a result that was attributed to the asymmetry of their energy band structure. The temperature dependence of the *IV* characteristics of SiO_2/HfO_2 gate stacks was next examined. These results revealed that the main conduction mechanism under substrate injection is the tunnelling effect through the SiO_2 layer and the Poole–Frenkel mechanism in the HfO_2 layer. Under gate injection, the Fermi level at the cathode lies below the energy levels of defects involved in the Poole–Frenkel effect within most of the gate voltage range, resulting in a weak temperature dependence of the gate current, consistent with the experimental results. Finally, the conduction mechanism through SiO_2/HfO_2 gate stacks with poly-Si gates was investigated. The absence of any polarity effect on the current–voltage characteristics of these structures, at room temperature as well as at high temperature, suggests that the poly-Si layer interacts with the HfO_2 layer, forming an $SiHf_xO_y$ layer at the top interface. The presence of this layer is thought to be responsible for the almost symmetric *IV* characteristics observed in these capacitors. The charge carrier conduction mechanism in these structures was suggested to be the tunnelling effect through the interfacial (top and bottom) layers, and the Poole–Frenkel effect in the HfO_2 layer.

Acknowledgments

The authors wish to thank their colleagues X. Garros, C. Leroux, F. Martin and G. Reimbold (CEA-LETI), P. Masson (L2MP-CNRS), X. Zu, M. Naili, S. De Gendt and M. M. Heyns (IMEC), A. Stesmans and V. V. Afanas'ev (University of Leuven) for their contribution to this work and for fruitful discussions. Part of this work was supported by the French Ministry of Research in the framework of the RMNT project 'KAPPA' (No 01V0893), the MEDEA project T653 (ALADIN) and International Sematech.

References

[1] Hillard R J, Mazur R G, Gruber G A and Sherbondy J C 1997 *Electrochem. Soc. Proc.* 97–12, 310
[2] Nicollian E H and Brews J R 1982 *MOS Physics and Technology* (New York: Wiley)
[3] Yang K J and Hu C 1999 *IEEE Trans. Electron Devices* **46** 1500
[4] Saga K and Hattori T 1996 *J. Electrochem. Soc.* **143** 3279
[5] Garros X, Leroux C and Autran J L 2002 *Electrochem. Solid-State Lett.* **5** F4
[6] Garros X, Leroux C, Blin D, Damlencourt J F, Papon A M and Reimbold G 2002 *ESSDERC Proc.*
[7] Autran J L unpublished

[8] Raynaud C, Autran J L, Masson P, Bidaud M and Poncet A 2000 *Mater. Res. Soc. Symp. Proc.* **592** 159

[9] Green M A 1990 *J. Appl. Phys.* **67** 2994

[10] Vankemmel R, Schoenmaker W and De Meyer K 1993 *Solid-State Electron.* **36** 1379

[11] Moglestue C 1986 *J. Appl. Phys.* **59** 3175

[12] Rodriguez S, Lopez-Villanueva J A, Melchor I and Carceller J E 1999 *J. Appl. Phys.* **86** 438

[13] http://schof.colorado.edu/~bart/book/effmass.htm#intro

[14] Kireev P 1975 *Physics of Semiconductor* (Moscow: Mir)

[15] http://www.ioffe.rssi.ru/SVA/NSM/Semicond/Si/bandstr.html#Masses

[16] Guillaumot B *et al* 2002 *IEDM Tech. Dig.* 355

[17] Campbell S A, Ma T Z, Smith R, Gladfelter W L and Chen F 2001 *Microelectron. Eng.* **59** 361

[18] Yamaguchi T, Satake H and Fukushima N 2001 *IEDM Tech. Dig.*

[19] Heiman F P and Warfield G 1965 *IEEE Trans. Electron Devices* 167

[20] Masson P, Autran J L, Houssa M, Garros X and Leroux C 2002 *Appl. Phys. Lett.* **81** 3392

[21] Masson P, Autran J L and Ghibaudo G 2001 *J. Non-Cryst. Solids* **280** 255

[22] Shockley W and Read W T 1952 *Phys. Rev.* **87** 62

[23] Houssa M, Naili M, Zhao C, Bender H, Heyns M M and Stesmans A 2001 *Semicond. Sci. Technol.* **16** 31

[24] Johnson N M 1988 *The Physics and Chemistry of SiO$_2$ and the Si–SiO$_2$ Interface* eds C R Helms and B E Deal (New York: Plenum) p 319

[25] Kao K C and Hwang W 1981 *Electrical Transport in Solids* (New York: Pergamon) p 237

[26] Sze S M 1981 *Physics of Semiconductor Devices* (New York: Wiley)

[27] Majkusiak B and Strojwas A 1993 *J. Appl. Phys.* **74** 5638

[28] Slavcheva G, Davies J H, Brown A R and Asenov A 2002 *J. Appl. Phys.* **91** 4326

[29] Houssa M, Afanas'ev V V, Stesmans A and Heyns M M 2000 *Appl. Phys. Lett.* **77** 1885

[30] Wilk G D, Wallace R M and Anthony J M 2000 *J. Appl. Phys.* **87** 484

[31] Copel M, Cartier E and Ross F M 2001 *Appl. Phys. Lett.* **78** 1607

[32] Lucovsky G and Phillips J C 2000 *Appl. Surf. Sci.* **166** 497

[33] Lucovsky G, Wu Y, Niimi H, Misra V and Phillips J C 1999 *J. Vac. Sci. Technol.* **B17** 1806

[34] Misra V, Lucovsky G and Parsons G 2002 *MRS Bull.* **327** 212

[35] Autran J L, Munteanu D, Dinescu R and Houssa M 2003 *J. Non-Cryst. Solids* **322** 219

[36] Autran J L, Munteanu D and Houssa M 2003 *Electrochem. Soc. Symp. Proc.* **2003-01** 383

[37] Houssa M, Naili M, Afanas'ev V V, Heyns M M and Stesmans A *Proc. 2001 VLSI-TSA Conf. (IEEE, Piscataway, 2001)* p 196

[38] Yang H Y, Niimi H and Lucovsky G 1998 *J. Appl. Phys.* **83** 2327

[39] Vogel E M, Ahmed K Z, Hornung B, Henson W K, McLarty P K, Lucovsky G, Hauser J R and Wortman J J 1998 *IEEE Trans. Electrons Devices* **45** 1350

[40] Fromhold A T 1981 *Quantum Mechanics for Applied Physics and Engineering* (New York: Dover)

[41] Depas M, Vermeire B, Mertens P W, Van Meirhaeghe R L and Heyns M M 1995 *Solid-State Electron.* **38** 1465

[42] Merzbacher E 1998 *Quantum Mechanics* (New York: Wiley)

[43] Afanas'ev V V, Houssa M, Stesmans A and Heyns M M 2001 *Appl. Phys. Lett.* **78** 3073

[44] Afanas'ev V V, Houssa M, Stesmans A and Heyns M M 2002 *J. Appl. Phys.* **91** 3079

[45] Xu Z, Houssa M, De Gendt S and Heyns M M 2002 *Appl. Phys. Lett.* **80** 1975

[46] De Salvo B, Ghibaudo G, Pananakakis G, Guillaumot G and Reimbold G 1999 *J. Appl. Phys.* **86** 2751

[47] Chanelière C, Autran J L, Devine R A B and Balland B 1998 *Mater. Sci. Eng. R* **22** 269

[48] Houssa M, Degraeve R, Mertens P W, Heyns M M, Jeon J S, Halliyal A and Ogle B 1999 *J. Appl. Phys.* **86** 6462

[49] Maiti C K, Varma S, Patil S, Chatterjee S, Dalapati G K and Samanta S K unpublished

SECTION 4

THEORY

Chapter 4.1

Defects and defect-controlled behaviour in high-κ materials: a theoretical perspective

*Marshall Stoneham, Alexander Shluger, Adam Foster,
Marek Szymanski,*

Introduction: why do defects matter?

To think of a gate dielectric as defect-free is science fiction. Such illusions may be understandable, in view of the remarkable quality achieved for silicon dioxide. Even though defects are certainly present in the oxide on silicon, its natural advantages are many. The oxide passivates; it has a wide bandgap and large band offsets; it offers good stoichiometry with neutral intrinsic defects (the neutrality of these native defects being both rare and important); it enables convenience in processing. Moreover, decades of applications have made it possible to avoid many likely problems. For silica, and only for silica at present, there is enormous knowledge of defects and their consequences, with indications of how to minimize the impact of these defects (see, for example, recent reviews in [1–4]).

Miniaturization is a powerful driving force in semiconductor technology. It brings problems of several quite different sorts. There are problems associated with materials processing, like lithography. There are problems with materials, such as the way that tunnelling sets limits on how thin a silicon dioxide layer can be made before power losses or breakdown become unacceptable. In particular, an ideal insulator should have both a low leakage current and a low loss tangent. The leakage current has two components. One comes from tunnelling, and is especially a problem when the real oxide thickness is very small. The second contribution is associated with defect-related channels, and involves dislocations, grain boundaries, or point defects. The loss tangent describes how well the polarization follows an applied field. For DRAM, when fields change on nanosecond timescales, the loss tangent should be less than 0.005. There are also problems with materials quality and its assessment. All high-κ materials will contain defects, some far worse than silicon dioxide. As films get thinner, it becomes more difficult to study these defects experimentally. Perhaps our most important message is that theory can give important input to studies of high-κ dielectrics, either as a complement to experiment or, at times, as an alternative. At its simplest, theory can give a framework in which to understand a series of experiments. The standard reaction/diffusion model (Deal-Grove) picture does just this, although it is very clear that the picture fails for very thin oxide [1]. At the next level, theory can scope a problem. Good quantitative predictions can determine which defects are likely to matter most, or whether interface relaxation is likely to modify band offsets. At this level, theory can help to choose between alternative descriptions suggested from experiment. For some questions, at least, theory can now give accurate predictions for aspects of dielectrics which are either inaccessible to experiment, or where experiment cannot be accurate. We shall address some of the questions, which can be answered by modelling (rather than by theory in general), with illustrative examples.

The specific problems caused by defects include these: they initiate degradation; they cause charge trapping, and they create fixed charge, which causes carrier scatter in the Si substrate. Performance will vary from one device to another, and even from one region to another over a wafer. Even for a single device, as the defects or their charge states evolve in time, so the dielectric properties will vary. Such problems are much more important in materials which are significantly non-stoichiometric, or when the constituent ions readily form more than one charge state. Further issues relate to degradation and failure as defects or charges move within the dielectric. For Ba/Sr titanate, there is 'resistive degradation,' apparently due to O vacancy motion. The motion of hydrogen in working devices can control the populations of fixed charges and electron traps.

Interfaces are a further issue. A thin film will often be inhomogeneous because of the different epitaxial constraints at the two interfaces. There will be effects on apparent band offsets from interface dipoles. In certain circumstances, these dipoles could be exploited to control band offsets, but this is not easy. Strain and defects can strongly affect the dielectric properties of thin films [5]. Other effects seem to be related to electrode roughness, possibly analogous to those seen for silica and for other oxide films.

The way the dielectric is made will strongly affect device reliability. Some of the requirements are listed in table 4.1.1. It will determine whether its microstructure is amorphous, polycrystalline, or epitaxial and to some extent its defectiveness. That it can exist in many closely related forms need not be a problem; indeed, the many structural forms of SiO_2 do not seem to have been an issue. However, it is very important that the dielectric reacts chemically neither with Si [6] nor with the gate contact. Likewise,

Table 4.1.1. Criteria for acceptability of high-κ oxides.

Requirement	Example	Comments
Basic properties as a dielectric	ε_0 in range 10–25; mainly ionic polarization; linear dielectric; band offsets acceptable	Criteria like these effectively eliminate cubic oxides like MgO, certain oxides like TiO_2, ferroelectrics, or II–VI or III–V semiconductors
Can the dielectric be processed without deterioration or interference with other materials?	Non-reactive; stress-free; accurately stoichiometric; homogeneous	Concerns about dislocations (crystalline oxides), alloy fluctuations, and diffusion of oxygen
Properties which affect performance in operation	Stability: little diffusion; low leakage current; low effective fixed charge; few defects which can change charge state to initiate damage	Problems of H motion; non-stoichiometry and dislocations usually problems

the interfaces should be largely stress-free during processing. If not, interfacial constraints may force either defect creation (dislocations, possibly point defects like P_b centres) or new phases, such as amorphous forms of usually crystalline oxides (MgO, Al_2O_3, ZrO_2, HfO_2). Such new phases may have non-optimal properties. If the dielectric is to be *amorphous*, then it must remain amorphous at processing temperatures, and also during the operating life. This appears to be satisfactory for alumina, for Si-doped Zr aluminate, and possibly other systems. A further condition might be required, since many glasses phase-separate (e.g. Al_2O_3/Y_2O_3 alloys, which have two liquid phases of different density [7]). Homogeneity of composition and of properties is essential, which includes maintaining stoichiometry accurately. For thin layers, the precise definition of stoichiometry is not always clear, and one might prefer to think in terms of a coordination criterion for a system like SiO_2.

In this chapter, we will review some of the main issues and recent calculations pertaining mainly to point defects and defect processes in prospective high-κ oxides. Several years of exploratory research and hundreds of publications have not led so far to a final consensus about the materials to be used in DRAM and gate applications. Some of the perspective materials are listed in table 4.1.2. The prevailing line of thought at the moment seems to favour 'something HfO_2 based' [8, 9]. As a candidate for gate dielectric, HfO_2 combines relatively high dielectric permittivity with low leakage current and reasonably high band offset with Si, limiting electron tunnelling. Although the situation remains fluid, HfO_2 and its silicates and

Table 4.1.2. Summary of classes of oxides as high-κ materials. The more important candidates are shown in boldface.

Value of ε_o	System	Comment
Very high (>35)	TiO_2 $SrTiO_3$, $BaTiO_3$	Ferroelectric systems are non-linear. Values above about 60 may be misleading because of non-uniformities
Useful 10–35	CaO, SrO, CdO, BaO, CoO. M_2O_3 (M = **Al**, Dy, Tm, Eu, La, Pr) $LaAlO_3$, $SrZrO_3$ MO_2 (M = Ge, **Zr**, **Hf**, Ce, U), **(Si/M)O$_2$ (M = Y, Hf, Zr)**	Some are very non-stoichiometric. Others have poor band offsets. There are issues of crystallinity. The maximum value of ε_0 for Si/Y oxide alloys is $\varepsilon_0 \sim 10$
Low <10	**Oxynitrides of silicon** MgO, NiO, M_2O_3 (M = Y, Er, Gd, Ho, Yb) SnO_2	**a-SiO$_2$** and **c-SiO$_2$** are in this range. Apart perhaps from oxynitrides, these alternative oxides probably do not offer sufficient improvement to justify a new technology

aluminates emerged as strong contenders and provide good examples of typical defect-related issues. Therefore, we will use them, where appropriate, to illustrate the present state of the art of calculations and main conclusions relevant to the performance of oxide materials. To some extent we will be following empirical approaches, which rely on knowledge of other materials in similar states to those to be used. Therefore, we will draw some analogies with other oxides, such as silica and MgO. The properties we are going to consider will relate mainly to non-stoichiometry and disorder of prospective oxides. Many other related issues, such as determination of band offsets, calculation of structural and dielectric properties of oxides, and interfaces with Si are considered in other chapters.

Our discussion of the theory starts by identifying some of the questions which are asked in relation to defects in the context of high-κ technology and microelectronics. First, there is the nature of the fixed charge. Which defects are likely to be present? Are they likely to be in undesirable charge states? And can one eliminate them via annealing? Charged defects can cause substantial random fields, which may alter breakdown thresholds locally. Typically, for 100 ppm of charged defects, there will be random fields of order $10^5 \, V \, cm^{-1}$. Second, which defects are involved in stress-induced leakage currents (SILCs)? In such cases, the defect offers a channel for electron transfer across the dielectric: an electron tunnels from one electrode into an oxide defect; there are then relaxation processes or a sequence of defect processes, and the electron can then tunnel to the other electrode. What can we say about the defects responsible, their electrical energy levels and relaxation energies (differences in energy for electron capture and emission)? Can we give useful estimates of electron and hole trapping cross-sections? Will there be any field-stimulated diffusion of defects? Third, what are the likely defects at interfaces? We shall have to make informed predictions as to the nature of the interface, and its abrupt or graded nature. This interface nature can have major effects on band offsets. Fourth, is the oxide crystalline, polycrystalline, or amorphous? Key related issues concern the roles of grain boundaries and dislocations, and the effects of disorder on diffusion, especially of H or O. Amorphous oxides will not have grain boundaries, but the intrinsic fluctuations will affect defect formation energies, mechanisms, and charge states.

Notes on techniques for modelling defects and defect processes in dielectrics

Theory, we have noted, can provide a framework of understanding, a scoping which lets one focus on the probable rather than the conceivable, and a route to make specific accurate predictions, which elude experiment. Possible defect types and their impact on device performance are listed in table 4.1.3. We can identify a number of basic questions for accurate predictive theory to address. The first concern is the stable structures and

Table 4.1.3. Possible defect types and their impact on device performance.

Origin of defect	Defect type	Possible role
Imprecise stoichiometry	Vacancies or interstitials, annealing induced	Fixed charge, SILC; motion under applied field
Growth-related impurities from fabrication or diffusion	Precursor related: e.g. Cl, H; source/drain related: e.g. B, P	Fixed charge; time-dependence of operating device
Features of polycrystallinity	Grain boundaries and associated defects, e.g. vacancies	Charge, diffusion
Amorphous structure	Structure-induced traps	Leakage current
Interface	Wrong bonds, relaxation with associated dipoles, roughness	Dipole, charge, electric field

electronic properties of Si/dielectric systems. This component includes calculations of band offsets; possible interface defects, such as dangling bonds, Si–Si bonds, hydrogen defects, etc; structures and properties of grain boundaries (which can, for example, accumulate vacancies); and electronic states due to disorder if the dielectric is amorphous. Second, there is a need to predict structures and electrical levels of those defects likely to be induced by growth and post-deposition annealing in different ambient conditions. These might include incorporation of boron, oxygen, nitrogen, hydrogen, and other species. Third, theory should determine the most stable charge states of interface and bulk defect species, and identify their possible role in static charging of as-grown and annealed oxides. Fourth, predictions can suggest candidates for transient traps capable of trapping and releasing electrons and holes, which are responsible for leakage current, and ideally find possible ways of eliminating them. This includes predicting defect diffusion and re-charging in applied electric fields. Theory should also help in identifying defects by predicting their vibrational, EPR, optical, and other properties.

Even for the best-studied Si/SiO_2 system, these issues have still to be treated fully, although some recent encouraging examples will be mentioned later. A major source of difficulty is that accurate quantitative computational methods have been achieved only recently. Many of the high-κ candidates listed in table 4.1.2 have been thoroughly studied for other classes of applications. Thus, there are relevant studies of electron and hole defects in MgO and other cubic oxides, ferroelectrics, alumina, ZrO_2, HfO_2, and in other materials (see, for example, reviews in [10–14]). At the present time, this area of research presents a remarkable situation where theory is capable of treating most of the issues outlined above, but where experimental data for some materials are either transient or still not available. The reliable and predictive theoretical screening of complex oxide materials needs a major investment of effort, and can take much longer than a pragmatic decision to reject a material for technological reasons. Consequently, computer experiments to investigate basic dielectric properties, processing, oper-ational, and other relevant performance measures of high-κ materials, are carried out mainly on model systems, such as ZrO_2, HfO_2, and $ZrSiO_4$. These systems provide a useful platform from which to guide decisions on other systems.

Some of the main features of existing computational methods used in these and similar studies are presented in table 4.1.4. The methods are thoroughly discussed in recent reviews [14–17] and will not be considered here in detail. Most of the studies to date of the properties of high-κ materials have been carried out so far using a periodic model and density functional theory (DFT) in plane wave basis sets (see, for example, detailed description of a method in [18]). The main advantage of this approach is that both ideal crystals and point defects can be treated on the same footing. If defects are charged or have a dipole moment, the interaction between periodically

Table 4.1.4. Comparison of computational methods used in oxide research.

Models	Methods	Properties
Molecular cluster: molecule represents local defect environment in a solid	DFT, and *ab initio* and semi-empirical Hartree–Fock theory in localized basis sets	Can predict local defect structure, vibrational, optical and EPR properties providing intermediate and long-range order are not important
Embedded cluster: local defect environment is treated quantum mechanically and the rest of the solid structure treated classically	DFT, and *ab initio* and semi-empirical Hartree–Fock theory in localized basis sets	Can predict local defect structure, ionization energies and electron and hole affinities, vibrational, optical and EPR properties and diffusion parameters. Can treat infinite amorphous structures
Periodic: defects are repeated periodically in a lattice	DFT, Hartree–Fock theory, tight-binding methods in plane wave and localized basis sets	Can predict local defect structure, electrical levels, vibrational and EPR properties and diffusion parameters providing the interaction between periodically translated defects is small. Treats amorphous structures as periodic arrangement of disordered cells

arranged defects could be a problem. Also, most of the available plane wave DFT methods do not take into account the non-local exchange interaction between electrons, and so strongly underestimate gaps between occupied and unoccupied electronic bands. This hampers reliable treatment of shallow defect states close to the bottom of conduction band. Our review below of existing calculations uses some of our calculations of defects in ZrO_2, HfO_2, and SiO_2 to illustrate some of the important points pertaining both to defect models and the accuracy of available techniques.

Most defect calculations have been carried out for bulk materials. There are very few studies of the electronic structure of Si/high-κ interfaces (see, for example [16, 17, 19]). Calculations of band offsets are usually based on estimates for bulk oxides, and do not usually address issues of interface dipoles. One of the features of many high-κ oxides, including hafnia, is that in the bulk crystalline form they form several polymorphs. When deposited on Si, in many cases they initially form amorphous films, which become polycrystalline after annealing. The suggestions regarding the structure of polycrystallites are sometimes controversial and certainly depend on annealing procedure (e.g. [20, 21]). In particular, both bulk hafnia and zirconia exist in three polymorphs at atmospheric pressure: at low temperatures hafnia is in the monoclinic C_{2h}^5 phase (space group $P2_1/c$), above 2000 K (1400 K for zirconia) in the tetragonal D_{4h}^{15} ($P4_2/nmc$) phase, and above 2870 K (2600 K for zirconia) in the cubic fluorite O_h^5 ($Fm3m$) phase. The monoclinic structure is the most stable, even for thin films grown on Si [22].

Zirconia, especially when stabilized in the cubic form, has wide uses, including fuel cells, oxygen sensors, and as a nuclear material [23], and so has attracted much scientific study. Although HfO_2 is homologous, and has very similar properties, it has been used less in applications and so has been studied less. There are several experimental studies of both zirconia [24–26] and hafnia [22, 27–30] in the monoclinic phase, but only very few for the high temperature phases [31]. Theoretical studies were scarce at first, focusing on the impurity-stabilized cubic zirconia phase [32–35]. Recent interest in these materials also led to an increase in theoretical studies, with state-of-the-art calculations for bulk zirconia [36–40] and hafnia [41–43].

Non-stoichiometry

Oxides can be non-stoichiometric in many ways [44]. Whereas SiO_2 composition is usually not far from stoichiometric with two oxygens per silicon (alternatively, O is bonded to two Si and each Si bonded to four O), other systems can deviate a lot from their nominal composition. Examples are systems for which cations can easily exist in more than one charge state (such as TiO_{2-x}) or where interstitial oxygen is readily formed (like SnO_2). Why might non-stoichiometry matter? First, the defects, which enable non-stoichiometry, often have charge carriers associated with them. These can give rise to charge transport (in some cases by activated small polaron transport) or to dielectric loss. Conduction along grain boundaries or dislocations may be especially worrying [45]. Second, these defects are involved in degradation processes, such as resistance degradation. This degradation may become more important for very thin films, since the dielectric will have statistical variations in composition, and some regions will be more vulnerable than others. In certain cases, doping can help: for $SrTiO_3$, for instance, doping with Er apparently suppresses O vacancies and reduces the rate of resistance degradation [46]. Third, there are likely to be sample-to-sample variations. These will arise both from the nature of the material as created and from changes during subsequent process steps.

To illustrate how the charging properties of defects can be modelled, let us consider in more detail oxygen vacancies and interstitial species in hafnia (HfO_2). Vacancies can be generated in hafnia films and bulk samples due to growth, deposition, and doping processes. Post-deposition annealing of films is an intrinsic part of any growth cycle and involves oxygen or nitrogen diffusion through the oxide, and the possible formation of interstitial oxygen or nitrogen species. Oxygen vacancies in silica have been implicated as possible electron and hole traps responsible for oxide charging and degradation. However, basic issues, such as the structure of oxygen vacancies and the nature of the diffusing oxygen species and the diffusion mechanisms in hafnia and zirconia, remain unknown. Experiments on zirconia [24–26] have suggested that oxygen incorporates and diffuses in atomic form. Once incorporated, the oxygen can act as an electron trap, changing its charge state and, therefore, its properties and interactions with the oxide and other defects. The performance of thin films in devices is likely to be strongly influenced by these defect processes.

All the calculations discussed below have been performed using the plane wave basis VASP code [47, 48], implementing spin-polarized DFT and the generalized gradient approximation (GGA) of Perdew and Wang known as GGA-II [49]. Ultrasoft Vanderbilt pseudopotentials [50, 51] were used to represent the core electrons. The pseudopotential for the hafnium atom was generated in the electron configuration $[Xe \cdot \{4f\}^{14}]\{5d\}^3 6s^1$ and that for the oxygen atom in $[\{1s\}^2]\{2s\}^2\{2p\}^4]$, where the core electron configurations are shown in square brackets. This method has been shown to give excellent agreement with experiment for bulk hafnium and hafnia properties [41] and be well suited to studying defects in this class of materials [36, 52]. All calculations were made using a 96-atom unit cell, which is generated by extending the 12-atom monoclinic unit cell by two in three dimensions. For this cell, the total energy was converged for a plane wave cut-off of 400 eV and 2 k-points in the first Brillouin zone. One oxygen atom was removed or added to this cell to model the vacancy or interstitial defect, correspondingly. A neutralizing background was applied for calculations of charged defects. The large size of the cell separates the periodic defect images by over 10 Å keeping the Coulomb interaction between charged defects in different periodic cells to below 0.1 eV [41].

Oxygen vacancies

In calculations, the neutral vacancy is generated by removing the corresponding neutral atom from its site in the relaxed perfect lattice supercell, followed by full atomic relaxation. Charged oxygen vacancies are also well known in many oxides, such as MgO and silica [2, 11]. Besides thermal processes and doping, they can be generated in thin hafnia films by electron and hole trapping from silicon. Once generated, they can take part in other electronic processes and serve as electron traps; therefore, we also consider the generation of charged vacancies by removal of electrons from the system. Systematic studies of oxygen vacancies in different charge states have been performed for bulk zirconia [36] and hafnia [41], and will be discussed below. However, oxygen vacancies have also been considered in normal [35] and Y-stabilized cubic [32] and tetragonal [53] ZrO_2, and also in detail for $ZrSiO_4$ [52].

Monoclinic zirconia and hafnia are characterized by three- and four-coordinated lattice oxygen sites, which both could potentially act as defect sites. Oxygen vacancies in monoclinic hafnia formed in three- and four-coordinated oxygen sites are called V_3 and V_4, respectively. The lattice relaxation around them involves small displacements of the nearest neighbour Hf ions. The displacements of these ions for both types of vacancy are in the range of 0.01–0.02 Å, which corresponds to 0.5–1.0% of the Hf–O bond length. These values are very similar to those obtained for oxygen vacancies in zirconia [36]. Such small displacements are characteristic for the F-centre type defects well studied in cubic ionic oxides, such as MgO [54, 55]. They correspond to almost full screening of the anion vacancy by the two remaining electrons, which are strongly localized around the vacancy site. The relaxation energies with respect to the perfect lattice are 0.09 and 0.06 eV, for V_3 and V_4, respectively. The vacancy formation energies are 9.36 and 9.34 eV, for V_3 and V_4, correspondingly. These are the energies required for removing an oxygen atom from a corresponding site to infinity outside the lattice. These values are also similar to those obtained for MgO [54, 55], silica [54, 56], and zirconia [36]. The formation of a vacancy introduces a new double-occupied level in the bandgap situated at 2.8 and 2.3 eV above the top of the valence band for V_3 and V_4, respectively. We note that these energies are defined as differences between Kohn–Sham levels.

The position of the bottom of the conduction band is known to be too low in these DFT calculations [57–59]. In consequence, an extra electron added to the neutral vacancy appears to be delocalized at the bottom of the conduction band, so the electron affinity of the neutral vacancy cannot be reliably established. On the other hand, removing electrons from relaxed neutral vacancies results in the formation of positively charged defects, V_3^+ and V_4^+. We should note that in these calculations electrons are removed to a common zero energy level. The charged vacancies are characterized by a much stronger lattice relaxation—the Hf ions surrounding the vacancy displace outwards by about 0.1–0.2 Å, which corresponds to 5–10% of the Hf–O separation. The corresponding relaxation energies are 0.65 and 0.61 eV for V_3^+ and V_4^+, respectively. The strong relaxation is caused by the fact that the neighbouring Hf ions lose part of the screening effect provided by the two electrons in the neutral vacancy case. The electron remaining in the V_3^+ vacancy is strongly localized inside the vacancy site, as can be seen in the spin density map shown in figure 4.1.1. The single-occupied state of this defect lies lower in the bandgap. Removal of yet another electron from the system leads to the formation of doubly positively charged vacancies, V_3^{2+} and V_4^{2+}, and even stronger relaxation of the neighbouring Hf ions.

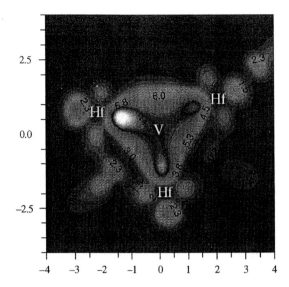

Figure 4.1.1. Calculated spin density in a plane through a V_3^+ oxygen vacancy and two surrounding hafnium atoms in hafnia. Density is in $0.1\,e\,\text{Å}^{-3}$.

The system total energy is much lower (0.44 and 0.76 eV, respectively) for charged vacancies created in the three- rather than in the four-coordinated site. Although the initial formation of a neutral vacancy is energetically balanced between the two sites, once electrons are removed, the V_3 defect is strongly favoured and vacancies are likely to diffuse to these sites.

To summarize, anion vacancies in zirconia and hafnia have similar properties to those of F-like centres. These centres are known to form in cubic alkali halides and oxides, such as MgO, CaO, and BaO [11]. In these ionic materials, the oxygen site is surrounded by six equivalent metal ions. A classical F-centre will have one (alkali halides) or two (cubic oxides) electrons in an s-type deep state in the bandgap. Formation of this centre is accompanied by small and fully symmetric (O_h point group) relaxation of the surrounding ions. A significant amount of the electron density is localized in the vacancy with considerable and equal contributions of six neighbouring metal ions (see figure 4.1.2(*a*)). Ionization of the F-centre leads to characteristic strong and, again, symmetric distortion of surrounding ions, where the nearest metal ions are moving outwards from the vacancy and 12 next nearest neighbour anions move inwards.

Another characteristic oxygen vacancy type is represented by what may be called a E′-like model. We should note that this notation is not fully consistent, as E′-centre notifies a special type of paramagnetic defect in SiO_2 associated with an oxygen vacancy. However, these defects have been studied very extensively and therefore we believe that using their name to represent a

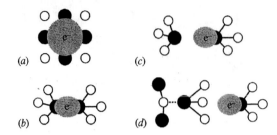

Figure 4.1.2. Schematic of different types of oxygen vacancy in insulators. Cations are shown in black and anions in white. (*a*) An F-like centre defect characterized by symmetric electron distribution centred in the vacancy (shown in grey); (*b*)–(*d*) various types of E′-centre model (see text for discussion).

class of models will not cause any problem. They are formed in different forms of SiO_2, GeO_2, and in silicates, where each oxygen ion is surrounded by only two nearest neighbour positive ions. They are known to exist in alkali silicates and can also be expected in some of the Hf and Zr silicates. Formation of a neutral oxygen vacancy in these materials is accompanied by the creation of a Me–Me bond and by strong displacements of these two Me ions towards each other (see figure 4.1.2(*b*)). The best-known example is a neutral vacancy in SiO_2 [2]. A paramagnetic positively charged vacancy has a different structure to that of the neutral vacancy and originally was called an E′-centre. When one electron is removed from the Me–Me bond, it becomes much weaker and can relax into two distinct configurations. One is still very similar to that of the neutral vacancy, where the remaining electron is delocalized over the two Me ions, but with a much longer Me–Me bond distance. Another one is asymmetric with an hole almost fully localized on one Me ion, which is accompanied by strong and asymmetric lattice distortion [60, 61] (see figure 4.1.2(*c*) and (*d*)). We should note that this description is valid mainly for crystalline silicates and germinates, and becomes more complicated in amorphous materials. However, it provides a generic model useful for discussion and classification in other materials.

Interstitial oxygen species

The incorporation energy of oxygen into the hafnia lattice can be calculated with respect to different processes involving different gas oxygen species [41]. If we consider the case where oxygen incorporates from a molecular source, we find defect energies of $+1.6\,eV$ for incorporation of atomic oxygen and $+4.2\,eV$ for molecular oxygen, favouring incorporation of two atoms over one molecule by 1 eV (oxygen in either form always favours incorporation at the more open three-coordinated lattice oxygen site [36, 41]). However, for incorporation from an atomic source, such as in ultraviolet ozone oxidation processes (see, for example [62]), the defect energy for atomic incorporation drops to $-1.3\,eV$ and it is now favoured over molecular incorporation by almost $7\,eV$. In both cases, it is clear that oxygen prefers to incorporate in an atomic form. This agrees with experimental results for similarly structured zirconia [24–26], and contrasts with the much more open silica structure where molecular incorporation is favoured [63].

Figure 4.1.3 shows a slice of the charge density through the neutral oxygen interstitial atom and original three-coordinated site. The defect forms a strong bond with lattice oxygen, since this is the closest source of electrons in the system, forming a negatively charged defect pair. We note that qualitatively the structure of this defect is very similar to that of an interstitial oxygen atom forming a peroxy bridge defect in silica [63]. In ionic materials like hafnia, the large Coulomb energies assist interstitial oxygen in gaining as much charge as possible to exploit the crystal potential. The DFT calculations described above predict very small and even negative vertical electron affinities for neutral and single negatively charged interstitial oxygen

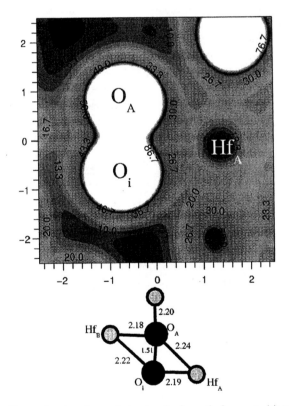

Figure 4.1.3. Charge density plot and structural schematic for neutral interstitial oxygen (O_i) in a plane through O_A, O_i, and O_B for equilibrium geometry near to a lattice oxygen (O_A). Charge density is in $0.1\,e\,\text{Å}^{-3}$.

species (see discussion below). However, large lattice relaxation favours the electron trapping and both single and double negatively charged relaxed interstitial oxygen species are predicted to be stable in zirconia and hafnia [36, 41].

Electron trapping

The charge state of defects will depend on band offset, voltage, temperature, and processing. Charge transfer between defects and the oxide/silicon conduction bands, and between defects themselves, has serious implications for device fabrication, as noted earlier. To define which charge state of an intrinsic defect (e.g. vacancy) or incorporated defect species is favoured, assuming the common case of oxide film grown on silicon, the source or sink of electrons is the bottom of the silicon conduction band at the interface. Hence, the chemical potential of the electron can be chosen at the bottom of the Si conduction band or at the silicon midgap energy [64]. Electrons are assumed to be able to tunnel elastically or inelastically [65] from/to these states to/from defect states in the oxide and create charged species.

Since the interface is not explicitly included in most calculations, one needs to use experimental information [66–68] (e.g. 3.1 eV for the valence band offset at the Si/HfO$_2$ interface, 1.1 eV for the bandgap of Si) to estimate the energy of an electron at the bottom of the conduction band of Si with respect to the theoretical zero energy level (see discussion in [36, 41, 63]). The calculated energies are readily adjusted if doping, temperature, or bias favours another source. One can define vertical (unrelaxed, i.e. nuclei remain in their initial states after electron is added) and relaxed defect electron affinities from the oxide conduction band. Using the electron affinities from the oxide conduction band and the experimental band offset information one can then place defect levels with respect to silicon bands (see figure 4.1.4). Thus, defined positions of unrelaxed electrical defect levels with respect to the top of the Si valence band can be used as indicators for the possibility of elastic tunnelling from Si into these states. For defects to serve as stepping-stones for electrons in leakage current and breakdown processes, defect relaxation energies should be of the order of the voltage drop across the oxide [69].

For defects in hafnia these energies have been studied in detail in [41]; here we discuss only those properties relevant to vacancies and atomic oxygen interstitials. Figure 4.1.4 shows that charged vacancies have positive electron affinities and are likely to be neutralized by electrons from Si. However, their relaxation energies are relatively small (of the order of 0.7 eV). Both the neutral and singly charged oxygen interstitials have large relaxed electron affinities, and would gain energy by trapping an electron from the conduction band. This is consistent with the drive for the neutral defect to gain electrons by bonding to the lattice oxygen, discussed in the previous section, and demonstrates that the ionicity of the crystal

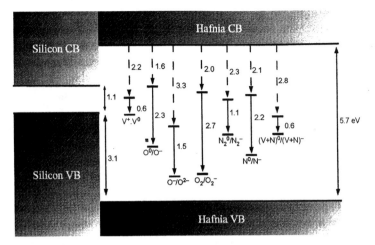

Figure 4.1.4. Energy level diagram showing the vertical (dashed lines) and relaxed (solid lines) electron affinities for various defects in monoclinic hafnia. All energies are in eV. Here superscript represents the overall charge of the supercell during the calculation (i.e. O^- is a calculation of an interstitial oxygen atom with an extra electron added). $(V+N)$ represents calculation of a nitrogen atom substituting an oxygen atom in the hafnia lattice [14].

energetically favours the existence of oxygen defects as O^{2-} (the negative U behaviour of oxygen is discussed in [36, 41]). The behaviour for nitrogen atomic and molecular species is very similar to that for oxygen, although overall the nitrogen species have a lower electron affinity [14].

Amorphous versus crystalline dielectrics

A number of factors favour an amorphous dielectric. The absence of dislocations or grain boundaries is especially important. Interface stresses can be taken up by modest topological variations in a random network, rather than through misfit dislocations (for the oxide on Si, it is possible that P_b centres play a role in accommodating the misfit, since they correlate in number with the interfacial stress [70]); a continuous random network tends to minimize electrically active defects. However, the randomness of structure will often give rise to shallow traps. Such traps can arise from potential fluctuations, or from composition fluctuations, such as regions of higher Hf concentration in Hf/SiO_2. The density may be hard to control, and composition fluctuations are likely (e.g. there may be regions relatively rich in Hf and regions relatively rich in Si in an Hf/Si oxide). There is evidence from dielectric loss that, for some thin film oxides of Ta, Zn, and Si, the system behaves like a Coulomb glass of random charged defects, and this may be related to such fluctuations [71].

There is a wealth of information about those compositions which are associated with glass formation in bulk materials. For a gate dielectric, the traditional network formers (like Si or Ge) clearly have a role, and the usual network modifiers (such as alkalis) are to be avoided. Transition metals have ambiguous roles in conventional glasses, and may act as modifiers or formers. For thin films, there is some evidence from alumina, a-Al_2O_3, that amorphous forms can be stabilized, presumably partly by the interfacial stress. Amorphous forms of MgO, ZrO_2, and HfO_2 have been reported. In the cases where the bulk form is normally crystalline, one should be cautious about the stability of an amorphous thin film and the possibility of crystallization over longer times or under electrical stress.

There are at least three basic fallacies concerning amorphous systems. First, it is wrong to assume there is only a single amorphous structure for a given composition, since the method of preparation can have significant effects. Second, it is wrong to think that the *mean* energies of defect formation or of trap ionization in an amorphous structure are sufficient to understand the behaviour: the form and tails of the distribution are equally important. Third, one should not assume that crystals and amorphous systems of the same composition have the same values for defect and trap energies. This third fallacy also applies to dielectric constants and band offsets, especially when the ionic polarization contributions are substantial. How a dielectric layer is created, manipulated, or shaped can be important. That these are fallacies is evident especially for SiO_2. Navrotsky's work [72] on the enthalpies and molar volumes for silicas shows significant ranges of values. At the molar volume of α-quartz, amorphous systems have an energy per molecular unit larger by about 0.25 eV; the lowest energy amorphous 'phases' have about 30% larger molar volume than the quartz. There is similar information for other glass systems. At the very least, the density must be defined for an amorphous system, not just the composition. More generally, amorphous materials depend on the way they are created, and especially on any thermal treatment.

Modelling defects in amorphous materials

Structural disorder in amorphous materials generally means that all sites are different. Understanding how this affects defect formation and properties is important fundamentally and for many technological applications. In most experimental studies the effect is hidden in broadening of the spectral features and only rarely disorder has been implicated in formation of different structural types of the same defect (e.g. E'_δ and E'_γ centres shown in figure 4.1.2(*b*)–(*d*) are two distinct types of structural relaxation of positive oxygen vacancies in silica). Predicting defect properties and relative abundance of different defect configurations generally requires considering not one or several, as in crystals, but a statistical ensemble of structural sites.

Yet additional factors include the history of the sample and the mechanism of defect formation [2]. This provides new challenges for both theoretical and experimental analysis of defects in amorphous materials.

How to correlate properties and different structural models of the same type of centre with local structural characteristics in amorphous material, such as bonds, rings, dihedral angles? If one could solve this problem, then relative concentrations of defects could be found simply by analysing the amorphous structure itself. Recent years have seen considerable progress in understanding the role of disorder in the defect properties of silica (see, for example [73–79]). Several approaches have been employed so far. On the theoretical front, a popular method has been to build a continuous random network model of amorphous structure and then produce defects in the selection of sites and build a distribution of defect parameters. It has been applied to study peroxy linkage defects [76] and oxygen diffusion [79], and recently to analyse different types of E'-centre in silica [74].

In our laboratory, Szymanski *et al* [76] have used a combination of molecular dynamics and density functional methods to create realizations of amorphous SiO_2 of density similar to that of thermal oxide. The analysis of the energies of oxidizing species (O and O_2 in their neutral, negative, and doubly negative charge states) showed striking results. First, the molecular species have a character very different from the atomic forms. Whereas O_2^0 is a relatively simple interstitial, O^0 forms a peroxy linkage (essentially, two oxygens in a bent structure between two silicons, similar to that shown in figure 4.1.3). This structural difference has the interesting consequence that the atomic form is very efficient at isotope exchange, whereas the molecular species only exchanges with network oxygens at special sites. Second, electrons with energies corresponding to the bottom of the Si conduction band could create ions, or induce reactions with ionic products. Third, there is a substantial variation of insertion energies (and indeed also of diffusion barriers) from site to site in the amorphous system. This variation has a significant component associated with *medium range order* and intrinsic topology-related strain fields in amorphous material, not just the configuration of the closest shells of ions. Fourth, the values of key energies for the oxidation process (e.g. the insertion energy for moving an oxygen molecule from the gas phase into an interstitial site) prove very different for the amorphous oxide and for quartz. The values for the amorphous oxide combine to give an activation energy for the oxidation process in good agreement with experiment. This work made the important point that earlier, simplistic ideas were misleading for key defects underpinning processing, performance, and degradation.

We may usefully compare the advantages and disadvantages of four classes of dielectric [80], namely (i) amorphous SiO_2; (ii) some other amorphous oxide, where we might think of a glassy Si-rich (Si/Hf) oxide; (iii) a polycrystalline oxide, where we might think of an Hf-rich (Si/Hf) oxide;

(iv) an epitaxial crystalline oxide on silicon (see table 4.1.5). In making the comparison, we shall assume a dielectric constant for the alternative oxide of about four times that of silica. Silica has some strikingly good features: its effectiveness in passivation, its consistency of stoichiometry; its band offsets for silicon, and its uses in processing, not to mention the wealth of many years of industrial experience. The fact that the mobile native defects are neutral is an advantage often unrecognized. Likewise, silica has problems: the dielectric constant is not large, the very thin oxide is vulnerable, perhaps even to a single critically placed defect, and tunnelling will be a problem. On the other hand, there are ways to minimize charged defects.

Interface-related defects

An effective gate dielectric needs adequate band offsets from those of Si [81]. The band offset is, however, a subtle quantity [82, 83], and will be affected by any interfacial dipoles. These account for discrepancies between predictions from band theory and electrochemical flat-band potentials. Whenever one deals with highly polar materials (like almost all high-κ materials), there is an ionic dipole layer which will affect offsets. For example, atoms placed on a surface which transfer some electronic charge to the substrate (like Cs, or like H on diamond) will decrease the electron affinity, even occasionally leading to negative electron affinities. However, space charge, or modified probabilities of different charge states of defects or impurities, can also contribute. There can also be an image charge effect, when the dielectric has a dielectric constant significantly different from that of Si. The dipoles can be quite large. Again, there are a very few predictions. For BaO/NiO, Stoneham and Tasker [84] predict a 2 eV potential barrier (corresponding to 4 eV for a $2+$ ion) stabilizing positive charges in the NiO. This suggests that *interface engineering* (such as having at least one layer of ions of chosen electric charge) might resolve the offset problem, even if this is easier said than done.

One factor in the interfacial dipole is certainly interfacial stress. Most oxides have a relatively poor mismatch with silicon. Basically, there are five main ways to take up the strain. For very thin crystalline films, elastic strain may suffice, as in strain-layer systems. For thicker crystalline films, misfit dislocations are expected. A third possibility is that there is a thin layer of a different phase. Point defects are a fourth possibility. It is known that the P_b centre concentration at the Si/SiO_2 interface correlates with interfacial stress [85], and it is plausible that P_b centre creation does relieve the stress to some degree [70]. Finally, topology changes in an amorphous oxide offer another way to reduce mismatch stress. One should bear in mind that the dielectric may be in a form which would be metastable in the bulk. The interface stress will certainly affect the dipole, especially for piezoelectric oxides. The stress can cause defect structures to develop in the oxide as oxidation or deposition

Table 4.1.5. Comparison of four different classes of dielectric.

Property	a-SiO$_2$	Glassy, Si-rich oxide, e.g. Si-rich (Si/Hf)O$_2$	Polycrystalline oxide, e.g. Hf-rich (Si/Hf)O$_2$	Epitaxial crystalline oxide
Passivation	Good. No misfit dislocations		Misfit and other dislocations are possible	Misfit and other dislocations possible
Stoichiometry	Good; uniform	Alloy fluctuations	Alloy fluctuations	Depends on oxide
Tunnelling	Problem for very thin films	Possible problems from interstitial O species	Possible problems from interstitial O species	
Other issues	Lithography can exploit properties of this oxide			Expensive should MBE be needed
Defect aspects	Neutral species common. Some control of the interface defects possible	Randomness in structure; alloy fluctuations lead to shallow traps	Alloy fluctuations lead to shallow traps. Grain boundaries and dislocations likely to cause problems. Mobile charged defects	Hard to avoid grain boundaries or dislocations

proceeds. Such structures may be the sites which nucleate damage under electrical stress.

Defect processes and the possible improvement of the dielectric by novel processing

Processes involving atomic reorganizations can be efficient at low energies with interstitial oxygen present, or with impurity hydrogens. One interesting class of process involves interstitial oxygen, its various charge states, and the molecular and atomic (peroxy-linkage) forms. Some of the processes need only electrons with energy corresponding to the bottom of the silicon conduction band. Whether or not interstitial oxygens have a direct role in breakdown is unclear; no such role has been demonstrated. However, there is the possibility that *tailoring* the interstitial oxygen species during the growth or annealing process might eliminate some of the more vulnerable features in the oxide. For example, discussions of the optical writing of Bragg gratings in silica optical fibres often suggest processes starting from Si–Si bonds (*not* O vacancies, but wrongly coordinated silicons). Or there may be defects in the oxide, which evolve from P_b centres at the Si/oxide interface on further oxidation (such defects cannot be eliminated except through reaction with single unpaired spins and a change of coordination). Could the right combination of atomic species be able to eliminate these defects? The answer may be yes [86], in that one has some control over both charge state and atomic/molecular form, especially if one chooses to apply electric fields during growth. The field can bias either electron tunnelling from the silicon, or the motion of negatively charged oxygen species. The timescales for these two processes differ, as do the signs of field, so the phasing and timings of the applied fields must be controlled, as well as their magnitudes. The basic ideas discussed below might prove helpful for other proposed dielectrics, although the processes may differ.

The actual defects present in a real system will depend on the processing, the source of any electrons, the applied voltage, and the temperature of that system. Starting from an initial distribution of defects in different charge states and electron transfer between defects, we can use the predictions of Foster *et al* [41] for various defects in different charge states as a guide to those defect combinations which are energetically more favourable. These results we can extend to consider defect levels with respect to the hafnia and silicon conduction bands to indicate likely electron and hole trapping in real devices.

Defect reactions

Studies of defect and impurity processes lead to four important points. First, there are various defect or impurity species which can change charge state, e.g. those involving oxygen molecules. Second, there are various electronically induced reactions, especially those which exchange H^+ and an hole. Third, in an amorphous dielectric, such reactions may be easier at certain sites, either because the energy needed is lower, or because there is a capture process which is favoured. In all these cases, one must distinguish between constant-geometry transitions (optical, most tunnelling) and the thermal transitions which involve relaxed states. Energies predicted for various reactions in [41] are summarized in table 4.1.6. These energies were calculated as differences in total energies of pairs of individual defects, and each pair has the same total charge state and number of atoms. Positive energies indicate that a reaction in the direction of the arrow is energetically favourable: positive energies mean exothermic reactions, negative mean endothermic. None of the reactions listed requires the total energy of delocalized states, e.g. O^+ and V^-. The energies presented in table 4.1.6 do not include the interaction between defects, which can be strong especially in close charged defect pairs.

Reactions 1 and 2 in table 4.1.6 show that charge transfer between oxygen vacancies and interstitials is favourable. A separated pair of doubly charged defects has 1.5 eV lower energy than the neutral pair. The associated Frenkel pair energies are 8.0 eV for formation of the neutral pair, 7.3 eV for the singly charged pair, and 5.8 eV for the doubly charged pair.

Table 4.1.6. Defect reactions and associated energies in hafnia. Here a superscript represents the overall charge of the supercell during the calculation (i.e. O^- is a calculation of an interstitial oxygen atom with an extra electron added). E_0^0 is the total energy of the ideal bulk monoclinic 96-atom cell, which is used to conserve particle numbers in the reactions. Here positive reaction energies refer to exothermic reactions.

No.	Reaction	Energy (eV)
1	$O^0 + V^0 \rightarrow O^- + V^+$	0.7
2	$O^- + V^+ \rightarrow O^{2-} + V^{2+}$	1.5
3	$O^0 + O^{2-} \rightarrow 2O^-$	-0.8
4	$V^{2+} + V^0 \rightarrow 2V^+$	0.2
5	$O_2^0 + V^0 \rightarrow O_2^- + V^+$	1.4
6	$O_2^- + V^+ \rightarrow O_2^{2-} + V^{2+}$	1.8
7	$O_2^{2-} + O_2^0 \rightarrow 2O_2^-$	-0.4
8	$2O^0 \rightarrow O_2^0 + E_0^0$	-1.0
9	$O^0 + O^- \rightarrow O_2^- + E_0^0$	-0.3
10	$2O^- \rightarrow O_2^{2-} + E_0^0$	0.8
11	$O_2^0 + V^0 \rightarrow O^0 + E_0^0$	9.0
12	$O_2^- + V^+ \rightarrow O^0 + E_0^0$	7.6
13	$O_2^{2-} + V^{2+} \rightarrow O^0 + E_0^0$	5.7

For both zirconia [36] and hafnia, reaction 3 predicts 'negative U' behaviour of an oxygen ion in hafnia: two isolated O^- species would transform into O^{2-} and O^0. The same is true for molecular species but with a much smaller energy gain: reaction 7 shows that the oxygen molecule also has 'negative U', and two O_2^- species would decay to form a doubly charged and a neutral molecule. The 'negative U' terminology comes from Anderson's model for semiconductors, where this effect is much more common than in insulators [87, 88]. Experimentally our result means that atomic and molecular oxygen species in hafnia prefer to stay diamagnetic, and will be difficult or impossible to detect by paramagnetic resonance. A similar process for vacancies in reaction 4 shows a balance between the different pairs for four-coordinated and a slight energy gain of 0.2 eV for three-coordinated to form two singly charged vacancies. Again, this is very similar to the results found in zirconia. However, in the concentration of paramagnetic V^+ vacancies will strongly depend on temperature. Molecular species also demonstrate a tendency for charge transfer between the vacancies and interstitial molecules. The combination of doubly charged molecule and vacancy pair is 3.2 eV lower in energy than the neutral pair (reactions 5 and 6).

Some reactions involve changes in defect type, rather than simply a change of charge state. In such reactions, two independent defects combine to form a different defect, e.g. it is energetically favourable for an oxygen molecule to separate into two interstitial atoms (reaction 8, discussed earlier). It is energetically favourable for two singly charged interstitial oxygen ions to recombine and form a doubly charged interstitial molecule (reaction 10). All combinations of molecule and vacancy defect pairs (reactions 11–13) are predicted to 'annihilate', leaving only a neutral oxygen interstitial. The energy gain is less for the more favourable doubly charged defect pair. Such reactions are likely to occur during annealing in an oxygen atmosphere.

Diffusion

How significant will diffusion be during processing [80]? For *processing conditions*, one might think of 15 s at 1050°C, whereas an *operating lifetime* might correspond to 6 years (2×10^8 s) at room temperature. If the relevant diffusion distance is 2 nm, then this means that one would prefer the diffusion constant during processing at 1050°C to be less than $0.3 \times 10^{-14} \, \text{cm}^2 \, \text{s}^{-1}$ (see the 'Electrolysis of a high-κ dielectric under an applied field' section for the constraints corresponding to operating conditions). Data for many oxides are available (e.g. p 258 of [89]). Many of the oxides shown fail, sometimes by cation motion (even Mg in MgO and Ca in CaO; Y in Y_2O_3 probably fails), sometimes by anion (O) motion (in calcium-stabilized zirconia (CSZ),

TiO_2, and Y_2O_3). Most non-stoichiometric oxides fail the criterion. Alumina seems safe so long as there are no grain boundaries, although Ag or Cu may diffuse fast enough to break the criterion [90]. It is possible that the criterion given is marginally too stringent, in that O in fused SiO_2 is close to failing.

Polycrystalline materials will contain grain boundaries and dislocations. These defects offer fast diffusion pathways and can trap charge [24, 25, 91–94]. They may give rise to field concentration and breakdown initiation. In polar solids (which these dielectrics are, cf the 'Introduction: why do defects matter?' section) diffusion is likely to involve ions. This suggests problems with changes of stoichiometry and the development of static charge. The electric fields associated with such charges can be large, if only local, and may have secondary effects, especially if the dielectric has a κ value very different from Si. Interfacial stress has to be taken up somehow. For very thin films, elastic deformation may suffice; in other cases, strain may lead to misfit dislocations or alternative phases. Epitaxial crystalline forms can be largely free of grain boundaries, and possibly even largely free from dislocations. However, as discussed below, one can expect fast oxygen diffusion even in ideal crystalline hafnia and zirconia.

The mechanism of diffusion of oxygen in oxides (and many materials in general) can be classified as occurring via either an exchange or interstitial process. The exchange mechanism involves the continuous replacement of a lattice site by the diffusing defect, and the lattice site then becoming the diffusing species. This mechanism is traditionally known as the 'interstitialcy' mechanism [95], but we will refer to it as 'exchange' for clarity. It is characteristic of diffusion of anions, for example, in oxides such as MgO [96] and fluorides such as CaF_2 [97]. In the interstitial mechanism, the defect diffuses through empty space between the lattice sites. This mechanism is characteristic of diffusion in oxides such as silica [86, 98]. The structure of hafnia is more complex than most of these classical cubic oxides, yet retains the same lack of interstitial space, therefore it is especially interesting to see which mechanism is energetically favoured. Also in some materials, such as MgO [99], the mechanism of diffusion is very dependent on the oxygen charge state.

In hafnia, we noted above that the neutral oxygen interstitial forms a strong bond with a lattice oxygen site (O_A in figure 4.1.3). This need to find an electron source dominates the diffusion process [100]. So oxygen diffusion in hafnia is governed by two competing phenomena: (i) the crystalline potential in hafnia means that oxygen defects are only stable as ions, and (ii) relaxation of atoms along the diffusion path. The barriers for each diffusing oxygen charge state are given in table 4.1.7. The neutral oxygen interstitial causes the least disruption of the surrounding crystal during diffusion, but its need to form an 'ion-pair' with a lattice oxygen produces a large barrier. The doubly charged oxygen is the most stable defect in hafnia film on silicon, yet

Table 4.1.7. Exchange (E_{ex}; the diffusing defect diffuses by regular replacement with lattice oxygen) and interstitial (E_{in}; the defect moves by a normal interstitial motion) activation barriers for different charge states of oxygen interstitial defects in hafnia. All values are in eV.

Defect	E_{ex}	E_{in}
O^0	0.8	1.3
O^-	0.3	1.1
O^{2-}	0.6	1.8

its large Coulomb interaction means it generates large displacements during diffusion. Hence, the singly charged defect proves to be the best balance—it is more independent of the lattice oxygens than the neutral species, but does not produce as large disruption of the crystal as the doubly charged defect. In terms of the general mechanism, the small space in between atoms in hafnia means that an exchange mechanism is favoured over the interstitial mechanism for all defect species, since it needs less lattice disruption (as shown in table 4.1.7). However, in the neutral case, displacements are usually modest, and the difference between the two mechanisms is much smaller than for the other defect species.

These results show that, although hafnia has a much more complex atomic structure than other, simpler ionic materials, its geometry shares a similar lack of interstitial space and the lattice exchange (or interstitialcy) remains the favoured diffusion mechanism. Barriers for interstitial oxygen diffusion in hafnia are small, and the defects will be very active, especially during the high temperature processing common in microelectronic processes. These barriers are much smaller than the measured activation energy of 2.3 eV for oxygen diffusion in m-zirconia [25], but this activation energy is dominated by the Schottky formation energies (about 2.2 eV [34]). The fact that oxygen is predicted to diffuse as a charged species suggests the possibility of using an applied electric field to influence the diffusion, and perhaps control defect concentrations. All oxygen species have large electron affinities, so they can all act as traps within a device, creating intrinsic electric fields and contributing to dielectric losses. Their control would be highly desirable for efficient device design.

One should also ask about that ubiquitous impurity, hydrogen. For all oxides studied, the H diffusion rate [101–107] is substantially in excess of the critical value [$LiNbO_3$, $LiTiO_3$, TiO_2 (parallel and perpendicular to the *c*-axis), a-SiO_2, MgO, MgO:Li, Al_2O_3, Al_2O_3:Mg, BeO, and also in certain ceramics (spinel $MgO \cdot Al_2O_3$, amorphous cordierite $2MgO \cdot 2Al_2O_3 \cdot 5SiO_2$)]. Thus, H can readily diffuse across a 2 nm film in 15 s at 1050°C.

Diffusion in amorphous systems

Extensive experiments on gas migration in glasses [108] show systematic trends in diffusion for rare gases and for hydrogen. In particular, inert gas mobility and solubility depend primarily on the accessible free volume of the glass structure. This simple concept rationalises the dependence on composition (network former and modifier changes), thermal history, and applied stress. A smaller role of phase separation has been identified, and should be considered with care in considering glassy systems like Hf/SiO_2. There are analogous processes exploiting interstitial ion diffusion and mobility, which are central to phenomena ranging from ionic conduction to Mallory bonding [109, 110]. We ask here whether similar ideas apply to a-SiO_2, for which many different amorphous silicas have been observed, varying in density and formation enthalpy [72]. Experimental data are sparse, mainly associated with molecular oxygen in amorphous gate dielectric oxides. However, theory [76, 86] has identified the most important features. Our own work shows that Shelby's rules need a simple but important modification, and we discuss how this might be verified.

Diffusion modelling, as discussed above, generally gives predictions for diffusion in *crystals* in good agreement with experiment. In crystals, the small number of critical processes can usually be identified with certainty. In amorphous systems, there are further issues. First, how is the amorphous system to be described? Is density enough, as [86] implies, or do ring statistics matter? Second, if we predict a realization of an amorphous structure, how can we be sure that the structure is realistic? This is especially a problem in an inhomogeneous amorphous system, such as a gate oxide, where the standard scattering methods are harder to apply. Mott [111] suggests that those sites which determine solubility of interstitial species may be relatively rare, whereas diffusion may have to sample sites which are energetically less favourable. Third, if we have a credible realization, how is it generalized to give statistically acceptable results? Also, as emphasized in the 'Amorphous versus crystalline dielectrics' section and the 'Modelling defects in amorphous materials' section, energetics of the same processes in amorphous and crystalline materials can be strikingly different, which can result even in an altogether different picture of diffusion. Results obtained for crystalline systems do not have to carry over to amorphous materials.

Some of these issues have been addressed in recent studies on oxygen diffusion in amorphous silica [63, 76, 86, 112–114]. There are usually two initial steps: the creation of a realization of an amorphous silica using quenched molecular dynamics [76], and a significant number of density functional calculations (DFT) for different geometries and defects present. In particular, Szymanski and co-workers [86] calculated a variety of stable sites and diffusion transition states for molecular oxygen in amorphous silica using DFT. These also underlay estimates of solubility and diffusion.

Diffusion is especially complicated, since the study of percolative diffusion requires analysis of the topology of amorphous realizations and the character of the important percolation paths. On this basis, one can estimate the activation energy for diffusion. The DFT calculations for transition states [113, 114] identify important effects of different sizes and shapes of network rings. The stable sites energies depend on the sizes and shapes of the voids. Figure 4.1.5 shows oxygen molecule energy at the stable sites (in the voids) and the transition states (in the rings). This energy shows a clear dependence on the free volume available to the molecule (here measured through the distance to the closest embedding network atom). An analysis of various energy contributions shows the influence of the intrinsic strain in the SiO_2 network on these energies (locally compressed or stretched parts of the network), and also the influence of the topology (locally soft versus hard embedding network). These results suggest a general rule that the diffusion will proceed mostly through the bigger voids and bigger rings in the SiO_2 network (the voids are connected by the passages through the bigger rings in a percolative manner). This contrasts with O_2 diffusion in α-quartz where the molecule can travel in the structural channels of the crystal with very low

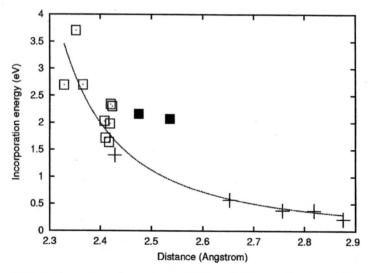

Figure 4.1.5. Incorporation energy of neutral O_2 into a-SiO_2 as a function of the distance between the centre of the molecule and the closest atom of the non-defective a-SiO_2 network. The stable configurations in the voids are marked by ' + ', whereas the transition states for diffusion through the rings are marked by empty squares (seven-, six-, and five-member rings are included). As a reference, the filled squares give values for a stable site and a transition state for O_2 diffusion in the big helix of α-quartz [112]. The incorporation energies exhibit clear dependence on the distances; the dotted line is a guide for the eye. The data for α-quartz do not follow the trend for amorphous silica due to higher density and crystalline topology of the medium.

activation energy since the channels provide little variation in the free volume available to the molecule.

Electrolysis of a high-κ dielectric under an applied field

The concern addressed here is the extent to which one need worry about electrolysis of the gate dielectric during the operational lifetime of a dielectric, i.e. ionic diffusion processes under applied fields [80]. There will be potential problems if defects or impurities have diffusion constants (or ionic mobilities) above some threshold.

In ordinary thermal diffusion with a diffusion constant D, there will be significant diffusion over distance L after a time of order L^2/D. If L is an oxide thickness, 2 nm, and τ is an operational lifetime of order 6 years (2.1×10^8 s), the largest acceptable D would be 2×10^{-22} cm^2 s^{-1}. In real devices, of course, an electric field E will be applied for some fraction of the time. This will cause ions present to drift with the velocity $v = \mu E q e$, with q the charge, e the proton charge, and μ the mobility, which is related to the diffusion constant D by the Nernst–Einstein relation. If the drift distance $v\tau$ is to be less than L, the upper bound on D is proportional to $LT/E\tau$ (note that this bound gives a slightly different dependence on L, T, and E than the L^2/τ of simple diffusion). With $T = 300$ K, $E = 10^6$ V cm^{-1}, and other values as before, the upper bound is $D = 2.9 \times 10^{-23}$ cm^2 s^{-1}. For ionic conduction to be negligible in a dielectric in an operating device, the diffusion constant must be an order of magnitude smaller than for simple diffusion (or the same as thermal diffusion if the field is only on for 10% of the time). If we assume $D = D_0 \exp(-\varepsilon/kT)$ with a pre-exponential factor of $D_0 = 10^{-4}$ cm^2 s^{-1}, then the device will only be safe against diffusive or electrochemical degradation for an activation energy ε greater than about 1 eV. Diffusion in dislocations or grain boundaries is especially likely to cause damage over an operational lifetime.

If diffusion is to occur over no more than 2 nm, the diffusion constant must be less than 2×10^{-22} cm^2 s^{-1}. For O diffusion, most oxides *except* clearly non-stoichiometric oxides are satisfactory. CSZ and Y_2O_3 fail; a-SiO_2 may fail marginally (corresponding to diffusion over about 50 nm in 10 years), but the extrapolation to low temperatures is unreliable. The situation is less clear for H diffusion. For some oxides [$LiNbO_3$, $LiTiO_3$, TiO_2 (both parallel and normal to the c-axis)] D is in excess of the critical value. For other oxides (MgO, MgO:Li, Al_2O_3, Al_2O_3:Mg, BeO) the diffusion rate is less than the critical value. Obviously, cases with *neutral* ions (like SiO_2 in normal situations) only thermal diffusion is an issue. Likewise, no field is applied when the rapid thermal anneal is done, so again thermal diffusion is all that matters. The estimates may be slightly pessimistic, in that a thicker oxide can be used for higher κ, so one might have real thickness of say 6–8 nm. However, ions may not need to traverse the whole thickness to cause problems. The conclusion is that electrolysis of the gate dielectric must be considered seriously whenever there are mobile ionic species, either intrinsic or common extrinsic ions.

Damage initiation

Defects can be created in operation only by having sufficient localized energy to achieve bond reorganization or the like, together with some means by which this energy can be used efficiently. One can understand such processes only by considering electronic excited states. Typically, an energy of the order of the bandgap might be available for defect creation [109]. However, this is often less than defect formation energies in perfect crystalline oxides. It is necessary to identify vulnerable regions, perhaps at grain boundaries or dislocations, where energy localization and modification are easy. For crystalline oxides, dislocations are especially important. It is observed that dislocation densities rise rapidly in MgO under electrical stress, for example. Dislocation motion, both climb and glide, is assisted by the presence of electron–hole pairs. Such recombination-enhanced phenomena (chapter 7 [109]) are seen in very many systems, from III–Vs to UO_2, and are presumably a part of dielectric degradation processes following damage initiation.

What are the atomic processes? A significant number of such processes have been identified especially for halides. A relevant example is α-quartz, where the self-trapped exciton has been studied in some detail (p 178 [109]), with extensive experimental validation of detailed calculations [115, 116]. Following bandgap excitation, energy is localized in the form of a self-trapped exciton, in which the oxygen moves very significantly away from its perfect-lattice site. There is a small but finite probability of defect production (oxygen vacancies and interstitials) and of the nucleation of amorphization (p 232 [109]; [117]). It is probable that similar processes are stimulated in the amorphous gate oxide a-SiO_2, and that these are among the processes which initiate breakdown. What is important about this intrinsic process is that it can be efficient in the use of energy, even if its overall probability is low. For very thin oxides, a breakdown *field* is inappropriate: even $5 \times 10^7 \, V \, cm^{-1}$ will give an electron only $10 \, eV$ on traversing a $2 \, nm$ film. Such an energy can create defects in silica, albeit with low probability. An elastic collision process would have essentially zero probability at such low energies.

The atomistic modelling of breakdown is at an early stage. Yet it is clear that it promises to link with phenomenological Monte Carlo models, in which defects form and finally link to give an easy transport of electrons across the gate dielectric. If it proves possible to pin down the defect phenomena in more detail, there is an opportunity for improved materials design for resistance to electrical degradation.

Final comments

New materials appear each year in the semiconductor industry. Some changes are evolutionary, affecting relatively limited areas. Others, like a decision to replace SiO_2 as the industry-standard gate dielectric, would have wide-ranging consequences, and so would depend on the industry's longer-term strategy. Strategic aims, whether related to speed, power consumption, miniaturization, or flexibility to make possible some unforeseen range of gadgets, can only be achieved if the right materials are available. Materials modelling can enable correct decisions to be taken. The prediction might be 'material X looks good' but even more important could be the prediction that 'favoured materials option Y will fail'. The combination of theory and experiment offers opportunities both for successful materials policies and for the avoidance of failure.

Acknowledgments

Some of this work was supported by the EPSRC and by the Fujitsu European Centre for Information Technology. We are grateful to EU project HIKE for financial support of this work. ASF wishes to thank the CSC, Helsinki for computational resources. The authors are grateful to J. Gavartin, P. Sushko, S. Mukhopadhyay, K. Tanimura, F. Lopez Gejo, V. Sulimov, R. Nieminen, G. Bersuker, and A. Korkin for valuable comments and discussions.

References

[1] Sofield C J and Stoneham A M 1995 *Semicond. Sci. Technol.* **10** 215
[2] Pacchioni G, Skuja L and Griscom D L (eds) 2000 *Defects in SiO₂ and Related Dielectrics: Science and Technology* (Dordrecht: Kluwer)
[3] Lenahan P M and Conley J J F 1998 *J. Vac. Sci. Technol.* **16** 2134
[4] Massoud H Z, Baumvol I J R, Hirose M and Poindexter E H 2000 *The Physics and Chemistry of SiO₂ and the Si–SiO₂ Interface* (Pennington, NJ: Electrochemical Society)
[5] Sirenko A A, Bernhard C, Golnik A, Clark A M, Hao J, Si W and Xi X X 2000 *Nature* **404** 373
[6] Hubbard K J and Schlom D C 1996 *J. Mater. Res.* **11** 2757
[7] Aasland S and McMillan P F 1994 *Nature* **369** 633
[8] Wilk G D, Wallace R M and Anthony J M 2000 *J. Appl. Phys.* **87** 484
[9] Wilk G D, Wallace R M and Anthony J M 2001 *J. Appl. Phys.* **89** 5243
[10] Neumark G F 1997 *Mater. Sci. Eng. Rep.* **21** 1
[11] Henderson B 1980 *CRC Crit. Rev. Solid State Mater. Sci.* **9** 1
[12] Kotomin E E, Maier J, Eglitis R I and Borstel G 2002 *Nucl. Instrum. Methods Phys. Res. B* **191** 22
[13] Kotomin E A and Popov A I 1998 *Nucl. Instrum. Methods Phys. Res. B* **141** 1

[14] Shluger A L, Foster A S, Gavartin J L and Sushko P V 2003 *Nano and Giga Challenges in Micro-Electronics* ed A Korkin (Berlin: Elsevier)

[15] Dovesi R, Orlando R, Roetti C, Pisani C and Saunders V R 2000 *Phys. Status Solidi (b)* **217** 63

[16] Kawamoto A, Jameson J, Cho K and Dutton R W 2000 *IEEE Trans. Electron Devices* **47** 1787

[17] Kawamoto A, Cho K and Dutton R 2002 *J. Comput.-Aided Mater. Des.* **8** 39

[18] Payne M C, Teter M P, Allan D C, Arias T A and Joannopoulos J D 1992 *Rev. Mod. Phys.* **64** 1045

[19] Kawamoto A, Cho K, Griffin P and Dutton R 2001 *J. Appl. Phys.* **90** 1333

[20] Neumayer D A and Cartier E 2001 *J. Appl. Phys.* **90** 1801

[21] Cho Y J, Nguyen N V, Richter C A, Ehrstein J R, Lee B H and Lee C C 2002 *Appl. Phys. Lett.* **80** 1249

[22] Balog M, Schieber M, Michiman M and Patai S 1977 *Thin Solid Films* **41** 247

[23] Ryshkewitch E and Richerson D W 1985 *Oxide Ceramics: Physical Chemistry and Technology* (Florida: Academic)

[24] Busch B W, Schulte W H, Garfunkel E, Gustafsson T, Qi W, Nieh R and Lee J 2000 *Phys. Rev. B* **62** R13290

[25] Brossman U, Würschum R, Södervall U and Schaefer H-E 1999 *J. Appl. Phys.* **85** 7646

[26] Martin D and Duprez D 1996 *J. Phys. Chem.* **100** 9429

[27] Ruh R and Corfield P W R 1970 *J. Am. Ceram. Soc.* **53** 126

[28] Adams D M, Leonard S, Russel D R and Cernik R J 1991 *J. Phys. Chem. Solids* **52** 1181

[29] Lakhlifi A, Leroux C, Satre P, Durand B, Roubin M and Nihoul G 1995 *J. Solid State Chem.* **119** 289

[30] Kingon A I, Maria J P and Streiffer S K 2000 *Nature* **406** 1032

[31] Wang J, Li H P and Stevens R 1992 *J. Mater. Sci.* **27** 5397

[32] Stapper G, Bernasconi M, Nicoloso N and Parrinello M 1999 *Phys. Rev. B* **59** 797

[33] Mackrodt W C and Woodrow P M 1986 *J. Am. Ceram. Soc.* **69** 277

[34] Dwivedi A and Cormack A N 1990 *Phil. Mag. A* **61** 1

[35] Králik B, Chang E K and Louie S G 1998 *Phys. Rev. B* **57** 7027

[36] Foster A S, Sulimov V B, Gejo F L, Shluger A L and Nieminen R M 2001 *Phys. Rev. B* **64** 224108

[37] Jomard G, Petit T, Pasturel A, Magaud L, Kresse G and Hafner J 1999 *Phys. Rev. B* **59** 4044

[38] Fabris S, Paxton A T and Finnis M W 2000 *Phys. Rev. B* **61** 6617

[39] Rignanese G M, Detraux F, Gonze X and Pasquarello A 2001 *Phys. Rev. B* **64** 134301

[40] Zhao X and Vanderbilt D 2002 *Phys. Rev. B* **65** 075105

[41] Foster A S, Gejo F L, Shluger A L and Nieminen R M 2002 *Phys. Rev. B* **65** 174117

[42] Demkov A A 2001 *Phys. Status Solidi B* **226** 57

[43] Zhao X and Vanderbilt D 2002 *Phys. Rev. B* **65** 233106

[44] Catlow C R A and Stoneham A M 1981 *J. Am. Ceram. Soc.* **64** 234

[45] Duffy D M and Stoneham A M 1983 *J. Phys. C* **16** 4087

[46] Kita R, Matsu Y, Masuda Y and Yano S 1999 *Mod. Phys. Lett. B* **13** 983

[47] Kresse G and Furthmüller J 1996 *Comp. Mat. Sci.* **6** 15

[48] Kresse G and Furthmüller J 1996 *Phys. Rev. B* **54** 11169

[49] Perdew J P, Chevary J A, Vosko S H, Jackson K A, Pederson M R, Singh D J and Fiolhais C 1992 *Phys. Rev. B* **46** 6671

[50] Vanderbilt D 1990 *Phys. Rev. B* **41** 7892

[51] Kresse G and Hafner J 1994 *J. Phys.: Condens. Matter* **6** 8245

[52] Crocombette J P 1999 *Phys. Chem. Miner.* **27** 138

[53] Eichler A 2001 *Phys. Rev. B* **64** 174103

[54] Pacchioni G, Ferrari A M and Ieranó G 1997 *Faraday Discuss.* **106** 155

[55] Kantorovich L N, Holender J and Gillan M J 1995 *Surf. Sci.* **343** 221

[56] Sulimov V B, Sushko P V, Edwards A H, Shluger A L and Stoneham A M 2002 *Phys. Rev. B* **66** 024108

[57] Lægsgaard J and Stokbro K 2001 *Phys. Rev. Lett.* **86** 2834

[58] Gavartin J L, Sushko P V and Shluger A L 2003 *Phys. Rev. B* **67** 035108

[59] Pacchioni G, Frigoli F, Ricci D and Weil J A 2001 *Phys. Rev. B* **63** 054102

[60] Feigl F J, Fowler W B and Yip K L 1974 *Solid State Commun.* **14** 225

[61] Rudra J K and Fowler W B 1987 *Phys. Rev. B* **35** 8223

[62] Ramanathan S, Wilk G D, Muller D A, Park C M and McIntyre P C 2001 *Appl. Phys. Lett.* **79** 2621

[63] Szymanski M A, Stoneham A M and Shluger A L 2000 *Microel. Reliab.* **40** 567

[64] Blöchl P E 2000 *Phys. Rev. B* **62** 6158

[65] Fowler W B, Rudra J K, Zvanut M E and Feigl F J 1990 *Phys. Rev. B* **41** 8313

[66] Alay J L and Hirose M 1997 *J. Appl. Phys.* **81** 1606

[67] Mihaychuk J G, Shamir N and van Driel H M 1999 *Phys. Rev. B* **59** 2164

[68] Afanas'ev V, Houssa M, Stesmans A and Heyns M M 2001 *Appl. Phys. Lett.* **78** 3073

[69] Warren W L, Lenahan P M, Robinson B and Stathis J H 1988 *Appl. Phys. Lett.* **53** 482

[70] Stoneham A M, Harker A H and Jain S C 1993 Unpublished

[71] Fleming R M, Lang D V, Jones C D W, Steigerwald M L and Kowach G R 2001 *Appl. Phys. Lett.* **78** 4016

[72] Navrotsky A 1987 *Diff. Diff. Data* **53/54** 61

[73] Griscom D L and Cook M 1995 *J. Non-Cryst. Solids* **182** 119

[74] Lu Z-Y, Nicklaw C J, Fleetwood D M, Schrimpf R D and Pantelides S T 2002 *Phys. Rev. Lett.* **89** 285505

[75] Donadio D, Bernasconi M and Boero M 2001 *Phys. Rev. Lett.* **87** 195504

[76] Szymanski A M, Shluger A L and Stoneham A M 2001 *Phys. Rev. B* **63** 224207

[77] Stirling A and Pasquarello A 2002 *Phys. Rev. B* **66** 245201

[78] Bakos T, Rashkeev S N and Pantelides S T 2002 *Phys. Rev. Lett.* **88** 055508

[79] Bongiorno A and Pasquarello A 2002 *Phys. Rev. Lett.* **88** 125901

[80] Stoneham A M 2002 *J. Non-Cryst. Solids* **303** 114

[81] Robertson J 2000 *J. Vac. Sci. Technol. B* **18** 1785

[82] Herring C and Nichols M H 1949 *Rev. Mod. Phys.* **21** 185

[83] van Vechten J H 1985 *J. Vac. Sci. Technol. B* **3** 1240

[84] Stoneham A M and Tasker P W 1984 *Mater. Res. Soc. Symp.* **40** 291

[85] Stesmans A 1993 *Phys. Rev. B* **48** 2410

[86] Stoneham A M, Szymanski M A and Shluger A L 2001 *Phys. Rev. B* **63** 241304(R)

[87] Anderson P W 1975 *Phys. Rev. Lett.* **34** 953

[88] Stoneham A M and Sangster M J L 1983 *Radiat. Eff.* **73** 267

[89] Stoneham A M and Smith L W 1991 *J. Phys.: Condens. Matter* **3** 225

[90] Atkinson A 1985 *Rev. Mod. Phys.* **57** 437

[91] Puls M P, Woo C H and Norgett M J 1977 *Phil. Mag.* **36** 1457

[92] Duffy D M 1987 *Current Issues in Solid State Science: Condensed Matter Structure* ed A M Stoneham (London: Institute of Physics)

[93] Bristowe P D and Domingos H S 2000 *Adv. Eng. Mater.* **2** 497

[94] Harding J H 2003 *Interface Sci.* **11** 81

[95] Lidiard A B 1957 *Handbuch der Physik XX* pp 246–349

[96] Brudevoll T, Kotomin E A and Christensen N E 1996 *Phys. Rev. B* **53** 7731

[97] Chadwick A V and Terenzi M 1985 *Defects in Solids: Modern Techniques*

[98] Hamann D R 1998 *Phys. Rev. Lett.* **81** 3447

[99] Brudevoll T, Kotomin E A and Christensen N E 1996 *Phys. Rev. B* **53** 7731

[100] Foster A S, Shluger A L and Nieminen R M 2002 *Phys. Rev. Lett.* **89** 225901

[101] Johnson O W, Pack S H and de Ford J W 1975 *J. Appl. Phys.* **46** 1026

[102] Bates J B, Wang J C and Perkins R A 1979 *Phys. Rev. B* **19** 4130

[103] Moulson A J and Roberts J P 1961 *Trans. Faraday Soc.* **57** 1208

[104] Gonzales R, Chen Y and Tsang K L 1982 *Phys. Rev. B* **26** 4637

[105] Gonzales R, Chen Y, Tsang K L and Summers G P 1982 *Appl. Phys. Lett.* **41** 739

[106] Engstrom H, Bates J B, Wang J C and Abraham M M 1980 *Phys. Rev. B* **21** 1520

[107] Fowler J D, Chandra D, Elleman T S, Payne A W and Verghesse K 1977 *J. Am. Ceram. Soc.* **60** 155

[108] Shelby J E 1974 *Gas Migration in Glass* SLL-74-5210 (Albuquerque: Sandia Laboratory)

[109] Itoh N and Stoneham A M 2000 *Materials Modification by Electronic Excitation* (Cambridge: Cambridge University Press)

[110] Stoneham A M and Tasker P W 1985 *J. Phys. C* **18** L543

[111] Mott N F 1981 *Proc. R. Soc. Lond. A* **376** 207

[112] Szymanski M A, Stoneham A M and Shluger A L 2001 *Solid-State Electron.* **45** 1233

[113] Szymanski M A, Stoneham A M and Shluger A L To be published

[114] Szymanski M A 2001 Modelling of Silicon Oxidation Processes *PhD Thesis* University of London

[115] Fisher A J, Hayes W and Stoneham A M 1990 *J. Phys.: Condens. Matter* **2** 6707

[116] Fisher A J, Stoneham A M and Hayes W 1990 *Phys. Rev. Lett.* **64** 2667

[117] Itoh N, Stoneham A M and Tanimura K 2000 *Structure and Imperfections in Amorphous and Crystalline Silica* eds R Devine, J P Duraud and E Dooryhee, pp 329–347

Chapter 4.2

Chemical bonding and electronic structure of high-κ transition metal dielectrics: applications to interfacial band offset energies and electronically active defects

Gerald Lucovsky and Jerry L Whitten

Introduction

The search for alternative high-κ dielectrics to replace thermally grown SiO_2 and Si oxynitride alloys as gate dielectrics in aggressively scaled Si complementary metal oxide semiconductor (CMOS) devices has focused on transition metal and lanthanide rare earth oxides and their respective silicate and aluminate alloys [1]. There are several factors that must be considered to make these replacements possible; these are based on more than 20 years of increases in integration, and the accompanying decreases in lateral dimensions and oxide physical and equivalent thickness in SiO_2 dielectrics. First, there must be significant reductions in direct tunnelling at operating bias levels by at least three to four orders of magnitude with respect to SiO_2 as the equivalent oxide thickness (EOT) is reduced to 1.0 nm and below. Additionally, the channel mobilities of electrons and holes must be approximately 90% of their values in SiO_2 devices. Finally, other device performance and reliability metrics must be essentially the same as SiO_2, which means that (i) interfacial and bulk defect densities such as interfacial traps, D_{it}, and fixed charge, Q_f, must be $< 10^{11}\, cm^{-2}$ and (ii) defect generation under accelerated stress bias testing must correspond to times to failure of about 10 years. In addition, there are many other issues relating to process integration; paramount among these are issues related to the stability of alternative dielectrics against chemical phase separation and/or crystallization below temperatures required for down-stream process steps including dopant activation in source and drain contacts at temperatures $\sim$ 900–1000°C. Other process integration issues include the eventual substitution of a pair of metal gate electrodes with Fermi level energies respectively equivalent to those of n + and p + doped polycrystalline Si. This integration issue will not be addressed in this chapter, although issues related to chemical bonding and stability at the interfaces between alternative dielectrics and metal gates are at least as challenging as those at the crystalline Si–dielectric interface that is one of the focal points of the chapter.

As noted above, one of the primary driving forces for alternative dielectrics is the reduction of direct tunnelling through gate dielectrics in which EOT is reduced to 1.0 nm and below. Since direct tunnelling is an exponential function of the physical thickness of the gate dielectric, this has focused the search for alternatives to SiO_2 on materials with dielectric constants, κ, in the range of 15–30 or approximately four to eight times the value of κ for SiO_2. Group IIIB and IVB transition metal, and lanthanide series rare earth oxides have dielectric constants in this range, as well as bandgaps that are generally expected to give conduction and valence band offset energies greater than about 1.5 eV. Since gate oxide capacitance scales directly with κ, these substitutions will potentially provide increases of up to four- to eightfold in physical thickness for the same EOTs as thermally grown SiO_2. However, since the tunnelling transmission probability also depends on the conduction band offset energy and the electron tunnelling mass, reductions in either or both of these could mitigate some of the gains associated with the anticipated increases in physical thickness [1, 2]. This mandates an increased understanding of the fundamental differences between the electronic structure of SiO_2 and the transition metal and rare earth alternative dielectrics, which contribute to both of the band offset energy and the tunnelling mass. From this point forward, no distinction will be made between rare earth lanthanides, and group IIIB transition metals, since they have similar electronic structures, and these will then be grouped under the designation *transition metal dielectrics*. In addition and equally important, *lower-κ* interfacial transition regions (nitrided SiO_2, or higher SiO_2 content silicate alloys) between the Si substrate and the gate dielectric, which result from processing, or which may be inherent due to interface bonding, can significantly reduce the physical thickness of the high-κ dielectric for a given value of EOT, and therefore have a marked effect on attainable reductions in direct tunnelling.

This chapter addresses chemical bonding and fundamental electronic structure of transition metal gate dielectrics, including amorphous morphology, and uses electronic structure of SiO_2 and Si–SiO_2 interfaces as a point of reference. The next section of the review identifies the fundamental and unique properties of thermally grown SiO_2 and Si–SO_2

interfaces in terms of chemical bonding and electronic structure, providing a background for understanding the qualitative and quantitative differences in bulk and interface properties associated with the introduction of deposited alternative dielectrics. The following sections address chemical bonding and electronic structure of alternative dielectrics, defining fundamental limitations for the replacement of thermally grown SiO_2, or equivalently, intrinsic factors that will contribute to the *inevitable end-of-the road-map* for the aggressive scaling of bulk CMOS integrated circuit devices.

SiO₂ and the Si–SiO₂ interface

Non-crystalline SiO_2, in particular bulk fused silica, holds a *very special* place among non-crystalline solids, and additionally, the interface between thermally grown thin film SiO_2 and Si is *unique* among semiconductor–native oxide interfaces. Fused silica glass is the prototypical non-crystalline solid with a continuous random network (CRN) structure, CRN [3 and references therein] and has been the testing and proving ground for x-ray, electron and neutron diffraction techniques that are used to obtain information about the local atomic bonding. Diffraction studies have been complemented by studies of vibrational properties by infrared, Raman and x-ray techniques [4]. This era of research was highlighted by the *ball and stick* model of Bell and Dean [5], and calculations based on the model are in excellent agreement with radial distribution functions deduced from diffraction studies providing an unambiguous identification of the local bonding geometry, including statistical distributions of bond-lengths and bond-angles. Calculations of vibrational properties based on the Bell and Dean model have identified empirical force constants that provide important information relative to glass formation and thin film annealing [6, 7]. However, only very recently, *ab initio* calculations, based on the local atomic structure deduced from analysis of diffraction studies, were able to provide a quantitative understanding of the mechanical and vibrational properties based on the fundamental electronic structure [8].

SiO_2 is the prototypical network amorphous solid [3], and has a CRN structural morphology in which the Si atoms are fourfold coordinated in a tetrahedral geometry, and the O atoms are twofold coordinated with bond angles in the range from about 130 to 170° [8]. The ideal network is chemically ordered containing only Si–O bonds. Extrinsic defect bonding configurations—which can be produced by x-ray and particle irradiation—have been proposed, and have been attributed to broken bonds, or wrong bonds, Si–Si, or equivalently O-atom vacancies, and interstitials [9]. The random aspects of the structure, which provide the configurational entropy required for glass formation, and metastability of the *CRN amorphous* morphology are essentially two in number: (i) the large distribution of

Si–O–Si bond angles, approximately ±10–20° [10, 11] and (ii) the randomness of the dihedral angle distribution, or four-body correlation functions.

There was considerable interest in the electronic structure of SiO_2 and the Si–SiO_2 interface that began in the early and mid-1970s, as Si integrated circuits started to emerge and herald an information revolution based on the processing and storage of high densities of digitally coded information. Two approaches were prevalent: (i) semi-empirical tight binding calculations performed on cluster Bethe lattices (CBLs) [12, 13] and (ii) calculations based on periodic structures with the local atomic structure of different crystalline phases of SiO_2 such as crystrobalite and α-quartz [14, 15]. A third approach, used less frequently, was based on the application of *ab initio* quantum chemistry calculations to molecules that included the local bonding arrangements of Si and O in SiO_2 [16, 17]. This approach has recently been refined by our group at NC State University and similar calculations have been extended to high-κ transition metal dielectrics as well [8].

The electronic structure calculations of [8] are *ab initio* in character, and employ variational methods in which an exact Hamiltonian is used. No core potential or exchange approximations are assumed [18, 19]. The Hamiltonian, H, is given in equation (4.2.1),

$$H = \sum_i (-1/2\nabla_i^2 + \sum_k (-Z_k/r_{ik})) + \sum_{i<j} 1/r_{ij} \qquad (4.2.1)$$

and contains kinetic energy ($1/2\nabla_i^2$), nuclear–electron attraction ($-Z_k/r_{ik}$) and electron–electron repulsion ($1/r_{ij}$) contributions. Equation (4.2.1) is formulated in atomic units (a.u.): 1 a.u. (distance) = 1 Bohr = a_0 = 0.5292 Å; 1 a.u. (energy) = 1 Hartree = e^2/a_0 = 27.21 eV.

The calculations are performed initially using a self-consistent field (SCF) Hartree–Fock method with a single determinant wave function represented by

$$\Psi = C\det(\phi_a(1)\phi_b(2)\phi_c(3)\phi_d(4)\cdots\phi_x(N)), \qquad (4.2.2)$$

where the functions ϕ are molecular orbitals containing spatial and spin components. This approach does not include effects due to electron correlation. Following this, there is a configuration interaction (CI) refinement of the bonding orbitals that is based on a multi-determinant expansion wave function,

$$\Psi = \sum_i c_i \Phi_i, \qquad (4.2.3)$$

where the Φ_i denote determinants of the form specified in equation (4.2.2) with different choices of occupied molecular orbitals [18–20]. Expansions of this type specifically include electron correlation effects. The expansion coefficients c_i are determined by an energy minimization algorithm. Since the Hamiltonian in equation (4.2.1) is exact through electrostatic interactions, the energy variational theorem is satisfied and any increases in total binding energy are then significant.

The SCF and CI calculations have been applied to the clusters shown in figures 4.2.1 and 4.2.2, which are based solely on the elements of short range order that are extracted from analysis of x-ray, electron and neutron diffraction: the Si–O bond-length, and the Si–Si and O–O inter-atomic distances that are *functionally equivalent* to the Si–O–Si and O–Si–O bond-angles. The local cluster in figure 4.2.1 is *embedded mathematically* into a larger network structure through a one-electron embedding potential $V(r)$ and basis functions $S1$ and $S2$, which are represented by Si* in the diagram. This termination is a significant aspect of the calculations since it means that the cluster contains no more information than is available from analysis of the diffraction data. In addition, the *linear termination* of the cluster by the Si* pseudo-atoms avoids complications due to steric effects resulting from electrostatic repulsions that depend on the angular orientation of terminating Si–OH groups as in [16, 17].

For Si bonded to O, $V(r) = 1/r$, relative to an origin 1.94 a.u. from the O nucleus, which is close to an OH bond length. Functions $S1$ and $S2$ are gaussian expansions of long and short range functions representing an sp³ hybrid orbital of Si. The mixing parameter, λ, in $S1 + \lambda\, S2$, is set so that: (i) all Si core energy levels are correct—this assures a correct overall charge distribution in the cluster—and (ii) the structure in figure 4.2.2 has a zero dipole moment. The cluster termination is similar in spirit to a cluster Bethe lattice calculation in that both approaches seek to terminate the local cluster to represent a continuation of the non-crystalline network morphology; otherwise, the approaches are completely different. The present work models the system as a finite cluster of atoms in which the terminating atoms, Si*, represent the next shell of SiO_n atoms. The Si* are chosen such that the correct dipole moment of the system, and uniformity in the charge distribution as measured by core eigenvalues, are achieved.

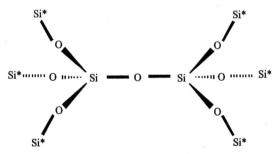

Figure 4.2.1. Schematic representation of the Si–O–Si terminated cluster used for the *ab initio* calculations of this paper. The Si–O–Si bond-angle, α, is 180° in this diagram, and will be varied from 120 to 150° for the calculations. The Si* represent an embedding potential such that Si core eigenvalues are correct.

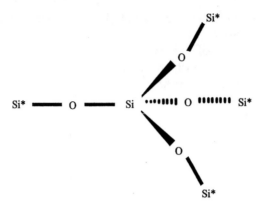

Figure 4.2.2. Schematic representation of a second Si–O–Si cluster that establishes the validity of the embedding potentials, Si*.

The total energy for the cluster in figure 4.2.1 has been calculated as a function of the Si–O–Si bond-angle α in figure 4.2.3 at three different computational levels: (i) SCF with d-polarization wavefunctions on the Si atoms, (ii) SCF + CI with no d-polarization terms and (iii) SCF + CI with d polarizations. The Si–O distances and angles for the central Si–O–Si group were optimized, and the values obtained are very close to the distances used for the next shell of atoms. Since the Hamiltonian is exact, increases in the total binding energy through a variational procedure represent an improved solution to the total energy calculation. This variational approach to binding energy optimization clearly establishes that the SCF + CI with d polarizations is the best approach for calculating the total binding energy, and therefore for other properties of these clusters as well.

The minimum in binding energy occurs at a Si–O–Si bond angle, α, of approximately 150°, very close to the average bond angle determined in [21] from recent diffraction studies, and in [11] from a new and improved analysis of the Bell and Dean model. The variation of the calculated binding energy around its minimum value of $\sim$150° in the calculations of [8] is in excellent agreement with the values obtained in [11, 21]. It is also consistent with a very small value of the empirically determined Si–O–Si three-body bond-bending force constant [3].

The electronic states that comprise the top of the valence band are non-bonding O 2p π states, and the states that form the lowest lying conduction band states are anti-bonding states with a Si 3s character [12–15]. There are also excitonic states at the band edge that have been identified both experimentally [22], and in empirical calculations [23]. The *ab initio* calculations suggest that they are transitions from the non-bonding O 2p π states at the top of the valence band, to anti-bonding O 2p π^* states that mix with the Si 3s* states. This explanation is consistent with measurements of the O-atom K_1 edge as determined from x-ray absorption spectroscopy studies [24].

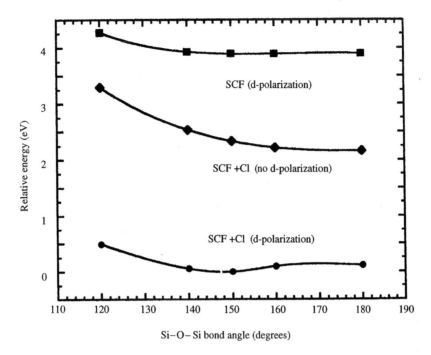

Figure 4.2.3. Calculated relative energy as a function of the Si–O–Si bond angle, α, for SCF with d polarization, SCF + CI (no d polarization), and SCF+ CI + d polarization.

The interface between crystalline Si and thermally grown, non-crystalline SiO_2 displays electrical properties, which have set the standard for metal–oxide–semiconductor (MOS) devices that include alternative gate dielectric materials and their interfaces. The conduction and valence band offset energies at Si–SiO_2 interfaces with respect to Si are 3.2 ± 0.1 and 4.5 ± 0.1 eV, respectively, and are sufficient to reduce thermionic emission currents at operating temperatures up to at least 150°C to below the levels of detection [25]. Densities of interface traps, D_{it}, and fixed charge, Q_f, can be reduced to levels below 5×10^{10} and 1×10^{11} cm^{-2}, respectively. As noted above, these defect levels, as well as defect generation under accelerated stress bias testing, have become the standards by which interfaces with other dielectrics are evaluated.

Incorporation of alternative deposited dielectrics other than thermally grown SiO_2 and thermally grown nitrided SiO_2 requires significant changes in gate stack processing and process integration. For thermally grown SiO_2, the Si–SiO_2 device interface is *continuously regenerated* as the SiO_2 film growth proceeds, and is therefore not formed at the initially processed surface of the Si substrate. The properties of this interface are determined by the oxidation temperature, as well as any other post-deposition thermal annealing and/or exposures. In contrast, for deposited dielectrics the interface may be formed (i) at the starting surface prior to film deposition, (ii) during the initial stages of deposition by interface reactions with the process gases or (iii) after deposition by a post-deposition anneal in an oxidizing ambient. Devices employing deposited SiO_2, and/or silicon nitride and oxynitrides, have required separate and independent steps for interface formation, low temperature film deposition and post-deposition annealing [26 and references therein], and it is anticipated that similar considerations will apply to transition metal oxides, silicates and aluminates as well [1].

The electronic structure at an ideal $Si–SiO_2$ interface is determined in part by the relative band alignments, and in particular by the conduction and valence band offset energies as noted above. In the calculations that follow, intrinsic interfacial dipoles due to differences in chemical potentials between Si and the dielectrics have been included in the band offset energies. However, there are departures from an otherwise ideal interface that are required to balance the intrinsic strain due to the molar volume mismatch between Si and SiO_2. Stated differently, departures from ideal abrupt interfaces are required to accommodate the large difference between the Si–Si bond length in crystalline Si, 0.235 nm, and the average Si–Si second neighbour distance in SiO_2, 0.31 nm. Accommodation of this strain requires a sub-oxide transition region between the crystalline Si substrate and the non-crystalline SiO_2 [27].

Synchrotron x-ray photoemission spectroscopy (SXPS) studies have demonstrated that in addition to the core shifts for Si 2p states in the Si substrate, and SiO_2, designated as Si^0 and Si^+, respectively, and the core shifts associated with ideal interface bonding, e.g. Si^+ for Si(111), there are additional features indicative of a transition region with *excess* sub-oxide bonding arrangements, i.e. Si^{2+} and Si^{3+} [28]. Two recent studies by Keister and co-workers [29, 30] have focused on several aspects of these sub-oxide transition regions for interfaces prepared by remote plasma processing including (i) changes in the excess sub-oxide bonding in the spectral regime between the Si substrate and the SiO_2 features as a function of annealing and (ii) differences in excess integrated sub-oxide bonding at nitrided interfaces. Figures 4.2.4 and 4.2.5 summarize these results for Si(111) and Si(100) interfaces. There is approximately a 25% reduction in the areal density of Si atoms in sub-oxide bonding arrangements for interfaces formed on Si(111) and Si(100) surfaces by remote plasma-assisted oxidation at 300°C, and then subjected to annealing in inert ambients at temperatures up to 900°C [28]. These additional spectral features fall into three spectral regions, that have been designated as Si^+, Si^{2+} and Si^{3+}, based on an assignment that assumes they present bonding arrangements with one, two and three oxygen atom neighbours, respectively [28]. Following this notation, the substrate Si feature is Si^0, and the SiO_2 feature is Si^{4+}. After a 900°C anneal, there is approximately one monolayer of Si atoms in sub-oxide bonding arrangements in

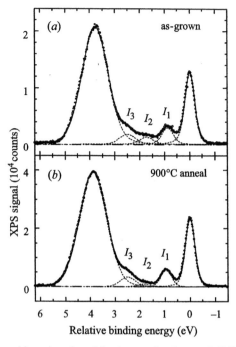

Figure 4.2.4. Spin–orbit stripped and background subtracted Si(2p) data for Si(111). (*a*) shows 'as-grown' sample (~1.2 nm), and (*b*) is for a piece of the same wafer, annealed *ex situ* to 900°C in Ar. The energy scale is referenced to the substrate Si feature (with four Si neighbours) at 0.0 eV, and the three features marked I_1, I_2 and I_3 are features assigned to Si with one, two and three oxygen neighbours. The broad feature centred at approximately 4 eV is the SiO$_2$ feature (with four O neighbours).

excess of what is expected for an ideal interface. As shown in figure 4.2.4 the largest spectral decrease for a Si(111)–SiO$_2$ interface is in the Si^{2+} oxidation state corresponding to a Si atom with two oxygen neighbours. This bonding configuration is not *native* to a Si(111) surface unless there are deviations in planarity such as surface steps. If the excess areal density of Si atoms is converted into an equivalent thickness that corresponds to an average SiO composition, then the physical thickness of this transition region is approximately 0.27 nm [31]. Controlled incorporation of nitrogen atoms in a concentration range up to about 1.5 monolayers has been achieved by a 300°C remote plasma-assisted, post-oxidation, interface nitridation step [32, 33]. Quantification of interfacial nitrogen was by secondary ion mass spectrometry (SIMS) and nuclear reaction analysis (NRA). For a nitrogen plasma processing time of 90 s, a concentration of $7.5 \pm 1 \times 10^{14}$ nitrogen atoms cm^{-2}, or equivalently one nitrogen atom per interface silicon atom on Si(100) was obtained. Additionally, the nitrogen concentration was found by analysis of SIMS data to be linear in processing time up to at least

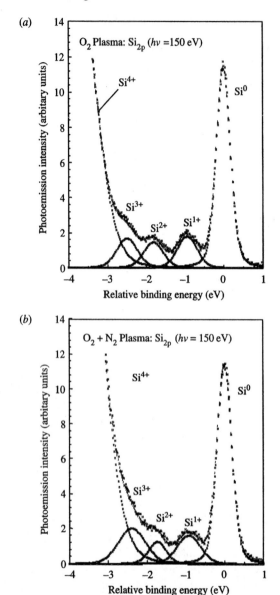

Figure 4.2.5. Spin–orbit stripped and background subtracted Si(2p) data for Si(100). (*a*) shows 'as-grown' sample (~2 nm), and (*b*) is for a piece of the same wafer, annealed *ex situ* to 900°C in Ar. The energy scale is referenced to the substrate Si feature (with four Si neighbours) at 0.0 eV, and the three features marked I_1, I_2 and I_3 are features assigned to Si with one, two and three oxygen neighbours. The broad feature centred at approximately 4 eV is the SiO_2 feature (with four O neighbours).

1.5 monolayers coverage. Analysis of XPS data on Si(100) surfaces with and without interface nitridation after a 900°C anneal indicate an increase in the excess sub-oxide/sub-nitride bonding that is consistent with a physically thicker interface transition region, 0.35 nm as compared to 0.27 nm [31]. This is consistent with the nitrogen incorporation process being a subcutaneous nitridation, rather than a N-atom–O-atom exchange reaction. The spectra shown in figure 3 of [30] indicate changes in the character of the N-atom 1s feature as a function of increasing interfacial nitrogen, with the largest change occurring for a concentration increase from approximately 1 to 1.5 monolayers. The position of this interfacial nitrogen atom feature is consistent with nitrogen atoms being bonded at the interface, and not within an interfacial oxide, e.g. as in an oxynitride alloy.

Studies using complementary spectroscopic techniques have also yielded qualitatively and quantitatively similar transition regions with sub-oxide bonding, and with approximately the same effective thickness of 0.3 nm. These include Auger electron spectroscopy [27], infrared absorption spectroscopy (IRAS) [34], high resolution electron energy loss spectroscopy (HREELS) [35] and medium energy ion scattering (MEIS) [36].

Alternative dielectrics

Zallen [3] has identified three different atomic scale amorphous morphologies for non-crystalline non-polymeric solids, (i) CRNs, as exemplified by SiO_2, with predominantly covalent bonding between the constituent atom pairs, (ii) modified continuous random networks (MCRNs) as exemplified by silicate alloys, in which metal atom ionic bonds *disrupt* and *modify* the covalently bonded CRN structure, and (iii) random close packed (RCP) nonperiodic solids comprised entirely of negative and positive ions.

In the CRN network, each atom is bonded according to its primary chemical valence; the Si atoms of SiO_2 are fourfold coordinated and the oxygen atoms are twofold coordinated. The randomness of the SiO_2 network, which is an important source of the configurational entropy that promotes the formation of CRNs with low densities of intrinsic bonding defects ($\leq 10^{16}$ cm^{-3}), derives from two sources, (i) a large spread in bond angle at the O-atom sites, $\sim 150° \pm 15°$ [11, 21], and (ii) a random distribution of dihedral angles [3].

Phillips has demonstrated that the perfection of CRNs relative to intrinsic bonding defects is correlated with the average number of bonds/atom, N_{av}, and the average number of bonding constraints/atom, C_{av} [6, 7]. It was demonstrated that an important criterion for ideal bulk glass formation is obtained by matching C_{av} to the dimensionality of the network; $C_{av} \sim 3$ for three-dimensional CRNs such as SiO_2. For CRNs in which the atoms are either two-, three- or fourfold coordinated, and the bonding geometries of the three- and fourfold coordinated atoms are non-planar, C_{av} is directly proportional to N_{av} by the following relationship:

$$C_{av} = 2.5 N_{av} - 3. \tag{4.2.4}$$

Equation (4.2.4) is derived from the relations between bonding coordination and the number of valence bond-stretching and bond-bending constraints that apply. If m is the coordination of one of the network atoms, and is greater than or equal to 2, then the number of stretching constraints per atom is $m/2$, and the number of bending constraints per atom is $2m - 3$. Based on equation (4.2.4), N_{av} for an ideal CRN in which the bonding is non-planar at three- and fourfold coordinated atomic sites is 2.4, as for the compound $As_2S(Se)_3$, and the chalcogenide-rich $Ge-S(Se)$ alloys, $GeS(Se)_4$ [6, 7]. If the atomic coordination is greater than or equal to 3, and the atom is in a planar bonding arrangement, e.g. B in B_2O_3, or N in Si_3N_4, then the number of bond-bending constraints per atom is reduced to $m - 1$. Additional considerations apply to terminal atoms with a coordination of 1; however, these are not relevant to the alternative gate dielectrics that are addressed in this review.

For SiO_2, $N_{av} = 2.67$, so that $C_{av} = 3.67$ and is sufficiently high for the creation of bulk defects at levels $> 10^{17}$ cm^{-3} [37]. However, since the bending force constant at the O-atom site is exceptionally weak, one bond-bending constraint per atom can be neglected so that C_{av} is then reduced to 3.0. This reduction explains the ease of glass formation for bulk SiO_2, and the unusually low density of intrinsic defects that are present in thermally grown and deposited SiO_2 thin films that have been annealed at temperatures of 900–1000°C [37]. The concentration of bulk and interfacial defects in over-constrained dielectrics has been shown to increase as the square of the difference between the number of bonds per atom and the number of bonds per atom in an idealized low defect density material, and is a factor in Si_3N_4 and silicon oxynitrides, both in bulk films and in their interfaces with Si and SiO_2 [38].

The *second* amorphous morphology includes silicate and aluminate alloys, and the elemental oxides Al_2O_3, TiO_2 and Ta_2O_5, which have modified CRN structures that include metal atom ions. In the silicate dielectrics, the covalently bonded network is *disrupted* by the introduction of ionic metals such as Na, Ca, Zr, etc. These atoms coordinate with O atoms at levels that exceed their chemical valence, and as a result dative bonds are formed between these metal atoms and the non-bonding pairs on the network O atoms. This increases the number of bonds per atom for oxygen, and decreases the bond-order at the metallic sites as well. The average bonding coordination of the oxygen atoms is typically increased from 2 in the CRN dielectrics to ~ 3 in the MCRN dielectrics. As an example, tetrahedrally coordinated Al groups in non-crystalline Al_2O_3 with a net negative charge comprise 75% of the oxide structure, and provide the network bonding component. Octahedral interstitial voids within this network are sites where Al^{3+} ions are incorporated through dative bonds with non-bonding orbitals of bridging oxygen atoms of the network [39, 40]. This increases the average coordination of the oxygen atoms from 2 to 3.

As a second example, the IR spectra of SiO_2-rich $(ZrO_2)_x(SiO_2)_{1-x}$ alloys, $x \lesssim 0.30$, as well as other transition metal silicate alloys, display vibrational features that are consistent with a modified network structure with an oxygen atom bond coordination < 3 [40, 41]. It has not yet been demonstrated that the bond-constraint approach of Phillips and Kerner [42] that has been successfully applied to silicate alloys with low concentrations of group I and II alloy atoms, e.g. Na and Ca, respectively, and for low concentrations of Al_2O_3 in SiO_2, is also applicable to non-crystalline Al_2O_3, TiO_2 and Ta_2O_5 and transition metal silicate and aluminate alloys.

The last group of non-crystalline oxides has an amorphous morphology that can be characterized as a random close packing of ions. It includes transition metal oxides, as well as Zr and Hf silicate alloys in which the ZrO_2 and HfO_2 fractions, x, are greater than 0.5. The average oxygen coordination in Hf and Zr oxide, as well as other group IIIB transition metal oxides, Y_2O_3, La_2O_3, etc, is approximately equal to 4.

A classification scheme based on bond ionicity distinguishes between the three different classes of non-crystalline dielectrics with different amorphous morphologies discussed above. Since there is an intimate relationship between electronegativity, X, and electronic structure reflected in the variation of X across the periodic table, X provides a useful and pragmatic metric for defining a bond ionicity that is directly correlated with the electronic structure of the constituent atoms [43]. A definition of bond ionicity, I_b, as introduced by Pauling is the basis for the classification scheme of this paper. Similar results can be obtained using other electronegativity scales such as the one proposed by Sanderson [44]. The approach below represents the first attempt at using bond ionicity scaling to discriminate between different bonding morphologies in non-crystalline elemental oxides and their alloys.

If $X(O)$ is the *Pauling electronegativity* of oxygen, 3.44, and $X(Si)$ is the corresponding electronegativity of silicon, 1.90, then the electronegativity difference between these atoms, ΔX, is 1.54. Applying Pauling's empirical definition of bond ionicity, I_b [43],

$$I_b = 1 - \exp(-0.25(\Delta X)^2), \qquad (4.2.5)$$

yields a value of I_b for Si–O bonds of $\sim 45\%$. The range of ΔX values of interest in this classification ranges from about 1.5 to 2.4, corresponding to a bond ionicity range from ~ 45 to 76%. For this range of ΔX, I_b is approximately a linear function of ΔX so that ΔX and I_b can be used as *functionally equivalent* scaling variables. It is important to understand that charge localization on the silicon and oxygen atoms, i.e. effective ionic charges, cannot be determined directly from these values of bond ionicity. The bonds in other good glass formers including oxides such as B_2O_3, P_2O_5, GeO_2 and As_2O_3, and chalcogenides such as $As_2S(Se)_3$ and $GeS(Se)_4$, are generally less ionic in character than SiO_2. For pseudo-binary oxide and chalcogenide alloys, e.g. $(SiO_2)_x(B_2O_3)_{1-x}$, and $(GeS_2)_x(As_2S_3)_{1-x}$, respectively, compositionally averaged values of ΔX are used.

The oxides and chalcogenides mentioned above form covalently bonded network structures in which each of the constituent atoms has a coordination that reflects its primary chemical valence; e.g. 2 for O, S and Se, 4 for Si and Ge, 3 for B and As, etc. This class of dielectrics is differentiated from other oxide dielectrics by a bond ionicity of less than about 47%, or equivalently a ΔX value less than about 1.6. This class of CRN dielectrics also includes Si_3N_4 and silicon oxynitride alloys.

The second class of non-crystalline dielectrics form MCRNs, which include ionic bonding arrangements of metal atoms that *disrupt* the network structure. This class of dielectrics is characterized by values of ΔX between about 1.6 and 2.0, or equivalently a bond ionicity between approximately 47 and 67%. The most extensively studied and characterized oxides in this group are the metal atom silicate alloys, e.g. SiO_2 alloyed with Na_2O, CaO, MgO, PbO, etc and quenched from the melt [3]. This class also includes deposited thin film Al_2O_3, TiO_2 and Ta_2O_5, and transition metal atom silicate alloys such as $(Zr(Hf)O_2)_x(SiO_2)_{1-x}$ in the composition range up to about $x \sim 0.3$ [40, 41]. The non-crystalline range of alloy formation in deposited thin films is generally significantly increased with respect to what can be obtained in bulk glasses that are quenched from the high temperature melt. The coordination of oxygen atoms in CRNs is typically 2, and increases to approximately 3 in the MCRNs (see table 4.2.1). As examples, the coordination of oxygen increases from 2.8 in thin film Ta_2O_5, to 3.0 in Al_2O_3, and increases from 2 to 3 in the group IVB silicate alloys as the ZrO_2 or HfO_2 fraction is increased from doping levels to $x = 0.5$.

The third class of non-crystalline oxides are those that have an RCP ionic amorphous morphology [3]. This class of oxides is defined by $\Delta X > 2$, and a Pauling bond ionicity of greater than approximately 67%. This group includes transition metal oxides deposited by low temperature techniques including plasma deposition and sputtering with post-deposition oxidation [1]. The coordination of the oxygen in these RCP structures is typically 4.

The coordination of oxygen atoms in the dielectrics addressed above scales monotonically with increasing bond ionicity. This heralds a fundamental relationship between charge localization on the oxygen atom and bonding coordination that has been confirmed by spectroscopic studies of Zr silicate alloys in which the coordination has been shown to vary linearly with alloy composition [41]. Finally, as noted above, other ionicity scales, such as the one proposed by Sanderson [44], give qualitatively similar results

Table 4.2.1. Electronegativity difference, ΔX, average bond ionicity, I_b, and metal and oxygen coordination for SiO_2 and high-κ alternative dielectrics.

Dielectric	ΔX	I_b	Metal/silicon coordination	Oxygen coordination
CRNS				
SiO_2	1.54	0.45	4	2.0
MCRNS				
Al_2O_3	1.84	0.57	4 and 6 (3:1)	3.0
Ta_2O_5	1.94	0.61	6 and 8 (1:1)	2.8
TiO_2	1.90	0.59	6	3.0
$(ZrO_2)_{0.1}(SiO_2)_{0.9}$	1.61	0.48	8 and 4	2.2
$(ZrO_2)_{0.23}(SiO_2)_{0.77}$	1.70	0.51	8 and 4	2.46
$(ZrO_2)_{0.5}(SiO_2)_{0.5}$	1.88	0.59	8 and 4	3.0
$(TiO_2)_{0.5}(SiO_2)_{0.5}$	1.72	0.52	6 and 4	2.5
$(Y_2O_3)_1(SiO_2)_2$	1.88	0.59	6 and 4	2.86
$(Y_2O_3)_2(SiO_2)_3$	1.93	0.61	6 and 4	3.0
$(Y_2O_3)_1(SiO_2)_1$	1.99	0.63	6 and 4	3.11
$(Al_2O_3)_4(ZrO_2)_1$	2.02	0.64	4 and 8	3.0
$(Al_2O_3)_3(Y_2O_3)_1$	1.97	0.62	4 and 6	3.0
Random ions				
HfO_2	2.14	0.68	8	4.0
ZrO_2	2.22	0.71	8	4.0
$(La_2O_3)_2(SiO_2)_1$	2.18	0.70	6 and 4	3.5[a]
Y_2O_3	2.2	0.7	6	4.0
La_2O_3	2.34	0.75	6	4.0

[a] The $(La_2O_3)_2(SiO_2)_1$ alloy contains both O ions and silicate groups, and as such the O atoms have local coordinations of 4 and 3, respectively.

for separation of non-crystalline dielectrics into the three amorphous morphologies discussed above.

Electronic structure of transition metal dielectrics

Figure 4.2.6 is a schematic molecular orbital energy level diagram for a group IV transition metal, e.g. Ti, in an octahedral bonding arrangement with six oxygen neighbours [45–47], $Ti(OH)_6^{3-}$. The symmetries and π or σ character of the calculated molecular orbitals are determined by the symmetry character of the group IVB and oxygen atomic states. The energies of the bonding and anti-bonding states are determined in the usual manner by the calculation of coulomb integrals [45, 46]. The top of the valence band is assigned to non-bonding p-state π orbitals of oxygen atoms, and the first two conduction bands are associated with transition metal atom d states.

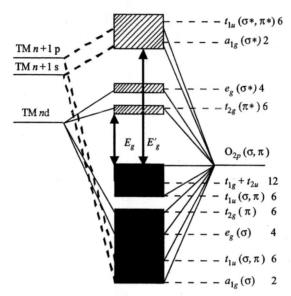

Figure 4.2.6. Molecular orbital energy level diagram for a group IV transition metal, e.g. Ti, in an octahedral bonding arrangement with six oxygen neighbours.

The next conduction band is derived from transition metal s states with $a_{1g}(\sigma^*)$ character. The energy separation between the top of the valence band and the $a_{1g}(\sigma^*)$ band edge defines an *ionic band gap* with essentially the same energy as that of non-transition metal insulating oxides including MgO or Al_2O_3, $\sim$8–9 eV. The ordering of the lowest conduction bands in crystalline TiO_2 has been verified by electron energy loss and x-ray spectroscopies, which confirm the relative sharpness of the $t_{2g}(\pi^*)$ and $e_g(\sigma^*)$ bands, the increased width of the $a_{1g}(\sigma^*)$ band, and an ionic band gap > 8 eV [47, 48].

There are several aspects of the energy band scheme in figure 4.2.6 that are important for band gap and conduction band offset scaling: (i) the symmetry character of the highest valence bonding states, non-bonding p states with an orbital energy approximately equal to the energy of the atomic 2p state, (ii) the weak π-bonding of the transition metal atoms establishes that the lowest anti-bonding d* state is close in energy to the atomic nd state of the transition metal atom and (iii) the energy separation between the nd and $n + 1s$ derived anti-bonding states is correlated with the difference between the atomic n d and $n + 1s$ states. Figure 4.2.7 contains plots of (*a*) the lowest optical band gap and (*b*) the conduction band offset energies, both from the papers of Robertson [49, 50], versus the absolute value of the energy of the transition metal atomic nd state in the $s^2 d^{\gamma-2}$ configuration appropriate to insulators, where $\gamma = 3$ for the group IIIB transition metals, Sc, Y and Lu(La), and the rare earth lanthanides, and $\gamma = 4$ for the group IVB transition metals Ti, Zr and Hf. The linearity of these

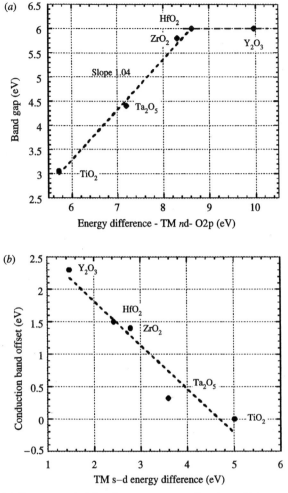

Figure 4.2.7. (*a*) Lowest optical band gap energy as function of the energy difference between the transition metal *n*d atomic state energy and O 2p atomic state energy. (*b*) Conduction band offset energy as function of the differences in energy between the atomic *n* + 1s and *n*d states. Both are for representative transition metal oxides.

plots supports the qualitative and quantitive *universality* of the energy band scheme of figure 4.2.6, including the band gap scaling slope of approximately one.

The band offset energy between the conduction band of Si and the empty anti-bonding or conduction band states of a high-κ gate dielectric in figure 4.2.7(*b*) is important in MOS device performance and reliability. It defines the barrier for direct tunnelling, and/or thermal emission of electrons from an n^+ Si substrate into a transition metal oxide. In alloys such as Al_2O_3–Ta_2O_5, or SiO_2–ZrO_2, it also defines the energy of localized transition metal trapping states relative to the Si conduction band [51–53]. The next sub-section extends the *ab initio* calculations of SiO_2 to local bonding arrangements of Zr atoms in ZrO_2 and Zr silicate alloys, $(ZrO_2)_x(SiO_2)_{1-x}$.

Following the methods of previous studies which have addressed the electronic structure of transition metal ion complexes [45–47], the *ab initio* results presented below are based on relatively small clusters (9–17 atoms in size) that include the transition metal atom at the centre, and H-atom terminated O-atom neighbours. The details of these calculations are being addressed in a paper which is in progress [54]. This review presents results of these calculations which have addressed: (i) the valence band electronic structure of TiO_2 and ZrO_2, (ii) transitions between 3p core states and antibonding 3d* and 4s* states in ZrO_2 and Zr silicate alloys, (iii) the O K_1 edge transitions in ZrO_2 and (iv) band edge transitions in ZrO_2. The transitions in (ii) are intra-atomic in character, whilst those in (iii) and (iv) include inter-atom contributions as well. Figure 4.2.8 summarizes the results of the ground state calculations including valence band structure, and figure 4.2.9 the intra- and inter-atomic transitions.

Figure 4.2.8 indicates ground state electronic structure for clusters containing fourfold coordinated Ti and Zr atoms, each bonded to four OH groups. The focus here is the atomic states of Ti and Zr atoms in $3d^24s^2$ and $4d^25s^2$ configurations, respectively, and the highest valence band O atom 2p states. The energies of the atomic 5s states of Zr and 4s states of Ti relative to the top of the valence band (arbitrarily set at 0.0 eV) are approximately the same, whereas the 3d states of Ti are about 2.7 eV closer to the valence band than those of Zr. The highest valence band states at 0.0 are associated with non-bonding O-atom 2p states with a π-symmetry character. The next four states in each of the valence bands are associated predominantly with O-atom σ and π states that are mixed with the respectively σ and π components of the transition metal d states, so that all five of the valence bands are highly localized on transition metal and O atoms within the cluster. The second and third valence band states have a π-bonding character and are derived from mixtures of occupied O-atom 2p and transition metal Zr 4d or Ti 3d π states, whilst the more closely spaced lower conduction band states are derived from mixtures of O-atom 2p σ states and Zr 4d, 5s, or Ti 3d, 4s σ states.

Figure 4.2.9 summarizes the intra- and inter-atomic energy transitions for ZrO_2. The calculations are non-relativistic and do not include the spin–orbit splittings of Zr 3p states in $3p_{1/2}$ and $3p_{3/2}$ components. The calculated Zr 4d* and 5s* final state energies for inter-atom Zr K_1 and $M_{2,3}$ transitions, in which the final states are localized on Zr atoms, are compared with the corresponding final state energies for the O K_1 edge, and band edge inter-atom transitions in which the Zr 4s* and 5s* states are mixed with 2p*

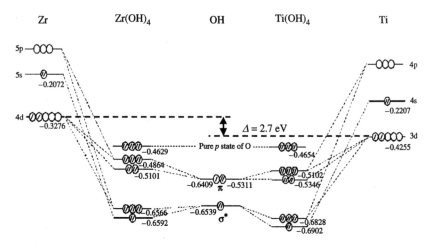

Figure 4.2.8. Ground state property energy levels for a nine-atom Ti(OH)$_4$ cluster, that includes atomic energies of Ti and the OH molecular group.

states and therefore more delocalized, in figure 4.2.9. The 4d*–5s* energy separations are respectively 14 ± 1 eV and 12.5 ± 0.5 eV for Zr K$_1$ and M$_{2,3}$ transitions, and decrease to 2.5 ± 0.3 for the O K$_1$ edge. Finally, for the band edge transition, the Zr 5s* state falls between the symmetry split 4d* state

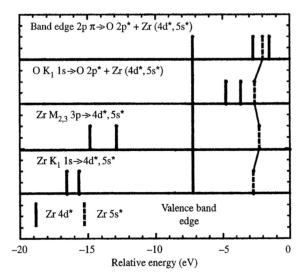

Figure 4.2.9. Calculated Zr 4d* and 5s* final state energies for Zr K$_1$, Zr M$_3$, O K$_1$ and band edge electronic transitions. The energy of the O 2p π non-bonding state, that defines the top of the valence band in ZrO$_2$, is included as reference energy.

doublet. In contrast, the symmetry splitting of the 4d* doublet is 1.5 ± 0.5 eV for all four transitions. As a point of reference, the figure includes the energy of the O 2p π non-bond state that forms the top of the valence band. The energy of the 5s* state is approximately the same for all four transitions, whilst the relative energies of the 4d* doublet vary by almost 10 eV. Most importantly, the similarity between the final state energies for O K$_1$ and band edge transitions establishes that the relative energies of the 4d* and 5s* states in the O K$_1$ transitions are a good indicator of the symmetry splittings of the band edge of 4d* states, but underestimate the extent of overlap between the 4d* and 5s* states.

Experimental studies of electronic structure

X-ray absorption spectra (XAS) for transition oxides and their pseudo-binary alloys and compounds are interpreted in terms of the *ab initio* calculations presented in the previous section [55, 57]. The Zr silicate alloys, including ZrO_2, and Ta and Hf silicate and aluminate alloys, including respectively Ta_2O_5 and HfO_2, were prepared by remote plasma enhanced chemical vapour deposition (RPECVD). This deposition process, and chemical and structural characterizations relevant to bonding coordination, have been addressed in other publications [41 and references therein].

Figures 4.2.10–4.2.13 summarize the results of XAS measurements performed on Zr silicate alloys, $(ZrO_2)_x(SiO_2)_{1-x}$, prepared by RPECVD. Figure 4.2.10 displays XAS spectra for alloys with $x \sim 0.05$, 0.2, 0.5 and 1.0 as obtained at the Brookhaven Laboratory National Synchrotron Light Source, NSLS [55, 56]. Spectral features in figure 4.2.10 have been assigned to dipole allowed transitions between $M_{2,3}$ 3p spin–orbit split $3p_{3/2}$ and $3p_{1/2}$ core states of the Zr atoms, and empty conduction band states derived from Zr-atom $N_{4,5}$ 4d* states (*a* and *b*, and *a'* and *b'*) and O$_1$ 5s*atomic states (*c* and *c'*). As noted above these transitions are intra-atomic in character. The transitions from the $3p_{3/2}$ states are the lower energy features, *a*, *b* and *c*, and the transitions from the $3p_{1/2}$ states are the higher energy features, *a'*, *b'* and *c'*. As-deposited alloys prepared by plasma deposition are non-crystalline and pseudo-binary with Si–O and Zr–O bonds, but no detectable Zr–Si bonds. After annealing at 900–1000°C in Ar, the $x = 0.5$ alloy phase separates into SiO_2 and crystalline ZrO_2; the $x = 1.0$ is also crystalline after annealing [41]. The separation in the $x = 0.05$ and 0.2 samples is not accompanied by crystallization of the ZrO_2 component. The energies of the features identified in the $M_{2,3}$ XAS spectra in figure 4.2.10 are independent, up to an experimental uncertainty of ±0.2 eV, of the alloy composition and the state of crystallinity. The independence of spectral features on alloy composition, and the relatively small changes that take place upon crystallization of a ZrO_2 phase after

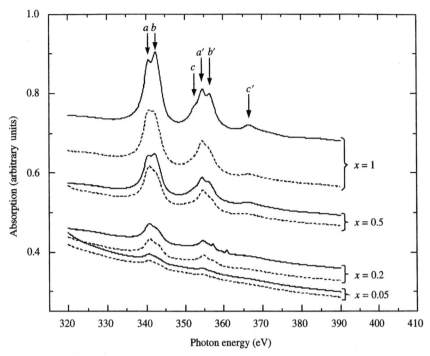

Figure 4.2.10. XAS as a function of photon energy for excitation from Zr $M_{2,3}$ p states in empty Zr 4d and 5s states of Zr silicate alloys. Before and after a rapid thermal anneal at 1000°C. The letters *a, b* and *c* and a', b' and c' designate the energy differences between the M_2 and M_3 p states, respectively, and the anti-bonding Zr states.

annealing, are consistent with the localization of the 4d* states on the Zr atoms, i.e. intra-atom transitions. The spectra for the $x = 0.05$ alloy, and crystalline (c-) ZrO_2 have been fitted by Lorentzian functions with Gaussian wings, and the effective line-widths (full width at half maximum (FWHM)) have been determined. The agreement between the frequencies obtained from the *ab initio* calculations and experiment is excellent as shown in figure 4.2.11 and is based on the deconvolution of the spectral features labelled *a, b* and *c* for ZrO_2. The dashed lines are calculated from the *ab initio* calculations of the preceding section in figure 4.2.9.

Figure 4.2.12 displays XAS spectra for the K_1 edge of oxygen in ZrO_2. The two lower features in the ZrO_2 spectrum are associated with transitions to the same Zr atom 4d* doublet in figures 4.2.10 and 4.2.11; however, the separation of these features is about 1 eV larger in figure 4.2.12, consistent with the *ab initio* calculations presented in figure 4.2.9. The experimental spectra also show the additional localized 5s* feature predicted by theory, as well as the Zr 5s*/O 2p* band.

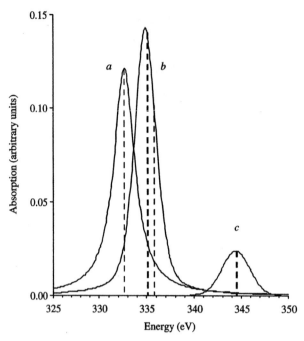

Figure 4.2.11. Deconvolution of *a*, *b* and *c* features for ZrO_2. The dashed lines indicate the relative transition energies obtained from the SCF/CI *ab initio* calculations.

Figure 4.2.12 displays XAS spectra for the K_1 edge of oxygen in ZrO_2. The lowest-energy spectral feature in this spectrum is associated with transitions to the Zr atom 4d* doublet in figures 4.2.10 and 4.2.11; however, the separation of these features is not as well resolved in this spectrum due to mixing with O 2p* states. The second feature is associated with localized 5s*, also mixed with O 2p*, states. The broader, highest-energy feature includes considerably more O 2p* character, but is also mixed with Zr 5s*.

Figure 4.2.13 displays the O K_1 edge for three Zr silicate alloys. The lowest-energy features in these spectra is associated with Zr 4d* states mixed with O 2p*. The higher-energy feature in these three alloys is assigned to the Si 3s* states mixed with O 2p*, and overlaps Zr 5s* states as well. These assignments, for spectra of Zr silicate alloys with *x* ranging from 0.35 to 0.6, indicates that (i) the difference in energy between the localized Zr atom 4d* feature, and the more extended Si 3s* states, is not dependent on the alloy composition, whereas (ii) the relative amplitudes of the Zr 4d* and Si 3s* features scale with alloy composition.

The spectra in figure 4.2.14(*a*) and (*b*) are qualitatively similar to those for Zr silicate alloys in figure 4.2.13. The Hf silicate spectra indicate that the energy difference between the Hf 5d* states, and the onset of transitions involving Si 3s* states, is independent of alloy composition.

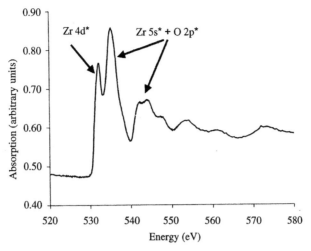

Figure 4.2.12. XAS K_1 spectrum for ZrO_2. The assignment of the first two spectral features is from figure 4.2.9.

Differences between the Hf silicate and aluminate spectra are consistent with an approximately 1 eV red shift of the Al_2O_3 conduction band states relative to those of SiO_2, as determined from band offset calculations and measurements.

Figure 4.2.15(*a*) and (*b*) compares the O-atom K_1 edges of Y_2O_3 and ZrO_2. The d* and s* states are spectroscopically distinct in ZrO_2, whilst in

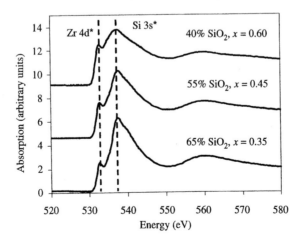

Figure 4.2.13. O K_1 spectra for Zr silicate alloys for $x = 0.35$, 0.45 and 0.60. The dashed lines indicate the energies of the Zr 4d* and Si 3s* spectral features.

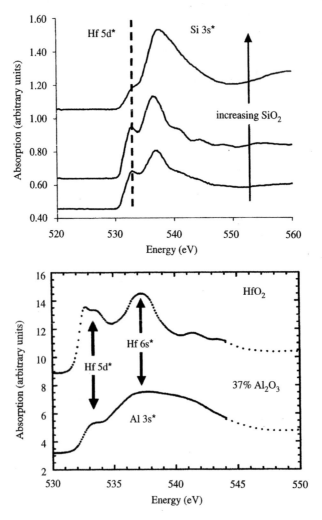

Figure 4.2.14. (*a*) XAS K₁ spectra for Hf silicate alloys. The assignments of the Hf 5d* and Si 3s* features parallel those in figure 4.2.13. The Hf 6s* feature is not marked in the figure, but is clearly evident about 540 eV in the HfO₂-richer alloys. (*b*) XAS K₁ spectrum for a Hf aluminate alloy. The Al 3s* feature also overlaps the Hf 6s* feature. The spectral overlap between the Al 3s* and Hf 5d* features in (*b*) is greater than for the Si 3s* and Hf 5d* features in (*a*).

Y_2O_3 there is considerable spectroscopic overlap. This is a manifestation of a smaller energy difference between the atomic 3d and 4s states of Y, ~1.5 eV, and the 4d and 5s states of Zr, ~2.8 eV. The extent of the overlap is a manifestation of solid state broadening of the respective d* and s* states.

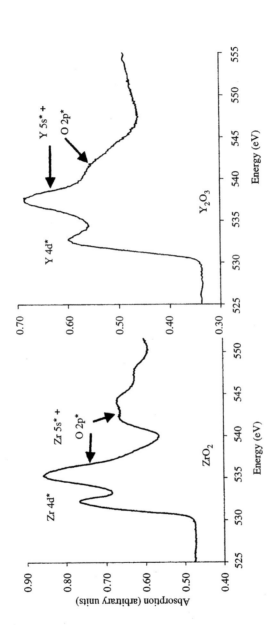

Figure 4.2.15. XAS O K$_1$ spectra for (*a*) ZrO$_2$ and (*b*) Y$_2$O$_3$. The spectral overlap between the 4d* and 5s* states in Y$_2$O$_3$ is greater than in ZrO$_2$, consistent with the respective differences between the energies of the corresponding atomic 4d and 5s states in the 5s^{2}4d^2 configurations of Y and Zr. The Zr 5s* states for both ZrO$_2$ and Y$_2$O$_3$ are mixed with O 2p states as indicated in the respective spectral plots.

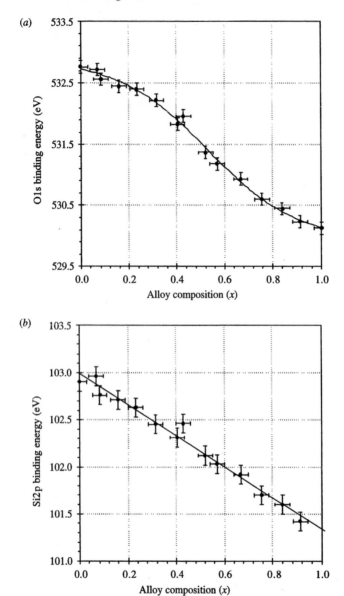

Figure 4.2.16. XPS chemical shifts of (*a*) O 1s, (*b*) Si 2p and (*c*) Zr $3d_{5/2}$ core levels from as-deposited (300°C) $(ZrO_2)_x(SiO_2)_{1-x}$ alloys as a function of composition, x.

This is also the origin of the departure from linearity in the plot of band gap versus atomic d-state energy (see figure 4.2.7(*a*)).

A comprehensive study of XPS and AES measurements is presented in [41] for Zr silicate alloys, $(ZrO_2)_x(SiO_2)_{1-x}$. Figure 4.2.16(*a*)–(*c*) summarizes

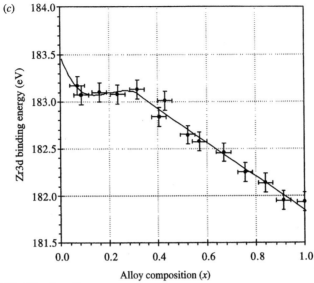

Figure 4.2.16. (Continued)

the results of XPS measurements of O 1s, Zr 3d and Si 2p core levels for the end-member oxides, SiO_2 and ZrO_2, and for 13 pseudo-binary oxide alloy compositions distributed approximately equally over the entire alloy composition range. These results are for as-deposited thin films. Studies of films annealed at 500°C in Ar display essentially the same spectra, whereas films annealed at 900°C show evidence for a chemical phase separation into SiO_2 and ZrO_2, independent of whether the phase separation is accompanied by crystallization of the ZrO_2 component.

Figure 4.2.16(a) indicates the compositional dependence of the O 1s binding energy. The sigmoidal character of the plot is a manifestation of mixed coordination for O atoms as anticipated by the discussion above relative to the classification scheme scaling between oxygen atom coordination and bond ionicity. The coordination of oxygen increases from 2 to 3 in the composition range from 2 in SiO_2, to 3 in the 50% ZrO_2 chemically ordered alloy that defines the stoichiometric silicate composition, $ZrSiO_4$. Derivative XPS spectra, displayed in [41], confirm that the sigmoidal dependence is due to mixed coordination. Finally, the total core level shift in the O 1s binding energy between SiO_2 and ZrO_2 is 2.45 eV.

Figure 4.2.16(b) and (c) displays similar spectra for the Si 2p and Zr $3d_{5/2}$ core levels. The Si 2p data in figure 4.2.16(b) show a linear dependence consistent with a single atomic coordination of 4, and a total core level shift of 1.85 eV between SiO_2 and ZrO_2. Note that the core level shifts in figure 4.2.16(a)–(c) are in the same direction, with the values for SiO_2 being more negative. As discussed in [41], this is consistent with partial charges calculated on the basis of electronegativity equalization [44].

The data for the compositional dependence of the Zr $3d_{5/2}$ core level show some additional structure for low values of x, < 0.4. Consider first the total change in binding energy across the alloy system; this is 1.85 eV, essentially the same as for the 2p Si core level. This means that the slopes of the plots in figure 4.2.16(b) and (c) in the linear regime are the same and consistent with the calculations of partial charges derived from application of the principle of electronegativity equalization [44]. The equivalence of these slopes is also consistent with the XAS data for Zr silicate alloys; it is equivalent to the 4d* anti-bonding states of Zr and the 3s* band peak of Si maintaining a constant energy separation as a function of alloy composition as shown in figure 4.2.13.

Finally, the departure from linearity for $x < 0.4$ in figure 4.2.16(c) has been assigned to changes in chemical bonding at the Zr site as a function of alloy composition [41]. The coordination of Zr has been assumed to be 8, independent of alloy composition; however, these eight oxygen atoms are inequivalent with respect to bonding neighbour coordination and electronic structure. The number of ionic Zr–O bonds associated with *network disruption* increases from four to eight with increasing x for alloys in the SiO_2-rich bonding regime. In this alloy regime, each O atom in a Si–O–Zr arrangement makes at least one Zr–O bond with a bond order of one, and there must be at least four of these bonds. The remainder of the eight-fold coordination is made up of donor–acceptor pair bonds with bridging O atoms of the non-disrupted portion of the SiO_2 CRN. These weaker bonds have been modelled in *ab initio* calculations as components of a dipolar electrostatic field, and alternatively, and equivalently can also be described as *donor–acceptor pair or dative bonds*. The donor–acceptor bonds are replaced by Si–O–Zr ionic bonding arrangements as x increases, and the network disruption increases. At a composition of $x = 0.5$, network disruption is essentially complete, and the O-atom coordination is 3, and the bond order of the Zr atoms is formally one-half with all bonds between eightfold coordinated Zr^{4+} ions and terminal O atoms of silicate ions, SiO_4^{4-}. Each of the terminal O atoms of a silicate ion makes bonds with two Zr^{4+} ions.

Chemical shifts of O_{KVV} and Zr_{MVV} transitions as a function of composition for derivative AES spectra are shown in figure 4.2.17(a) and (b). They show nearly identical sigmoidal behaviours that are qualitatively different and *complementary* to the XPS chemical shifts of the Si 2p and Zr $3d_{5/2}$ core levels shown respectively in figure 4.2.16(b) and (c). The compositional dependence of the *peak kinetic energy* values displays a similar sigmoidal non-linear dependence; however, for the O_{KVV} data the spread in energies between the end-member compounds at $x = 0.0$ and $x = 1.0$ is about 1 eV less due to the compositional dependence of the spectral line-width. Finally, due to spectral overlap between the Zr_{MVV} and Si_{LVV}

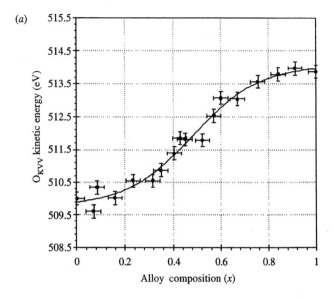

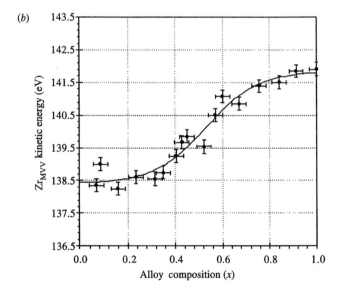

Figure 4.2.17. AES chemical shifts of (*a*) O_{KVV} and (*b*) Zr_{MVV} kinetic energies in as-deposited $(ZrO_2)_x(SiO_2)_{1-x}$ alloys as a function of composition. The plots in (*a*) and (*b*) are for the *highest energy peaks* in the respective AES derivative spectra. The solid lines are polynomial fits that are intended to emphasize the sigmoidal character of the compositional dependence.

features in the AES spectra, it was not possible to track the compositional dependence of the AES Si_{LVV} feature.

The chemical shifts of the Auger electron kinetic energies for O_{KVV} and Zr_{MVV} transitions in the as-deposited films are consistent with changes in the calculated partial charges and their effects on the O and Zr core state energies, i.e. the kinetic energies of the Auger electrons increase with increasing x reflecting the *decreases in the negative* XPS binding energies, i.e. shifts to less negative values. The differences between the XPS and AES spectral features derive from differences between the XPS and AES processes. Following [58], the AES electrons of figure 4.2.17 originate in the valence band, whereas the XPS electrons of figure 4.2.16 originate in the respective core states with no valence band participation.

The XPS and AES results are now combined with determinations of valence bond offset energies for SiO_2 and ZrO_2 [56] to generate an empirical model for the compositional variation of valence band offset energies with respect to Si. The O_{KVV} transition in a-SiO_2 has been investigated theoretically, and it has been shown that the highest kinetic energy AES feature is associated with two electrons being released from the non-bonding O 2p π states at top of the valence band; one of these is the AES electron, and the second fills the O 1s core hole generated by electron beam excitation [58]. Based on this mechanism, the XPS and AES results are integrated into a model that provides an estimate of valence band offsets with respect to Si as a function of alloy composition. For an *ijk* AES *A*-atom transition, the kinetic energy of the AES electron, $E_K(A,ijk)$, is related to the XPS binding energies, $E_B(A,i)$, $E_B(A,j)$, and $E_B(A,k)$, and a term $\Omega(A)$ includes all final state effects:

$$E_K(A,ijk) = E_B(A,i) - E_B(A,j) - E_B(A,k) - \Omega(A). \qquad (4.2.6)$$

Applied to the O_{KVV} transition, $A = O$, $i = K$ (O 1s) and $j, k = L = O$ (2p π non-bonding). Equation (4.2.6) is the basis for an empirical model for the energy of the Zr silicate valence band edge with respect to vacuum, and then with respect to Si, both as functions of the alloy composition. If $E_{BE}(O 1s)$ is the XPS binding energy, and $E_{KE}(O_{KVV})$ is the average kinetic energy of the Auger electron with respect to the top of the valence band edge, then the offset energy, $V_{OFFSET}(x)$, is given by

$$V_{OFFSET}(x) \sim -0.5A[E_B(O\ 1s) - E_k(O_{kvv})] + B, \qquad (4.2.7)$$

where A and B are determined from the experimental valence band offsets of 4.6 eV for SiO_2 and 3.1 eV for ZrO_2 [56]. This model is presented in figure 4.2.18, and the sigmoidal shape is determined by the relative compositional dependences of the XPS (O 1s) and AES (O_{KVV}) results in figures 4.2.16(*a*) and 4.2.17(*a*). The analysis has also been applied to the Zr_{MVV} AES and Zr $3d_{5/2}$ XPS results of figures 4.2.16(*c*) and 4.2.17(*b*), and gives essentially the same compositional dependence as is displayed in

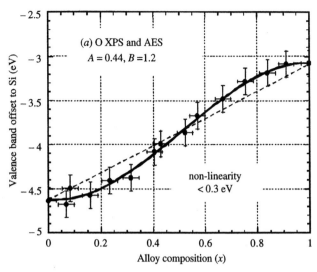

Figure 4.2.18. Values of the valence band offset energies relative to the valence band of crystalline Si at $\sim -5.2\,\text{eV}$ as calculated from the two-parameter empirical model of equation (4.2.7). The plots are derived from O-atom XPS and AES data. The sigmoidal dependence results from differences between the compositional dependences of the respective XPS and AES results used as input, and not on empirical constants.

figure 4.2.18, but with different empirical constants, A' and B'. The weakly sigmoidal dependence is a manifestation of the discreteness of the O-atom coordinations as function of the alloy composition, a mixture of two- and threefold for $x < 0.0$, and three- and fourfold for $x > 0.5$.

Figure 4.2.19 contains plots of the average conduction and valence band offset energies of Zr silicate alloys. This plot combines XAS results of figure 4.2.13 with the model of equation (4.2.7), and the experimentally determined band gaps for SiO_2, $\sim 9\,\text{eV}$, and ZrO_2, $\sim 5.6\,\text{eV}$. This approach demonstrates that essentially all of the band gap variation occurs in the valence band offsets, so that the offset energies of the respective Zr 4d* states and Si 3s* states are constant to $< \pm 0.2\,\text{eV}$ with respect to the conduction band edge of Si. The contributions of these Zr and Si states to the conduction band density of states are proportional to their alloy concentrations, with *qualitative differences* in these states playing a significant role in determining direct tunnelling currents.

Studies of the electrical properties of Ta aluminate alloys have provided another example of fixed energy differences between transition metal atom d* states, and s* conduction band states of the aluminium oxide alloy constituent. This is discussed in detail in [52] and for alloys of Al_2O_3 and Ta_2O_5 and in [53] for alloys of Al_2O_3 and HfO_2.

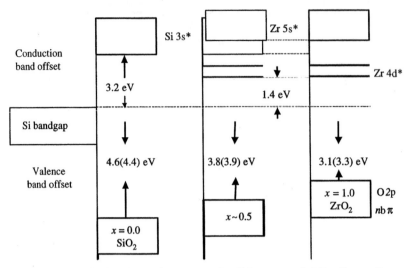

Figure 4.2.19. Band edge electronic structure for SiO_2, an $x = 0.5$ Zr silicate alloy and for ZrO_2.

Interface electronic structure: direct tunnelling in silicate alloys

This section applies the energy band scheme of figure 4.2.19 for Zr silicates to direct tunnelling in Hf(Zr) silicate and Y(La,Lu) silicate alloys, demonstrating that reductions in conduction band offset energies and tunnelling electron masses in transition metal (and rare earth) silicates can in some systems mitigate increases in physical thickness in the higher-κ oxide members, and minimize direct tunnelling in mid-alloy range silicates instead of end member transition metal oxides.

The exact solution for the tunnelling transmission through a rectangular barrier is obtained by matching boundary conditions for the wave function and its derivative [60]. The tunnelling transmission probability, T, is a function of the barrier height, E_b, the effective electron mass, m_{eff}^*, and the physical thickness of the barrier, t_{phys}, and is given by

$$T = \exp(-2A\, t_{phys}[E_b m_{eff}^*]^{1/2}), \qquad (4.2.8)$$

where A is a constant. The relationship of equation (4.2.8) states that T is an exponential function of the ratio of the film thickness to the *effective* de Broglie wavelength in the barrier region, squared. This expression is applicable at the onset of direct tunnelling where the bias voltage across a dielectric film is very much smaller than the barrier height. This solution has been extended to silicate alloys, $(A_n O_m)_x (SiO_2)_{1-x}$ ($n = 1$ and $m = 2$ for Hf(Zr) oxide and $n = 2$ and $m = 3$ for the group IIIB and lanthanide oxides), through an effective medium approximation in which t_{phys}, E_b and m^*_{eff} are assumed to vary linearly with alloy composition

$$t_{phys}(x) = t_{phys}(SiO_2)(1 + x(\kappa(j) - \kappa(SiO_2))/\kappa(SiO_2)),$$
$$E_b(x) = E_b(SiO_2) - x(E_b(SiO_2) - E(j)), \qquad (4.2.9)$$
$$m^*_{eff}(x) = m^*_{eff}(SiO_2) - x(m^*_{eff}(SiO_2) - m^*_{eff}(j))$$

where j identifies the appropriate high-κ oxide. Substitution of the relationships in equation (4.2.9) into equation (4.2.8), and defining a, b and c, as $a = (\kappa(j) - \kappa(SiO_2))/\kappa(SiO_2)$, $b = (E_b(j) - E_b(SiO_2))/E_b(SiO_2)$ and $c = (m^*_{eff}(j) - m^*_{eff}(SiO_2))/m^*_{eff}(SiO_2)$ provides a *short-hand* notation of identifying the value of x, that corresponds to the minimum tunnelling current in a given silicate alloy system. The tunnelling transition probability has a minimum, when the derivative of $(1 + ax)((1 - bx)(1 - cx))^{0.5} = 0$. Values of a, b, c and x_{min} are given in table 4.2.2.

The values of a are obtained from nominal values of κ for the respective end members, 3.8 for SiO_2, 7.6 for Si_3N_4, and 20 for the transition metal oxides. Values of E_b have been obtained either from Robertson's calculations [49, 50], or from spectroscopic studies of Miyazaki [56, 59], and are 3.2 eV for SiO_2, 2.2 eV for Si_3N_4, 1.4 eV for $HfO_2(ZrO_2)$ and 2.3 eV for Y_2O_3. Values of m^*_{eff} have been obtained from experimental studies and/or band gap scaling with a simplified expression for the variation of κ^2 in the band gap region as given by a Franz two-band model [61, 62]; SiO_2, $0.55m_0$, Si_3N_4, $0.25m_0$, HfO_2, $0.15m_0$, and Y_2O_3, $0.20m_0$. Assuming equal conduction and valence band mass, $m_v = m_c = m_b$, the tunnelling effective mass, m^*_{eff}, at energy E, relative to the valence band of a non-crystalline dielectric is given by

$$(m^*_{eff}E)^{-1} = (m_c(E - E_c))^{-1} + (m_v(E_v - E))^{-1}. \qquad (4.2.10)$$

The values of m^*_{eff} obtained in this way scale monotonically with the difference in energy between the band gap and conduction band offset energy as shown in figure 4.2.20.

As indicated in table 4.2.2, this quantum mechanical model for tunnelling barrier transmission in equation (4.2.8) predicts that minimum

Table 4.2.2. Values of a, b and c, and x_{min} for Si oxynitride, and Hf(Zr) and Y silicate alloys.

Alloy system	a	b	c	x_{min}
Si oxynitride	1	0.38	0.55	0.58
Hf(Zr) silicate	4	0.56	0.73	0.65
Y(La) silicate	4	0.28	0.55	>1.0

tunnelling current for Si oxynitrides, $(Si_3N_4)_x(SiO_2)_{1-x}$, should occur in the mid-alloy range, $x \sim 0.6$, as has been demonstrated in [63]. The model predicts a similar behaviour for HfO_2, also confirmed by experiment in [64]. Finally, the predictions of this model for the group IIIB transition metal and the rare earth lanthanide silicates have yet to be tested; however, results presented by Iwai at IEDM 2002 indicate very low tunnelling currents in Gd_2O_3, consistent with the predictions of this model calculation.

Figures 4.2.21 and 4.2.22 present, respectively, the compositional dependence of (i) *normalized* tunnelling transmission probabilities based on values presented in table 4.2.2 and calculated from equation (4.2.8) and (ii) the tunnelling current density at an oxide bias of 1 V as calculated using the WKB relationship of [65]. The plots in figure 4.2.21 are proportional to tunnelling transmission probabilities, T^*, calculated from a normalized form of equation (4.2.8)

$$T^* = \exp(-B\kappa[E_b - m^*_{\mathrm{eff}}]^{1/2}), \qquad (4.2.11)$$

where κ, E_b and m^*_{eff} have been normalized to value of 1 at $x = 0.0$, and B includes the physical thickness, and a number of constants. The values of x at which the tunnelling transmission probability is a minimum are in excellent agreement with the approximate values in table 4.2.2 for Si oxynitrides and Hf silicates, and are in very good agreement for the Y silicates, where the approximation gives a value greater than 1 ($\sim$1.08), and the plot gives a value between 0.9 and 1.0.

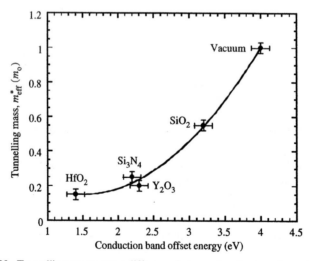

Figure 4.2.20. Tunnelling mass versus difference between band gap and conduction band offset energy.

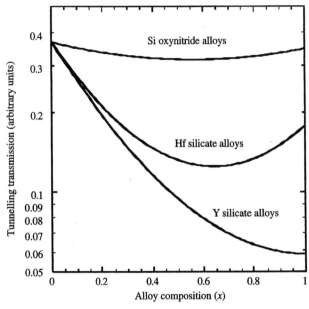

Figure 4.2.21. Normalized tunnelling transmission versus alloy composition.

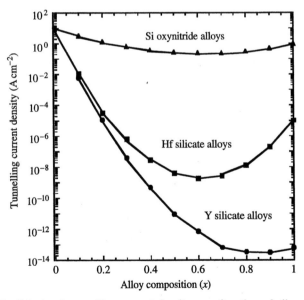

Figure 4.2.22. Calculated tunnelling current density as a function of alloy composition for an EOT = 2.0 nm for Si oxynitrides and 1.5 nm for the silicates, and at an oxide bias of 1 V.

The plots in figure 4.2.22 are for the tunnelling current in n+ Si–dielectric–N+ poly-Si at an oxide bias in excess of 1 V above flat band for substrate accumulation. The calculation takes into account the potential drops across the poly-Si and the channel region, so that there is a potential drop of 1 V across the dielectric for the gate potential used in the evaluation of the current density [65]. The doping concentration in the substrate was $2.5 \times 10^{17} \mathrm{cm}^{-3}$ and that in the poly-Si, $9.0 \times 10^{19} \mathrm{cm}^{-3}$. The values of the computed tunnelling current density are independent of these values for n+ and N+ because of the corrections made for the potential drops in the poly-Si and channel regions of the dielectric stack. The similarity in the shapes of the curves in figures 4.2.21 and 4.2.22, and in the alloy composition at which the tunnelling current is at a minimum, indicates the validity of the analytical approach in which the applied bias across the oxide was assumed to be very small relative to the barrier height.

Figures 4.2.23 and 4.2.24 compare experimental results for MOSCAPs with Si oxynitride alloys [63] and Hf silicate alloys [64], respectively. The agreement between the calculated compositional dependence and the experimental points is generally very good. There are issues that relate to the interfacial oxide/silicate layers and the way they must be included in these comparisons that must be addressed in comparisons between experiments and theoretical models. However, the point to be emphasized here is not so much the comparisons of the magnitudes of the direct tunnelling currents, but rather the compositions at which the minimum tunnelling occurs. The experimental data clearly indicate that these minima are in the mid-alloy

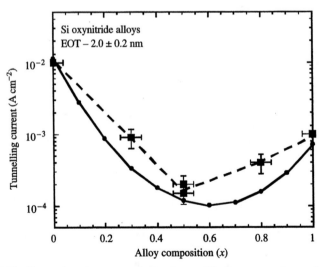

Figure 4.2.23. Comparison between calculated tunnelling current and experiment for Si oxynitride alloys.

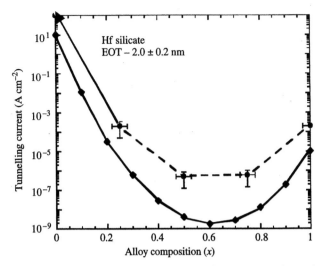

Figure 4.2.24. Comparison between calculated tunnelling current and experiment for Hf silicate alloys.

composition range for both the Si oxynitride and Hf silicate alloys, providing direct experimental evidence in excellent agreement with the quantum mechanical model of figure 4.2.21, and good quantitative agreement with the tunnelling calculations presented in figure 4.2.22.

The composition variations of κ, m^*_{eff} and E_b have been assumed to be linear functions of the alloy composition for the Si oxynitride and transition metal and rare earth silicate alloys that have been discussed above. There are at least two factors that can lead to non-linear behaviours for κ: (i) differences in the densities of the alloy system end members and (ii) dielectric constant enhancements due to changes in the detailed nature of the chemical bonding in group IVB silicates in the low composition alloy range [41, 66]. Inclusion of either, or both of these will change the quantitative behaviour of the tunnelling current at given composition, but not the overall trends that have been illustrated in figures 4.2.21 and 4.2.22. In particular, the compositions at which minimum tunnelling current occurs in the Si oxynitrides and Hf silicates will remain in the mid-alloy composition range, and if anything will be shifted to slightly lower values of x by the dielectric constant enhancement of [66].

Linearity is assumed for m^*_{eff}; however, the end member values themselves are at best empirically determined parameters. Their values can be refined by a more detailed evaluation of the direct tunnelling in the end member Si_3N_4, and transition metal and rare earth oxides by including the effects of interfacial transition regions. See as an example the calculations in [67, 68] which address tunnelling in stacked gate dielectrics that include interfacial transition regions. It is not likely that refinements to these masses will markedly affect the relative behaviours shown in figures 4.2.21 and 4.2.22 with respect to the value of x corresponding to a tunnelling current minimum.

The linear dependence of band offset energies for transition metal and rare earth lanthanide alloys is supported by empirical calculations based on direct spectroscopic studies [49, 50]. These studies indicate that the energies of the transition metal and rare earth lanathide d states in their respective silicate and aluminate alloys relative to the conduction band of the Si substrate remain essentially constant as a function of alloy composition. Additionally, the same studies indicate that the energy differences between these d states and the conduction band states of the SiO_2 (or Al_2O_3) alloy component is also independent of alloy composition. The only issue is then the compositional averaging used in the calculations described above.

The spectroscopic results in [41, 69] are supported by a more general aspect of alloy bonding that is revealed in studies of the alloy dependence of the core shifts of the Si (or Al) and transition metal (or rare earth) atoms [41]. The principle of electronegativity equalization as applied to these pseudo-binary alloys predicts that partial charges of these respective alloy atoms track together as a function of alloy composition [44]. The charge transfer that gives rise to these systematic changes is the dominant contributor to changes in core level binding energies, and therefore explains the parallel linear shifts of these experimentally determined energies. Parallel tracking of partial charges and core levels is functionally equivalent to energy differences between transitional metal d-state and Si (or Al) s-state conduction bands maintaining a constant separation as a function of alloy composition [41, 69]. Coupled with systematic changes in the valence band offset energy and band gaps, it provides a firm foundation for the linear scaling that has been applied.

The occurrence of transition regions is well established for gate dielectrics involving transition metal oxides, silicates and aluminate alloys [1]. This has two effects on the tunnelling results above: (i) it should be factored in the extraction of tunnelling parameters from experimental data and (ii) it must be addressed in the calculations of the tunnelling current. The effects of these transition regions have been identified in [2], and the relevant plot is presented in figure 4.2.25. This is a plot of EOT versus physical thickness of the high-κ dielectric for abrupt interfacial transitions, and for the inclusion of a nitrided Si oxide transition region that contributes about 0.35 nm to EOT. The primary effect of the transition region is to reduce the physical thickness of the high-κ dielectric by an amount Δt_{phys}, given by

$$\Delta t_{phys} = EOT_{tr}(\kappa(j)/\kappa(SiO)_2), \qquad (4.2.13)$$

where EOT_{tr} is the transition region contribution to EOT, ~ 0.35 nm, $\kappa(j)$ is the dielectric constant of the alternative dielectric and $\kappa(SiO_2)$ is the dielectric constant of SiO_2, ~ 3.8, so that $\Delta t_{phys} \sim 0.092 \, \kappa(j)$ in units of nm. If this correction is applied, then the very low direct tunnelling reported in

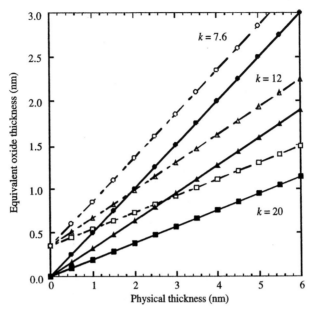

Figure 4.2.25. EOT versus physical thickness for dielectrics with $\kappa = 7.6, 12$ and 20 with and without an interfacial transition region, $EOT_{tr} = 0.35\,nm$.

Pr_2O_3 may be taken as evidence for the validity of this approach to rare earth lanthanide oxides, and in particular to the compositional dependence shown in figure 4.2.22 [70].

Interfacial defects

Interfaces between transition metal gate dielectrics have shown fixed charge and traps at levels considerably higher than $Si-SiO_2$ or nitrided $Si-SiO_2$ interfaces. Defects at the $Si-SiO_2$ interfaces have been assumed to be an 'intrinsic' limitation associated with interfacial bonding constraints, which derive from two sources [27]: the mismatch in molar volumes between Si and SiO_2, and additionally the differences in bond rigidity on either side of the interface. These two factors promote at least one excess layer of sub-oxide bonding, and it has been proposed that this layer establishes the limiting concentrations of defects.

The higher densities of defects at transition metal dielectric interfaces with Si are yet to be explained. However, studies by electron spin resonance (ESR) have demonstrated that the dangling bond density is higher at these interfaces [71] (see also the chapter of Stesmans and Afanas'ev in the present book). This suggests that other bonding defects may also increase proportionally. Even though there has been much speculation about the bonding environments associated with D_{it} and Q_f at Si–SiO$_2$ interfaces, at this point in time, no definitive and unambiguous assignment can be made. In addition, studies of Si–high-κ dielectric interfaces by SXPS have indicated Si sub-oxide bonding [72], so that this transition region may indeed be the source of the interfacial trapping states and the fixed charge as well.

There has been considerable debate about the origin of decreases in channel electron and hole mobilities in FETs with transition metal gate dielectrics. Three possible mechanisms have been proposed, interface trapping of charged carriers [73], scattering due to fixed charge [74] and scattering due to highly polar vibrational modes [75], and these will now be addressed.

Mobilities have been extracted from FET transport experiments by assuming that the charge available for transport was determined solely by the capacitance and the gate voltage, including its variation with distance along the length of the channel. The validity of this assumption has recently been questioned by the Ma group at Yale [74]. Their approach to these issues is based on a large amount of experimental evidence that indicates that values of D_{it} extracted from C–V data for capacitors with transition metal and Al$_2$O$_3$ dielectrics are about an order of magnitude larger, $\sim 10^{11}\,\mathrm{cm}^{-2}\,\mathrm{eV}^{-1}$, than those at Si–SiO$_2$ interfaces [1]. Their analysis of FET data indicates that if trapping is taken into account the values of the channel mobility are increased, and are closer to the values obtained in devices with SiO$_2$ dielectrics for the same substrate doping.

Additionally, it has been proposed that fixed charge is an intrinsic property of a *heterovalent* interface between a covalent substrate and a significantly more ionic dielectric [75]. This model is supported by the differences in sign between fixed charge at Al$_2$O$_3$–Si interfaces, negative, and the high-κ dielectric–Si interfaces, positive. However, it is difficult to account for the magnitude of the fixed charge, $\sim$ mid $10^{12}\,\mathrm{cm}^{-2}$, in terms of a specific bonding arrangement, and its electronic states (see figure 4.2.26(*a*) and (*b*)). These charge densities are sufficiently high to contribute to mobility reductions; however, calculations must include the scattering to dopants in the channel region. This must be increased as devices are scaled and in effect mitigates some of the reduction expected from charge carrier scattering. For example, a doping density in the channel corresponds to a charged centre scattering of $10^{12}\,\mathrm{cm}^{-2}$.

The third source of charge carrier scattering is associated with remote phonon scattering and been addressed in detail by Max Frischetti at IBM [76]. Low wavenumber vibrations in transition metal oxides, silicates and aluminates, which contribute to the high values of κ in these dielectrics, are sufficient to contribute to mobility reductions.

It is important to understand that from the perspective of device performance, it is the current drive per unit channel length that is the important parameter, and this must be comparable to values for SiO$_2$ dielectrics. The values of drive current for NMOS- and PMOSFETs must be

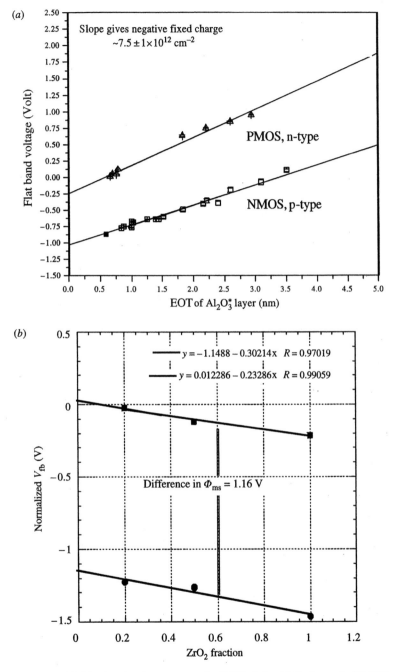

Figure 4.2.26. (*a*) Flat band voltage versus the EOT of the Al₂O₃ layers in NMOS and PMOS caps. (*b*) Flat band voltage versus alloy composition for a fixed EOT in Zr silicate alloys.

sufficient to maintain scaling rules for increased integration. This raises another issue, that has not yet been resolved, namely, *asymmetries* in the respective current drives. This means that combinations of fixed charge, and interface trapping that determine the drive current, may be Fermi level dependent and therefore make it difficult to meet the scaling for both NMOS- and PMOSFETs.

Narrowing the field of high-κ alternative dielectrics

The results described above have already been implemented in *first generation alternative dielectrics*, i.e. Si oxynitride alloys, by at least two major semiconductor chip makers in the United States. The choice of the dielectric for second generation has focused in the short term on HfO_2, and in the longer term on group IIIB and rare earth lanthanide oxides. The results of this study suggest that Hf (also Zr) silicate alloys with about 20–30% HfO_2 (ZrO_2) would be equally effective in reducing direct tunnelling, and present several other advances as well. These include: (i) increased thermal stability against crystallization and chemical phase separation in general, (ii) reduced ionic conductivity, (iii) reduced hydrophilic behaviour (water incorporation), (iv) reduced interfacial fixed charge and trapping and (v) increased effectiveness for bulk film nitridation. Since similar problems relative to thermal stability and crystallization and to hydrophilic behaviour apply to group IIIB and lanthanide rare earth oxides, and since the model calculations indicate quantitatively similar tunnelling current in their silicate alloys and Hf silicates, the alternative gate dielectrics of choice may well be Hf (or Zr) alloys as originally suggested by Wilk, Wallace and Anthony [1], and independently by our group at NC State University [76].

Summary

This chapter has first discussed the chemical bonding and electronic structure of SiO_2 and Si–SiO_2 interfaces in order to highlight the differences with respect to transition metal and rare earth dielectrics and their interfaces with Si. Changes in the local bonding have been introduced through a classification scheme based on bond ionicity, which has identified O-atom bonding coordination as a key factor. This has provided a link to *ab initio* theory. The correspondence between the results of *ab initio* calculations and XAS and other spectroscopies validates this approach, and allows the following conclusions to be made:

(i) localization of Zr (and other group IIIB, IVB and VB transitional metal) d* states explains the absence of any dependence of Zr M_2, to $N_{4,5}$ and O_1 transition energies on Si alloy content, and crystallinity;

(ii) the non-linear behaviour of the XPS O 1s BE is due to the discreteness of O-atom bonding coordinations; the non-linearity in the Zr $3d_{5/2}$ BEs for $x < 0.4$ is consistent with a ZrO_2-disrupted CRN in which donor–acceptor electrostatic bonds contribute only at low x;

(iii) the different non-linear behaviours for Zr $3d_{5/2}$ and O 2p BEs support a model for the microscopic origin of the enhanced dielectric constants reported for low concentration Zr (and Hf) silicate alloys [66];

(iv) non-linear behaviours of the O_{KVV} and Zr_{MVV} AES kinetic energies reflect changes in the energy of the top of the O 2p non-bonding valence band states with respect to vacuum; these are correlated with systematic changes in the alloy mix of O-atom bonding coordinations, and contribute to the variation in valence band offset energies between the silicate dielectrics and the Si substrate;

(v) the combination of XPS and AES establish that (a) the bandgaps of Zr(Hf) silicate alloys decrease monotonically, but non-linearly between those of SiO_2 ($\sim 9\,eV$) and ZrO_2 (~ 5.5–$5.6\,eV$), (b) that conduction band offset energies associated with the lowest Zr and Si anti-bonding states are essentially independent of x to at most $< \pm 0.2\,eV$, but (c) that valence band offsets decrease monotonically with increasing O-atom coordination, N: from $\sim 4.6\,eV$ in SiO_2 ($N = 2$), to $3.7\,eV$ in the stoichiometric silicate, $x = 0.5$ ($N = 3$), and to $3.1\,eV$ in ZrO_2 ($N = 4$); and

(vi) these band offset energies have been incorporated into an empirical direct tunnelling model in which coupling of tunnelling electrons to the extended Si 3s* and localized Zr(Hf) 4(5)d* band is parametrized through different empirically determined effective masses, $\sim 0.55m_0$ and $\sim 0.15m_0$, respectively.

Finally, this chapter has also presented a discussion of the composition dependence of direct tunnelling currents in high-κ pseudo-binary alloys. This includes Si oxynitride alloys, $(Si_3N_4)_x(SiO_2)_{1-x}$, Hf(Zr) silicate alloys, $(Hf(Zr)O_2)_x(SiO_2)_{1-x}$, as well as Y silicate alloys, $(Y_2O_3)_x(SiO_2)_{1-x}$, that are also representative of La, Lu and rare earth silicates. A quantum mechanical model based on three parameters, the barrier height, E_b, the effective electron mass, m_{eff}^*, and the physical thickness of the barrier, t_{phys}, has been applied. It was demonstrated that the minimum direct tunnelling current does not necessarily occur at the end member high-κ dielectric with the largest dielectric constant, κ, and the physical thickness of the film, but instead can occur in the mid-alloy range. This behaviour has been calculated for the Si oxynitride alloys, and for Hf(Zr) silicate alloys as well. In particular, increases in physical thickness, t_{phys}, with increasing κ in these alloys are mitigated by reductions in tunnelling barrier heights, E_b, e.g. conduction band offset energies, and effective masses of tunnelling electrons, m^*_{eff}. The corresponding decreases are reduced in group IIIB and lanthanide silicate alloys and the minimum tunnelling current in these alloys occurs in the end-member high-κ oxides.

The calculations are based on a WKB formalism using an effective medium approximation for the compositional variation of the three parameters. These are assumed to vary linearly with alloy composition. Experimental results for Si oxynitride alloys and for Hf silicate alloys are in excellent agreement with the model calculations. In addition, the model explains the very low value of direct tunnelling in Pr_2O_3.

Several refinements to the model are discussed including (i) non-linear behaviours in the three-parameter representation, and more importantly, (ii) the inclusion of interfacial transition regions in the analysis of tunnelling results as well as in the calculations.

The tunnelling model raises important issues relative to narrowing the field of alternative high-κ dielectrics, with Zr and Hf silicate alloys emerging as the leading candidates. This conclusion is based not only on the criterion of minimum direct tunnelling for a given EOT, but also includes considerations of (i) increased thermal stability against crystallization and chemical phase separation in general, (ii) reduced ionic conductivity, (iii) reduced hydrophilic behaviour (water incorporation), (iv) reduced interfacial fixed charge and trapping and (v) increased effectiveness for bulk film nitridation.

Finally, considerably more research is required to understand the local bonding arrangements that give rise to interface traps and fixed charge at Si–high-κ dielectric interfaces. If these are intrinsically higher than those at Si–SiO_2 interfaces by factors as large as 10, then this raises some important issues relative to device scaling. For example, if decreases in EOT are accompanied by decreases in current drive associated with an intrinsic combination of trapping, fixed charge and remote phonon scattering that is associated with the increased ionicity of the transition metal high-κ dielectrics, then this is will require a re-evaluation of how far roadmap scaling based on increases in capacitance can go, and whether there application specific dielectrics. For example, reductions in direct tunnelling and off state current are more important in mobile devices than in high performance devices, and this may identify niches for transition metal silicates in spite of reductions in current drive.

Acknowledgments

Research supported by the Office of Naval Research, the Air Office of Scientific Research and the SEMATECH/ SRC Front End Processes Center

References

[1] Wilk G, Wallace R M and Anthony J M 2001 *J. Appl. Phys.* **89** 5243

[2] Lucovsky G 2001 *Extended Abstracts of the 6th Workshop on Formation, Characterization, and Reliability of Ultrathin Silicon Oxides*, Atagawa Heights, Japan 26–27 January p 5

[3] Zallen R 1983 *The Physics of Amorphous Solids* (New York: Wiley) chapter 2

[4] Galeener F L and Lucovsky G 1976 *Phys. Rev. Lett.* **37** 1474

[5] Bell R J and Dean P 1970 *Discuss. Faraday Soc.* **50** 55; *Amorphous Materials* 1972 ed R W Douglas (London: Wiley–Interscience) p 443

[6] Phillips J C 1979 *J. Non-Cryst. Solids* **34** 153

[7] Phillips J C 1981 *J. Non-Cryst. Solids* **43** 37

[8] Whitten J L, Zhang Y, Menon M and Lucovsky G 2002 *J. Vac. Sci. Technol.* **B20** 1710

[9] Griscom D L 1978 *The Physics of SiO_2 and Its Interfaces* ed S T Pantelides (New York: Pergamon) p 232

[10] Mozzi R L and Warren B E 1969 *J. Appl. Crystallogr.* **2** 164

[11] Robertson L and Moss S 1988 *J. Non-Cryst. Solids* **106** 330

[12] Laughlin R B and Joannopoulos J D 1977 *Phys. Rev.* **B16** 2942

[13] Chadi D J, Laughlin R B and Joannopoulos J D in reference [9] p 55

[14] Batra I P in reference [9] p 65

[15] Pantelides S T and Harrison W A 1976 *Phys. Rev.* **B13** 2667

[16] Revesz A G and Gibbs G V 1980 *The Physics of MOS Insulators* eds G Lucovsky, S T Pantelides and F L Galeener (New York: Pergamon) p 92

[17] Lucovsky G and Yang H 1997 *J. Vac. Sci. Technol.* **A15** 836

[18] Whitten J L and Yang H 1995 *Int. J. Quantum Chem. Quantum Chem. Symp.* **29** 41

[19] Whitten J L and Yang H 1996 *Surf. Sci. Rep.* **24** 55

[20] Whitten J L and Hackmeyer M 1969 *J. Chem. Phys.* **51** 5584

[21] Neuefeind J and Liss K D 1996 *Ber. Bunsen. Phys. Chem.* **100** 1341

[22] Philipp H R 1966 *Solid State Commun.* **44** 73

[23] Pantelides S T in reference [9] p 80

[24] Senemaud C and Costa Lima M T in reference [9] p 85

[25] DiStanfano T H in reference [9] p 362

[26] Lucovsky G 1999 *IBM J. Res. Dev.* **43** 301 and references therein

[27] Lucovsky G, Yang H, Niimi H, Keister J W, Rowe J E, Thorpe M F and Phillips J C 2000 *J. Vac. Sci Technol.* **B18** 1742

[28] Himpsel F J, McFeely F R, Taleb-Ibrahimi A, Yarmoff Y A and Hollinger G 1988 *Phys. Rev.* **B38** 6084

[29] Keister J W, Rowe J E, Kolodzie J J, Niimi H, Tao N-S, Madey T E and Lucovsky G 1999 *J. Vac. Sci. Technol.* **A17** 1250

[30] Keister J W, Rowe J E, Lee Y-M, Niimi H, Lucovsky G and Lapeyre G J 2000 *J. Vac. Sci. Technol.* **B18**

[31] Yang H, Niimi H, Keister J W, Lucovsky G and Rowe J E 2000 *IEEE Electron Device Lett.* **21** 76

[32] Lee D R, Lucovsky G, Denker M R and Magee C 1995 *J. Vac. Sci. Technol.* **A13** 1671

[33] Niimi H and Lucovsky G 1999 *J. Vac. Sci. Technol.* **A17** 3185; **B17** 2610

[34] Weldon M, Queeny K T, Chabal Y J, Stefanov B and Raghavachai K 1999 *J. Vac. Sci. Technol.* **B17** 1795

[35] Muller D E *et al* 1999 *Nature* **399** 758

[36] Feldman L C, Stensgaard I, Silverman P J and Jackman T E in reference [9] p 339

[37] Lucovsky G and Phillips J C 1998 *J. Non-Cryst. Solids* **227** 1221

[38] Lucovsky G, Wu Y, Niimi H, Misra V and Phillips J C 1999 *Appl. Phys. Lett.* **74** 2005

[39] Lucovsky G, Rozaj-Brvar A and Davis R F 1983 *The Structure of Non-Crystalline Materials 1982* eds P H Gaskell, J M Parker and E A Davis (London: Taylor and Francis) p 193

[40] Rayner B, Niimi H, Johnson R, Therrien R, Lucovsky G and Galeener F L 2001 *AIP Conf. Proc.* **550** 149

[41] Rayner G B Jr, Kang D, Zhang Y and Lucovsky G 2002 *J. Vac. Sci. Technol.* **B20** 1748

[42] Phillips J C and Kerner X 2001 *Solid State Commun.* **117** 47

[43] Pauling L 1948 *The Nature of the Chemical Bond* 3rd edn (Ithaca, NY: Cornell University Press) chapter 2

[44] Sanderson R T 1971 *Chemical Bonds and Bond Energy* (New York: Academic)

[45] Gray H B 1962 *Electrons and Chemical Bonding* (New York: Benjamin) chapter 9

[46] Ballhausen C J and Gray H B 1964 *Molecular Orbital Theory* (New York: Benjamin) chapter 8

[47] Cox P A 1992 *Transition Metal Oxides* (Oxford: Oxford Science Publications)

[48] Grunes L A, Leapman R D, Walker C D, Hoffman R and Kunz A B 1982 *Phys. Rev.* **B25** 7157

[49] Robertson J and Chen C W 1999 *Appl. Phys. Lett.* **74** 1164

[50] Robertson J 2000 *J. Vac. Sci. Technol.* **B18** 1785

[51] Johnson R S, Lucovsky G and Hong J 2001 *J. Vac. Sci. Technol.* **A19** 1353

[52] Johnson R S, Lucovsky G and Hong J 2001 *J. Vac. Sci. Technol.* **B19** 1606

[53] Johnson R S, Hinkle C L, Hong J G and Lucovsky G 2002 *J. Vac. Sci. Technol.* **B20** 1126

[54] Lucovsky G, Whitten J L and Zhang Y 2002 *Solid State Electron.* **46** 1687

[55] Lucovsky G, Rayner G B Jr, Kang D, Appel G, Johnson R S, Zhang Y, Sayers D E, Ade H and Whitten J L 2001 *Appl. Phys. Lett.* **79** 1775

[56] Miyazaki S, Narasak M, Ogasawaga M and Hirose M 2001 *Microelectron. Eng.* **59** 373

[57] Lucovsky G, Zhang Y, Rayner G B *et al* 2002 *J. Vac. Sci. Technol.* **B20** 1739

[58] Raymaker D E, Murday J S, Turner N H, Moore C and Legally M G in reference [9] p 99

[59] Miyazaki S and Hirose M 2000 *AIP Conf. Proc.* **550** in reference [16] p 89

[60] Schiff L I 1955 *Quantum Mechanics* (New York: McGraw-Hill) chapter 5

[61] Franz W 1965 *Handbuch der Physik* ed S Flugge (Berlin: Springer) vol 18 p 155

[62] Maserjian J 1974 *J. Vac. Sci. Technol.* **11** 996

[63] Yang H and Lucovsky G 1999 *IEDM* 245

[64] Kim I, Han S, Hong J G, Osburn C and Lucovsky G unpublished

[65] Schuegraf K F, King T-J and Hu C-M 1992 *VLSI Symposium*

[66] Lucovsky G and Rayner G B Jr 2000 *Appl. Phys. Lett.* **77** 2912

[67] Yang H-Y, Niimi H and Lucovsky G 1998 *J. Appl. Phys.* **83** 2327

[68] Vogel E M *et al* 1998 *IEEE Trans. Electron Dev.* **45** 1350

[69] Yu H Y *et al* 2002 *Appl. Phys. Lett.* **81** 376

[70] Osten H H *et al* 2001 *Extended Abstracts International Workshop on Gate Insulator(s)* Nov 1–2 p 100

[71] Stesmans A and Afanas'ev V V 2002 *J. Vac. Sci. Technol.* **B20** 1720

[72] Ulrich M D, Johnson R S, Hong J G, Rowe J E and Lucovsky G 2002 *J. Vac. Sci. Technol.* **B20** 1732

[73] Zhu M, Ma T P, Tamagawa T, Kim J, Carruthers R, Gibson M and Furukawa T 2000 *IEDM Digest of Technical Papers* 463

[74] Lucovsky G, Phillips J C and Thorpe M F 1998 *Characterization and Metrology for ULSI Technology: 1998 International Conference* eds D G Seiler, A C Diebold, W M Bullis, T J Shaffner, R McDonald and E J Walters (Woodbury, NY: American Institute of Physics) p 273

[75] Frischetti M V, Neumayer D A and Cartier E A 2002 *Proceedings of ISCSI IV* Karuizawa, Japan 21–25 October

[76] Lucovsky G, Wolfe D, Flock K, Therrien R, Johnson R, Rayner B, Gunther L, Brown N and Claflin B 1999 *Mater. Res. Soc. Symp. Proc.* **567** 343

Chapter 4.3

Electronic structure and band offsets of high-dielectric-constant gate oxides

J Robertson and P W Peacock

Introduction

The decrease of the dimensions of metal oxide semiconductor (MOS) transistors has led to the need for alternative, high-dielectric-constant (κ) oxides to replace silicon dioxide as their gate dielectric [1–3]. Silicon dioxide layers thinner than 1.6–2 nm have a leakage current over $1\,\mathrm{A\,cm^{-2}}$ due to direct tunnelling through the oxide [4], which is too large for portable devices. As tunnelling decreases exponentially with thickness, the tunnelling current can be reduced by using thicker layers of high-κ oxides, which have the same equivalent capacitance or equivalent silicon dioxide thickness 'EOT'.

A variety of oxides have been proposed, and they must satisfy certain criteria to be acceptable as gate dielectrics [1]. The most important criterion is that the oxides should be stable and compatible with the processing, by not reacting with the Si channel to form an SiO_2 or silicide interlayers [5], by not crystallizing in the case of amorphous dielectrics and in the case of silicates by not phase separating into component oxides. The material should also be able to block boron penetration from p-type poly-silicon gates. A second criterion is that the oxide should provide a high-quality electrical interface, giving high carrier mobility in the channel and a low defect density with a low trapped charge and small gate threshold voltages. Our main focus is the third criterion, that they should act as potential barriers to both electrons and holes [6–8].

Band offsets

The barrier at each band or 'band offset' should be over about 1 eV for both conduction and valence bands in order to inhibit Schottky emission of

372

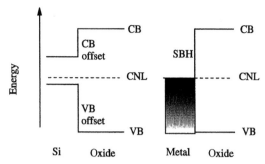

Figure 4.3.1. Band offsets and Schottky barriers.

electrons or holes into their bands. The band offset is the alignment of bands between the Si and the oxide, as shown in figure 4.3.1. There is a band offset at the conduction band and one at the valence band.

It is unclear that some of the oxides can act as potential barriers, as some have quite small band gaps. SiO_2 has a wide gap of 9 eV, so it has large band offsets for both electrons and holes. On the other hand, $SrTiO_3$ has a band gap of only 3.2 eV, so its bands must be aligned almost symmetrically with respect to Si for both barriers to be 1 eV. It turns out that the conduction band offset is smaller than the valence offset in most high-κ oxides, so this limits the leakage current. Generally, a band offset is a function of the oxide band gap and the asymmetry of the band alignment. We will see that the asymmetry of alignment depends on details of the electronic structure. The band gap is a fundamental property of the oxide. It is interesting that the band gap tends to vary inversely with the dielectric constant, as seen in figure 4.3.2. This means there is a trade-off between their κ and their band gap.

Recently, Yeo *et al* [9] produced a figure of merit for gate dielectrics in terms of their resistance to direct tunnelling. Direct tunnelling dominates if

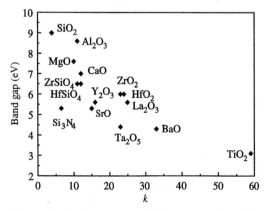

Figure 4.3.2. Variation of dielectric constant with band gap of candidate oxides.

the barrier has a square shape under applied voltage, rather than triangular as in Fowler–Nordheim tunnelling. From the WKB approximation, the tunnel current density is then proportional to

$$J = J_0 \exp(-2kt)$$

where k is the decay constant of the electron wavefunction in the oxide and t is the oxide thickness. If the applied gate voltage is less than the barrier height ϕ, k is given by

$$e\phi \sim \frac{\hbar^2 k^2}{2m^*}$$

The equivalent oxide thickness (EOT) t_{ox} is related to the actual oxide thickness t by

$$\frac{\kappa}{t} = \frac{\kappa(SiO_2)}{t_{ox}}$$

where the dielectric constant of SiO_2 is 3.9. Thus, Yeo *et al* [9, 10] found that the tunnelling current is given by

$$J = J_0 \exp\left(-\frac{8\sqrt{\pi e}}{3h}(m^* \phi)^{1/2} \kappa t_{ox}\right)$$

Thus, $(m^* \phi)^{1/2}\kappa$ is a figure of merit for direct tunnelling. The larger $(m^* \phi)^{1/2}\kappa$ is, the lower is the tunnelling current through the oxide. The barrier height ϕ is the most important factor in this because it can vary from 0 to 3 V. However, the tunnelling effective carrier mass m^* also matters. This factor is of order unity, but it can vary typically from 0.5 to 2. It can be found by fitting the tunnelling current for a known oxide thickness to the theoretical equation, if the potential barrier ϕ is known. Note that the tunnelling mass can differ numerically from the band edge effective mass [11].

Oxides have recently been compared by plotting their leakage current at 1 V applied versus the EOT, as in figure 4.3.3. We see that the leakage for SiO_2 increases roughly exponentially as the EOT decreases. HfO_2 is seen to be about four orders of magnitude lower current at an EOT of 1 nm [12], while La_2O_3 or Y_2O_3 are some orders of magnitude lower still [13] (figure 4.3.3).

Experimental methods to determine band offsets

The band offsets of the various oxides can be determined experimentally, by photoemission or by internal photoemission (figure 4.3.4). They can also be calculated from the electronic band structures.

The photoemission method is a direct method to determine band offsets. However, it determines the VB offset, while the CB offset is the more critical

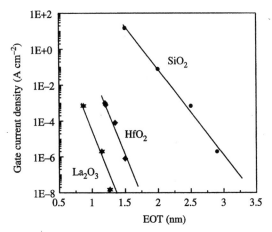

Figure 4.3.3. Tunnelling current densities at 1 V applied versus EOT for SiO_2, HfO_2, and La_2O_3 [9,13,14].

parameter. The CB offset also requires a value of the band gap. The first photoemission method is to measure evolution of the valence band spectrum by ultraviolet photoemission spectroscopy (UPS), as the oxide is deposited onto a silicon substrate (figure 4.3.4(*a*)). The valence band edge of Si becomes replaced by a deeper edge of the oxide. The valence band maxima are located by extrapolating the spectra in each case. This gives a direct measurement of the valence band offset.

A more accurate method involves a comparison to the core levels. The valence band edge of Si and of the oxide are assumed to lie at a fixed energy above their respective bulk core levels. This is measured on bulk samples

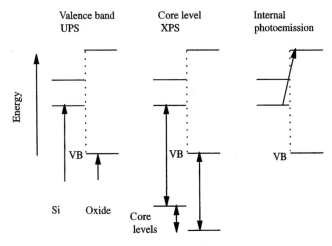

Figure 4.3.4. Measurement method of band offsets.

(figure 4.3.4(*b*)). The experiment then measures the difference of the Si and oxide core levels as a thin overlayer of oxide covers the Si. The valence band offset can then be obtained by difference.

The third method is internal photoemission (figure 4.3.4(*c*)), as used by Afanas'ev *et al* [14]. This is actually a photoconductivity experiment. Electrons from the Si valence band are excited to the oxide conduction band as a function of the incident photon energy. The onset of photoconductivity corresponds to the energy difference from the Si valence band edge to the oxide conduction band. The opposite transition from the Si oxide valence band to the Si conduction band can be inhibited by various means, such as the lower hole mobility in oxides.

Electronic structure of some high-κ oxides

In practice, many of the band offsets were first calculated from the electronic structure. In order to understand how we calculate the band offsets, we first describe the electronic structure of the most important high-κ oxides [15].

The band structures and density of states (DOS) were calculated using the local density approximation (LDA) and the pseudopotential method using the CASTEP code [16]. The CASTEP code uses a plane wave basis set and Vanderbilt ultra-soft pseudopotentials. A pseudopotential usually retains only the valence orbitals. Some core states which are not very deep should be included explicitly [17], such as the outer d core states in Zr and Hf, and the 5p states in Ba. A Monkhorst–Pack grid of special k points was used to generate the DOS. A plane wave cut-off of 380 eV is used. The exchange–correlation potential of the electrons was calculated using the generalized gradient approximation (GGA) variant of the LDA.

The LDA is well known to underestimate the band gap of semiconductors by 30–50%. The bands can be corrected to a good approximation by a rigid upward shift of the conduction bands, to fit the experimental value of the band gap. The *corrected* bands are displayed here. The uncorrected LDA and the experimental band gaps are given in table 4.3.1.

For reference, we first show SiO_2. It is a covalent random network in which Si is four-fold bonded to oxygen and O is two-fold bonded to Si. Its band gap is 9 eV wide (figure 4.3.5). Its valence band maximum consists of non-bonding oxygen π states with a large effective mass. The conduction band minimum is a broad minimum of Si 3s states with an effective mass of about 0.5. The ionicity of each Si–O bond is about 50%. The high-κ oxides are largely ionic [18].

Al_2O_3 is the only s–p bonded oxide currently considered as a candidate gate oxide. It has a moderate dielectric constant of 11. It is amorphous, in which the Al coordination is mainly six-fold and some four-fold. As a crystal,

Table 4.3.1. Tabulation of band gaps, experimental value of electron affinity (EA), experimental ε_∞, the S factor, and a comparison of the calculated values of change neutrality levels (CNLs) and conduction band offsets, found by the previous TB and the present LDA methods, respectively.

	Gap (eV) uncorrected LDA	Gap (eV) exp	EA (eV) exp	ε_∞	S	CNL, TB (eV)	CNL, LDA (eV)	CB offset, TB (eV)	CB offset, LDA (eV)
SiO_2		9	0.9	1.5	0.83	3.3			
Ta_2O_5		4.4	3.3	4.84	0.4	5.5	6.6	0.3	2.4
Al_2O_3	6.5	8.8	1[a]	3.4	0.63	2.4	2.7	2.8	2.2
Y_2O_3	3.3	6	2[a]	4.4	0.46	2.4	2.5	2.3	2.3
La_2O_3	3.7	6	2[a]	4	0.53	3.6	3.3	2.3	1.6
ZrO_2	3.4	5.8	2.5	4.8	0.41	3.7	4	1.4	1.3
HfO_2	3.4	5.8	2.5	4	0.53	2.6	1.7	1.5	0.4
$SrTiO_3$	1.7	3.3	3.9	6.1	0.28	3.7	2.7	-0.1	1.6
$SrZrO_3$	3.3	5.3	2.6	4	0.53		3.8	0.8	1.0
$LaAlO_3$	3.1	5.6	2.5[a]	4[a]	0.53				
$ZrSiO_4$	4.7	6.5[a]	2.4[a]	3.8	0.56	3.6	4.0	1.5	1.3

[a] Estimates.

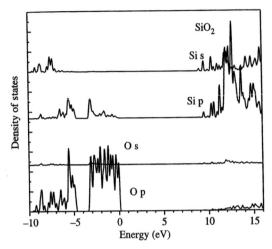

Figure 4.3.5. DOS of SiO_2.

Al_2O_3 occurs in a number of polymorphs. α-Al_2O_3 or sapphire is hexagonal, in which Al is six-fold coordinated [19]. γ-Al_2O_3 is a cubic spinel structure, in which Al is both four- and six-fold coordinated. The structure also contains hydrogen [20].

The bands were calculated for the α-Al_2O_3 and are shown in figure 4.3.6. The band gap is 8.8 eV and direct, as previously found by French [19] and Ching [21]. The conduction band shows a broad minimum at Γ of Al 3s states. The upper valence bands are oxygen non-bonding p states. The overall valence band is 7 eV wide. The partial DOS shows considerable charge transfer between Al and O states, and the valence band is mainly oxygen-like. The charge transfer is larger than in SiO_2.

The simplest alternative gate oxide is the group IVA oxide ZrO_2. ZrO_2 films are amorphous at room temperatures, but crystallize relatively easily. ZrO_2 is stable in the monoclinic structure at room temperature, it transforms to the tetragonal structure above 1170°C and it can be stabilized in the cubic fluorite structure by the addition of Y [22]. In cubic ZrO_2, Zr has eight oxygen neighbours and each oxygen has four Zr neighbours, while in monoclinic or tetragonal ZrO_2, each Zr atom has seven oxygen neighbours.

The band structure of cubic ZrO_2 in figure 4.3.7 shows an indirect gap of 5.8 eV, the experimental value [22]. Our calculated bands are similar to those found by others [22, 23]. The valence band is 6 eV wide, and it has a maximum at X formed from O p states. The conduction band minimum is a Γ_{12} state of Zr 4d orbitals. The Zr d states are split by the crystal field into a lower band of e symmetry (d_{z^2}) states and an upper band of t_2 (d_{xy}) states 5 eV higher (at Γ), with the Zr 4s states lying at 9 eV in between. The partial DOS shows considerable charge transfer, with the valence band being strongly O p states, and the conduction band Zr d states, with 30% admixture.

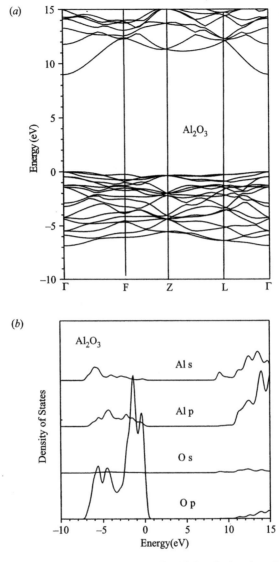

Figure 4.3.6. Band structure and partial DOS of α-Al$_2$O$_3$. The band gap is corrected to the experimental value, 8.8 eV.

Structurally, HfO$_2$ is similar to ZrO$_2$. Electronically, the band structure of HfO$_2$ in figure 4.3.8 is very similar to ZrO$_2$ [24], except that the crystal splitting of the Hf 5d states in the conduction band is larger. Its band gap is also 5.8 eV [25]. Lowering the symmetry of ZrO$_2$ and HfO$_2$ narrows the band gap, broadens the valence states slightly, and intermixes the d states in the conduction band, so that there is no longer a simple crystal field gap [22].

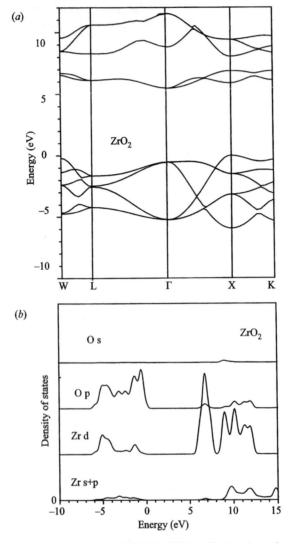

Figure 4.3.7. Band structure and partial DOS of ZrO_2. The band gap is corrected to the experimental value, 5.8 eV.

Crystalline La_2O_3 has the La_2O_3 structure in which La is seven-fold coordinated, with four short La–O bonds and three longer bonds [26]. The band structure of La_2O_3 is shown in figure 4.3.9. The band gap is indirect and expected to be 6 eV. The valence band maximum is at Γ and the valence band is now only 3.5 eV wide, narrower than in ZrO_2. This indicates a higher ionicity than in ZrO_2. The conduction band shows a weak minimum at K due

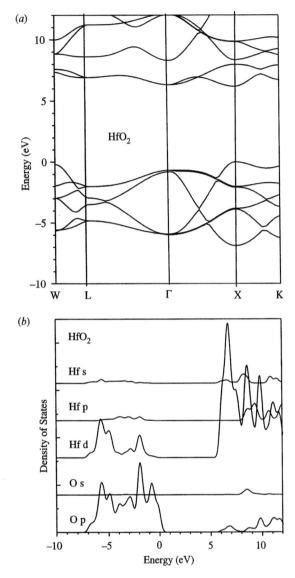

Figure 4.3.8. Band structure and partial DOS of HfO$_2$. The band gap is corrected to the experimental value.

to La d states. The DOS of La$_2$O$_3$ in figure 4.3.9(*b*) shows that the valence band is strongly localized on O p states and the conduction band in La d states with some La s,p states starting at 8 eV. La$_2$O$_3$ can be taken as a model for other lanthanide oxides.

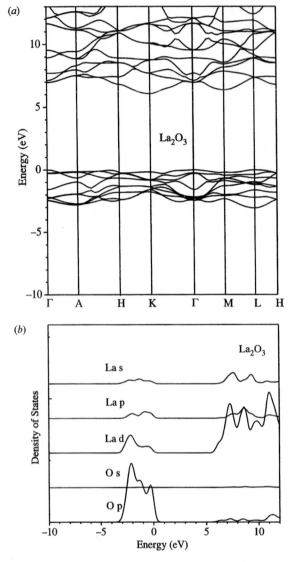

Figure 4.3.9. Band structure and partial DOS of La_2O_3. The band gap is corrected to the experimental value.

Of the group IIIA metal oxides, Y_2O_3 has the cubic bixbyite (defect spinel) structure [26]. This has a large 56-atom unit cell in which there are two types of Y site, both seven-fold coordinated. These coordinations will carry over roughly into an amorphous phase. We have calculated the bands of Y_2O_3 in the La_2O_3 structure, because it has a much smaller unit cell and the metal site is also seven-fold coordinated. The bands are shown in figure 4.3.10.

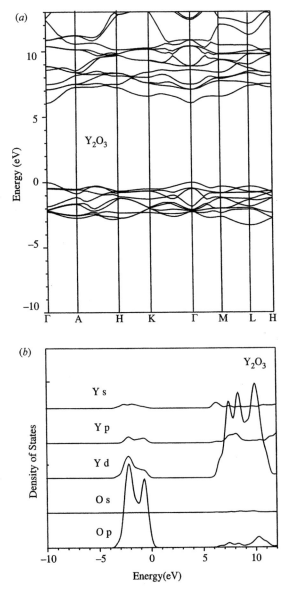

Figure 4.3.10. Band structure and partial DOS of Y_2O_3. The band gap is corrected to the experimental value.

The gap is direct at Γ and is about 6 eV. The valence band is again only 3 eV wide. The partial DOS in figure 4.3.10(*b*) shows the valence band is largely of O p states. The conduction band minimum is at Γ and has mixed Y d,s character. The major part of the conduction band, however, is of Y d states. This is similar to the results of Ching [27] using local orbitals.

The largest class of oxides of potential interest has the cubic perovskite structure, ABO_3. This is partly because of the ability to grow epitaxial $SrTiO_3$ layers on Si [28, 29]. Many of the perovskites are ferroelectric and so distort into lower-symmetry tetragonal or rhombohedral phases, but the cubic phase displays their essential electronic structure. The smaller transition metal ion occupies the B site, which is octahedrally coordinated by six oxygens. The oxygens are bound to two B ions, while the A ion is surrounded by 12 oxygen ions. Figure 4.3.11 shows the band structure of $BaTiO_3$ and figure 4.3.11(b) shows the partial DOS. $SrTiO_3$ is very similar, and is as found by others [30, 31]. The valence band has nearly degenerate maxima of Γ_{15} and R_{15} states and the valence band is 4.2 eV wide. The conduction band minimum is a Γ'_{25} state. The direct band gap at Γ is 3.3 eV wide. The lowest conduction bands are Ti d_{xy} t_2 states followed by the Ti d_{z^2} states. The next states above 7 eV are Ti p states followed by Ba s states. The A ion states (Ba or Sr) are well away from the band gap, and the ion can be considered to be passive and essentially fully ionized. On the other hand, the Ti–O bond is polar, but only about 60% ionic.

The band structure of the perovskite zirconate $BaZrO_3$ is shown in figure 4.3.12. This differs from the titanate in that the Zr d states lie 2 eV higher than the Ti d states, and the Zr d states have a larger crystal field splitting. The higher Zr d states open up the gap in $BaZrO_3$ to 5.7 eV, 2 eV wider than in $BaTiO_3$. The conduction band minimum is Γ'_{25}. The higher Zr d states also result in less repulsion of the O p states of the same symmetry, so this modifies dispersion in the valence band, compared to $BaTiO_3$. The zirconate is also more ionic than the titanate. Overall, the calculated bands of the perovskites are very similar to those found earlier by King-Smith and Vanderbilt [31], only shown over a wider energy range here.

$LaAlO_3$ is another perovskite oxide, but it is unusual in that the transition metal La occupies the A site and Al occupies the octahedral B site. $LaAlO_3$ is typical of aluminates, which are of interest as they have larger dielectric constants than the silicates. It is lattice matched to Si(110). The band structure and partial DOS of $LaAlO_3$ is shown in figure 4.3.13. The band gap is found to be 5.6 eV by ellipsometry [25].

$ZrSiO_4$ is typical of the transition metal silicates. They are of interest for gate dielectric application because of their greater glass-forming tendency than the simple oxides, despite their lower dielectric constant. $ZrSiO_4$ has the body-centred tetragonal structure. The Zr and Si atoms are organized in chains. Each Zr atom has eight O neighbours, four in its chain, and four in adjacent chains. Each Si has four O neighbours in a tetrahedral arrangement. Each O is bonded to two Zr and one Si atom. Its band structure and partial DOS is shown in figure 4.3.14. The band gap is taken to be about 6.5 eV. The valence band is 7 eV wide and has a maximum at Γ, as found by Rignanese *et al* [32]. The conduction bands form two blocks. The lower conduction

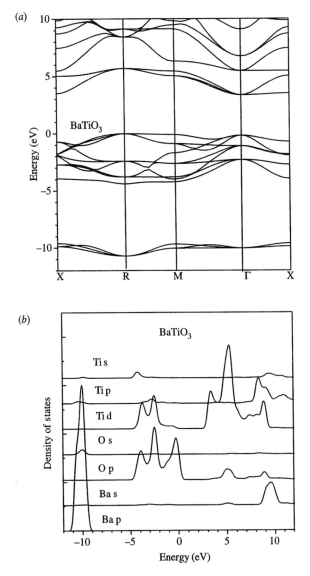

Figure 4.3.11. Band structure and partial DOS of BaTiO$_3$. The band gap is corrected to the experimental value.

bands due to Zr d states lie between 6.5 and 8 eV, followed by a set of bands due to Si–O antibonding states mixed with further Zr d states. Thus, the conduction band DOS can be considered to be the sum of ZrO$_2$-like and SiO$_2$-like components. HfSiO$_4$ is essentially similar to ZrSiO$_4$.

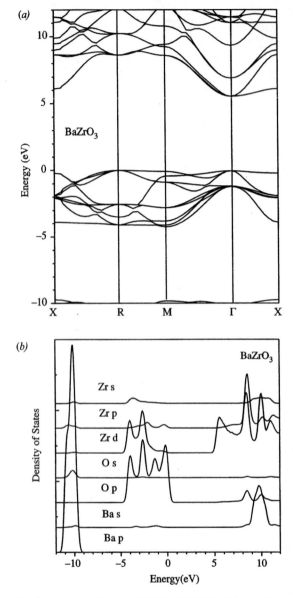

Figure 4.3.12. Band structure and partial DOS of $BaZrO_3$. The band gap is corrected to the experimental value.

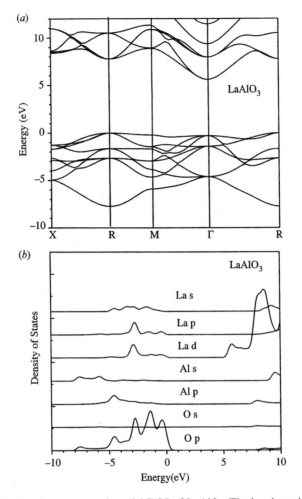

Figure 4.3.13. Band structure and partial DOS of LaAlO₃. The band gap is corrected to 5.6 eV.

Calculation of band offsets

Method

The band offsets of an oxide on silicon can be found by treating the oxide as a wide-band-gap semiconductor. It is then the band offset between two semiconductors. The band offset is closely related to the barrier height between the oxide and a metal, the oxide's Schottky barrier height.

The band offset between two semiconductors depends on the energy levels of the two semiconductors and the presence of any charge transfer

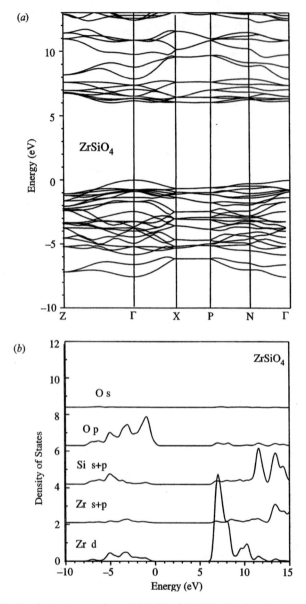

Figure 4.3.14. Band structure and partial DOS of $ZrSiO_4$. The band gap is corrected to the experimental value, of about 6.5 eV.

across the interface, which would create an interface dipole (figure 4.3.15). In the absence of charge transfer, the band offset is given by placing the energies of each semiconductor on a common energy scale—for example with respect to the vacuum level. This is called the electron affinity rule [33], which says

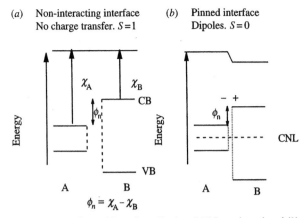

(*a*) Non-interacting interface   (*b*) Pinned interface
No charge transfer. $S=1$ Dipoles. $S=0$

$\phi_n = \chi_A - \chi_B$

Figure 4.3.15. Charge transfer and interface dipoles. (*a*) Non-pinned and (*b*) pinned limits.

that the conduction band offset is given by the difference in electron affinities (EAs). Experimental EAs would be used. Alternatively, the band energies could be calculated by the TB method using orbital energies of free atoms, as by Harrison [34]. This places each semiconductor on a common energy scale of the free atoms.

In practice, there is charge transfer across the interface. Consider first the Schottky barrier case. The charge transfer will create a dipole layer at the interface (figure 4.3.15). The charge transfer is between the metal Fermi level and the intrinsic interface states of the semiconductor [35]. The Schottky barrier height for electrons ϕ_n between a semiconductor S and a metal M is given by

$$\phi_n = S(\Phi_M - \Phi_S) + (\Phi_S - \chi_s) \qquad (4.3.1)$$

Here, Φ_M is the metal work function, Φ_S is the charge neutrality level (CNL) of the semiconductor, and χ_S is the electron affinity (EA) of the semiconductor.

The parameter S in equation (4.3.1) is a dimensionless pinning factor, which describes whether the barrier is 'pinned' or not. S varies from $S = 1$ in unpinned Schottky limit, equivalent to the electron affinity rule, to $S = 0$ for the Bardeen limit, which is pinned by a high density of interface states.

S is given in the linear approximation by [35]

$$S = \frac{1}{1 + \frac{e^2 N \delta}{\varepsilon_0}} \qquad (4.3.2)$$

where e is the electronic charge, ε_0 is the permittivity of free space, N is the density of the interface states per unit area and δ is their extent into the semiconductor.

There are several theories of Schottky barrier pinning [36–42]. We take the intrinsic model of virtual gap states (VGSs) or metal-induced gap states

[41–48]. The VGS can be thought of as the dangling bond states of the broken surface bonds of the semiconductor dispersed across its band gap, or alternatively as the evanescent states of the metal wavefunctions continued into the forbidden energy gap of the semiconductor. Φ_S is the CNL of the interface states. The CNL is like a Fermi level for interface states; it is the energy near mid-gap to which the interface states are filled on a neutral surface. The charge transfer at a Schottky barrier tends to align the Fermi level of the metal to the CNL of the semiconductor, as shown in figure 4.3.1. All the energies in equation (4.3.1) are measured from the vacuum level, except ϕ_n, which is measured from the band edge.

It was originally believed that S depends on the semiconductor ionicity [49]. It is now known that S depends on the electronic part of the dielectric constant ε_∞ [50]. Monch found that S empirically obeyed [47]

$$S = \frac{1}{1 + 0.1(\varepsilon_\infty - 1)^2} \tag{4.3.3}$$

while Tersoff [46] suggests $S = 0.5/\varepsilon_\infty$.

The band offset of two semiconductors is defined in the same way, as the energy difference between the conduction bands or the valence bands. We can use the VGS model, and then for two semiconductors, a and b, the electron barrier ϕ_n is the conduction band offset, given by

$$\phi_n = (\chi_a - \Phi_{S,a}) - (\chi_b - \Phi_{S,b}) + S(\Phi_{S,a} - \Phi_{S,b}) \tag{4.3.4}$$

S is the pinning parameter of the wider-gap semiconductor, that is the oxide. In the case of strong pinning, $S = 0$, the CNLs of each semiconductor line up, while if there is no pinning, $S = 1$, the band offset is given by the electron affinity rule. A wide comparison [51] of the band offsets of zinc-blende semiconductors found that the charge neutrality model with $S = 0$ gives a good description. The S parameter is calculated here from equation (4.3.3) and ε_∞ using the experimental value of refractive index, $\varepsilon_\infty = n^2$.

The method used here is to calculate the band offset between these two semiconductors, Si and oxide, using the method of virtual induced gap states and CNLs. The advantage of this method is that the CNLs are properties which are defined by the band structure of the bulk materials, which is applicable for both covalent and ionic bonding, and does not require an explicit description of the atomic bonding at the interface. This is an advantage because the oxides are often amorphous and so it is difficult to define the interface bonding accurately. Band offsets at zinc-blende semiconductors have also been calculated by methods using explicit atomic models of the interface, and then to calculate the offset in terms of the potential step at the interface, using the core levels as reference, following the work of Baldereschi *et al* [52]. Alternatively, the potential step can be calculated from the bonding using the model solid method of van de Walle [53, 54]. These methods need a calculation for each interface structure. These direct calculations allow one to study the variation of interface dipoles with interface structure and oxide termination.

The CNL method is used here. The CNL can be evaluated as the energy at which the Green function of the band structure, integrated over all bands and over k points in the Brillouin zone, is zero [55]

$$G(E) = \int_{BZ} \int_{-\infty}^{\infty} \frac{N(E')dE'}{E - E'} = 0 \qquad (4.3.5)$$

This integral can be replaced by a sum over special points of the zone [54]. For TB bands, there is a finite number of bands, whereas for pseudopotential bands we fix a finite upper limit in integral (equation (4.3.5)).

Results

Table 4.3.1 gives the calculated energy of the CNL for each compound with respect to the valence band maximum. The calculation uses equation (4.3.5) and the LDA band energies, with the conduction band energies shifted upwards by the scissors correction to give the experimental band gaps.

The band offsets on Si are then calculated, and given in table 4.3.1. The offsets are summarized in figure 4.3.16. The S values are derived from experimental values of the refractive index [56], which are also given in table 4.3.1. The experimental band gaps are used. The offsets also need

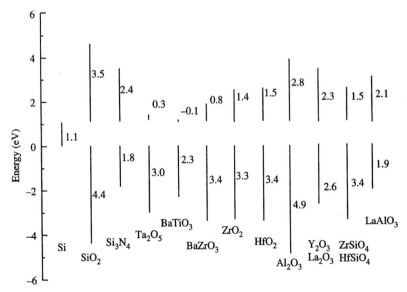

Figure 4.3.16. Summary of calculated band offsets.

experimental values of EA. These are taken from photoemission or electrochemical data [57–59].

The band-offset values are reasonably similar to those found earlier by TB [7]. The CNLs lie slightly lower in the gap for Al_2O_3, $BaTiO_3$ and $BaZrO_3$ than in the TB calculations, and so they give a slightly larger conduction band offset. The change in offset is less than the change in CNL, due to the finite S values. The CNLs lie at similar energies in both calculations for La_2O_3, Y_2O_3, ZrO_2 and $ZrSiO_4$.

The differences arise for two reasons. First, the conduction bands found by LDA are more dispersive than the TB bands. Second, the LDA calculation of CNLs includes more conduction bands in equation (4.3.5), and it does not cut off after including the small valence basis. This has the effect of lowering the CNL. This can be significant.

Experimental band offsets

A number of the band offsets have since been measured experimentally, and were found to be in surprisingly close agreement with the predicted values. The calculated and experimental values of conduction band offset on Si are compared in table 4.3.2. An offset of $SrTiO_3$ of 0 eV was found by photoemission by Chambers *et al* [60, 61], in close agreement to our calculated value. Their measured offset is not altered by doping of their $SrTiO_3$ overlayer; this just alters the position of the surface Fermi level.

The offset of Al_2O_3 on Si was found to be 2.8 eV by DiMaria [62], 2.8 eV by Ludeke [63], and 2.2 eV by Afanas'ev [14]. Our calculated value of 2.4 eV from LDA is in good agreement with these values. The lower offset value

Table 4.3.2. Comparison of the calculated conduction band offset (by LDA method) and experimental values for various gate oxides, by various authors.

	Calculated, LDA (eV)	Experiment (eV)	References
SiO_2		3.1	[74]
Ta_2O_5	0.35	0.3	Miyazaki [64]
$SrTiO_3$	0.4	< 0.1	Chambers [60]
ZrO_2	1.6	1.4	Miyazaki [65]
		2.0	Afanas'ev [14]
		1.4	Rayner [66]
HfO_2	1.3	1.3	Sayan [67]
		2.0	Afanas'ev [68]
		1.1	Zhu [69]
Al_2O_3	2.4	2.8	Ludeke [55]
		2.2	Afanas'ev [14]

found by Afanas'ev *et al* [14] was mainly due to the much lower band gap found by them, 7.0 eV, than the 8.8 eV found by the others. The gap increases when the sample is annealed [64].

For Ta_2O_5, Miyazaki [64, 65] found a conduction band offset of 0.3 V, similar to the TB estimate [7].

Miyazaki [65] found the conduction band offset of ZrO_2 to be 1.4 eV, using photoemission to find the valence band offset. This is similar to the 1.6 eV calculated here and the 1.4 eV from TB. Afanas'ev *et al* [14] found the conduction band offset of ZrO_2 directly, using internal photoemission from the Si valence band to the ZrO_2 conduction band. Their value of 2.0 eV is larger than found by the photoemission method. Rayner *et al* [66] found a VB offset of 3.1 eV by photoemission, equivalent to a CB offset of 1.6 eV for a gap of 5.8 eV.

For HfO_2, Sayan *et al* [67] found a CB offset of 1.3 eV assuming a gap of 5.8 eV. This compares with the calculated 1.3 eV here or 1.5 eV by TB. Afanas'ev *et al* [68] found 2.0 eV for HfO_2 using internal photoemission. Zhu *et al* [69] obtain a CB offset of 1.1 eV, which is close to that found here. However, this measurement is derived from the tunnelling current and uses the Schottky limit to extract a barrier. We know that the Schottky limit is not valid for HfO_2 as $S \sim 0.5$, unlike in SiO_2.

Rayner [66] found a VB offset of about 4 eV for a $ZrSiO_4$ alloy. This is equivalent to a CB offset of 1.4 eV if we take the band gap as 6.5 eV. This CB offset is close to our calculated 1.3 eV by LDA and 1.5 eV by TB.

There is currently no direct experimental determination of the CB offset for La oxide. Nevertheless, the fact that the leakage current density in figure 4.3.3 is lower than that of HfO_2 suggests that La oxide has the larger CB offset, as predicted in table 4.3.1.

Recently, Osten *et al* [70, 71] studied the case of epitaxial Pr_2O_3 on Si(001). They found a valence band offset of about 1.1 eV and a conduction band offset of 0.5–1.5 eV. The CB offset varied with annealing conditions, which they attribute to a change in oxygen bonding and the interface dipole. Large changes in interface dipoles have been seen in $Si:TiO_2$ interfaces [72].

Overall, there is good agreement between experiment and calculated values for the various alternative gate oxides. Surprisingly, the TB predictions are often closer to experiment than the new LDA values.

Lucovsky [73] has recently shown how the CB offsets tend to follow the differences in s–d atomic energies in transition metals, because the band gap between metal s and oxygen 2p states remains rather constant. This allows a method to scan oxides more quickly.

Why are some conduction band offsets so small? The CNL energy is driven by competing forces [8]. From equation (4.3.5) and figure 4.3.17, we see that a large density of valence states pushes the CNL up, and a large density of conduction states pushes it down. Now, valence states are O 2p states and conduction states are metal d or s,p states. The ratio of these states is just the

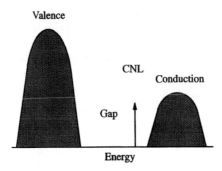

Figure 4.3.17. How the valence and conduction DOS pushes the CNL in the gap.

ratio of oxygen to metal atoms–the oxide stoichiometry. Thus, the CNL depends on metal valence. A high metal valence pushes the CNL up and a low valence pushes it down. Hence La_2O_3 has a lower CNL than ZrO_2. Interestingly, $SrTiO_3$ has a higher CNL than TiO_2, despite Ti having the same valence, because there is the extra oxygen for the Sr. The Sr states are well above the gap so they do not affect the CNL much. This is why $SrTiO_3$ has a small CB offset unless explicit moves are made to introduce an extra dipole layer.

There are two ways to get a larger conduction band offset, either have a smaller metal valence, by using metals from column 2, 3 or 4, and avoid perovskites which have a passive A site ion which donates more electrons, or use a wide oxide band gap in the first place. The latter means using transition metals with a high d state, like metals from columns IIIB or IVB, or 4d or 5d metals instead of 3d metals. This selects oxides of Zr, Hf, La, Y, Al, Gd, or Pr. It is interesting that the offset constraint selects the same oxides as the stability constraint. This is because a high metal d state gives a large formation energy, due to a large charge transfer energy.

References

[1] Wilk G, Wallace R M and Anthony J M 2001 *J. Appl. Phys.* **89** 5243
[2] Kingon A I, Maria J P and Streiffer S K 2000 *Nature* **406** 1032
[3] Wallace R M 2002 *MRS Bull.* **27** (3)
[4] Lo S H, Buchanan D A, Taur Y and Wang W 1997 *IEEE Electron Device Lett.* **18** 209
[5] Hubbard H J and Schlom D G 1996 *J. Mater. Res.* **11** 2757
[6] Robertson J and Chen C W 1999 *Appl. Phys. Lett.* **74** 1168
[7] Robertson J 2000 *J. Vac. Sci. Technol. B* **18** 1785
[8] Robertson J 2002 *MRS Bull.* 217
[9] Yeo Y- C, King T J and Hu C 2002 *Appl. Phys. Lett.* **81** 2091
[10] Lee W C and Hu C 2001 *IEEE Trans. Electron Device Lett.* **48** 1366
[11] Weinberg Z A 1982 *J. Appl. Phys.* **53** 5052
[12] Lee S J *et al* 2000 *Tech. Digest. IEDM*

[13] Iwai H Private communication
[14] Afanas'ev V V, Houssa M, Stesmans A and Heyns M M 2001 *Appl. Phys. Lett.* **78** 3073
[15] Peacock P W and Robertson J 2002 *J. Appl. Phys.* **92** 4712
[16] Payne M C, Teter M P, Allan D C, Arias T A and Joannopoulos J D 1992 *Rev. Mod. Phys.* **64** 1045
[17] Wei S H and Zunger A 1987 *Phys. Rev. Lett.* **59** 144
[18] Lucovsky G 2001 *J. Vac. Sci. Technol. A* **19** 1553
[19] French R H 1990 *J. Am. Ceram. Soc.* **73** 477
[20] Sohlberg K, Pennycook S J and Pantleides S T 1999 *J. Am. Chem. Soc.* **121** 7493
[21] Xu Y N and Ching W Y 1991 *Phys. Rev. B* **43** 4461
[22] French R H, Glass S J, Ohuchi F S, Xu Y N and Ching W Y 1994 *Phys. Rev. B* **49** 5133
[23] Kralik B, Chang E K and Louie S G 1998 *Phys. Rev. B* **57** 7027
[24] Demkov A A 2001 *Phys. Stat. Solidi b* **226** 57
[25] Lim S G, Kriventsov S, Jackson T N, Haeni J H, Schlom D G, Balbashov A M, Uecker R, Reiche P, Freeouf J L and Lucovsky G 2002 *J. Appl. Phys.* **91** 4500
[26] Wyckoff R 1964 *Crystal Structures* vol 2 (New York: Wiley)
[27] Ching W Y and Xu Y N 1990 *Phys. Rev. Lett.* **65** 895
[28] McKee R A, Walker F J and Chisholm M F 1998 *Phys. Rev. Lett.* **81** 3014
[29] Yang G Y, Finder J M, Wang J, Wang Z L, Droopad R, Eisenbeiser K W and Ramesh R 2002 *J. Mater. Res.* **17** 204
[30] Mattheis L F 1972 *Phys. Rev. B* **6** 4718
[31] King-Smith R D and Vanderbilt D 1994 *Phys. Rev. B* **49** 5828
[32] Rignanese G M, Gonze X and Pasquarello A 2001 *Phys. Rev. B* **63** 104305
[33] Anderson R L 1962 *Solid State Electron.* **5** 341
[34] Harrison W A 1977 *J. Vac. Sci. Technol.* **14** 1016
[35] Rhoderick E H and Williams R H 1988 *Metal Semiconductor Contacts* (Oxford: Oxford University Press)
[36] Cowley A W and Sze S M 1965 *J. Appl. Phys.* **36** 3212
[37] Spicer W E, Lindau I, Skeath P R, Su C Y and Chye P W 1980 *Phys. Rev. Lett.* **44** 420
[38] Allen R E and Dow J D 1982 *Phys. Rev. B* **25** 1423
[39] Brillson L J 1994 *Surf. Sci.* **299** 909
[40] Flores F and Tejedor C 1987 *J. Phys. C* **20** 145
[41] Mönch W 1987 *Phys. Rev. Lett.* **58** 1260
[42] Tejedor C, Flores F and Louis E 1977 *J. Phys. C* **10** 2163
[43] Louie S G, Chelikowsky J R and Cohen M L 1977 *Phys. Rev. B* **15** 2154
[44] Tersoff J 1984 *Phys. Rev. Lett.* **52** 465
[45] Tersoff J 1984 *Phys. Rev. B* **30** 4874
[46] Tersoff J 1985 *Phys. Rev. B* **32** 6968
[47] Mönch W 1994 *Surf. Sci.* **300** 928
[48] Monch W 1996 *Appl. Surf. Sci.* **92** 367
[49] Kurtin S, McGill T C and Mead C A 1969 *Phys. Rev. Lett.* **22** 1433
[50] Schluter M 1978 *Phys. Rev. B* **17** 5044; *Thin Solid Films* 1982 **93** 3
[51] Yu E T, McCaldin J O and McGill T C 1992 *Solid State Phys.* **46** 1
[52] Baldereschi A, Baroni S and Resta R 1988 *Phys. Rev. Lett.* **61** 734
[53] van de Walle C G 1989 *Phys. Rev. B* **39** 1871
[54] Franciosi A and van de Walle C G 1996 *Surf. Sci Rep.* **25** 1
[55] Cardona M and Christensen N E 1987 *Phys. Rev. B* **35** 6182

[56] Palik E D 1985 *Handbook of Optical Properties of Solids* vols 1–3 (New York: Academic)

[57] Scott J F 1998 *Ferroelect. Rev.* **1** 1

[58] Hartmann A J, Lamb R M, Scott J F, Johnston P N, ElBouanani M, Chen C W and Robertson J 1998 *J. Korean Phys. Soc.* **32** S1329

[59] Schmickler W and Schultze J W 1986 *Modern Aspects of Electrochemistry* vol 17 ed J M O'Bockris (London: Plenum)

[60] Chambers S A, Liang Y, Yu Z, Droopad R, Ramdani J and Eisenbeiser K 2000 *Appl. Phys. Lett.* **77** 1662

[61] Chambers S A, Liang Y, Yu Z, Droopad R and Ramdani J 2001 *J. Vac. Sci. Technol. A* **19** 934

[62] Maria D J 1974 *J. Appl. Phys.* **45** 5454

[63] Ludeke R, Cuberes M T and Cartier E 2000 *Appl. Phys. Lett.* **76** 2886

[64] Miyazaki S 2001 *J. Vac. Sci. Technol. B* **19** 2212

[65] Miyazaki S 2001 *Appl. Surf. Sci.* **190** 66

[66] Rayner G B, Kang D, Zhang Y and Lucovsky G 2002 *J. Vac. Sci. Technol. B* **20** 1748

[67] Sayan S, Grafunkel E and Suzer S 2002 *Appl. Phys. Lett.* **80** 2135

[68] Afanasev V V, Stesmans A, Chen F, Shi X and Campbell S A 2002 *Appl. Phys. Lett.* **81** 1053

[69] Zhu W J, Ma T P, Tamagawa T, Kim J and Di Y 2002 *IEEE Electron Device Lett.* **23** 97

[70] Osten H J, Liu J P and Mussig H J 2002 *Appl. Phys. Lett.* **80** 297

[71] Fissel A, Dabrowski J and Osten H J 2002 *J. Appl. Phys.* **91** 8986

[72] Fulton C C, Lucovsky G and Nemanich R J 2002 *J. Vac. Sci. Technol. B* **20** 1726

[73] Lucovsky G, Zhang Y, Rayner G B, Appel G, Ade H and Whitten J L 2002 *J. Vac. Sci. Technol. B* **20** 1739

[74] Keiser J W, Rowe J E, Kolodziej J J, Niimi H, Madey T E and Lucovsky G 1999 *J. Vac. Sci. Technol. B* **17** 1831

Chapter 4.4

Reduction of the electron mobility in high-κ MOS systems caused by remote scattering with soft interfacial optical phonons

Massimo V Fischetti, Deborah A Neumayer
and Eduard A Cartier

Introduction

Insulators with a large static dielectric constant (usually referred to as 'high-κ insulators') are currently being considered as possible replacements for SiO_2, because of the necessity of increasing the gate capacitance of Si metal–oxide–semiconductor field-effect transistors (MOSFETs), while avoiding the problems that arise when the SiO_2 thickness is reduced below the 1.5–1.0 nm range, as demanded by device scaling [1, 2]. At least at present, these efforts are still mainly aimed at improving the chemical and physical properties of the insulating materials. Yet, in this paper, we point out an intrinsic, possibly unavoidable, and unwanted property of these materials, namely, the fact that their high dielectric constant may necessarily cause a reduction of the electron mobility in the inversion layer of Si MOSFETs. The dielectric constant of a (non-metallic) solid results from the contribution of the ionic and the electronic polarization. The latter is roughly inversely proportional to the square of the direct band gap of the solid, averaged over the Brillouin zone. Insulators, by definition, have large band gaps, so that there is little one can do to increase the electronic polarization and a larger (static) dielectric constant can only stem from a larger ionic polarization, often due to highly polarizable ('soft') metal–oxygen bonds. Associated with soft bonds are low-energy optical phonons. By contrast, the 'hard' Si–O

397

bonds in SiO$_2$ yield a reduced ionic polarization. Associated with 'hard' bonds are 'stiff' optical phonons.

In 1972, Wang and Mahan [3] showed that electrons in the inversion layer at the interface between a semiconductor of optical permittivity ϵ_s^∞ and a dielectric of static and optical permittivities ϵ_{ox}^0 and ϵ_{ox}^∞, respectively, can couple with the surface-optical (SO) modes (arising at the insulator/Si interface from the longitudinal-optical (LO) modes of the insulator) with a coupling strength proportional to

$$\hbar\omega_{SO}\left[\frac{1}{\epsilon_s^\infty + \epsilon_{ox}^\infty} - \frac{1}{\epsilon_s^\infty + \epsilon_{ox}^0}\right]. \qquad (4.4.1)$$

Here $\hbar$ is the reduced Planck constant and ω_{SO} is the frequency of the SO insulator phonon, given by:

$$\omega_{SO} = \omega_{TO}\left[\frac{\epsilon_{ox}^0 + \epsilon_{Si}^\infty}{\epsilon_{ox}^\infty + \epsilon_{Si}^\infty}\right]^{1/2}. \qquad (4.4.2)$$

Equation (4.4.1) is physically equivalent to the well-known Fröhlich electron/LO-phonon scattering strength, proportional to

$$\hbar\omega_{LO}\left[\frac{1}{\epsilon^\infty} - \frac{1}{\epsilon^0}\right], \qquad (4.4.3)$$

in a material with static and optical permittivities ϵ^0 and ϵ^∞, respectively, and LO-phonon frequency ω_{LO}. In equation (4.4.3) the difference between the inverse of ϵ^0 and of ϵ^∞ is proportional to the squared amplitude of the dipole field solely due to the oscillating ionic polarization of the material; that is, to the coupling between electrons and the bulk LO phonons. Equation (4.4.1) results from the same physics, but the dipole field is modified by 'image-charge effects' at the insulator/semiconductor interface, affecting the decay of the dipole field of the insulator phonons away from the bulk of the insulator into the semiconductor inversion layer. The effect of this scattering mechanism, called 'remote phonon scattering', on hot-electron transport in the Si/SiO$_2$ system was later studied by Hess and Vogl [4] and by Moore and Ferry [5], and its effect on the effective electron mobility by one of us (Fischetti) [6]. For the Si/SiO$_2$ system, and restricting our attention—now and throughout the rest of the chapter—to the electron mobility, remote scattering does not play a major role. There are two reasons for this. First, the ionic polarizability of SiO$_2$ is not very large, because of the hard nature of the Si–O bond. While this results in a small static dielectric constant, it also results in a small difference between ϵ_{ox}^0 and ϵ_{ox}^∞, and so in a small coupling constant for electron/remote-phonon scattering. Second, the stiff Si–O bond also results in a large LO (and SO) phonon energy ($\hbar\omega_{LO} \approx 0.15\,\text{eV}$). Electrons at thermal energies (which should be considered when interested in their Ohmic mobility) cannot emit excitations of such a large energy, and at room temperature there are too few thermally excited phonons to be absorbed. Note that another bulk SiO_2 phonon with $\hbar\omega_{LO} \approx 63$ meV could potentially have a larger effect, as far as these energetic considerations are concerned, but its oscillator strength is too small; if it were not so, SiO_2 would have a significantly larger κ.

Considering now the case of high-κ insulators, their high-frequency dielectric response is mainly electronic—since heavier and 'slower' ions cannot respond fully at sufficiently large frequencies—and so it is not too different from that of SiO_2. In contrast, the large ionic response dominates at low frequency. This does indeed yield a larger static dielectric constant, but also yields both a large difference between ϵ_{ox}^0 and ϵ_{ox}^∞, and so a large scattering strength, equation (4.4.1), and a low SO-phonon frequency, which allows frequent emissions and absorption processes by thermal electrons. The net result is that the very same physical properties that are responsible for the higher κ of the insulator are also likely to yield (with some important exceptions we shall consider later) a degradation of the effective electron mobility in the inversion layer of MOS systems using the high-κ insulator. We should remark that there is nothing novel about these arguments. Hess and Vogl [4] were already very well aware of these ideas in 1979, when they concluded their article with the optimistic remark: 'In passing, we note that a reduction of the ionic polarizability of SiO_2, or better of the difference $\epsilon_{ox}^0 - \epsilon_{ox}^\infty$, would reduce the electron–phonon coupling, [...] and correspondingly, enhance the field dependent electron mobility in MOS transistors'. Unfortunately in our context, we must move in the opposite direction, from SiO_2 to higher-κ materials with a higher difference $\epsilon_{ox}^0 - \epsilon_{ox}^\infty$, thus achieving the opposite effect of depressing the mobility.

It is instructive to start by giving a rough idea about the size of the effect we are considering. In figure 4.4.1 we plot the effective electron mobility in the inversion layer at the interface between Si and an infinitely thick film of several insulators we have considered (SiO_2, HfO_2, ZrO_2, $ZrSiO_4$, AlN, and Al_2O_3). Full details will be given below. For now it suffices to say that the triangular-well approximation has been used to treat the inversion layer, an anisotropic and nonparabolic band-structure model has been used to account for (anisotropic) scattering with acoustic phonons, as described in [7], surface roughness has been accounted for empirically using Matthiessen's rule, and scattering with remote SO modes has been treated using Fermi's golden rule with the Wang–Mahan matrix element proportional to the scattering strength given in equation (4.4.1). While this model is excessively oversimplified for the reasons stated below, it shows that effects as large as a factor of two or more can be expected.

In principle, the results shown in figure 4.4.1 are only suggestive of what we should expect. A more accurate assessment of the importance of remote phonon scattering in realistic high-κ MOS systems requires that we account

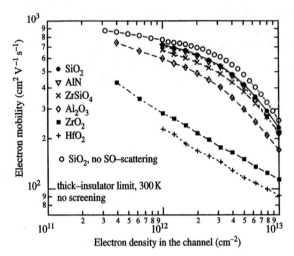

Figure 4.4.1. Effective electron mobility in Si inversion layers of MOS systems with the insulators indicated. A triangular-well approximation has been used to model the subband structure of the inversion layer. Anisotropic scattering with acoustic phonons and remote scattering with surface optical phonons has been accounted for (when indicated). Scattering with surface roughness has been added empirically using Matthiessen's rule and fitting the roughness parameters to match the experimental 'universal' mobility for the Si/SiO$_2$ system at an electron sheet density of 10^{13} cm^{-2}. The limit of infinite insulator thickness has been taken, and no additional dielectric effects (screening by substrate and gate electrons, plasmon−phonon coupling) have been considered here.

for two additional complications: the coupling between surface/interface optical modes and the two-dimensional electron plasma at the insulator/ semiconductor interface, and the coupling between interfacial optical and plasma modes at the substrate/insulator and at the gate/insulator interfaces. In thin-insulator structures, this coupling alters significantly the dispersion of the excitations and their coupling with the electrons in the channel. The coupling between substrate- and gate-interface plasmons has been investigated before, finding that a significant gate Coulomb drag yields by itself a reduction of the electron mobility for SiO$_2$ layers thinner than about 2–1.5 nm [7]. Here we must extend the treatment by including the coupling of surface plasmons to SO modes, by accounting for electron scattering with the resulting phonon-like modes, and considering MOS systems with various thicknesses of different insulators of practical technological importance. Anticipating our main result, the proximity of the heavily doped gate has the beneficial effect of screening to a large extent the interaction between electrons and interface optical modes at the smallest insulator thickness considered for all but two (HfO$_2$ and ZrO$_2$) of the high-κ insulators we have considered, and at sufficiently large electron density in the depleted poly-Si gate.

This chapter is organized as follows: In the section 'Interface modes' we present our theoretical scheme. In the section 'Effective mobility of the 2D electron gas' we present our results. In particular, the section 'Insulator parameters' presents a discussion of the non-trivial problem of selecting physical quantities (LO and/or transverse-optical, or TO, phonon energies, dielectric constants, and oscillator strengths) of the insulating films, comparing information available in the literature with data extracted from Fourier transform infrared (FTIR) spectroscopy. Finally, in the section 'Effect of a silicon dioxide interfacial layer' we present some estimates about the role played by an interfacial SiO_2 layer and conclude in the section 'Discussion and conclusions'. The role of remote phonon scattering on hot-electron (i.e. non-Ohmic) transport will not be investigated here, although we should expect significant effects, along the line of a previous investigation of long-range Coulomb effects on the transconductance of Si n-MOSFETs [8].

Interface modes

Dispersion

Here, as in [7], we shall consider a structure consisting of degenerately doped n-type Si (representing the poly-Si gate) in the half-space $z < 0$ (gate), an SiO_2 or high-κ insulating layer for $0 \leq z < t$, and a p-type Si substrate filling the half-space $z \geq t$. The latter is assumed to be electrically inverted, and so it is treated as a two-dimensional electron gas (2DEG). We shall denote by $\epsilon_g(\omega)$, $\epsilon_{ox}(\omega)$, and $\epsilon_s(Q,\omega)$ the dielectric functions of the gate, insulator, and substrate, respectively (all in the long-wavelength limit discussed below). We denote by $\mathbf{Q}$ and $\mathbf{R}$ the two-dimensional wave vector and coordinate vector in the (x,y)-plane of the interfaces, respectively.

We are only interested in the longitudinal electric eigenmodes of the system, since transverse modes (given by poles of the total dielectric response) correspond to a vanishing electric field, and so to a vanishing coupling with the charge carriers. These are transverse-magnetic solutions (TM or p waves) of Maxwell's equations. As described in [7], we can safely work in the non-retarded limit. Thus, the 'usual' boundary conditions require that the components of the electric field on the plane of the interfaces be continuous across the two interfaces at $z = 0$ and $z = t$, and similarly for the component of the displacement field normal to the plane of the interfaces. We can expand the electrostatic potential at frequency ω in its Fourier components as:

$$\phi(\mathbf{R},z,t) = \sum_{\mathbf{Q}} \phi_{Q,\omega}(z) \, e^{i\mathbf{Q}\cdot\mathbf{R}} e^{i\omega t}. \qquad (4.4.4)$$

Here and in the following it must be understood that only the real part of the complex exponentials must be retained. Assuming an isotropic dielectric response everywhere, and thanks to the cylindrical symmetry of the problem, $\phi_{Q,\omega}(z)$ depends only on the magnitude of the wave vector $\mathbf{Q}$. Thus, we are led to finding the solution of the Laplace equation which in Fourier space reads as:

$$\frac{d^2\phi_{Q,\omega}(z)}{dz^2} - Q^2\phi_{Q,\omega}(z) = 0. \tag{4.4.5}$$

The boundary conditions at the interfaces imply that a physically acceptable solution of equation (4.4.5) exists provided we satisfy the secular equation:

$$\epsilon_{ox}(\omega)^2 + \epsilon_{ox}(\omega)[\epsilon_g(\omega) + \epsilon_s(Q,\omega)]\coth(Qt) + \epsilon_g(\omega)\epsilon_s(Q,\omega) = 0. \tag{4.4.6}$$

The solutions of this equation yield the dispersion of the modes, $\omega(Q) = \omega_Q^{(i)}$, where the index i runs over the branches of the modes. The solution $\phi_{Q,\omega}(z)$ has the form:

$$\phi_{Q,\omega_Q^{(i)}}(z) = \begin{cases} a_{Q,\omega_Q^{(i)}}\, e^{Qz} & (z < 0) \\ b_{Q,\omega_Q^{(i)}}\, e^{-Qz} + c_{Q,\omega_Q^{(i)}}\, e^{Qz} & (0 \le z < t) \\ d_{Q,\omega_Q^{(i)}}\, e^{-Qz} & (z \ge t) \end{cases}, \tag{4.4.7}$$

where:

$$b_{Q,\omega_Q^{(i)}} = \frac{\epsilon_{ox}(\omega_Q^{(i)}) - \epsilon_g(\omega_Q^{(i)})}{2\epsilon_{ox}(\omega_Q^{(i)})} a_{Q,\omega_Q^{(i)}}, \tag{4.4.8}$$

$$c_{Q,\omega_Q^{(i)}} = \frac{\epsilon_{ox}(\omega_Q^{(i)}) + \epsilon_g(\omega_Q^{(i)})}{2\epsilon_{ox}(\omega_Q^{(i)})} a_{Q,\omega_Q^{(i)}}, \tag{4.4.9}$$

$$d_{Q,\omega_Q^{(i)}} = \frac{\epsilon_{ox}(\omega_Q^{(i)}) - \epsilon_g(\omega_Q^{(i)})}{\epsilon_{ox}(\omega_Q^{(i)}) + \epsilon_s(Q,\omega_Q^{(i)})} a_{Q,\omega_Q^{(i)}}. \tag{4.4.10}$$

The determination of multiplicative constant $a_{Q,\omega_Q^{(i)}}$ will be discussed later in the section 'Scattering strength'.

The selection of a model dielectric response for the insulator is a quite delicate issue. In principle, we should account for polarization effects due to a multitude of optical modes, functions not only of the chosen materials, but also of their chemical composition (stoichiometric or not, depending on deposition and annealing conditions), on their allotropic forms (amorphous or, if crystalline, on their crystallographic structure), etc. In order to keep the model tractable, we consider only two bulk optical modes, obtained by averaging over direction (e.g. over the A_{2u} and E_u modes for bc tetragonal

ZrO$_2$ or ZrSiO$_4$), by considering only the two modes exhibiting the largest oscillator strength, by lumping 'bands' of modes into two groups, or by combining of all of these approximations[1]. Thus, we assume an ionic dielectric function of the 'oscillator' form:

$$\epsilon_{\text{ox}}(\omega) = \epsilon_{\text{ox}}^{\infty} + (\epsilon_{\text{ox}}^{i} - \epsilon_{\text{ox}}^{\infty})\frac{\omega_{\text{TO2}}^2}{\omega_{\text{TO2}}^2 - \omega^2} + (\epsilon_{\text{ox}}^{0} - \epsilon_{\text{ox}}^{i})\frac{\omega_{\text{TO1}}^2}{\omega_{\text{TO1}}^2 - \omega^2}, \quad (4.4.11)$$

where ϵ_{ox}^{0} and $\epsilon_{\text{ox}}^{\infty}$ are the static and optical permittivity of the insulator, respectively (so that $\kappa = \epsilon_{\text{ox}}^{0}/\epsilon_0$, where ϵ_0 is the permittivity of vacuum) and ω_{TO1} and ω_{TO2} are the angular frequencies of the only two TO-phonon modes we shall consider in the insulator. We assume $\omega_{\text{TO1}} \lesssim \omega_{\text{TO2}}$. Finally, ϵ_{ox}^{i} is the insulator permittivity describing the dielectric response at some intermediate frequency ω_{int} such that $\omega_{\text{TO1}} \lesssim \omega_{\text{int}} \lesssim \omega_{\text{TO2}}$. Physically, it is related to the oscillator strength of each mode and it must be determined from the energy splitting between longitudinal and transverse optical modes via the Lyddane–Sachs–Teller (LST) relation (or its trivial extension in the case of two optical modes), which allows us to rewrite equation (4.4.11) as:

$$\epsilon_{\text{ox}}(\omega) = \epsilon_{\text{ox}}^{\infty} \frac{(\omega_{\text{LO2}}^2 - \omega^2)(\omega_{\text{LO1}}^2 - \omega^2)}{(\omega_{\text{TO2}}^2 - \omega^2)(\omega_{\text{TO1}}^2 - \omega^2)}, \quad (4.4.12)$$

where the frequency of the two LO modes is given by the generalized LST relation:

$$\omega_{\text{LOi}}^2 = \frac{1}{2\Delta}[b \pm (b^2 - 4\Delta c)^{1/2}] \quad (i = 1,2), \quad (4.4.13)$$

with:

$$\Delta = \epsilon_{\text{ox}}^{\infty},$$
$$b = \Delta(\omega_{\text{TO1}}^2 + \omega_{\text{TO2}}^2) + (\epsilon_{\text{ox}}^{i} - \epsilon_{\text{ox}}^{\infty})\omega_{\text{TO2}}^2 + (\epsilon_{\text{ox}}^{0} - \epsilon_{\text{ox}}^{i})\omega_{\text{TO1}}^2, \quad (4.4.14)$$
$$c = (\Delta + \epsilon_{\text{ox}}^{0} - \epsilon_{\text{ox}}^{\infty})\omega_{\text{TO1}}^2\omega_{\text{TO2}}^2.$$

For the electronic response of the gate we take the usual long-wavelength expression:

$$\epsilon_{\text{g}}(\omega) = \epsilon_{\text{Si}}^{\infty}\left(1 - \frac{\omega_{\text{p,g}}^2}{\omega^2}\right), \quad (4.4.15)$$

where $\omega_{\text{p,g}}^2 = e^2 N_{\text{g}}/(\epsilon_{\text{Si}}^{\infty} m_{\text{g}})$ is the bulk plasma frequency of the polycrystalline-Si gate with an electron density N_{g} (obtained from some suitable average of the electron density over the depletion layer of the poly-Si gate, as

---

[1] We neglect here additional complications, which may arise when dealing with layers of thickness comparable to the lattice constant or the size of the molecular bonds, such as localized/quantized phonon modes and electronic states.

discussed below in the section 'The model'), with an effective mass m_g (=0.32 m_0, where m_0 is the electron mass), and optical permittivity ϵ_{Si}^∞. Finally, for the inverted substrate we assume:

$$\epsilon_s(Q,\omega) = \epsilon_{Si}^\infty \left[1 - \frac{\omega_{p,s}(Q)^2}{\omega^2} \right], \tag{4.4.16}$$

where $\omega_{p,s}(Q)^2 = \left[\sum_\nu e^2 n_\nu Q / (\epsilon_{Si}^\infty m_\nu) \right]^{1/2}$ is the plasma frequency of the 2DEG, n_ν and m_ν being the electron density and conductivity mass in each of the occupied subbands labelled by the index ν.

Equation (4.4.6) is an algebraic equation of sixth degree in ω^2, and we shall label its six positive solutions as $\omega_Q^{(i)}$. Two of these solutions (which we shall label with the indices $i = 5, 6$) are associated with a small scattering field and will be ignored. Indeed, for small values of Qt, they behave like bulk TO modes and couple poorly with the electrons in the inversion layer. At large values of Qt, instead, they are mainly localized at the 'far' gate/insulator interface—thus yielding a scattering strength depressed by a factor $\sim e^{-2Qt}$— with frequencies approaching the frequencies of the bare SO modes at that interface. The remaining four solutions (which we shall label with the index i running from 1 through 4, ordered so that $\omega_Q^{(1)} \geq \omega_Q^{(2)} \geq \omega_Q^{(3)} \geq \omega_Q^{(4)}$) represent coupled interface plasmon–phonon modes.

Two issues regarding these modes must be addressed before we can evaluate their scattering strength—how to estimate their separate phonon and plasmon content, and how to handle them in a regime in which Landau damping dominates.

The first issue can be addressed by extending the result of Kim and co-workers [9] to the case of interest here. The gate-plasmon content of mode i will be defined as:

$$\Pi^{(G)}(\omega_Q^{(i)}) \approx \left| \frac{(\omega_Q^{(i)2} - \omega_Q^{(-g,1)2})(\omega_Q^{(i)2} - \omega_Q^{(-g,2)2})(\omega_Q^{(i)2} - \omega_Q^{(-g,3)2})}{(\omega_Q^{(i)2} - \omega_Q^{(j)2})(\omega_Q^{(i)2} - \omega_Q^{(k)2})(\omega_Q^{(i)2} - \omega_Q^{(l)2})} \right|, \tag{4.4.17}$$

where the indices (i,j,k,l) are cyclical. The 'approximate' sign above results from having neglected the two solutions mentioned above. Similarly, considering the three solutions $\omega_Q^{(-s,\alpha)}$ ($\alpha = 1,3$), obtained from the secular equation (4.4.6) by ignoring the plasma response of the 2DEG in the substrate (that is, by replacing $\epsilon_s(Q,\omega)$ with ϵ_{Si}^∞), we define the substrate-plasmon content of mode i as:

$$\Pi^{(S)}(\omega_Q^{(i)}) \approx \left| \frac{(\omega_Q^{(i)2} - \omega_Q^{(-s,1)2})(\omega_Q^{(i)2} - \omega_Q^{(-s,2)2})(\omega_Q^{(i)2} - \omega_Q^{(-s,3)2})}{(\omega_Q^{(i)2} - \omega_Q^{(j)2})(\omega_Q^{(i)2} - \omega_Q^{(k)2})(\omega_Q^{(i)2} - \omega_Q^{(l)2})} \right|. \tag{4.4.18}$$

It can be verified that equations (4.4.17) and (4.4.18) satisfy the normalization conditions:

$$\sum_{i=1}^{4} \Pi^{(G)}(\omega_Q^{(i)}) = 1, \quad \sum_{i=1}^{4} \Pi^{(S)}(\omega_Q^{(i)}) = 1. \qquad (4.4.19)$$

Having ignored the solutions $\omega_Q^{(5)}$ and $\omega_Q^{(6)}$ forces us to approximate the phonon content of each mode as follows. From equations (4.4.17) and (4.4.18) it follows that the total phonon content of mode i will be:

$$\Phi(\omega_Q^{(i)}) = 1 - \Pi^{(G)}(\omega_Q^{(i)}) - \Pi^{(S)}(\omega_Q^{(i)}). \qquad (4.4.20)$$

In order to define separate phonon-1 and phonon-2 contents, we also consider the three solutions $\omega_Q^{(-\text{TO1},\alpha)}$ ($\alpha = 1, 3$), obtained from the secular equation (4.4.6), but now ignoring the response of phonon 1—that is, by replacing $\epsilon_{ox}(\omega)$ with $\epsilon_{ox}^{\infty}(\omega_{\text{LO2}}^2 - \omega^2)/(\omega_{\text{TO2}}^2 - \omega^2)$—and the three solutions $\omega_Q^{(-\text{TO2},\alpha)}$ ($\alpha = 1, 3$) similarly obtained by ignoring the response of the TO-mode 2 by setting in equation (4.4.6) $\epsilon_{ox}(\omega) \rightarrow \epsilon_{ox}^{\infty}(\omega_{\text{LO1}}^2 - \omega^2)/(\omega_{\text{TO1}}^2 - \omega^2)$. Therefore, the relative phonon-1 content of mode i will be:

$$R^{(\text{TO1})}(\omega_Q^{(i)}) \approx \left| \frac{(\omega_Q^{(i)2} - \omega_Q^{(-\text{TO1},1)2})(\omega_Q^{(i)2} - \omega_Q^{(-\text{TO1},2)2})(\omega_Q^{(i)2} - \omega_Q^{(-\text{TO1},3)2})}{(\omega_Q^{(i)2} - \omega_Q^{(j)2})(\omega_Q^{(i)2} - \omega_Q^{(k)2})(\omega_Q^{(i)2} - \omega_Q^{(l)2})} \right|, $$
$$(4.4.21)$$

(where, as before, i, j, k, l are cyclical) so that, finally, the TO-phonon-1 content of mode i will be:

$$\Phi^{(\text{TO1})}(\omega_Q^{(i)}) \approx \frac{R^{(\text{TO1})}(\omega_Q^{(i)})}{R^{(\text{TO1})}(\omega_Q^{(i)}) + R^{(\text{TO2})}(\omega_Q^{(i)})} [1 - \Pi^{(G)}(\omega_Q^{(i)}) - \Pi^{(S)}(\omega_Q^{(i)})], $$
$$(4.4.22)$$

and similarly for $\Phi^{(\text{TO2})}(\omega_Q^{(i)})$. Once more, it has been verified numerically that these definitions satisfy the additional normalization conditions:

$$\sum_{i=1}^{4} \Phi^{(\text{TO1})}(\omega_Q^{(i)}) = 1, \quad \sum_{i=1}^{4} \Phi^{(\text{TO2})}(\omega_Q^{(i)}) = 1, \qquad (4.4.23)$$

and, for each mode i:

$$\Pi^{(G)}(\omega_Q^{(i)}) + \Pi^{(S)}(\omega_Q^{(i)}) + \Phi^{(\text{TO1})}(\omega_Q^{(i)}) + \Phi^{(\text{TO2})}(\omega_Q^{(i)}) = 1. \qquad (4.4.24)$$

At sufficiently short wavelengths plasmons cease to be proper excitations. In our context this may happen when the gate-plasma-like solution $\omega_Q^{(i_g)}$ (where usually $i_g = 1$ at large enough N_g) enters the single-particle continuum in the gate, the substrate-plasma-like solution enters the single-particle continuum in the substrate, or both. In order to account for

Landau damping, albeit approximately, we proceed as follows. Whenever the substrate-plasmon-like excitation $\omega_Q^{(i_s)}$ (where, usually, $i_s = 4$) enters the single-particle continuum of the 2DEG evaluated in the extreme quantum limit (i.e. $\omega_Q^{(i_s)} \leq [(\hbar Q)/(2m_t)](Q + 2K_F)$, where $K_F = (\pi n_s)^{1/2}$, $m_t = 0.19 m_0$ being the transverse effective mass), we consider only the three solutions $\omega_Q^{(-s,i)}$ ($i = 1,3$) given above. These represent the three coupled gate-plasmon/insulator-TO modes which exist when the substrate plasma is absent. The plasmon/phonon content and scattering strength for these modes are obtained in a way completely analogous to the one discussed so far. Similarly, when the frequency of the gate-plasmon-like excitation enters the single-particle continuum of the gate (that is, $\omega_Q^{(i_g)} \leq (\hbar Q)/(2m_g)(Q + 2k_F)$, where k_F is the zero-temperature Fermi wave vector of the electron gas in the gate, $k_F = (\pi^2 N_g/2)^{1/3}$, and the index i_g takes a value of 1 or 2, depending on the electron density in the gate, frequency of the high-energy TO mode, ω_{TO2}, and dielectric constants of the material considered), we consider only the three solutions $\omega_Q^{(-g,i)}$ ($i = 1,3$) representing the three coupled substrate-plasmon/insulator-TO modes which exist when the gate plasma does not respond. In this case, only phonon-like scattering with these modes is considered. Equation (4.4.33) describes the surface-phonon scattering field, setting $\epsilon_{gate}(\omega) = \epsilon_{Si}^\infty$ in $\epsilon_{TOT,high}^{(TOi)}(\omega)$ and $\epsilon_{TOT,low}^{(TOi)}(\omega)$ to reflect the absence of the gate plasma.

Finally, when both the gate- and the substrate-plasmon-like dispersions are within their respective Landau-damping regions, we consider only the two phonon-like modes of frequencies $\omega_Q^{(SOi)}(i = 1, 2)$, whose associated scattering field is given by equation (4.4.33) with $\epsilon_{gate}(\omega) = \epsilon_{substrate}(Q, \omega) = \epsilon_{Si}^\infty$ employed in $\epsilon_{TOT,high}^{(TOi)}(\omega)$ and $\epsilon_{TOT,low}^{(TOi)}(\omega)$.

Scattering strength

The amplitude $a_{Q,\omega}$ of the field, equation (4.4.7), can be determined using the semiclassical approach originally proposed by Stern and Ferrel [10] which we also followed in [7][2] and described in a simple case in Appendix A of [8]. We first consider the time-averaged total (electrostatic, including self-energy) energy, $\langle W_Q^{(i)} \rangle$, associated with the field $\phi_Q^{(i)}(\mathbf{R}, z, t)$ caused by the excitation of mode i oscillating at the frequency $\omega_Q^{(i)}$. (The bra–kets $\langle \cdots \rangle$ denote time average.) Let us write the electrostatic potential at a given wavelength as:

$$\phi_Q^{(i)}(\mathbf{R}, z, t) = \phi_{Q,\omega_Q^{(i)}}^{(i)}(z) \cos(\mathbf{Q} \cdot \mathbf{R} - \omega_Q^{(i)} t). \qquad (4.4.25)$$

---

[2]As explained in [7], scattering with the substrate-plasmon component of the excitations is assumed to have no effect on the electron momentum relaxation rate, and so on the mobility, since it involves no direct loss of momentum by the 2DEG.

Since phonons and plasmons in the harmonic and linear-response approximations, respectively, are represented as harmonic oscillations, the time-averaged total energy associated with these excitations is simply twice the time-averaged potential energy, $\langle U_Q^{(i)} \rangle$. This, in turn, is the electrostatic (self-) energy of the interface polarization charge density $\rho_Q^{(i)}(\mathbf{R}, z, t)$ in the presence of the potential $\phi_Q^{(i)}(\mathbf{R}, z, t)$ caused by the interface charge itself. We may express this potential energy in two alternative equivalent ways: From expression (4.4.7) for the potential, the density of the polarization charge associated with mode i is localized at the two interfaces and can be obtained from the Poisson equation $\rho_Q^{(i)}(\mathbf{R}, z, t) = -\nabla \cdot [\epsilon(\omega_Q^{(i)}; z) \nabla \phi_Q^{(i)}(\mathbf{R}, z, t)]$ (where the z-dependence in $\epsilon(\omega_Q^{(i)}; z)$ reflects the fact the we must use the appropriate dielectric functions across the interfaces):

$$\rho_Q^{(i)}(\mathbf{R}, z, t) = \{\delta(z)[\epsilon_{\text{gate}}(\omega_Q^{(i)}) a_{Q, \omega_Q^{(i)}} + \epsilon_{\text{insulator}}(\omega_Q^{(i)}) (b_{Q, \omega_Q^{(i)}} - c_{Q, \omega_Q^{(i)}})]$$
$$+ \delta(z - t)[\epsilon_{\text{insulator}}(\omega_Q^{(i)}) (c_{Q, \omega_Q^{(i)}} e^{Qt} - b_{Q, \omega_Q^{(i)}} e^{-Qt})$$
$$+ \epsilon_{\text{substrate}}(Q, \omega_Q^{(i)}) d_{Q, \omega_Q^{(i)}} e^{-Qt}]\} Q \cos(\mathbf{Q} \cdot \mathbf{R} - \omega_Q^{(i)} t), \qquad (4.4.26)$$

having introduced the new functions $\epsilon_{\text{gate}}(\omega)$, $\epsilon_{\text{insulator}}(\omega)$, and $\epsilon_{\text{substrate}}(Q, \omega)$ which must be chosen in a way consistent with the component of the polarization charge $\rho_Q^{(i)}(\mathbf{R}, z, t)$ we are considering, as discussed below. Therefore, for the energy $\langle W_Q^{(i)} \rangle$ we can write:

$$\langle W_Q^{(i)} \rangle = 2 \langle U_Q^{(i)} \rangle = \frac{2}{\Omega} \left\langle \int_\Omega d\mathbf{R} \int_{-\infty}^{\infty} dz \phi_Q^{(i)}(\mathbf{R}, z, t) \rho_Q^{(i)}(\mathbf{R}, z, t) \right\rangle, \qquad (4.4.27)$$

where Ω is a normalization area. Alternatively, using Green's identity and accounting for the discontinuity of the electric and displacement fields across the interfaces, we can express $\langle W_Q^{(i)} \rangle$ in terms of the electrostatic energy of the field $\mathbf{E}_Q^{(i)} = -\nabla \phi_Q^{(i)}$:

$$\langle W_Q^{(i)} \rangle = \frac{2}{\Omega} \left\langle \int_\Omega d\mathbf{R} \int_{-\infty}^{\infty} dz\, \epsilon(\omega_Q^{(i)}; z) |\mathbf{E}_Q^{(i)}(\mathbf{R}, z, t)|^2 \right\rangle. \qquad (4.4.28)$$

From equations (4.4.8)–(4.4.10) and either using equation (4.4.27) or performing the integrals in equation (4.4.28) using equation (4.4.7), we obtain:

$$\langle W_Q^{(i)} \rangle = Q \epsilon_{\text{TOT}}(Q, \omega_Q^{(i)}) \left[\frac{\epsilon_{\text{ox}}(\omega_Q^{(i)}) - \epsilon_{\text{g}}(\omega_Q^{(i)})}{\epsilon_{\text{ox}}(\omega_Q^{(i)}) + \epsilon_{\text{s}}(Q, \omega_Q^{(i)})} \right]^2 a_{Q, \omega_Q^{(i)}}^2 e^{-2Qt}. \qquad (4.4.29)$$

Here the 'total' effective dielectric function of the substrate coupled to the gate and the insulating layer has been defined as:

$$\epsilon_{\text{TOT}}(Q, \omega) = \epsilon_{\text{gate}}(\omega) \left[\frac{\epsilon_{\text{ox}}(\omega) + \epsilon_{\text{s}}(Q, \omega)}{\epsilon_{\text{ox}}(\omega) - \epsilon_{\text{g}}(\omega)} \right]^2 e^{2Qt}$$

$$+ \epsilon_{\text{insulator}}(\omega) \left\{ \left[\frac{\epsilon_{\text{ox}}(\omega) + \epsilon_{\text{s}}(Q, \omega)}{2\epsilon_{\text{ox}}(\omega)} \right]^2 (e^{2Qt} - 1) \right.$$

$$\left. + \left[\frac{\epsilon_{\text{ox}}(\omega) - \epsilon_{\text{s}}(Q, \omega)}{2\epsilon_{\text{ox}}(\omega)} \right]^2 (1 - e^{-2Qt}) \right\} + \epsilon_{\text{substrate}}(Q, \omega), \quad (4.4.30)$$

having made repeated use of the relation (4.4.6) to reach one of the many possible equivalent algebraic forms. The semiclassical nature of the argument enters the final step of setting the quantity $\langle W_Q^{(i)} \rangle$ equal to the zero-point energy, $\hbar \omega_Q^{(i)}/2$, of the quantized excitation. This finally determines the 'normalization constant', $a_{Q,\omega_Q^{(i)}}$, and thus the amplitude of the scattering field in the substrate ($z \geq t$):

$$\phi_{Q,\omega_Q^{(i)}}^{(i)} = \left[\frac{\hbar \omega_Q^{(i)}}{2Q\epsilon_{\text{TOT}}(Q, \omega_Q^{(i)})} \right]^{1/2} e^{-Q(z-t)}, \quad (4.4.31)$$

up to the appropriate Bose factors of the excitations, $n_Q^{(i)1/2}$ and $(1 + n_Q^{(i)})^{1/2}$, which multiply the scattering potential for absorption and emission processes, respectively.

The choice of the dielectric functions $\epsilon_{\text{gate}}(\omega)$, $\epsilon_{\text{insulator}}(\omega)$, and $\epsilon_{\text{substrate}}(Q, \omega)$, which appear in the expression for $\epsilon_{\text{TOT}}(Q, \omega)$, is a crucial element of our discussion. Whenever we are interested in determining the potential energy due to a particular type of response of the system (ionic or electronic), we cannot include this response in these dielectric functions. For example, by setting $\epsilon_{\text{gate}}(\omega) = \epsilon_{\text{g}}(\omega)$, $\epsilon_{\text{insulator}}(\omega) = \epsilon_{\text{ox}}(\omega)$, and $\epsilon_{\text{substrate}}(Q, \omega) = \epsilon_{\text{s}}(Q, \omega)$, we effectively lump the entire dielectric response, ionic and electronic, into the dielectric functions, and we expect that the potential energy of 'whatever response is left' (none, in this case) in the field and charge, ϕ_Q and ρ_Q, should vanish. Indeed when so doing, the resulting $\epsilon_{\text{TOT}}(Q,\omega)$ vanishes for $\omega = \omega_Q^{(i)}$, the equation $\epsilon_{\text{TOT}}(Q, \omega) = 0$ being equivalent to the secular equation (4.4.6). So, when taking $\epsilon_{\text{gate}}(\omega) = \epsilon_{\text{Si}}^{\infty}$, $\epsilon_{\text{substrate}}(Q, \omega) = \epsilon_{\text{Si}}^{\infty}$, and $\epsilon_{\text{insulator}}(\omega) = \epsilon_{\text{ox}}^{0}$ we consider only the plasmon contribution to the polarization charges. Indeed, the response of the insulator lattice is removed from the electrostatic field by being lumped into the insulator permittivity when setting $\epsilon_{\text{insulator}}(\omega) = \epsilon_{\text{ox}}^{0}$, while the electronic response is removed from the dielectric functions of the gate and substrate, and is included directly into the amplitude of the electrostatic field and polarization charge, ϕ_Q and ρ_Q. In this case, equation (4.4.31) represents the amplitude of the field induced by plasma excitations. Thus, defining as $\epsilon_{\text{TOT}}^{(\text{PL})}(Q, \omega_Q^{(i)})$ the total plasma dielectric function so obtained, scattering between electrons in the substrate and gate plasmons is described by the effective scattering field:

$$\phi_{Q,\omega_Q^{(i)}}^{(i,g,PL)}(z) = \left[\frac{\hbar\omega_Q^{(i)}}{2Q\epsilon_{TOT}^{(PL)}(Q,\omega_Q^{(i)})} \Pi^{(G)}(\omega_Q^{(i)}) \right]^{1/2} e^{-Q(z-t)}. \tag{4.4.32}$$

Scattering with the field induced by the polarization charges of the insulator lattice (i.e. with the optical phonons in the insulator) can be evaluated in a way essentially identical to the approach followed by Kittel [11] to evaluate the Fröhlich coupling in bulk polar materials. The only difference between Kittel's and our approach consists in following Stern and Ferrel [10] in evaluating the ground-state energy semiclassically, rather than from second-order perturbation theory. In order to isolate the contribution of each phonon independently and consider only the lattice polarization, the squared, time-averaged amplitude of the scattering field is computed by lumping the electronic response into the dielectric functions of the gate and substrate, while letting one phonon (say, phonon 2 to fix the ideas) respond, but first by 'freezing' the other mode (TO1) and then by considering its full response. The difference between the two squared amplitudes so obtained constitutes the effect of the ionic polarization charge associated solely with optical mode 1. To be explicit, in our case the amplitude of the field (equation (4.4.31)) when only phonon 2 responds is obtained by setting $\epsilon_{gate}(\omega) = \epsilon_g(\omega)$ (response of the gate plasmons lumped into the gate dielectric function), $\epsilon_{substrate}(Q,\omega) = \epsilon_s(Q,\omega)$ (response of the substrate plasmons lumped into the dielectric function of the inversion layer), and setting $\epsilon_{insulator}(\omega) = \epsilon_{ox}^\infty(\omega_{LO2}^2 - \omega^2)/(\omega_{TO2}^2 - \omega^2)$ (phonon 2 responds at the frequency ω, while phonon 1 does not respond). Let $\epsilon_{TOT,high}^{(TO1)}$ be the resulting effective dielectric function. In contrast, when phonon 1 is allowed to respond fully, we have $\epsilon_{insulator}(\omega) = \epsilon_{ox}^\infty[(\omega_{LO2}^2 - \omega^2)/(\omega_{TO2}^2 - \omega^2)](\omega_{LO1}/\omega_{TO1})^2$ (which reduces to ϵ_{ox}^0 in the simpler case of insulators exhibiting only one TO-mode), the full response of phonon 1 now being accounted for by the term $(\omega_{LO1}/\omega_{TO1})^2$. Let $\epsilon_{TOT,low}^{(TO1)}$ denote the resulting dielectric function. Thus, the interaction between electrons in the inversion layer and the TO1-phonon content of mode i will be described by the scattering field:

$$\phi_{Q,\omega_Q^{(i)}}^{(i,PH1)}(z) = \left\{ \frac{\hbar\omega_Q^{(i)}}{2Q} \left[\frac{1}{\epsilon_{TOT,high}^{(TO1)}(Q,\omega_Q^{(i)})} - \frac{1}{\epsilon_{TOT,low}^{(TO1)}(Q,\omega_Q^{(i)})} \right] \Phi^{(TO1)}(\omega_Q^{(i)}) \right\}^{1/2} e^{-Q(z-t)}. \tag{4.4.33}$$

The scattering strength with phonon mode 2 can be trivially obtained by swapping indices 1 and 2 in the discussion above.

Discussion

We summarize graphically the results of this section by showing in figures 4.4.2 and 4.4.3 the significant properties of the interface modes for the $Si/SiO_2/Si$ and the $Si/ZrO_2/Si$ systems, respectively, as a function of the in-plane wave vector Q. These two MOS systems are the extreme cases of a low-κ (SiO_2) and a high-κ (ZrO_2) material, the $Si/SiO_2/Si$ system exhibiting some of the stiffest modes, the $Si/ZrO_2/Si$ some of the softest optical modes. The curves in the figures have been obtained using electron concentrations in the gate and in the Si substrate and an 'equivalent' insulator thickness, t_{eq}, (defined as $t\epsilon^0_{ox}/\epsilon^0_{SiO_2}$), which are representative of typical situations. The subband model employed has been described in section 'Interface modes' of [7]; We have employed an anisotropic, non-parabolic band-structure, used a triangular-well approximation for the potential in the inversion layer, and embraced the long-wavelength approximation for the dielectric function discussed above, also ignoring intersubband-plasmons.

In figures 4.4.2 and 4.4.3 we show in (*a*) the dispersion of the modes and in (*b*) the total scattering strength for each mode. The scattering strength with the phonon-like component of each mode i, $\Lambda^{(i)}_{SO}(Q)$, has been defined, according to equation (4.4.33), by summing the scattering strength of both TO-modes α:

$$\Lambda^{(i)}_{SO}(Q) = \sum_{\alpha=1}^{2} \Lambda^{(i)}_{SO,\alpha}(Q)$$

$$= \epsilon_0 \sum_{\alpha=1}^{2} \left| \frac{\hbar\omega^{(i)}_Q}{2} \left[\frac{1}{\epsilon^{(TO\alpha)}_{TOT,high}(Q,\omega^{(i)}_Q)} - \frac{1}{\epsilon^{(TO\alpha)}_{TOT,low}(Q,\omega^{(i)}_Q)} \right] \Phi^{(TO1)}(\omega^{(i)}_Q) \right|,$$

$$(4.4.34)$$

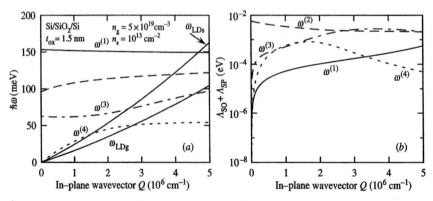

Figure 4.4.2. Calculated dispersion (*a*) and total scattering strength (*b*) for the insulator–optical-phonon/substrate-and-gate-plasmons interface modes for the $Si/SiO_2/Si$ system. In (*a*), the curves labelled by ω_{LDs} and ω_{LDg} identify the boundary of the substrate and gate Landau-damping regions, respectively. Modes 5 and 6 have been ignored.

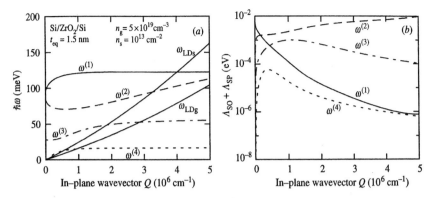

Figure 4.4.3. As in figure 4.4.2, but for the Si/ZrO₂/Si system. Note that the two highest energy modes are plasmon-like.

and similarly for the scattering strength for the gate-plasmon content of mode i, $\Lambda_{SP}^{(i)}(Q)$, defined, according to equation (4.4.32), as:

$$\Lambda_{SP}^{(i)}(Q) = \epsilon_0 \left| \frac{\hbar\omega_Q^{(i)}}{2\epsilon_{TOT}^{(PL)}(Q,\omega_Q^{(i)})} \Pi^{(G)}(\omega_Q^{(i)}) \right|. \qquad (4.4.35)$$

(In both equations (4.4.34) and (4.4.35) we have included a factor ϵ_0 to give the scattering strengths an 'intuitive' dimension of energy.) Note that comparing equation (4.4.33) with equation (4.4.34) and equation (4.4.32) with equation (4.4.35), the (squared) scattering amplitude will be modulated by a factor Q^{-1}—which is compensated by the density-of-states factor Q while integrating over final scattering states. The scattering field will also exhibit the exponential decay $\exp[-Q(z-t)]$ into the substrate, away from the substrate/insulator interface.

Landau damping has been ignored in these figures. However, the two curves labelled ω_{LDg} and ω_{LDs} identify the region of strong damping of the gate and substrate plasma, respectively. As explained above, Landau damping is approximately accounted for by ignoring the substrate plasma for values of Q such that $\omega^{(4)} \gtrsim \omega_{LDs}$ (in both figures) and by ignoring the response of the gate electrons whenever $\omega^{(2)} \gtrsim \omega_{LDg}$ (in figure 4.4.2) or $\omega^{(1)} \gtrsim \omega_{LDg}$ (in figure 4.4.3).

In figure 4.4.2(a), the mode labelled $\omega^{(1)}$ appears clearly to be mostly a phonon-like mode, originating from the high-frequency SiO₂ TO mode at about 0.135 eV. The mode $\omega^{(2)}$ is mainly a gate-plasmon mode, its coupling to the high-frequency phonon mode increasing at shorter wavelengths. The second, low-energy SiO₂ TO-mode at about 0.06 eV is strongly coupled to the substrate plasmons. Indeed, the modes labelled $\omega^{(3)}$ and $\omega^{(4)}$ result from this strong coupling: the former is mostly phonon-like at small Q, but it quickly becomes mainly a substrate-plasma mode as Q grows, while the mode $\omega^{(4)}$ shows the opposite behaviour.

Figure 4.4.3 conveys essentially the same information, but it shows how the stiffest mode, labelled $\omega^{(1)}$, is now mainly gate-plasmon-like, mode $\omega^{(2)}$ is mainly phonon-like at small Q, substrate-plasmon-like at shorter wavelength. The mode labelled $\omega^{(3)}$ is mainly phonon-like, but its phonon content switches from the low-energy (small-Q) to the high-energy (large-Q) insulator phonon mode, crossing the substrate-plasma mode at intermediate wavelength. The mode labelled $\omega^{(4)}$ starts as substrate-plasma-like at low Q, but it becomes almost completely phonon-like at larger Q.

Although not shown in the figures, the SiO_2 system ($\omega^{(2)}$) shows a larger gate-plasmon scattering strength than for the ZrO_2 system ($\omega^{(1)}$). As explained before [7], this is simply due to the closer proximity of the gate in the SiO_2 system. Conversely, looking at the scattering strength of the $\omega^{(2)}$-mode, for example, at the largest values of Q in the undamped region, the ZrO_2 system exhibits a stronger scattering strength with phonon-like modes. Finally phonon-like scattering with modes $\omega^{(3)}$ (in figure 4.4.2) and $\omega^{(2)}$ (in figure 4.4.3) is significantly enhanced by the phonon–plasmon coupling. This effect results from the anti-screening properties of the electron gas(es). Whenever the frequency of an excitation is larger than the frequency of the electron plasma, the coupling strength with the excitation is enhanced, while Landau damping gains strength. Indeed, this effect is significant well within the region in which we must account for Landau damping. In our case, the situation is noticeably complicated by the presence of two plasmas (the gate and the substrate) and by two optical phonons. As we shall see below discussing the effective electron mobility, each of the phonon-like excitations may be screened by one plasma and anti-screened by the other.

Effective mobility of the 2D electron gas

The model

The calculation of the effective electron mobility in the inversion layer of the MOS systems considered here has been performed using the approximations and models described in detail in section III of [7]. We have employed equation (63) of that reference, using the total relaxation time computed by adding relaxation rates due to electron scattering with intravalley, intra- and intersubband acoustic phonons using an anisotropic deformation potential (equation (86) of [7]), scattering with intervalley phonons as described in section III D of that reference, and considering scattering with the coupled plasmon/insulator-phonon interface modes employing equation (77) of [7], but substituting the 'effective' field amplitude $|\mathscr{A}_Q|^2$ with the scattering strengths $\Lambda_{SO,\alpha}^{(i)}(Q)/Q$ and $\Lambda_{SP}^{(i)}(Q)/Q$ given by equations (4.4.34) and (4.4.35) above. The Si inversion layer has been modelled using the triangular-well approximation, employing a number of subbands sufficient to account correctly for absorption processes by thermal electrons.

The complexity and computer requirements of the numerical procedure limit the maximum number of subbands (and so the minimum electron density) which we could consider. Even so, some 'numerical noise' is evident in our results. The dispersion of the interfacial modes has been obtained by solving the secular equation (4.4.6) for a set of tabulated values Q_i and storing its roots in look-up tables. In general there will be four of them in the undamped region of low Q, three or only two as Landau damping enters the picture at larger values of Q. We also tabulate, for each Q and branch i, the amplitude of the scattering fields and the group velocity of the modes, required to evaluate the Jacobian factor g (in the notation of [7]). A linear interpolation of these functions of the magnitude of the momentum-transfer Q has been performed during the numerical integration of the SO-limited relaxation time.

The electron concentration, N_g, entering the evaluation of the gate plasma has been estimated by first integrating numerically the Poisson equation in the gate for a given electron concentration, n_s, in the inversion layer of the Si substrate. We have then either employed the electron concentration at the insulator/gate interface, or, instead, the average quantity

$$\langle N_g(Q) \rangle = \frac{\int_{z_{max}}^{0^-} N(z) e^{Qz} \, dz}{\int_{z_{max}}^{0^-} e^{Qz} \, dz}. \tag{4.4.36}$$

This expression has a heuristic justification. The potential associated with the interface excitations has the form given by equation (4.4.7), exhibiting an exponential decay e^{Qz} for $z \leq 0$. Therefore, the 'effective' average electron density seen in the gate by an interface excitation will approach the gate donor density, N_{Dg}, at long wavelength and $N(z = 0^-)$ at short wavelength. Equation (4.4.36) empirically captures this behaviour. Note that this Q-dependent average must be employed to compute the dispersion of the interface modes, since it appears via $\omega_{p,g}$, and so via $\epsilon_g(\omega)$, in the secular equation (4.4.6).

Finally, scattering with interface roughness has been included using Matthiessen's rule and adding to the calculated mobility $\mu_{PH,SO,SP}$—including scattering with Si phonons (PH), coupled SO, and interface plasmon (SP) modes—the contribution of a surface-limited mobility, μ_{SR}, of the form $\mu_{SR} = \mu_0 (10^{13}/n_s)^2$, with n_s measured in cm^{-2}. The constant μ_0 has been determined by fitting the resulting 'total' mobility $\mu_{tot} = [1/\mu_{PH,SO,SP} + 1/\mu_{SR}]^{-1}$ calculated for thick (5 nm) SiO$_2$ systems at $n_s = 10^{13}$ cm^{-2} to the experimental value in lightly doped substrates of about 308 cm^2 V^{-1} s^{-1} (see [13]). The resulting value was $\mu_0 = 1.473 \times 10^3$ cm^2 V^{-1} s^{-1} when using

$N_g = N(z = 0^-)$, and $\mu_0 = 1.167 \times 10^3 \, \text{cm}^2 \, \text{V}^{-1} \, \text{s}^{-1}$ when using $N_g = \langle N_g(Q) \rangle$. Converting the latter value to commonly used expressions, it implies a surface-roughness-limited mobility about a factor of two larger than what we obtained in the past [6] (using Monte Carlo simulations, and also used in figure 4.4.1, which assumed the values $\Lambda = 1.3 \, \text{nm}$ and $\Delta = 0.48 \, \text{nm}$ for the Ando parameters [12]), and within 10% of a typical empirical fit [14] to the effective electron mobility.

Insulator parameters

We have considered MOS systems with six different insulators: SiO_2, Al_2O_3, AlN, ZrO_2, HfO_2, and $ZrSiO_4$. These materials cover a range of parameters (dielectric constants, phonon energies, etc) wide enough to give an idea of the qualitative behaviour of the electron mobility as a function of the physical properties of the insulator.

In order to select the parameters required to evaluate the electron/SO-mode scattering strength, it is convenient to rewrite equation (4.4.11) in the following more general form accounting for N_{TO} TO modes:

$$\epsilon_{ox}(\omega) = \epsilon_{ox}^{\infty} + \epsilon_0 \sum_{\alpha=1}^{N_{TO}} \frac{f_\alpha \omega_{TO,\alpha}^2}{\omega_{TO,\alpha}^2 - \omega^2}. \qquad (4.4.37)$$

We can relate this expression to its alternative form, equation (4.4.11), by rewriting the oscillator strength f_α of the TO mode α in terms of the 'intermediate' dielectric constants $\epsilon^{(\alpha)} \approx \epsilon_{ox}(\omega_\alpha)$, where $\omega_{\alpha-1} \leq \omega_{TO,\alpha} < \omega_\alpha$ for ω_α ordered so that $\omega_{\alpha-1} < \omega_\alpha$:

$$f_\alpha = \frac{\epsilon^{(\alpha-1)} - \epsilon^{(\alpha)}}{\epsilon_0}. \qquad (4.4.38)$$

Here $\epsilon^{(\alpha=0)} = \epsilon_{ox}^0$ and $\epsilon^{(\alpha=N_{TO})} = \epsilon_{ox}^{\infty}$. For a single TO mode and, approximately, for TO modes widely separated energetically, the LST relation provides the LO/TO splitting:

$$\omega_{LO,\alpha}^2 \approx \frac{\epsilon^{(\alpha-1)}}{\epsilon^{(\alpha)}} \omega_{TO,\alpha}^2. \qquad (4.4.39)$$

For materials exhibiting two or more TO modes at nearby frequencies, the energy of the LO modes must be determined by the generalized LST relation $\epsilon_{ox}(\omega) = 0$, such as equation (4.4.14) valid in the case of two TO phonons. In order to determine completely the frequency dependence of $\epsilon_{ox}(\omega)$ in the model form (4.4.37), for each bulk mode α we need knowledge of two of the quantities $\omega_{TO,\alpha}$, $\omega_{LO,\alpha}$, and $\epsilon^{(\alpha)}$ (or, equivalently, f_α). Experimentally, infrared (IR) absorption experiments can provide mainly information on $\omega_{TO,\alpha}$, while Raman and electron tunnelling spectroscopy

[15] can also provide direct information about the LO frequency, $\omega_{LO,\alpha}$. The relative amplitude of each peak in the IR spectra can be correlated with the oscillator strength f_α. In addition, one could rely on theoretical calculations. Unfortunately, things are more complicated. Our ultimate goal is the calculation of the electron mobility, possibly comparing our results with experimental data. However, as thin-insulator MOS structures are typically manufactured with processes which must be compatible with the current Si technology, the structure and composition of the grown or deposited insulator can often be inferred only indirectly. The dielectric response of any given material may depend on its morphology: for example, AlN exhibits different properties in its wurtzite and zinc-blende structures [16]. It can also depend on its closeness to the ideal chemical composition: incompletely oxidized Al_2O_3, for example, shows additional modes, possibly related to unoxidized Al ions [17]. Undesired, but so far unavoidable, 'native' interfacial layers (SiO_2 when dealing with oxides, Si_3N_4 when dealing with nitrides) can mask the response of the 'pure' dielectric under study, as discussed in the section 'Effect of a silicon dioxide interfacial layer'. Finally, the information available in the literature is incomplete, occasionally inconsistent. Here we shall rely on both experimental data and theoretical results, and we shall compare this input with FTIR spectra we have obtained, and attempt to obtain a consistent picture. Some of these have already been reported, including experimental details [18]. Additional spectra are shown in figure 4.4.4. Table 4.4.1 summarizes the values we have employed.

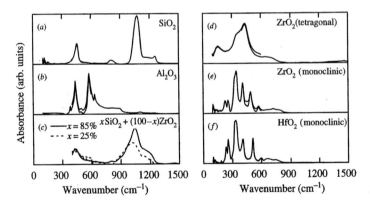

Figure 4.4.4. FTIR spectra for some of the insulators considered. In all spectra (except for $xSiO_2 + (100 - x)ZrO_2$ with $x = 85\%$ in the bottom panel) two curves are shown, one in the far-IR range of $50-600\,cm^{-1}$ and one in the mid-IR range of $400-4000\,cm^{-1}$. In the bottom frame ('nominally' $ZrSiO_4$ obtained from chemical-solution deposition of $xSiO_2 + (100 - x)ZrO_2$) spectra for $x = 85\%$ and $x = 25\%$ are shown, to illustrate the decay of the SiO_2 mode at about $1080\,cm^{-1}$ with decreasing x.

Table 4.4.1. Parameters used to compute the electron–phonon coupling in polar insulators.

Material quantity (units)	SiO_2[a]	Al_2O_3[b]	AlN[c]	ZrO_2[d–f]	HfO_2[f,g]	$ZrSiO_4$[h]
ϵ_{ox}^0 (ϵ_0)	3.90	12.53	9.14	24.0[i]	22.00	11.75
ϵ_{ox}^i (ϵ_0)	3.05	7.27	7.35	7.75	6.58	9.73
ϵ_{ox}^∞ (ϵ_0)	2.50	3.20	4.80	4.00	5.03	4.20
ω_{TO1} (meV)	55.60	48.18	81.40	16.67	12.40	38.62
ω_{TO2} (meV)	138.10	71.41	88.55	57.70	48.35	116.00
α_1	0.0248	0.0788	0.0248	0.2504	0.3102	0.0322
α_2	0.0113	0.0814	0.0423	0.0779	0.0362	0.2942

The frequency of the optical phonons and/or the dielectric functions ϵ^0, ϵ^i, and ϵ^∞ are taken from the literature, when available, and the unavailable data are obtained using the LST relation. When more than two modes are present, only the two strongest modes (based on the magnitude of the LO–TO energy splittings) have been considered. When anisotropic quantities are given— such as in [22] and [34] for the energies of E_T and A_{2T} (bc tetragonal) or E_u and A_{2u} (hexagonal) modes propagating in directions perpendicular and parallel to the crystal c-axis, respectively, and also for the elements of the dielectric tensor—a simple average has been taken. The Fröhlich coupling constants for each mode are also indicated. Note the large coupling constants for the low-energy modes in ZrO_2 and HfO_2, which are indeed the materials yielding the lowest mobility. [a] [4, 22] and references therein. [b] [23]. [c] [16]. [d] [26]. [e] [27]. [f] [25]. [g] [31]. [h] [34] and references therein. [i] [29].

In the table, we also show the dimensionless coupling constant

$$\alpha_i = \frac{e^2}{4\pi\hbar}\left(\frac{m_t}{2\hbar\omega_{SOi}}\right)^{1/2}\left(\frac{1}{\epsilon_{Si}^\infty + \epsilon_{ox}^\infty} - \frac{1}{\epsilon_{Si}^\infty + \epsilon_{ox}^0}\right), \qquad (4.4.40)$$

for each of the two modes i, which corresponds to the dimensionless Fröhlich coupling constant usually defined in bulk materials. These coupling constants have been obtained using for the energy of the SO phonons the approximate expressions given by equation (4.4.2), with the appropriate optical, static, and 'intermediate' dielectric constants. Note how well these values correlate with the mobility shown in figure 4.4.1. In particular, the high values of the coupling constants relative to the low-energy modes in HfO_2 and ZrO_2 hint very directly at the importance of remote scattering with SO modes in MOS systems using these materials.

The FTIR spectra of SiO_2 were obtained from thermally grown SiO_2. The α-Al_2O_3 FTIR spectra were obtained from a chemical-solution-deposited film annealed at 1200°C for 60 min in oxygen. The FTIR spectra of tetragonal and monoclinic ZrO_2 were obtained from chemical-solution-deposited films annealed at 500 and 900°C, respectively. The FTIR spectra of $xSiO_2 + (100 - x)ZrO_2$ with $x = 85$ and 25% were obtained from a chemical-solution-deposited film annealed at 700°C. Additional details regarding deposition, phase formation, and FTIR analysis of the ZrO_2, HfO_2 and $xSiO_2 + (100 - x)ZrO_2$ films can be found in the literature [18]. The monoclinic HfO_2 spectra were obtained from a film chemical-vapour deposited at 700°C.

SiO_2. The SiO_2 FTIR spectrum shown in figure 4.4.4 exhibits two strong peaks at 1076 and 461 cm^{-1}, corresponding to TO-modes at 133.4 and 57.2 meV, associated with an asymmetric stretching of the SiO_4 unit and a bending of the Si–O–Si bond, respectively. We neglect a weak third mode at 806 cm^{-1} ($\approx$100 meV), due to a symmetric stretching mode of the Si–O–Si bond. The 'shoulders' at 1255 and 532 cm^{-1} are related to the corresponding LO modes. These values are in good agreement with the experimental energies reported by Hess and Vogl [4]—from [19–21]. Also the LO/TO splittings are consistent with the oscillator strengths reported in the literature, but those derived from the areas under the FTIR peaks appear to give a stronger high-energy mode, the strength of the 1076 cm^{-1} mode being about five times greater than the strength of the low-energy phonon. We have decided to use the most common values for the oscillator strengths reported in the literature. Indeed, recent calculations of the Raman-active intensities in α-quartz, based on a first-principle density functional approach, give a variety of modes [22]. Averaging the two strongest transverse modes over symmetry directions (the A_{2T} and E_T modes at the Γ point) gives two modes at about 1100 and 450–480 cm^{-1}. A similar average over the longitudinal modes (A_{2L} and E_L) provides the LO/TO energetic splitting and, via the LST relation equations (4.4.39) and (4.4.38), a ratio of 3:4 for the oscillator strengths of the modes.

Al_2O_3. Two peaks are clearly visible in the FTIR spectrum, at 579 and 437 cm^{-1} for a film deposited at 600°C and annealed in oxygen at 1200°C for 60 min to ensure complete oxidation. The areas under the peaks yield a ratio 56:44 for their respective oscillator strengths. High-resolution energy loss spectroscopy in thin Al_2O_3 films provides two sets of modes, in the plane of the film and off plane [23]. The in-plane TO modes (at 578 and 390 cm^{-1}) are in fair agreement with our FTIR results. Chen and co-workers [17] see a variety of modes as a function of annealing conditions of the thin films, typically grouped into three LO bands around 400–430, 600–655, and 850–895 cm^{-1}. The low-energy band is attributed to excess (unoxidized) Al. The remaining two bands are in satisfactory agreement with the TO energies and oscillator strengths which can be derived from the FTIR spectrum and [23]. The selection reported in table 4.4.1 can be viewed as a satisfactory compromise.

AlN. For AlN we employ the theoretical results by Ruiz and co-workers [16] (an *ab initio* Hartree–Fock study of the hexagonal (wurtzite) phase of AlN) and by Gorczyca *et al* [24] (a muffin-tin analysis of both the wurtzite and zinc-blende phases). In the hexagonal phase two almost degenerate TO modes (at around 660 and 715 cm^{-1}, with oscillator strengths in a 59:41 ratio) originate from a single cubic TO mode at about $650\,cm^{-1}$, as a result of the doubling of the available optical modes moving from the cubic to the hexagonal structure. For the wurtzite phase, the modes reported in table 4.4.1 have been obtained by averaging the modes over the various allowed symmetries. The particular structure selected (wurtzite or zinc-blende) is largely immaterial as far as electron scattering is concerned, since the total oscillator strength carried by the almost-degenerate modes in the hexagonal phase corresponds approximately to the oscillator strength of the single mode in the cubic phase. Finally, we have neglected a weak low-energy mode at $250\,cm^{-1}$ seen in Raman spectra, reported in [16] and [24].

ZrO₂. Desgreniers and Lagarec [25] have published Raman spectra for polycrystalline (cotunnite phase) HfO_2 and ZrO_2. For the latter insulator, they have observed two TO modes at wavenumbers of about 390 (an oscillation of the Zr–O bond) and $100\,cm^{-1}$, with the corresponding LO modes at about 430 and $170\,cm^{-1}$. Raman spectra by Morell and co-workers [26] for Y- and Ca-stabilized ZrO_2 give two LO modes at 620 and $160\,cm^{-1}$, the former mode possibly influenced by the dopants. Lattice-dynamics calculations for cubic and tetragonal lattices [27] give three transverse modes at wavenumbers 164 (E_u), 339, and 467 (A_{2u}) cm^{-1}, with corresponding longitudinal frequencies at 232, 354, and $650\,cm^{-1}$. These values, in rough agreement with the dispersions calculated by Mirgorodsky *et al* [28] for cubic and tetragonal ZrO_2, show that the 339(TO)/354(LO) cm^{-1} A_{2u} mode is quite weak. In figure 4.4.4 we show FTIR spectra we have obtained for both tetragonal and monoclinic ZrO_2. The former exhibits a weak peak at $161\,cm^{-1}$, a weak shoulder at about $300\,cm^{-1}$, a signature of the weak $339\,cm^{-1}$ mode reported in [27, 28], and a stronger structure around $439\,cm^{-1}$. These are in good agreement with both the experimental [25] and the theoretical [27, 28] frequencies we have just discussed. The spectrum relative to tetragonal ZrO_2 is quite similar, while showing sharper peaks. Therefore, we have employed the LO energies reported in [27], but have lowered the TO frequency of the low-energy mode to account for the higher static dielectric constant ($\sim24\epsilon_0$) observed in thin films [29]. Note that in FTIR spectra oscillations at wavenumbers below about $100\,cm^{-1}$ remain elusive, as impurities (dopants) in the Si substrate render it opaque to the IR radiation.

HfO₂. The dominant high-frequency mode seen in the Raman spectra of [25] (TO at $395\,cm^{-1}$ with corresponding LO at $450\,cm^{-1}$)—a vibration of the Hf–O bond—is also seen in the FTIR spectrum of this material in the monoclinic phase, as a double peak around $337–409\,cm^{-1}$. It corresponds to one of the modes of the monoclinic structure also reported in [30]. The low-frequency mode seen in the spectrum of [25] (TO at $115\,cm^{-1}$, LO at $210\,cm^{-1}$), despite its strength, is not easily visible in IR spectroscopy, as explained above. Thus, we have embraced essentially unaltered the results of [25], using the values for the refractive index from [32] and [33], and ignoring the weaker modes at 235 and 256 cm^{-1} seen in the spectrum of figure 4.4.4.

ZrSiO$_4$. FTIR spectra of chemical-solution-deposited xSiO$_2$ + (100 − x) ZrO$_2$ (with x in per cent) films, such as those shown in figure 4.4.4 or in [18], usually show two TO bands (a strong one at 430–460 cm^{-1}, a weaker one around 810–930 cm^{-1}). A strong signal around 1080 cm^{-1} can be attributed to residual 'unconverted' SiO$_2$, since, as shown by the dashed line in the bottom spectrum of figure 4.4.4, its intensity decreases with decreasing x. Averaging over the frequencies of the A_{2u} and E_u modes calculated for bc tetragonal ZrSiO$_4$ by Rignanese and co-workers [34], the two strongest modes appear to be at approximately 310 and 940 cm^{-1} (with corresponding LO wavenumbers at 410 and 1060 cm^{-1}). The relative oscillator strengths are approximately in the ratio 73:27. The low-energy mode can be assigned to an oscillation of the Zr–O bond, the high-energy mode to a vibration of the Zr–O–Si bond. Using the dielectric constants reported in [34] and the index of refraction from [35], we obtain the values shown in table 4.4.1.

Results

Before presenting the results of our calculations including the full dielectric response of the gate and substrate electron plasmas and their coupling with the optical modes of the insulating layer, it is interesting to revisit figure 4.4.1. The results shown in this figure illustrate the effect of the polar field of the optical modes of the insulators on the mobility of the electrons in the inversion layer, as determined by the parameters discussed in the previous section and by the simple Wang–Mahan scattering strength given by equation (4.4.1). Let us recall that in these calculations, screening effects of the electrons in the inversion layer are ignored, as well as screening by the (infinitely far) gate.

SiO$_2$ (dots, solid line) is moderately affected by the presence of the SO modes, as a comparison with the curve calculated by neglecting them (open circles, dashed line) reveals. As stated in our introductory section, the stiffness of the Si–O bond results in a high-frequency LO mode which couples poorly with thermal electrons, and a mode of lower frequency with a small coupling constant, equation (4.4.1). Thus, SO modes have a very small effect, of about 5%, on the electron mobility. AlN has a somewhat larger dielectric constant (9.14) and in its wurtzite phase exhibits two almost degenerate modes at energies still larger than the thermal temperature. As for SiO$_2$, the electron mobility is only moderately affected by the presence of these phonons. But as soon as we consider materials with soft metal–oxygen bonds, the dielectric constant rises, and so does also the Wang–Mahan coupling. In addition, modes of lower energy—usually caused by oscillations of the oxygen ion in metal–O bonds—emerge and couple very effectively with thermal electrons. The insulators with the highest κ (ZrO$_2$ and HfO$_2$)

are negatively affected by the presence of low-energy modes and by the larger electron/SO-phonon coupling constant, exhibiting the lowest mobility over the entire range of electron sheet densities. The dependence of the mobility on n_s is almost completely dominated by scattering with the SO modes and by the electronic form-factor (overlap intergral) entering the expression for the relaxation rate associated with this process (see equation (79) of [7]). Al_2O_3 and $ZrSiO_4$ are intermediate materials, both as far as their mobility and their dielectric constant are concerned.

In figures 4.4.5 and 4.4.6 we now show the effective mobility accounting for the plasmon/TO-phonon coupling, and for the screening (or anti-screening) effects of the gate and substrate plasma. We show results relative to two SiO_2-equivalent insulator thicknesses: infinitely thick insulators, and 0.5 nm. In figure 4.4.5, we have employed the gate/insulator surface electron concentration to determine the bulk plasma frequency of the gate, as determined by the solution of the Poisson equation. In figure 4.4.6, instead, we have employed the Q-dependent average, $\langle N_g(Q) \rangle$, given by equation (4.4.36) above. We see a few major differences between figure 4.4.1 on the one hand, and figures 4.4.5 and 4.4.6 on the other: screening of the electron–SO

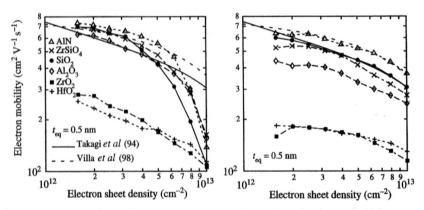

Figure 4.4.5. Calculated effective electron mobility in the inversion layer of MOS systems with various insulators for the 'SiO$_2$-equivalent' thicknesses of 0.5 (left) and 5.0 nm (right). The Kubo–Greenwood expression for the mobility has been used by employing an anisotropic momentum relaxation time accounting for scattering with bulk Si phonons, gate/insulator interface plasmons (SP), and SO insulator phonons. The effect of scattering with interface roughness at the substrate/insulator interface has been included by fitting the experimental mobility for the SiO$_2$-based system at an electron density of 10^{13} cm^{-2} and using Matthiessen's rule. The plasma response of the depleted poly-Si gate has been assumed to be given by the electron concentration at the gate/insulator interface determined by a numerical solution of the Poisson equation. In both frames the 'universal mobility curve' by Takagi *et al* [13] (thick solid line) and an empirical fit from [14] (thick dashed line) show the range of experimental value. All data refer to systems with negligible scattering with dopants in the substrate or charges in the insulator.

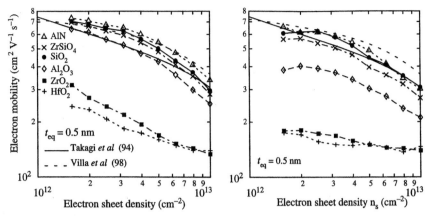

Figure 4.4.6. As in figure 4.4.5, but with the plasma response of the depleted poly-Si gate now computed using a wave-vector-dependent average electron concentration in the depletion layer of the gate.

scattering by the substrate electrons themselves results in a higher mobility for ZrO_2 and HfO_2 at large n_s: in these materials, the mobility is dominated by SO scattering; the relevant SO modes have relatively low energy, so their influence can be effectively screened by 2D plasmons of higher frequency. At lower sheet densities, however, the frequency of the interface excitations becomes larger than the plasma frequency of the 2DEG in the substrate, and anti-screening takes effect, boosting the scattering rate and lowering the mobility with respect to the unscreened value shown in figure 4.4.1. At the smaller t_{eq}, however, another interesting effect emerges: screening by the electrons in the gate. For sufficiently small electron sheet densities in the substrate, $n_s \leq 5 \times 10^{12}$ cm^{-2}, SiO_2, AlN, Al_2O_3, and $ZrSiO_4$ exhibit mobilities approaching the value limited only by scattering with Si phonons. Even the mobilities of ZrO_2 and HfO_2 improve at these small densities, as a result of the competition between gate screening and substrate anti-screening.

The difference between the results shown in figure 4.4.5 and those shown in figure 4.4.6 is not qualitative, but only quantitative. Since the choice of a Q-dependent average $\langle N_g(Q) \rangle$ results in a larger gate plasma frequency over most of the interesting range of values for Q, scattering with gate plasmons and the reduction of the SO frequency is less pronounced in the dispersions used in figure 4.4.6. This results in large effective mobilities at large values of n_s.

In figures 4.4.7 and 4.4.8 we show the dependence on the thickness of the insulator of the various components of the mobility. Note in figure 4.4.7 that the SP-limited mobility is quite large even for SiO_2 and at the smallest t_{eq} investigated, because both of Landau damping and of the large

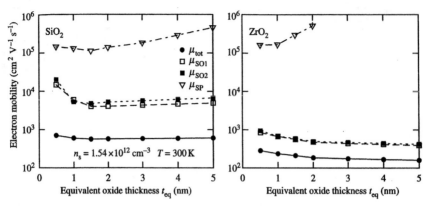

Figure 4.4.7. Calculated components of the total mobility (μ_{tot}) at an electron sheet density in the substrate of 1.54×10^{12} cm^{-2} as a function of the 'SiO$_2$-equivalent' thickness t_{eq} for the SiO$_2$- (left) and ZrO$_2$-based MOS structures (right). The curves labelled μ_{SO1} and μ_{SO2} refer to the components of the mobility limited by scattering with the TO1 and TO2 phonon-like components of the interface modes. The curve labelled μ_{SP} has been obtained from the total mobility and the SO-limited mobilities using Matthiessen's rule. The total mobility accounts for scattering with bulk Si phonons, SO- and SP-limited processes, but it does not account for scattering with interface roughness.

gate-plasma frequency at the small electron density in the inversion layer (and so larger electron density in the gate) assumed in the figure ($n_s = 1.54 \times 10^{12}$ cm^{-2}) [7]. Note also the screening effect of the gate plasma on the SO-limited mobility at small t_{eq}. In contrast, at the larger electron density employed in figure 4.4.8 ($n_s = 10^{13}$ cm^{-1}) we see all of the effects which our previous discussions, here and in [7], had anticipated: scattering with (gate) surface plasmons is negligible (i.e. small enough not to contribute significantly to the total mobility) in all materials, except

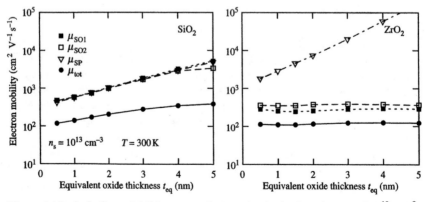

Figure 4.4.8. As in figure 4.4.7, but at an electron density in the substrate of 10^{13} cm^{-2}.

obviously SiO_2. In contrast, this 'advantage' of the high-κ materials is unfortunately more than compensated by a much stronger scattering with the TO components of the interface excitation. The SO-limited mobility decreases at small t_{eq}—as also evident in figure 4.4.5—because of the anti-screening effect of a strongly depleted gate; when using the gate/insulator interface electron concentration, the strongly depleted gate will exhibit a lower plasma frequency, and so it will be unable to screen the SO component of the scattering field, actually anti-screening it. Finally, in figure 4.4.9 we show the dependence of the SO-limited electron mobility on the static dielectric constant of the insulator. We have chosen a relatively small value for the electron concentration, in order to minimize the effect of scattering with surface roughness and with the gate-plasmon component of the interface excitations. Thus, the SO-limited component of the mobility is the major correction to the Si-phonon-limited component in the absence of Coulomb scattering with dopants and insulator charges. Note how the mobility decreases monotonically as κ increases, thanks to the softer oxygen bonds. AlN is indeed the single exception, thanks to the higher energy of the nitrogen-related optical phonons.

Effect of a silicon dioxide interfacial layer

In this final section, we discuss the effect caused by the presence of a thin layer of SiO_2 between the Si substrate and the high-κ dielectric on the electron mobility in the Si inversion layer. In particular, we have in mind the *beneficial* effect of the interfacial layer in MOS systems based on materials (such as HfO_2 or ZrO_2) exhibiting a large ionic polarizability, so

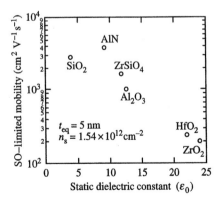

Figure 4.4.9. Calculated SO-limited component of the electron mobility in the inversion layer of MOS systems with various insulators at an electron density of $1.54 \times 10^{12}\,cm^{-2}$ as a function of the static dielectric constant of the insulator.

that removing the high-κ layer farther away from the Si substrate should reduce the strong interaction with the SO modes of the high-κ dielectric. Therefore, the formation of a thin SiO_2 layer may be not only hard to avoid during the growth/deposition/annealing of the high-κ insulator, but also desirable. To formulate more precisely our expectations, note that the length-scale relevant for the calculation of the electron mobility is the Fermi wavelength of the 2DEG, $\lambda_F \sim K_F^{-1} \sim n_s^{-1/2}$. Since the scattering potential decays with increasing distance z from the high-κ insulator as $\exp(-Qz)$, the effects of an SiO_2 interfacial layer of thickness t_{ox} will be: (a) turning on the interaction with SiO_2 SO modes, which we have seen is quite small, and (b) reducing the scattering strength of the high-κ SO modes by a factor $\sim \exp(-2K_F t_{ox})$. At small electron sheet densities n_s, the small value of the Fermi wave vector means that unreasonably thick oxides are required in order to boost the electron mobility. Not so at large n_s (and so at large K_F), in which case even a 0.5–1.0 nm thick SiO_2 layer can have a significant beneficial effect.

The analysis of the coupled plasmon–phonon modes of the full Si-gate/high-κ/SiO_2/Si system is quite cumbersome. Following a trivial generalization of the procedure described in the section 'Dispersion', the secular equation—whose solutions yield the dispersion of the modes—takes the form of a 16th-degree algebraic equation in ω^2. Even using the long-wavelength approximations for the dielectric functions, the problem is extremely cumbersome, the 16 positive solutions representing the dispersion of the 16 coupled modes resulting from two TO-like modes in each insulating film, and 12 surface modes (at large Q identifiable as groups of four modes mainly localized at each one of the three interfaces).

Here our aim is just to investigate qualitatively the effects caused by the interfacial layer. Thus, we reduce the complexity of the problem by making the following approximations.

1. A comparison between figure 4.4.1 and figures 4.4.5 or 4.4.6 shows that the 'infinitely thick insulator limit' captures the most important qualitative (and even quantitative) aspects of the problem. Thus, it seems appropriate to consider the simpler infinitely thick high-κ/SiO_2/Si system and ignore electronic screening effects. Thus, the dispersion of the modes is given by the solutions of the secular equation

$$\epsilon_{ox}(\omega)^2 + \epsilon_{ox}(\omega)[\epsilon_\kappa(\omega) + \epsilon_{Si}^\infty] \coth(Qt_{ox}) + \epsilon_{ox}(\omega)\epsilon_{Si}^\infty = 0. \quad (4.4.41)$$

This is exactly equation (4.4.6) with the role of the gate now played by the high-κ dielectric and with $\epsilon_s(Q,\omega)$ replaced by ϵ_{Si}^∞.

2. Low-energy SO modes are most important in determining the electron mobility. Thus, we approximate the response of the two insulators considering only their low-energy TO phonons and employ dielectric functions of the form:

$$\epsilon_{\text{ox}}(\omega) = \epsilon_{\text{ox}}^i \frac{\Omega_{\text{LO}}^2 - \omega^2}{\Omega_{\text{TO}}^2 - \omega^2}, \quad \epsilon_\kappa(\omega) = \epsilon_\kappa^i \frac{\omega_{\text{LO}}^2 - \omega^2}{\omega_{\text{TO}}^2 - \omega^2}, \tag{4.4.42}$$

where $\Omega_{\text{LO}} = (\epsilon_{\text{ox}}^0/\epsilon_{\text{ox}}^i)^{1/2}\Omega_{\text{TO}}$ and $\omega_{\text{LO}} = (\epsilon_\kappa^0/\epsilon_\kappa^i)^{1/2}\omega_{\text{TO}}$ are the longitudinal frequencies of the low-energy optical modes of the SiO_2 and high-κ layers, respectively.

3. Since we are most interested in understanding how 'low-mobility' materials behave in the presence of the interfacial oxide layer, and since these materials usually exhibit very soft optical modes, we typically have $\omega_{\text{TO}} \ll \Omega_{\text{TO}}$, as seen in table 4.4.1 for HfO_2 and ZrO_2. Therefore, the SiO_2 and high-κ modes become largely decoupled. In the fully decoupled limit, excellent approximations of the three solutions of equation (4.4.41) are given by:

$$\omega_Q^{(\kappa)} \approx \left(\frac{\epsilon_\kappa^0 + \Delta_Q^{(\kappa)}}{\epsilon_\kappa^i + \Delta_Q^{(\kappa)}} \right)^{1/2} \omega_{\text{TO}}, \tag{4.4.43}$$

where $\Delta_Q^{(\kappa)} = \epsilon_{\text{ox}}^0[\epsilon_{\text{ox}}^0 + \epsilon_{\text{Si}}^\infty \coth(Qt_{\text{ox}})]/[\epsilon_{\text{Si}}^\infty + \epsilon_{\text{ox}}^0 \coth(Qt_{\text{ox}})]$, and

$$\omega_Q^{(\pm)} \approx \left(\frac{\epsilon_{\text{ox}}^0 + \Delta_Q^{(\pm)}}{\epsilon_{\text{ox}}^i + \Delta_Q^{(\pm)}} \right)^{1/2} \Omega_{\text{TO}}, \tag{4.4.44}$$

where $2\Delta_Q^{(\pm)} = (\epsilon_\kappa^i + \epsilon_{\text{Si}}^\infty)\coth(Qt_{\text{ox}}) \pm \{[(\epsilon_\kappa^i + \epsilon_{\text{Si}}^\infty)\coth(Qt_{\text{ox}})]^2 - 4\epsilon_{\text{Si}}^\infty\epsilon_\kappa^i\}^{1/2}$. The solution $\omega_Q^{(\kappa)}$ represents the SO mode associated with the TO phonon of the high-κ film. For small Qt_{ox} the solution $\omega_Q^{(+)}$ approaches Ω_{TO}, thus being essentially a bulk SiO_2 TO mode, while in the limit of large Qt_{ox} it approaches the frequency of the SiO_2 mode at the 'far' high-κ/SiO_2 interface. In either limit, this mode couples only weakly with the electrons in the inversion layer and—while retaining it— could safely be neglected, similarly to the modes labelled $\omega_Q^{(5)}$ and $\omega_Q^{(6)}$ in the section 'Dispersion'. Finally, the solution $\omega_Q^{(-)}$ is the SO mode at the Si/SiO_2 interface associated with the SiO_2 TO mode. All modes, $\omega_Q^{(\kappa)}$ and $\omega_Q^{(\pm)}$, exhibit a very weak dependence on Q. We shall ignore their dispersion and employ their short-wavelength limits, $\omega_Q^{(\kappa)} \approx \omega_{\text{TO}}[(\epsilon_\kappa^0 + \epsilon_{\text{Si}}^\infty)/(\epsilon_\kappa^i + \epsilon_{\text{Si}}^\infty)]^{1/2}$, $\omega_Q^{(+)} \approx \Omega_{\text{TO}}[(\epsilon_{\text{ox}}^0 + \epsilon_\kappa^\infty)/(\epsilon_{\text{ox}}^i + \epsilon_\kappa^\infty)]^{1/2}$, and $\omega_Q^{(-)} \approx \Omega_{\text{TO}}[(\epsilon_{\text{ox}}^0 + \epsilon_{\text{Si}}^\infty)/(\epsilon_{\text{ox}}^i + \epsilon_{\text{Si}}^\infty)]^{1/2}$. The scattering strength associated with the high-κ mode can be obtained as in the section 'Scattering strength':

$$\Lambda^{(\kappa)}(Q) = \frac{\hbar\omega_Q^{(\kappa)}}{2} \left[\frac{1}{\epsilon_{\text{TOT}}^{\kappa,\text{hi}}(Q)} - \frac{1}{\epsilon_{\text{TOT}}^{\kappa,\text{lo}}(Q)} \right], \tag{4.4.45}$$

where

$$\epsilon_{\text{TOT}}^{\kappa,\text{hi/lo}}(Q) = \epsilon_\kappa^{i/0}\left[\frac{\epsilon_{\text{ox}}(\omega)+\epsilon_{\text{Si}}^\infty}{\epsilon_{\text{ox}}(\omega)-\epsilon_\kappa(\omega)}\right]^2 e^{2Qt_{\text{ox}}} + \epsilon_{\text{ox}}(\omega)\left\{\left[\frac{\epsilon_{\text{ox}}(\omega)+\epsilon_{\text{Si}}^\infty}{2\epsilon_{\text{ox}}(\omega)}\right]^2 (e^{2Qt_{\text{ox}}}-1)\right.$$

$$\left.+\left[\frac{\epsilon_{\text{ox}}(\omega)-\epsilon_{\text{Si}}^\infty}{2\epsilon_{\text{ox}}(\omega)}\right]^2 (1-e^{-2Qt_{\text{ox}}})\right\}+\epsilon_{\text{Si}}^\infty, \tag{4.4.46}$$

where the last expression is evaluated for $\omega = \omega_Q^{(\kappa)}$. Similarly, for the scattering strength of the modes $\omega_Q^{(\pm)}$ we get:

$$\Lambda^{(\pm)}(Q) = \frac{\hbar\omega_Q^{(\pm)}}{2}\left[\frac{1}{\epsilon_{\text{TOT}}^{\text{ox,hi}}(Q)} - \frac{1}{\epsilon_{\text{TOT}}^{\text{ox,lo}}(Q)}\right], \tag{4.4.47}$$

where

$$\epsilon_{\text{TOT}}^{\text{ox,hi/lo}}(Q) = \epsilon_\kappa(\omega)\left[\frac{\epsilon_{\text{ox}}(\omega)+\epsilon_{\text{Si}}^\infty}{\epsilon_{\text{ox}}(\omega)-\epsilon_\kappa(\omega)}\right]^2 e^{2Qt_{\text{ox}}} + \epsilon_{\text{ox}}^{i/0}\left\{\left[\frac{\epsilon_{\text{ox}}(\omega)+\epsilon_{\text{Si}}^\infty}{2\epsilon_{\text{ox}}(\omega)}\right]^2 (e^{2Qt_{\text{ox}}}-1)\right.$$

$$\left.+\left[\frac{\epsilon_{\text{ox}}(\omega)-\epsilon_{\text{Si}}^\infty}{2\epsilon_{\text{ox}}(\omega)}\right]^2 (1-e^{-2Qt_{\text{ox}}})\right\}+\epsilon_{\text{Si}}^\infty, \tag{4.4.48}$$

where $\epsilon_\kappa(\omega)$ and $\epsilon_{\text{ox}}(\omega)$ are evaluated at $\omega = \omega_Q^{(\pm)}$. Note that, as expected, $\Lambda^{(\kappa)}(Q) \sim e^{-2Qt_{\text{ox}}}$ while $\Lambda^{(\pm)}(Q) \to 0$ as $Qt_{\text{ox}} \to 0$.

4. Finally, we consider only one subband in the inversion layer. Therefore, the electron mobility can be obtained using equations (68) and (85) of [7] (the latter equation unfortunately being mistyped in that reference: the factor $b^6 e^{-2Qt_{\text{ox}}}$ in the integrand should be replaced by unity), by substituting $|\mathscr{A}_Q|^2$ with $[\Lambda^{(\kappa)}(Q)+\Lambda^{(+)}(Q)+\Lambda^{(-)}(Q)]/Q$ in equation (4.4.85).

We show in figure 4.4.10 our results for the HfO$_2$/SiO$_2$/Si system, similar results having been obtained also for the ZrO$_2$/SiO$_2$/Si system. The top frame shows the enhancement of the SO-limited mobility resulting from the presence of the SiO$_2$ interfacial layer, i.e. the ratio between the SO-limited mobility calculated for the HfO$_2$/SiO$_2$/Si system with an SiO$_2$ layer of thickness t_{ox}, and the mobility calculated using the same approximations for the 'pure' infinitely thick HfO$_2$/Si system. As expected, a 1.0 nm thick interfacial SiO$_2$ layer boosts the SO-limited mobility by a factor of more than four at the largest electron density considered (10^{13} cm^{-2}), but an SiO$_2$ layer as thick as 2.0 nm is required to obtain the same enhancement at the lowest density (10^{12} cm^{-2}). The bottom frame illustrates the dependence of the overall mobility (accounting also for scattering with Si phonons and surface

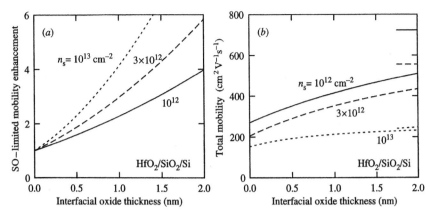

Figure 4.4.10. (*a*) Calculated enhancement of the SO-limited mobility in the inversion layer of an $HfO_2/SiO_2/Si$ system as a function of thickness of the interfacial SiO_2 layer for the three indicated values of the electron concentration in the inversion layer. An infinitely thick HfO_2 layer and only the low-energy TO modes in the insulators have been considered. For a given thickness of the interfacial oxide, the enhancement is larger at large electron densities, since the length-scale is set by the Fermi wavelength of the two-dimensional electron gas. (*b*) Calculated total electron mobility (including scattering with Si phonons and interfacial roughness), as in (*a*). The horizontal lines at the far right are the asymptotic limits of infinite oxide thickness (i.e. the mobility for the pure Si/SiO_2 interface). As in (*a*), in order to approach the mobility at the Si/SiO_2 interface, a thicker SiO_2 interfacial layer is required at lower electron sheet density.

roughness, included empirically using Matthiessen's rule). The horizontal lines at the far right indicated the asymptotic limit of an infinitely thick SiO_2 layer. Once more, at the largest density the mobility of the pure infinitely thick SiO_2/Si system is recovered quickly even for a thin interfacial layer. Not so at the lowest density: even in the presence of a 1.0 nm thick interfacial oxide layer, we remain almost a factor of two below the 'desired' SiO_2/Si limit.

Discussion and conclusions

The mobilities shown in figures 4.4.5 and 4.4.6 show a clear trend, emphasized in figure 4.4.9. It appears that the price one must pay for a higher κ is a reduced electron mobility. Among the materials we have investigated, metal oxides appear to be the worst, because of the soft modes caused by the oscillation of the oxygen ions, while AlN, and to some extent $ZrSiO_4$, show significant promise, albeit with the caveats we shall mention below.

Several sources of uncertainty affect the quantitative accuracy of our results: first and foremost is the choice of parameters for the insulators. This concerns both the overall quality of the parameters listed in table 4.4.1 for 'ideal' materials, as well as their applicability to 'real' insulators, almost invariably of a 'non-ideal' composition and structure. This has been already discussed above and will be emphasized again in the following paragraph. Second, the difference between the results shown in figure 4.4.5 and those shown in figure 4.4.6 clearly points to the importance of knowing accurately the electron density in the depletion layer of the gate. Again, this source of uncertainty affects not only the ideal calculations we have performed, but also the 'real world' complications we should expect; the poly-crystalline structure of the Si gate, for example, will undoubtedly result in an electron concentration exhibiting inhomogeneities not only in the z-direction, but also on the plane of the interface. Dopant segregation at grain boundaries is another possible cause of inhomogeneities. Nevertheless, our results stress at least qualitatively (but, hopefully, also quantitatively) the major role played by the gate both in screening the electron–SO interaction, and in triggering the gate/substrate Coulomb drag studied in [7]. In particular, we should note that the use of metal gates should be beneficial in both cases, by inducing a more complete screening of the surface-optical modes and by reducing the plasmon-mediated (long-range) component of the Coulomb drag. Finally, we have approximated the potential in the inversion layer as a triangular well. This approximation is likely to be satisfactory at the large electron sheet densities of interest, but it will cause additional inaccuracies in the opposite limit of low n_s.

It would be interesting to support the results of our calculation with experimental evidence. Unfortunately, we have already alluded to the many experimental and processing complications, which hamper a fair comparison. One could claim that, at least at present, the use of high-κ insulators has indeed resulted in disappointing performance in those few instances in which high-κ-based MOS transistors have been made in order to measure effective electron mobilities. Ragnarsson *et al* [36] have reported a peak mobility of $266 \, \text{cm}^2 \, \text{V}^{-1} \, \text{s}^{-1}$ for an nMOSFET fabricated using aluminium gates and Al_2O_3 films of 'equivalent' thickness $t_{eq} \approx 2.9 \, \text{nm}$. Similarly, Qi and co-workers [37] have measured low mobilities when using 1.6 nm thick $ZrSiO_4$ and 2.5 nm-thick ZrO_2 films, the former appearing to be about 40% better. Even lower values have been observed for other $ZrSiO_4$ films [29], and even for AlN films [38]. While some of these observations seem to agree quite well with our results (AlN being an exception we shall discuss shortly), it should be kept in mind that our calculations assume an ideal scenario: perfectly stoichiometric films with no charges, electron traps (and the associated hysteresis) or interfacial layers. In contrast, the structure or even the composition of the insulator itself is often unknown with the required accuracy. Charging effects, almost always seen, make it for a difficult, often impossible, accurate determination of the mobility (since an accurate determination of n_s becomes a hard task). Moreover, interfacial SiO_2 (or Si_3N_4) layers are almost always present. On the one hand, this changes substantially the theoretical picture, with the additional complication arising from the coupling of the optical modes of two insulators and the presence of an additional interface (in the section 'Effect of a silicon dioxide interfacial layer' we estimate these effects in a simplified situation). On the other hand, the structural property of the interface, and not SO scattering, may dominate the experimental situation. This is probably the case for AlN-based MOSFETs, in which the Si_3N_4 interfacial layer, with the well known associated electron traps and instabilities, may completely mask the effects we are trying to observe. In conclusion, it is fair to say that, at least at present, we only have 'suggestive' and 'circumstantial' experimental evidence supporting our results, but no quantitative conclusions should be drawn.

References

[1] Kotecki D E 1996 *Integr. Ferroelectrics* **16** 1
[2] Ohmi T, Sekine K, Kaihara R, Sato Y, Shirai Y and Hirayama M 1999 *Material Research Society, Symposium Proceedings* vol 567 eds H R Huff, C A Ritter, M L Green, G Lukovsky and T Hattori (Warrendale, PA: MRS) p 3
[3] Wang S Q and Mahan G D 1972 *Phys. Rev. B* **6** 4517
[4] Hess K and Vogl P 1979 *Solid State Commun.* **30** 807
[5] Moore B T and Ferry D K 1980 *J. Appl. Phys.* **51** 2603
[6] Fischetti M V and Laux S E 1993 *Phys. Rev. B* **48** 2244
[7] Fischetti M V 2000 *J. Appl. Phys.* **89** 1232
[8] Fischetti M V and Laux S E 2000 *J. Appl. Phys.* **89** 1205
[9] Kim M E, Das A and Senturia S D 1978 *Phys. Rev. B* **18** 6890
[10] Stern E A and Ferrel R A 1960 *Phys. Rev.* **120** 130
[11] Kittel C 1963 *Quantum Theory of Solids* (New York: Wiley)
[12] Ando T, Fowler A B and Stern F 1982 *Rev. Mod. Phys.* **54** 437
[13] Takagi S, Toriumi A, Iwase M and Tango H 1994 *IEEE Trans. Electron Device* **41** 2357
[14] Villa S, Lacaita A, Perron L M and Bez R 1998 *IEEE Trans. Electron Device* **45** 110
[15] See, for example, Balk P, Ewert S, Shmidtz S and Steffen A 1991 *J. Appl. Phys.* **69** 6510
[16] Ruiz E, Alvarez S and Alemany P 1994 *Phys. Rev. B* **49** 7115
[17] Chen P J, Colaianni M L and Yates J T 1990 *Phys. Rev. B* **41** 8025
[18] Neumayer D A and Cartier E 2001 *J. Appl. Phys.* **90** 1801
[19] Kleinman D A and Spitzer W G 1962 *Phys. Rev.* **125** 16
[20] Scott J F and Porto S P S 1967 *Phys. Rev.* **161** 903
[21] Galeener F L and Lucovsky G 1976 *Phys. Rev. Lett.* **37** 1474
[22] Umari P, Pasquarello A and Dal Corso A 2001 *Phys. Rev. B* **63** 94305
[23] Frederick B G, Apai G and Rhodin T N 1991 *Phys. Rev. B* **44** 1880
[24] Gorczyca I, Christensen N E, Peltzer y Blanca E L and Rodriguez C O 1995 *Phys. Rev. B* **51** 11936
[25] Desgreniers S and Lagarec K 1999 *Phys. Rev. B* **59** 8467
[26] Morell G, Katiyar R S, Torres D, Paje S E and Llopis J 1997 *J. Appl. Phys.* **81** 2830
[27] Pecharromán C, Ocaña M and Serna C J 1996 *J. Appl. Phys.* **80** 3479
[28] Mirgorodsky A P, Smirnov M B, Quintard P E and Merle-Méjan T 1995 *Phys. Rev. B* **52** 9111

[29] Callegari A C Private communication
[30] Philippi C M and Mazdiasni K S 1971 *J. Am. Ceram. Soc.* **54** 254
[31] Kang L, Lee B H, Qi W-J, Jeon Y, Nieh R, Gopalan S and Lee J C 2000 *IEEE Electron Dev. Lett.* **21** 181
[32] Alvisi M, Scaglione S, Martelli S, Rizzo A and Vasanelli L 1999 *Thin Solid Films* **354** 19
[33] Nishide T, Honda S, Matsuura M and Ide M 2000 *Thin Solid Films* **371** 61
[34] Rignanese G-M, Gonze X and Pasquarello A 2001 *Phys. Rev. B* **63** 104305
[35] Barrie J D and Fletcher E L 1996 *J. Vac. Sci. Technol. A* **14** 795
[36] Ragnarsson L-Å, Guha S, Bojarczuk N A, Cartier E, Fischetti M V, Rim K, Karasinski J *Electrical Characterization of Al₂O₃ n-Channel Metal–Oxide–Semiconductor Field-Effect Transistors with Aluminium Gates* 2001 unpublished
[37] Qi W-J *et al* 2000 *Symp. on VLSI Technol. Dig.* 40
[38] L-Å Ragnarsson Private communication

Chapter 4.5

Ab initio calculations of the structural, electronic and dynamical properties of high-κ dielectrics

Gian-Marco Rignanese, Xavier Gonze and Alfredo Pasquarello

Introduction

The accurate prediction of materials properties is one of the pivotal goals of computational condensed matter physics. In this framework, the density functional theory (DFT) has emerged as an extremely successful approach. The success of DFT not only encompasses standard bulk materials but also complex systems such as proteins and carbon nanotubes.

In the framework of the quest for an alternative high-κ material to conventional SiO_2 as the gate dielectric in MOS devices, first-principles calculations can provide insight into the nanoscopic behaviour of novel materials without requiring empirical data. This is particularly relevant for the early stages of research when relatively few experimental data are available. DFT is appropriate to study ground state properties of the electronic system. In this chapter, we focus on structural, vibrational and dielectric properties, which all relate to the ground state of the electronic system and are thus well described within DFT.

Despite its predictive accuracy, DFT calculations have one important limitation associated with their high computational cost, which limits both the length and time scales of the phenomena which can be modelled. With the most widespread DFT approach based on plane-wave basis sets and pseudopotentials, it is currently possible to treat systems containing up to hundreds of atoms. For our application to high-κ materials, it is important to note that materials containing transition-metal and first-row elements (e.g. oxygen) generally present an additional difficulty when treated with plane-wave basis sets. In fact, their valence wave functions are generally strongly localized around the nucleus and may require a large number of basis functions to be described accurately, thus further limiting the size of the system that can be investigated.

This chapter is dedicated to the first-principles study of the Hf–Si–O and Zr–Si–O systems which have drawn considerable attention as alternative high-κ materials. Indeed, the metal oxides HfO_2 and ZrO_2 as well as the silicates $HfSiO_4$ and $ZrSiO_4$ in the form of amorphous films are stable in direct contact with Si up to high temperature, which is highly desirable to avoid the degradation of the interface properties by formation of a low-κ interfacial layer. In fact, the Hf–Si–O and Zr–Si–O phase diagrams present a large phase field of stable silicates and the static permittivity ϵ_0 increases continuously with the amount of Hf and Zr incorporated into the silicate film. In order to be able to control this process, it is highly desirable to develop an understanding of how the permittivities of Hf and Zr silicates are affected by the underlying nanoscopic structure.

This chapter is organized as follows. In the 'Theoretical background' section, we present a brief summary of the DFT and the equations related to the properties that will be presented in the subsequent sections. We also give some technical details about the calculations. The 'Crystalline oxides' section is dedicated to the study of structural, vibrational and dielectric properties of hafnia (HfO_2) and zirconia (ZrO_2). Both the cubic and tetragonal phases are considered. The differences and the analogies between the two phases and between hafnia and zirconia are highlighted. In the 'Crystalline silicates' section, we present structural and electronic properties of hafnon ($HfSiO_4$) and zircon ($ZrSiO_4$). We compare their Born effective charge tensors and discuss the phonon frequencies at the Γ point. The dielectric permittivity tensors are analysed in detail. The fifth section is devoted to the study of amorphous silicates. We introduce a scheme which relates the dielectric constants to the local bonding of Si and M = (Hf, Zr) atoms. This scheme is based on the definition of parameters characteristic of the basic structural units (SUs) formed by Si and M = (Hf, Zr) atoms and their nearest neighbours, and allows us to avoid heavy large-scale calculations, which are beyond current computational capabilities. Applied to amorphous Zr silicates, our scheme provides a good description of the measured dielectric constants, both of the optical and the static ones. Finally, in the last section, we give the conclusions.

Theoretical background

Ground state properties

The main idea of DFT is to describe an interacting system of fermions through the electron density rather than through the many-body wave function. For N electrons in a solid obeying the Pauli principle and interacting via the Coulomb potential, this means that the basic variable of the system depends only on three (the spatial coordinates x, y and z) rather than $3N$ degrees of freedom.

In 1964, Hohenberg and Kohn [1] demonstrated that the ground state of the electron system is defined by the electron density which minimizes the total energy. Furthermore, they showed that all the other ground state properties of the system (e.g. the lattice constant, the cohesive energy, etc) are functionals of the ground state electron density. This means that, once the ground state electron density is known, all the other ground state properties follow (in principle, at least).

In 1965, Kohn and Sham [2] showed that this variational approach leads to equations of a very simple form:

$$(T + v_{KS}[n])|\psi_\alpha\rangle = (T + v_{ext} + v_H[n] + v_{xc}[n])|\psi_\alpha\rangle = \epsilon_\alpha|\psi_\alpha\rangle, \qquad (4.5.1)$$

nowadays known as the Kohn–Sham equations. These effectively single-particle eigenvalue equations are similar in form to the time-independent Schroedinger equation, T being the kinetic energy operator and v_{KS} the potential experienced by the electrons. The latter is generally split into a part which is external to the electronic system v_{ext}, for instance the electron–ion interaction, and a part describing the electron–electron interactions. For convenience, the latter is further split into the Hartree potential v_H and the exchange–dcorrelation potential v_{xc}, whose form is, in general, unknown.

The ground state energy of the electronic system is given by:

$$E_{el}\{\psi_\alpha\} = \sum_\alpha^{occ} \langle\psi_\alpha|T + v_{ext}|\psi_\alpha\rangle + E_{Hxc}[n], \qquad (4.5.2)$$

where E_{Hxc} is the Hartree and exchange–correlation energy functional of the electron density $n(\mathbf{r})$ with $\delta E_{Hxc}/\delta n = v_H + v_{xc}$, and the summation runs over the occupied states α. The occupied Kohn–Sham orbitals are subject to the orthonormalization constraints,

$$\int \psi_\alpha^*(\mathbf{r})\psi_\beta(\mathbf{r})\,d\mathbf{r} = \langle\psi_\alpha|\psi_\beta\rangle = \delta_{\alpha\beta}, \qquad (4.5.3)$$

where α and β label occupied states. The density is generated from

$$n(\mathbf{r}) = \sum_\alpha^{occ} \psi_\alpha^*(\mathbf{r})\psi_\alpha(\mathbf{r}). \qquad (4.5.4)$$

Nowadays, DFT is considered as the method of choice for simulating solids and molecules from first principles. For a review of DFT applications, we recommend the article of Pickett [3]. The interested reader may find more technical details about DFT in the review article of Payne *et al* [4].

Response properties

In this brief presentation, we focus on the *responses* of solid systems to two classes of perturbations: (a) collective displacements of atoms characterized by a wavevector $\mathbf{q}$ (phonons) and (b) homogeneous static electric fields. These responses can also be calculated within DFT using various methods, as reviewed by Baroni *et al* [5].

The method that we adopt in the calculations presented here relies on a variational approach to density-functional perturbation theory: a complete description can be found in [6, 7]. The first paper [6] describes the computation of the first-order derivatives of the wave functions, density and self-consistent potential with respect to the perturbations mentioned above; while the second paper [7] is dedicated to the second-order derivatives. We adopt the same notations as in those references to introduce the properties that are studied in the subsequent sections. In particular, κ and α run over the atoms in the unit cell and over the three Cartesian directions, respectively; $\tau_{\kappa\alpha}$ denote the equilibrium positions.

The squares of the phonon frequencies $\omega_{m\mathbf{q}}^2$ at $\mathbf{q}$ are obtained as eigenvalues of the dynamical matrix $\tilde{D}_{\kappa\alpha,\kappa'\beta}(\mathbf{q})$, or as solutions of the following generalized eigenvalue problem:

$$\sum_{\kappa'\beta}\tilde{C}_{\kappa\alpha,\kappa'\beta}(\mathbf{q})U_{m\mathbf{q}}(\kappa'\beta) = M_\kappa\omega_{m\mathbf{q}}^2 U_{m\mathbf{q}}(\kappa\alpha), \tag{4.5.5}$$

where M_κ is the mass of the ion κ, and the matrix $\tilde{C}$ is connected to the dynamical matrix $\tilde{D}$ through:

$$\tilde{D}_{\kappa\alpha,\kappa'\beta}(\mathbf{q}) = \tilde{C}_{\kappa\alpha,\kappa'\beta}(\mathbf{q})/(M_\kappa M_{\kappa'})^{1/2}. \tag{4.5.6}$$

The matrix $\tilde{C}_{\kappa\alpha,\kappa'\beta}(\mathbf{q})$ is the Fourier transform of the matrix of the inter-atomic force constants. It is related to the second-order derivative of the total energy with respect to collective atomic displacements [7].

The limit $\mathbf{q}\to\mathbf{0}$ must be performed carefully [7] by the separate treatment of the macroscopic electric field associated with phonons in this limit. A bare dynamical matrix at $\mathbf{q}=\mathbf{0}$ is first computed, then a non-analytical part is added, in order to reproduce correctly the $\mathbf{q}\to\mathbf{0}$ behaviour along different directions:

$$\tilde{C}_{\kappa\alpha,\kappa'\beta}(\mathbf{q}\to\mathbf{0}) = \tilde{C}_{\kappa\alpha,\kappa'\beta}(\mathbf{q}=\mathbf{0}) + \tilde{C}_{\kappa\alpha,\kappa'\beta}^{\text{NA}}(\mathbf{q}\to\mathbf{0}). \tag{4.5.7}$$

The expression of the non-analytical part will be given later on in this section.

For insulators, the dielectric permittivity tensor is the coefficient of proportionality between the macroscopic displacement field and the macroscopic electric field, in the linear regime:

$$\mathcal{D}_{\text{mac},\alpha} = \sum_\beta \epsilon_{\alpha\beta} \mathcal{E}_{\text{mac},\beta}. \tag{4.5.8}$$

It can be obtained as

$$\epsilon_{\alpha\beta} = \frac{\partial \mathcal{D}_{\text{mac},\alpha}}{\partial \mathcal{E}_{\text{mac},\beta}} = \delta_{\alpha\beta} + 4\pi \frac{\partial \mathcal{P}_{\text{mac},\alpha}}{\partial \mathcal{E}_{\text{mac},\beta}}. \tag{4.5.9}$$

In general, the displacement $\mathcal{D}_{\text{mac}}$, or the polarization $\mathcal{P}_{\text{mac}}$, will include contributions from ionic displacements. In the presence of an applied field of high frequency, the contribution to the dielectric permittivity tensor resulting from the electronic polarization, usually denoted $\epsilon^\infty_{\alpha\beta}$, dominates. This ion-clamped dielectric permittivity tensor is related to the second-order derivatives of the energy with respect to the macroscopic electric field [7]. Later on in this section, we will consider the supplementary contributions to the polarization coming from the ionic displacements.

For insulators, the Born effective charge tensor $Z^*_{\kappa,\beta\alpha}$ is defined as the proportionality coefficient relating, at linear order, the polarization per unit cell, created along the direction β, and the displacement along the direction α of the atoms belonging to the sublattice κ, under the condition of zero electric field. The same coefficient also describes the linear relation between the force on an atom and the macroscopic electric field:

$$Z^*_{\kappa,\beta\alpha} = \Omega_0 \frac{\partial \mathcal{P}_{\text{mac},\beta}}{\partial \tau_{\kappa\alpha}(\mathbf{q}=\mathbf{0})} = \frac{\partial F_{\kappa,\alpha}}{\partial \mathcal{E}_\beta}, \tag{4.5.10}$$

where Ω_0 is the volume of the primitive unit cell. The Born effective charge tensors are connected to the mixed second-order derivative of the energy with respect to atomic displacements and macroscopic electric field [7].

Finally, we discuss two phenomena that arise from the same basic mechanism: the coupling between the macroscopic electric field and the polarization associated with the $\mathbf{q} \to \mathbf{0}$ atomic displacements. The Born effective charges are involved in both cases.

First, in the computation of the low-frequency (infrared) dielectric permittivity tensor, one has to include the response of the ions. Their motion will be triggered by the force due to the electric field, while their polarization will be created by their displacement.

At the lowest order of approximation in the theory, the macroscopic frequency-dependent dielectric permittivity tensor $\epsilon_{\alpha\beta}(\omega)$ is calculated as follows:

$$\epsilon_{\alpha\beta}(\omega) = \epsilon^\infty_{\alpha\beta} + \frac{4\pi}{\Omega_0} \sum_m \frac{S_{m,\alpha\beta}}{\omega_m^2 - \omega^2}, \tag{4.5.11}$$

where the mode-oscillator strength $S_{m,\alpha\beta}$ is defined as:

$$S_{m,\alpha\beta} = \left(\sum_{\kappa\alpha'} Z^*_{\kappa,\alpha\alpha'} U^*_{m\mathbf{q}=0}(\kappa\alpha') \right) \left(\sum_{\kappa'\beta'} Z^*_{\kappa',\beta\beta'} U_{m\mathbf{q}=0}(\kappa'\beta') \right). \qquad (4.5.12)$$

A damping factor might be added to equation (4.5.11) in order to take into account anharmonic effects, and fit frequency-dependent experimental data. For our purpose, such a damping factor can be ignored.

At zero frequency, the static dielectric permittivity tensor is usually denoted $\epsilon^0_{\alpha\beta}$; it is obtained by:

$$\epsilon^0_{\alpha\beta} = \epsilon^\infty_{\alpha\beta} + \sum_m \Delta\epsilon_{m,\alpha\beta} = \epsilon^\infty_{\alpha\beta} + \frac{4\pi}{\Omega_0} \sum_m \frac{S_{m,\alpha\beta}}{\omega^2_m}. \qquad (4.5.13)$$

In parallel to this decomposition of the static dielectric tensor, one can define a mode-effective charge vector:

$$Z^*_{m,\alpha} = \frac{\sum_{\kappa\beta} Z^*_{\kappa,\alpha\beta} U_{m\mathbf{q}=0}(\kappa\beta)}{\left(\sum_{\kappa\beta} U^*_{m\mathbf{q}=0}(\kappa\beta) U_{m\mathbf{q}=0}(\kappa\beta) \right)^{1/2}}. \qquad (4.5.14)$$

This vector is related to the global polarization resulting from the atomic displacements for a given phonon mode m. The non-zero components indicate the directions in which the mode is infrared active.

Second, in the computation of phonons in the long-wavelength limit, a macroscopic polarization and electric field can be associated with the atomic displacements. At the simplest level of theory, the phonon eigenfrequencies then depend on the direction along which the limit is taken as well as on the polarization of the phonon. This gives birth to the LO–TO splitting, and to the Lyddane–Sachs–Teller relation [7].

For insulators, the non-analytical, direction-dependent part of the dynamical matrix $\tilde{C}^{NA}_{\kappa\alpha,\kappa'\beta}(\mathbf{q}\to 0)$ is given by:

$$\tilde{C}^{NA}_{\kappa\alpha,\kappa'\beta}(\mathbf{q}\to 0) = \frac{4\pi}{\Omega_0} \frac{\left(\sum_\gamma q_\gamma Z^*_{\kappa,\gamma\alpha} \right) \left(\sum_{\gamma'} q_{\gamma'} Z^*_{\kappa',\gamma'\beta} \right)}{\sum_{\alpha\beta} q_\alpha \epsilon^\infty_{\alpha\beta} q_\beta}. \qquad (4.5.15)$$

Hence, once the dynamical matrix at $\mathbf{q}=0$ as well as $\epsilon^\infty_{\alpha\beta}$ and the Born effective charge tensors are available, it is possible to compute the LO–TO splitting of phonon frequencies at $\mathbf{q}=0$.

Technical details

All our calculations are performed using the ABINIT package, developed by the authors and collaborators [8]. The exchange–correlation energy is evaluated within the local density approximation (LDA) to DFT, using the Perdew–Wang parametrization [9] of Ceperley–Alder electron-gas data [10].

Only valence electrons are explicitly considered using pseudopotentials to account for core–valence interactions. We use norm-conserving pseudopotentials [11, 12] with Hf(5s, 5p, 5d, 6s), Zr(4s, 4p, 4d, 5s), Si(3s, 3p) and O(2s, 2p) levels treated as valence states.

The wave functions are expanded in plane waves up to a kinetic energy cut-off of 30 Ha. The chosen kinetic energy cut-off and k-point sampling of the Brillouin zone ensure convergence of all the calculated properties.

Crystalline oxides

Introduction

Hafnia (HfO_2) and zirconia (ZrO_2) have many similar physical and chemical properties. These similarities result from the structural resemblance between the two oxides, which can in turn be explained by the chemical homology of Hf and Zr.

The electron configuration of hafnium is $4f^{14}5d^26s^2$ while it is $4d^25s^2$ for zirconium. In the periodic table, the inner transition (rare-earth) elements immediately preceding Hf add electrons to the inner 4f shell from element no 58, cerium, to no 71, lutetium. Because the nuclear charge increases while no additional outer shells are filled, there is a contraction in the atomic size. Consequently, element no 72, hafnium, has a slightly smaller atomic size than element no 40, zirconium, the group IVB element in the preceding. This results in the so-called lanthanide contraction. The atomic radii of Hf and Zr are close to each other: 1.44 and 1.45 Å, respectively [13]. They also have quasi-identical ionic radii (M^{4+}), 0.78 for Hf and 0.79 Å for Zr, respectively [14]. Their electronegativity values are almost equal, 1.23 for hafnium and 1.22 for zirconium [15]. All this explains the origin of the similarity between HfO_2 and ZrO_2.

Hafnia and zirconia undergo polymorphic transformations with changes in external parameters. At high temperature, the compounds are highly defective and their structure is fluorite type ($Fm\bar{3}m$). The decreasing temperature induces a cubic to tetragonal ($P4_2/nmc$) phase transition (c–t) at about 2650°C for HfO_2 [16] and about 2350°C for ZrO_2 [17]. This transition is followed by a tetragonal to monoclinic ($P2_1/c$) martensitic phase transition (t–m) at about 1650°C for hafnia [18] and about 1150°C for zirconia [19].

The crystalline structure may also depend on the presence of dopants (MgO, CaO, Y_2O_3). For instance, an addition of 3% (wt) Y_2O_3 stabilizes the tetragonal form of ZrO_2 at room temperature [20]. Finally, the contribution of the surface energy also influences the structural stability of nanocrystallites. A crystallite of 30 nm or less is now believed to stabilize the tetragonal form of ZrO_2 at room temperature [21], while for HfO_2 the critical size is about 10 nm [22].

The structural, electronic and dynamical properties of hafnia and zirconia have been the object of several first-principles studies [23–27]. For the sake of brevity, our results are only presented here for the cubic and tetragonal phases. For the monoclinic phase, we refer the interested reader to the work of Zhao and Vanderbilt [26, 27].

Structural properties

The cubic and tetragonal crystalline structures of HfO_2 and ZrO_2 are illustrated in figure 4.5.1. The cubic phase takes the fluorite structure (space group $Fm\bar{3}m$), which is fully characterized by a single lattice constant a. The M = (Hf, Zr) atoms are in a face-centred-cubic structure and the O atoms occupy the tetrahedral interstitial sites associated with this fcc lattice. The unit cell contains one formula unit of MO_2 with M = (Hf, Zr). The tetragonal phase (space group $P4_2/nmc$) can be viewed as a distortion of the cubic structure obtained by displacing alternating pairs of O atoms up and down by an amount Δz along the z direction, as shown in figure 4.5.1, and by applying a tetragonal strain. The resulting primitive cell is doubled compared to the cubic phase, including two formula units of MO_2. The tetragonal structure is completely specified by two lattice constants (a and c) and the dimensionless ratio $d_z = \Delta z/c$ describing the displacement of the O atoms. The cubic phase can be considered as a special case of the tetragonal structure with $d_z = 0$ and $c/a = 1$ (if the primitive cell is used for the tetragonal phase, $c/a = \sqrt{2}$).

In table 4.5.1, our calculated structural parameters for the cubic and tetragonal phases of HfO_2 and ZrO_2 are compared with the experimental

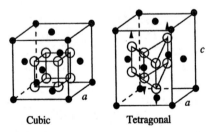

Cubic   Tetragonal

Figure 4.5.1. Structures of the cubic and tetragonal phases of HfO_2 and ZrO_2. The O atoms are in white while M = (Hf, Zr) atoms are in grey. For clarity, the Zr–O bonds are not indicated. For the tetragonal phase, the arrows indicate the displacements of oxygen pairs relative to the cubic structure.

Table 4.5.1. Structural parameters for the cubic (C) and tetragonal (T) phases of HfO_2 and ZrO_2.

	HfO_2		ZrO_2	
	Theoretical	Experimental	Theoretical	Experimental
C				
a	5.11	5.08	5.01	5.09
Volume	33.36	32.77	31.44	32.97
$d(M-O)$	2.21	2.20	2.17	2.20
T				
a	5.11	5.15	5.02	5.05
c	5.17	5.29	5.09	5.18
d_z	0.0310	–	0.0400	0.0574
Volume	33.75	35.08	32.07	33.04
$d(M-O)$	2.13	–	2.07	2.05
	2.32	–	2.31	2.39

The length unit is the Å. The experimental results for HfO_2 are taken from [28], while those for ZrO_2 are obtained by extrapolation to zero temperature using the thermal expansion data of Aldebert and Traverse [19].

values [19, 28]. The agreement is very good: the errors on the lattice constants and the volumes are smaller than 2%, as is typical for LDA calculations. The largest discrepancy is for d_z in t-ZrO_2 (the small displacement from the cubic phase localization): our value is about 30% smaller than the experimental data, but it is in excellent agreement with the results of other first-principles calculations. The discrepancy with experiment is probably due to the fact that our calculations are performed at zero temperature.

The bond lengths are also found to be in good agreement with experimental data, as indicated in table 4.5.1. In both the cubic and tetragonal structures, the M = (Hf, Zr) atoms are eightfold coordinated and the O atoms are fourfold coordinated. However, in the tetragonal phase, four O atoms are closer to the M = (Hf, Zr) atom than the other four (8 and 10% difference in the Hf–O and Zr–O bond lengths, respectively).

Born effective charge tensors

In table 4.5.2, we report non-vanishing components of the calculated Born effective charge tensors for M = Hf, Zr) and O atoms in the cubic and tetragonal phases of hafnia and zirconia. Due to the symmetry of the cubic phase, the Born effective charge tensors of M = (Hf, Zr) and O atoms are diagonal and isotropic. The value of Z^* is anomalously large for M = (Hf, Zr)

Table 4.5.2. Non-vanishing components of the calculated Born effective charge tensors for M = (Hf, Zr) and O atoms in the cubic (C) and tetragonal (T) phases of HfO$_2$ and ZrO$_2$.

Atom	HfO$_2$			ZrO$_2$		
C						
M	(+5.58	+5.58	+5.58)	(+5.74	+5.74	+5.74)
O	(−2.79	−2.79	−2.79)	(−2.87	−2.87	−2.87)
T						
M	(+5.57	+5.57	+5.24)	(+5.74	+5.74	+5.14)
O	(−3.22	−2.35	−2.62)	(−3.51	−2.24	−2.57)

The tensors are diagonal and only the principal elements are given.

atoms compared to the nominal ionic charge Z = 4. This behaviour has also been observed in the case of PbZrO$_3$ [29], and indicates a mixed covalent–ionic bonding. In the tetragonal structure, the symmetry imposes that the Born effective charge tensor of M = (Hf, Zr) atoms is diagonal and only has two independent components: parallel ($Z_\parallel^*$) and perpendicular ($Z_\perp^*$) to the c-axis. The value of $Z_\perp^*$ is identical to the one calculated for the cubic phase, while $Z_\parallel^*$ is 6 and 10% smaller for HfO$_2$ and ZrO$_2$, respectively. The Born effective charge tensor of O atoms is also diagonal, but with three independent components. It is quite anisotropic compared to the cubic phase. Such a strong anisotropy of the Born effective charge tensor for O atoms has already been observed in SiO$_2$-stishovite [30] and TiO$_2$-rutile [31].

It is interesting to note that the Born effective charges of c-HfO$_2$ are about 3% smaller (in absolute value) than those of c-ZrO$_2$. The comparison between the Z*-values of t-HfO$_2$ and t-ZrO$_2$ is also very instructive. In directions perpendicular to the c-axis, the Born effective charges of the M=(Hf, Zr) atoms compare in the same way as for the cubic phase, while the comparison for the Born effective charges of O atoms shows an anisotropy in t-ZrO$_2$ stronger than that in t-HfO$_2$ by about 30% (the values of $Z_\perp^*$ for t-HfO$_2$ are comprised between those of t-ZrO$_2$). In the direction parallel to the c-axis, the Born effective charges in t-HfO$_2$ are larger than in t-ZrO$_2$ by about 2%, showing an opposite trend with respect to the comparison for the cubic phase. The slightly different behaviour between hafnia and zirconia can be related to the differences in the inner electronic shells between Hf and Zr, which lead to different polarizabilities. This discussion shows the interest of analysing a dynamical property such as the Born effective charge tensors, which is able to highlight subtle differences between two very similar systems.

Phonon frequencies at the Γ point

The theoretical group analysis predicts the following irreducible representations of optical and acoustical zone-centre modes for the cubic phase:

$$\Gamma = \underbrace{F_{2g}}_{\text{Raman}} \oplus \underbrace{F_{1u}}_{\text{IR}} \oplus \underbrace{F_{1u}}_{\text{Acoustic}}$$

and for the tetragonal phase:

$$\Gamma = \underbrace{A_{1g} \oplus 2B_{1g} \oplus 3E_g}_{\text{Raman}} \oplus \underbrace{A_{2u} \oplus 2E_u}_{\text{IR}} \oplus \underbrace{A_{2u} \oplus E_u}_{\text{Acoustic}} \oplus \underbrace{B_{2u}}_{\text{Silent}}.$$

Because of the non-vanishing components of the Born effective charge tensors, the dipole–dipole interaction must be properly included in the calculation of the interatomic force constants [7, 32, 33]. In particular, the dipole–dipole contribution is found to be responsible for the splitting at the Γ point between the longitudinal and transverse optic (LO and TO, respectively) modes F_{1u} in the cubic phase, and E_u (perpendicular to the c-axis) and A_{2u} (parallel to c-axis) in the tetragonal phase.

In the absence of experimental data for the cubic phase of both materials and for the tetragonal phase of hafnia, the following discussion will first focus on the tetragonal phase of zirconia. A comparison will then be made between the calculated results for the various phases of both materials.

In the experimental Raman spectra, six lines corresponding to the six active modes have been observed for pure t-ZrO$_2$ at high temperature [34, 35] and for samples stabilized by dopants [34, 36–38]. In the case of pure t-ZrO$_2$, the Raman spectra are found to be very similar except for a slight down-shift of the frequencies, which was attributed to the increase in lattice constant with dopant concentration and to temperature effects [34]. In the absence of t-ZrO$_2$ single crystals of good quality, a reliable assignment of those lines could not be made. However, the symmetry classification proposed by Feinberg and Perry [34] is widely used in the literature. It is reported in table 4.5.3, together with the measured phonon frequencies for their yttria-stabilized t-ZrO$_2$ sample. Theoretical studies have been performed using lattice-dynamical models, and phonon frequencies in relatively good agreement with the experimental values have been predicted [39–41]. Consequently, the symmetry assignments given by Feinberg and Perry [34] were criticized and a second set of assignments was proposed. In particular, Mirgorodsky *et al* [40, 41] argued that the A_{1g} mode should be at lower frequency to account for the change in the dynamical properties of ZrO$_2$. More recently, using crystallite size effects to stabilize the tetragonal phase, Bouvier and Lucazeau [42] obtained experimental Raman spectra of pure t-ZrO$_2$ at room temperature. They proposed a third assignment of the vibration modes on the basis of a linear chain model: compared to Feinberg

Table 4.5.3. Fundamental frequencies of the cubic (C) and tetragonal (T) phases of HfO_2 and ZrO_2 (in cm^{-1}) with their symmetry assignments.

	HfO$_2$	ZrO$_2$	
Mode	Theoretical	Theoretical	Experimental
C			
Raman			
F_{2g}	579	596	
Infrared			
F_{1u} (TO)	285	280	
F_{1u} (LO)	630	677	
T			
Raman			
A_{1g}	218	259	266 (E_g)	269 (E_g)
$B_{1g}(1)$	244	331	326	319
$B_{1g}(2)$	582	607	616 (A_{1g})	602 (A_{1g})
$E_g(1)$	110	147	155 (B_{1g})	149
$E_g(2)$	479	474	474	461
$E_g(3)$	640	659	645	648 (B_{1g})
Infrared			
A_{2u} (TO)	315	339	320	339
A_{2u} (LO)	621	664	–	650 (E_u)
E_u (TO1)	185	153	140	164
E_u (LO1)	292	271	–	232
E_u (TO2)	428	449	550	467
E_u (LO2)	669	734	–	734 (*m*-ZrO$_2$)
Silent			
B_{2u}	665	673		

For tetragonal zirconia, the experimental values of the Raman modes are taken from [34, 42] (in the last two columns, respectively). For the infrared modes, the data for *t*-ZrO$_2$ are from [37, 39] (in the last two columns, respectively).

and Perry [34], it consists in an interchange of the B_{1g} mode and the highest E_g mode (see table 4.5.3).

As for the infrared (IR) spectra of *t*-ZrO$_2$, experiments have been carried out on crystals [43], doped powders [34, 37, 44] and undoped powders stabilized by their small particle size [39]. But, in the absence of *t*-ZrO$_2$ single crystals of good quality, a general agreement has not been reached so far. The two most recent studies [37, 39], which are reported in table 4.5.3, agree to assign the E_u TO modes to the lines at about 150 and $500 \, cm^{-1}$. These assignments were also confirmed by calculations [39–41]. However, the situation is more confused for the A_{2u} mode. To properly fit their reflectance spectra, Pecharromán *et al* [39] had to introduce three oscillators in addition to the two E_u modes. They placed the A_{2u} at 339 cm^{-1} and attributed the extra modes at 580 and 672 cm^{-1} to a secondary oscillator and to the presence of monoclinic zirconia, respectively. Hirata *et al* [37] also mentioned a broad band located at 320 cm^{-1}, but they attributed it to a B_u mode associated with traces of monoclinic zirconia.

Our calculated phonon frequencies and symmetry assignments are reported in table 4.5.3, where they are compared with those of various experimental results for t-ZrO$_2$. For the phonon frequencies, our results are globally in better agreement with experimental data than those of previous theoretical studies. Our symmetry assignments meet all the requirements discussed in the literature, solving the existing contradictions and clarifying some important issues.

For the Raman spectra of t-ZrO$_2$, our calculation presents an rms absolute deviation of 8.4–9.6 cm^{-1}, and an rms relative deviation of 2.7–2.6% with respect to the experimental data of Feinberg and Perry [34] and those of Bouvier and Lucazeau [42]. Our symmetry assignments reconcile the arguments developed by Mirgorodsky *et al* [40, 41] with those proposed by Bouvier and Lucazeau [42]. On the one hand, the A_{1g} mode is at lower frequency compared to the assignments by Feinberg and Perry, in agreement with the arguments of Mirgorodsky *et al* [40, 41]. On the other hand, the B_{1g} is found to have high frequency in accordance with Bouvier and Lucazeau [42].

For the IR-active frequencies of t-ZrO$_2$, our calculation presents an rms absolute deviation of 18.8 and 59.5 cm^{-1}, and an rms relative deviation of 7.6 and 12.3%, with respect to the experimental data of Pecharromán *et al* [39] and Hirata *et al* [37], respectively. We find an LO–TO splitting for the A_{2u} mode of 325 cm^{-1}, much larger than the 15 cm^{-1} found by Pecharromán *et al* [39]. Our result is consistent with the large difference between ϵ_∞ and ϵ_0 as discussed in the 'Dielectric permittivity tensors' section. As a result, we propose that the LO peaks at 650 and 734 cm^{-1} should be attributed to the A_{2u} and the second E_u modes.

In table 4.5.4, we indicate the relationship between the phonon modes of the cubic and tetragonal phases in the case of zirconia. Cubic zone boundary (X point) modes become zone-centre (Γ point) modes in the tetragonal structure. Note in particular that the unstable X_2^- zone-boundary mode in the cubic phase transforms into a stable zone-centre A_{1g} phonon in the tetragonal form.

It is also very interesting to compare the phonon frequencies calculated for HfO$_2$ and ZrO$_2$ (see table 4.5.3). There are several possible origins for the variations that are observed: structural changes (e.g. the volume), change of the mass ratio Hf/Zr = 1.96, and differences in interatomic force constants. Given the small structural changes reported in table 4.5.1, we suspect that their effect should be very small. In order to check this, we compute the

Table 4.5.4. Relationship of phonon modes for cubic and tetragonal phases.

	Cubic				Tetragonal	
	Γ		X		Γ	
Acoustic	F_{1u}	0			A_{2u}	0
					E_u	0
			X_4^-	361	B_{1g}	331
			X_5^-	141	E_g	147
Optic	F_{2g}	596			B_{1g}	607
					E_g	659
			X_2^-	$191i$	A_{1g}	259
			X_5^-	568	E_g	474
	F_{1u}	280			A_{2u}	339
					E_u	153
			X_1^+	697	B_{2u}	673
			X_5^+	325	E_u	449

The calculated frequencies are given in cm^{-1}. For the infrared modes, only the TO frequencies are reported.

phonon frequencies for hafnia assuming that the interatomic force constants are the same as those for zirconia, while the volume is allowed to vary. This analysis shows that the structural changes play a very minor role, in agreement with our intuition. As for the role of the mass ratio, it can be clearly evidenced in the modes in which the $M = (Hf, Zr)$ atoms move significantly more than O atoms: the $B_{1g}(1)$ and $E_g(1)$ modes of the tetragonal phase for which the frequencies vary by about 35%. In contrast, the modes in which the $M = (Hf, Zr)$ atoms are not involved (i.e. F_{2g} in the cubic phase, and A_{1g} and B_{2u} in the tetragonal phase), as well as those for which the O atoms move significantly more than the $M = (Hf, Zr)$ atoms (i.e. F_{1u} in the cubic phase, and $B_{1g}(2)$, $E_g(2)$ and (3), A_{2u}, and $E_u(1)$ and (2) in the tetragonal phase), should have frequencies very similar in HfO_2 and ZrO_2. This is indeed what is observed in most of the cases. However, there are some significant exceptions for the tetragonal phase: A_{1g} and $E_u(1)$ for which the frequencies vary by 19 and 17%, respectively. These are cases in which the effects due to differences in the interatomic force constants are dominant.

Dielectric permittivity tensors

In the cubic phase, the electronic (ϵ_∞) and static (ϵ_0) permittivity tensors are diagonal and isotropic. Due to the symmetry of the tetragonal crystal, these tensors are still diagonal, but have two independent components $\epsilon_\parallel$ and $\epsilon_\perp$, parallel and perpendicular to the c-axis, respectively. In table 4.5.5, the

Table 4.5.5. Electronic and static dielectric tensors for the cubic (C) and tetragonal (T) phases of HfO_2 and ZrO_2.

	HfO$_2$			ZrO$_2$		
C						
ϵ_∞	5.37			5.74		
$\Delta\epsilon$	20.80			27.87		
ϵ_0	26.17			33.61		
T						
	$\parallel$	$\perp$		$\parallel$	$\perp$	
ϵ_∞	5.13	5.39		5.28	5.74	
$\Delta\epsilon_1$	14.87	22.34		15.03	35.48	
$\Delta\epsilon_2$		5.08			6.91	
ϵ_0	20.00	32.81		20.31	48.13	

The contributions of the different phonon modes to the static dielectric tensor are also indicated. For the cubic phase, the tensors are diagonal and isotropic. The phonon mode contributions to $\epsilon_0^{\parallel}$ come from the IR-active F_{1u} mode. For the tetragonal phase, the tensors are also diagonal but they have different components parallel ($\parallel$) and perpendicular ($\perp$) to the c-axis. The phonon mode contributions to $\epsilon_0^{\parallel}$ come from the IR-active A_{2u} mode, while the contributions to $\epsilon_0^{\perp}$ come from the two IR-active E_u modes.

calculated values of ϵ_∞ and ϵ_0 are reported for the cubic and the tetragonal phases of hafnia and zirconia. In the tetragonal phase, the ϵ_∞ tensor is only slightly anisotropic with about 5 and 10% difference between the parallel and perpendicular values for t-HfO$_2$ and t-ZrO$_2$, respectively. In contrast, the ϵ_0 tensor is highly anisotropic: the value of ϵ_0 in the direction parallel to the c-axis is 1.6 and 2.4 times smaller than that in the perpendicular direction for t-HfO$_2$ and t-ZrO$_2$, respectively. While the values of ϵ_∞ for the cubic and tetragonal phases are very close, there is a significant difference in the values of ϵ_0.

A direct comparison of the calculated dielectric tensors with experimental values is very difficult since there are very few data available in the literature, especially for hafnia. The main problem encountered in the experimental determination of the dielectric properties is that good quality single crystals are not available. For the tetragonal phase, the results obtained for undoped powders stabilized by their small particle size must be analysed in the framework of effective medium theory [45]. As a result, a unique value of ϵ is found without distinction between the directions parallel and perpendicular to the c-axis. In order to compare our results with experimental data, we average the values parallel and perpendicular to the c-axis:

$$\bar{\epsilon} = \frac{2\epsilon_\perp + \epsilon_\parallel}{3}.$$

This average does not really have any physical meaning, and therefore the comparison is rather qualitative.

For hafnia, we are only aware of measurements of ϵ_0. Our calculated values of 26.17 for the cubic phase, and $\bar{\epsilon}_0 = 28.54$ for the tetragonal phase significantly overestimate the values of 16 [46] and 20 [47] obtained in recent measurements. This overestimation is considerably higher than what can be expected from our density functional approach and the origin of this difference remains poorly understood. For the cubic phase, our results agree within 1% with those obtained by Zhao and Vanderbilt [26] using similar methods. However, for the tetragonal phase, our calculations disagree significantly with those of Zhao and Vanderbilt [26]. In fact, we find a ratio of 1.6 between the values of the ϵ_0 tensor in directions parallel and perpendicular to the c-axis, to be compared with the value of 8.6 reported by Zhao and Vanderbilt [26]. We note that the value of $\bar{\epsilon}_0 = 70$ proposed by the latter authors appears excessively high in view of the dielectric constant of the cubic phase (~26.17) and the trends observed for zirconia (see below).

For zirconia, an experimental value of $\epsilon_\infty = 4.8$ is reported in the literature for c-ZrO$_2$ [43, 48], while measured values for t-ZrO$_2$ range between 4.2 [39] and 4.9 [49]. Our theoretical values ($\epsilon_\infty = 5.74$ and $\bar{\epsilon}_\infty = 5.59$ for the cubic and tetragonal phases, respectively) are larger than the experimental ones by about 10–15%, as often found in the LDA to DFT. For ϵ_0, the experimental values found in the literature vary from 27.2 [50] to 29.3 [51] for c-ZrO$_2$, and from 34.5 [51] to 39.8 [51] for t-ZrO$_2$. For the cubic phase, our calculated value $\epsilon_0 = 33.61$ is somewhat larger than experimental estimates, whereas, for the tetragonal phase, our calculated average $\bar{\epsilon}_0 = 38.86$ falls in the range of the experimental data.

For a deeper analysis of the static dielectric tensor, we can rely not only on the frequencies of the IR-active modes, but also on the corresponding eigendisplacements and Born effective charges. Indeed, the static dielectric tensor can be decomposed in the contributions of different modes as indicated in equation (4.5.13).

The contribution of the individual modes $\Delta\epsilon_m$ to the static dielectric constants is presented in table 4.5.5. For each IR-active mode, the relevant component of the oscillator strength tensor is reported in table 4.5.6. This tensor is isotropic for the F_{1u} mode in the cubic phase, while in the tetragonal phase we indicate the parallel–parallel component for the A_{2u} mode, and the perpendicular–perpendicular component for the E_u modes. We also give the magnitude of the mode-effective charge vector defined by equation (4.5.14) which is parallel and perpendicular to the tetragonal axis for A_{2u} and E_u modes, respectively, while it has an arbitrary orientation for the F_{1u} mode. The atomic motions for these vibrational modes have been described in detail in the literature [39, 52, 53].

In table 4.5.6, the $E_u(1)$ mode in the tetragonal phase has the lowest oscillator strength (S_m) and the lowest mode-effective charge (Z_m^*). However, it also has the lowest frequency (see table 4.5.3), which results in the largest contribution to the static dielectric constant in table 4.5.5. The F_{1u} mode in the cubic phase has similar characteristics: the oscillator strength is quite small (though larger than for the $E_u(1)$ mode) but it also has a low frequency (though not as low as for the $E_u(1)$ mode). The resulting contribution is of the same order of magnitude as the $E_u(1)$ mode in the tetragonal phase (though rather smaller). Comparatively, the $E_u(2)$ mode in the tetragonal phase gives a much smaller (but not negligible) contribution despite its larger oscillator strength and mode-effective charge.

In fact, the frequency factor plays a crucial role in equation (4.5.13). The A_{2u} has the largest oscillator strength (about twice that of the $E_u(1)$ mode) and the largest mode-effective charge. However, its frequency is about twice that of the $E_u(1)$ mode, and its contribution to the static dielectric constant is roughly half that of the $E_u(1)$ mode. This difference between A_{2u} and E_u modes explains why the ϵ_0 tensor is highly anisotropic, while the ϵ_∞ tensor is only slightly anisotropic.

The same argument holds to rationalize the differences observed in the static dielectric tensor between the tetragonal and the cubic phases. Indeed, as already mentioned, the oscillator strength and the mode-effective charge of the F_{1u} mode are comparable to those of the E_u modes of the tetragonal phase, while the frequency of the F_u mode is 1.7 times larger than that of the $E_u(1)$ mode. As a result, the static dielectric constant is noticeably smaller in the cubic case.

In table 4.5.6, it can be observed that the oscillator strengths and the mode-effective charges are smaller for HfO_2 than for ZrO_2. This can be

Table 4.5.6. Components of mode-effective charge vectors Z_m^* and oscillator strength tensor S_m for each of the IR-active modes of the cubic (C) and tetragonal (T) phases of HfO_2 and ZrO_2.

	HfO_2		ZrO_2	
	Z_m^*	S_m	Z_m^*	S_m
C				
F_{1u}	5.82	6.31	6.42	7.65
T				
A_{2u}	7.71	11.10	8.14	12.28
$E_u(1)$	5.75	5.76	5.95	5.91
$E_u(2)$	5.91	7.03	6.99	9.95

The description of the reported vector and tensor components corresponding to the different modes is given in the text. The components of the mode-effective charge vectors are given in units of $|e|$, where e is the electronic charge. The oscillator strengths are given in 10^{-4} atomic units (1 a.u. $= 0.342\,036\,m^3\,s^{-2}$).

related to the behaviour of the Born effective charges $Z^*_{\kappa,\alpha\alpha'}$ and the eigendisplacements $U_m(\kappa\alpha)$, the two quantities that appear in the definitions of $S_{m,\alpha\beta}$ and $Z^*_{m,\alpha}$ given in equations (4.5.12) and (4.5.14). On the one hand, as discussed in the 'Born effective charge tensors' section, the Born effective charges are globally smaller in HfO_2 than in ZrO_2. On the other hand, the displacements of Hf atoms are smaller than those of Zr atoms simply because they are heavier (as discussed in the 'Phonon frequencies at the Γ point' section).

If one now considers the contributions to the static dielectric constant reported in table 4.5.5, it appears clearly that for HfO_2 the contributions are smaller than for ZrO_2. However, despite the fact that in all cases the oscillator strengths are smaller for hafnia than for zirconia, two different situations can be distinguished depending on the behaviour of the phonon frequencies. On one hand, for the $E_u(1)$ mode, the frequency for HfO_2 is larger than for ZrO_2. In this case, the contribution for ZrO_2 is noticeably larger (about 60%) than for HfO_2. For the F_{1u} in the cubic phase, the situation is very similar though the frequency does not change very much. On the other hand, for the A_{2u} mode, the frequency changes in the opposite way. As a result, the increase by 6% of the oscillator strengths is almost completely compensated by the rise of 7% in the frequency: in the end, there only remains a 1% difference between the contributions for HfO_2 and ZrO_2. For the $E_u(2)$ mode, the rise of 5% in the frequency only slightly attenuates the 15% increase of the oscillator strengths.

Crystalline silicates

Introduction

Due to the chemical homology of Hf and Zr discussed in the 'Introduction' section, hafnon ($HfSiO_4$) and zircon ($ZrSiO_4$) resemble each other in many physical and chemical properties. Their similarities are such that there is complete miscibility between $ZrSiO_4$ and $HfSiO_4$ [54]. In addition to their importance as potential alternative gate dielectrics, hafnon and zircon are of geological significance. They both belong to the orthosilicate class of minerals, which can be found in igneous rocks and sediments. Zircon is used as a gemstone, because of its good optical quality, and resistance to chemical attack. In the earth's crust, hafnon and zircon are host minerals for the radioactive elements uranium and thorium. Therefore, they have been widely studied in the framework of nuclear waste storage.

In a recent paper [55], we have studied the structural, electronic and dynamical properties of zircon using first-principles calculations. In this section, we present a comparison between hafnon and zircon.

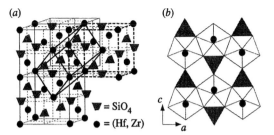

Figure 4.5.2. Structure of hafnon and zircon. (*a*) The individual SiO_4 units are represented schematically by the grey tetrahedra, while M = (Hf, Zr) atoms are indicated by black spheres. The two sets of dashed lines and heavy lines outline the body-centred-tetragonal unit cell and the primitive cell, respectively. (*b*) Besides the SiO_4 units, the MO_8 triangular dodecahedra with the M atoms in their centre are also drawn.

Structural properties

Hafnon and zircon have a conventional unit cell which is body-centred tetragonal (space group $I4_1/amd$, no 141) and contains four formula units of $MSiO_4$ with M = (Hf, Zr), as illustrated by the dashed lines in figure 4.5.2(*a*). A primitive cell containing only two formula units of $MSiO_4$ can also be defined, as indicated by the heavy lines in figure 4.5.2(*a*).

The structure of hafnon and zircon may be viewed as consisting of $(SiO_4)^{4-}$ anions and M^{4+} cations with M = (Hf, Zr), as illustrated by the grey tetrahedra and the black spheres in figure 4.5.2(*a*). This is consistent with the larger bond length (about 25%) of the M−O compared to the Si−O bond. The experimental data describing the structure of hafnon [54] and zircon [56] are reported in table 4.5.7.

Alternatively, as presented in figure 4.5.2(*b*), a different view may be adopted in which $HfSiO_4$ and $ZrSiO_4$ consist of alternating (discrete) SiO_4 tetrahedra and MO_8 units, sharing edges to form chains parallel to the *c*-direction. Note that in these MO_8 units four O atoms are closer to the M atom than the other four (about 6% difference in the M−O bond length; see table 4.5.7).

The positions of the M = (Hf, Zr) and Si atoms are imposed by symmetry: they are located at $(0, \frac{3}{4}, \frac{1}{8})$ and $(0, \frac{1}{4}, \frac{3}{8})$ on the 4a and 4b Wyckoff sites, respectively. The O atoms occupy the 16h Wyckoff sites $(0, u, v)$, where u and v are internal parameters.

Table 4.5.7 summarizes our results obtained after structural and atomic relaxation. The calculated lattice constants a and c, as well as the internal parameters u and v, are found to be in excellent agreement with their corresponding experimental values [54, 56]. Interatomic distances and angles are within 1 or 2% of the experimental values. This accuracy is to address in a meaningful way the dynamical and dielectric properties.

Table 4.5.7. Structural parameters of $HfSiO_4$ and $ZrSiO_4$.

| | $HfSiO_4$ | | $ZrSiO_4$ | |
	Theoretical	Experimental	Theoretical	Experimental
a	6.61	6.57	6.54	6.61
c	5.97	5.96	5.92	6.00
u	0.0672	0.0655	0.0645	0.0646
v	0.1964	0.1948	0.1945	0.1967
Volume	130.42	128.63	126.60	131.08
$d(Si-O)$	1.62	1.61	1.61	1.62
$d(M-O)$	2.14	2.10	2.10	2.13
	2.27	2.24	2.24	2.27
$\angle(O-Si-O)$	97°	97°	97°	97°
	116°	117°	116°	116°

The length unit is the Å. The experimental data are taken from [54] for $HfSiO_4$, and from [56] for $ZrSiO_4$.

Electronic structure

In figure 4.5.3, we present the calculated electronic density of states (DOS) for hafnon and zircon. The complete electronic band structure for $ZrSiO_4$ along several directions in the Brillouin zone can be found elsewhere [55]. For $HfSiO_4$, the electronic band structure is very similar apart from the position of the Hf 5s and 5p bands, as explained hereafter.

We clearly distinguish four groups in the DOS of the valence bands, of which the three lowest ones are rather peaked (small dispersion of the bands), indicative of a weak hybridization. The DOS of hafnon (zircon) exhibits a very sharp peak at -60.2 eV (-47.1 eV) attributed to the Hf 5s (Zr 4s) states, corresponding to two flat bands in the band structure [55]. The peak at -29.8 eV for hafnon (-25.5 eV for zircon) is related to the Hf 5p (Zr 4p) states: it includes six electrons per unit cell. The O 2s peak (eight electrons per unit cell) is located between -18.0 and -16 eV for both hafnon and zircon.

By contrast, the fourth group (24 electrons per unit cell) has a much wider spread of 8 eV. These states have mainly an O 2p character with some mixing of Si and M = (Hf, Zr) orbitals. This mixed covalent–ionic bonding of $HfSiO_4$ and $ZrSiO_4$, appearing in this group of valence bands, should be kept in mind when interpreting the Born effective charge tensors.

Born effective charge tensors

In the hafnon and zircon structures, the local site symmetry of M = (Hf, Zr) and Si atoms is rather high ($\bar{4}m2$). The Born effective charge tensors of

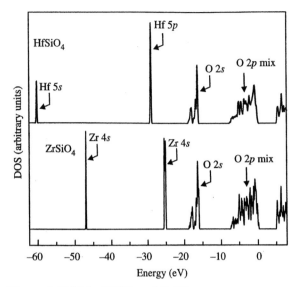

Figure 4.5.3. Electronic DOS for $HfSiO_4$ and $ZrSiO_4$.

M = (Hf, Zr) and Si atoms are diagonal and have only two independent components: parallel and perpendicular to the tetragonal axis, $Z^*_\parallel$ and $Z^*_\perp$, respectively. The Born effective charge tensors of M = (Hf, Zr) and Si atoms are reported in table 4.5.8.

We note that $Z^*_\perp$ for M = (Hf, Zr) is anomalously large compared to the nominal ionic charge of the hafnium and zirconium ions $Z = +4$. A similar behaviour was also observed in the case of $PbZrO_3$ [29] and of hafnia and zirconia, as discussed in 'Crystalline oxides' section. A detailed analysis of the physics of Born effective charges in the case of perovskite ferroelectrics (like $PbZrO_3$) ascribed this effect to a mixed covalent–ionic bonding [57]. In the 'Electronic structure' section, we have seen the occurrence of M–O 2p hybridization. Thus, the physical interpretation of this phenomenon is probably similar to the case of perovskite ferroelectrics. The other component of the M = (Hf, Zr) Born effective charge tensor ($Z^*_\parallel$) is also larger than the nominal ionic charge, although the effect is not as pronounced.

For the silicon atom, there are also some (weaker) deviations with respect to the nominal value ($Z = +4$), one component being slightly larger, and one being definitely lower. These are not very different from those observed in tetrahedrally bonded silica polymorphs, like quartz [58], in which each O atom is strongly bonded to two Si atoms, or in the more compact polymorph of silica, stishovite [30], in which each O atom has three close Si neighbours.

Note that $Z^*_\perp$ is about 3% smaller for hafnium in $HfSiO_4$ than for zirconium in $ZrSiO_4$, similarly to what is observed in hafnia and zirconia,

Table 4.5.8. Non-vanishing components of the calculated Born effective charge tensors for M = (Hf, Zr), Si and O atoms in $HfSiO_4$ and $ZrSiO_4$.

Atom	$HfSiO_4$			$ZrSiO_4$		
M	(+5.28	+5.28	+4.68)	(+5.41	+5.41	+4.63)
Si	(+3.18	+3.18	+4.35)	(+3.25	+3.25	+4.42)
O	$\begin{pmatrix} -1.15 & 0 & 0 \\ 0 & -3.08 & -0.19 \\ 0 & -0.35 & -2.26 \end{pmatrix}$			$\begin{pmatrix} -1.15 & 0 & 0 \\ 0 & -3.17 & -0.16 \\ 0 & -0.34 & -2.25 \end{pmatrix}$		
	[−1.15	−3.16	−2.18]	[−1.15	−3.23	−2.19]

For M = (Hf, Zr) and Si atoms, the tensors are diagonal and only the principal elements are given. For O atoms, the full tensor is reported and the principal elements of its symmetric part are indicated between brackets.

as discussed in the 'Born effective charge tensors' section. The Born effective charge of Si atoms for directions perpendicular to the tetragonal axis shows a very similar behaviour: it is about 2% smaller in hafnon than in zircon. For the Born effective charge in a direction parallel to the c-axis, we find for Si atoms the same trend as for perpendicular directions, but the opposite one for M = (Hf, Zr) atoms, the Born effective charges for Hf in hafnon being about 1% higher than for Zr in zircon.

The local site symmetry of the O atoms has only a mirror plane. As a consequence, the Born effective charge tensors of O atoms are not diagonal, and depend on five independent quantities. We examine the tensor for the O atom located at $(0, u, v)$, which is reported in table 4.5.8. The Born effective charge tensors of the other oxygen atoms can be obtained using the symmetry operations. For this particular atom, the mirror plane is perpendicular to x. Note that Z_{yz}^* and Z_{zy}^* are different, but rather small, making the Born effective charge tensor almost diagonal. They appear in the mirror plane, where one O–Si bond and two O–M bonds (one long and one short) are present. One can compute the projection of the Born effective charge on these directions: for the O–Si bond, the projection is −2.30 in $HfSiO_4$ (−2.29 in $ZrSiO_4$, while it is −3.16 (−3.23) for the shorter O–Hf (O–Zr) bond, and −2.97 (−3.02) for the longer bond. In this plane, the magnitude of the Born effective charge components is larger than the nominal ionic charge of oxygen ($Z = -2$). Following an alternative approach to the characterization of the anisotropy of this tensor, we select its symmetric part and diagonalize it. The principal values are given in table 4.5.8 and the principal direction associated with the largest principal value forms an angle of about 14° with the y-axis. Both analyses give the same type of anisotropy.

Such a strong anisotropy of the Born effective charge tensor for O atoms, with one component of magnitude much smaller than two and much smaller than the two others, has already been observed in SiO_2-stishovite [30] and TiO_2-rutile [31]. By contrast, in tetrahedrally bonded silica, there are *two* components of magnitudes much smaller than two. Thus, at the level of the Born effective charges, the ionic–covalent bonding of O atoms to M = (Hf, Zr) and Si atoms in $HfSiO_4$ and $ZrSiO_4$ is closer to stishovite than to quartz, in agreement with a naive bond-counting argument. Models of amorphous silicates MSi_xO_y should take into account this difference, and might be classified according to the anisotropy of the O Born effective charges. One expects that for a small content of M = (Hf, Zr) the quartzlike behaviour dominates, while for M atomic fractions closer to that of hafnon and zircon the stishovite-like behaviour becomes stronger.

Note finally that the Born effective charges for O atoms are very similar in $HfSiO_4$ and $ZrSiO_4$. The only significant difference is for the second principal value, which is 2% smaller in hafnon than in zircon.

Phonon frequencies at the Γ point

We also compute the phonon frequencies at the Γ point of the Brillouin zone for hafnon and zircon. The theoretical group analysis predicts the following irreducible representations of optical and acoustical zone-centre modes:

$$\Gamma = \underbrace{2A_{1g} \oplus 4B_{1g} \oplus B_{2g} \oplus 5E_g}_{\text{Raman}} \oplus \underbrace{3A_{2u} \oplus 4E_u}_{\text{IR}} \oplus \underbrace{A_{2u} \oplus E_u}_{\text{Acoustic}}$$

$$\oplus \underbrace{B_{1u} \oplus A_{2g} \oplus A_{1u} \oplus 2B_{2u}}_{\text{Silent}}.$$

Because of the non-vanishing components of the Born effective charge tensors, the dipole–dipole interaction must be properly included in the calculation of the interatomic force constants [7, 32, 33]. In particular, the dipole–dipole contribution is found to be responsible for the splitting between the longitudinal and transverse optic (LO and TO, respectively) modes E_u (perpendicular to c) and A_{2u} (parallel to c) at the Γ point.

In table 4.5.9, the calculated phonon frequencies are compared with experimental values. For hafnon, experimental data are only available for Raman modes, in the form of two sets of measurements [59, 60]. Since the agreement with both sets of data is excellent, we report here only the most recent data [60]. For zircon, both Raman and IR-active modes have been studied experimentally [61], the IR data being confirmed by more recent experiments [62, 63].

Overall, the agreement between theory and experiment is excellent, with an rms absolute deviation of $4.1\,\text{cm}^{-1}$ for $HfSiO_4$ ($9.4\,\text{cm}^{-1}$ for $ZrSiO_4$), and an rms relative deviation of 4.2% (2.5%). We obtain four Raman active

Table 4.5.9. Fundamental frequencies of $HfSiO_4$ and $ZrSiO_4$ (in cm^{-1}) with their symmetry assignments.

Mode	HfSiO$_4$		ZrSiO$_4$	
	Theoretical	Experimental	Theoretical	Experimental
Raman				
$A_{1g}(1)$	462	450	442	439
$A_{1g}(2)$	970	984	971	974
$B_{1g}(1)$	162	157	225	214
$B_{1g}(2)$	395	401	397	393
$B_{1g}(3)$	638	620	632	–
$B_{1g}(4)$	1016	1020	1017	1008
B_{2g}	247	267	252	266
$E_g(1)$	161	148	194	201
$E_g(2)$	204	212	225	225
$E_g(3)$	369	351	375	357
$E_g(4)$	530	–	536	547
$E_g(5)$	923	–	923	–
Infrared				
A_{2u} (TO1)	312		348	338
A_{2u} (LO1)	423		476	480
A_{2u} (TO2)	598		601	608
A_{2u} (LO2)	656		646	647
A_{2u} (TO3)	983		980	989
A_{2u} (LO3)	1095		1096	1108
E_u (TO1)	252		285	287
E_u (LO1)	313		341	352
E_u (TO2)	395		383	389
E_u (LO2)	409		420	419
E_u (TO3)	420		422	430
E_u (LO3)	461		466	471
E_u (TO4)	873		867	885
E_u (LO4)	1023		1029	1035
Silent				
B_{1u}	107		120	
A_{2g}	233		242	
A_{1u}	383		392	
$B_{2u}(1)$	573		566	
$B_{2u}(2)$	945		943	

The experimental values are taken from [60] for $HfSiO_4$ (Raman modes only), and from [61] for $ZrSiO_4$.

modes that could not be detected experimentally: two for hafnon (at $530\,\mathrm{cm}^{-1}$ $[E_g(4)]$ and $923\,\mathrm{cm}^{-1}$ $[E_g(5)]$) and two for zircon (at $632\,\mathrm{cm}^{-1}$ $[B_{1g}(3)]$ and $923\,\mathrm{cm}^{-1}$ $[E_g(5)]$).

We also obtain silent modes, inactive for both IR and Raman experiments. They are found to range from 107 to $945\,\mathrm{cm}^{-1}$ in $HfSiO_4$, and from 120 to $943\,\mathrm{cm}^{-1}$ in $ZrSiO_4$. Two of these (B_{1u} and A_{2g}) are very soft, and correspond, in a first approximation, to vibration modes in which the SiO_4 tetrahedra rotate as a unit [61] (in the u and g modes the tetrahedra move with opposite phases).

In the 'Phonon frequencies at the Γ point' section, we pointed out three origins for the variations of the frequencies in Hf- and Zr-based oxides: the structural changes, the mass ratio Hf/Zr equal to 1.96 and the differences in interatomic force constants. It is quite interesting to compare on the same basis the phonon frequencies calculated for $HfSiO_4$ and $ZrSiO_4$ (see table 4.5.9).

By performing a similar analysis as for HfO_2 versus ZrO_2, we find that the structural changes play a very minor role, in agreement with the intuition resulting from the very small variations observed in table 4.5.7. The effect of the mass ratio is clear for the $B_{1g}(1)$ mode in which the M = (Hf, Zr) atoms move significantly more than O atoms: the frequency increases by about 28% in $ZrSiO_4$. In contrast, the frequencies should not vary much from $HfSiO_4$ to $ZrSiO_4$ for modes in which the M = (Hf, Zr) atoms are not involved (i.e. all the silent modes, $A_{1g}(1)$ and (2), and B_{2g}), as well as for those in which the O atoms move significantly more than the M = (Hf, Zr) atoms. In most of these cases, this is indeed what is observed; in a few cases, however, the differences in the interatomic force constants dominate, e.g., for the B_{1u} mode for which the frequency increases by about 11%.

Dielectric permittivity tensors

Due to the tetragonal symmetry of the hafnon and zircon crystals, the electronic (ϵ_∞) and static (ϵ_0) permittivity tensors have two independent components $\epsilon_\parallel$ and $\epsilon_\perp$ parallel and perpendicular to the c-axis, respectively. The calculated values of ϵ_∞ and ϵ_0 are reported in table 4.5.10.

For zircon, values of 10.69 (3.8) [62] and 11.25 (3.5) [63] are reported for the static (electronic) dielectric permittivity in the directions parallel and perpendicular to the tetragonal axis, respectively. Our theoretical values are larger than the experimental ones by about 10%, as often found in the LDA to DFT. For hafnon, we were not able to find accurate measurements in the literature: for hafnium silicates, values ranging from 11 to 25 have been reported.

The contribution of the individual modes $\Delta\epsilon_m$ to the static dielectric constant, as defined in equation (4.5.13), is also indicated in table 4.5.10. The largest contribution comes from the lowest frequency mode. The

Table 4.5.10. Electronic and static dielectric tensors of $HfSiO_4$ and $ZrSiO_4$.

| | HfSiO$_4$ | | ZrSiO$_4$ | |
	∥	⊥	∥	⊥
ϵ_∞	4.11	3.88	4.26	4.06
$\Delta\epsilon_1$	4.93	4.38	5.90	5.16
$\Delta\epsilon_2$	0.81	0.75	0.52	1.31
$\Delta\epsilon_3$	0.80	0.35	0.85	0.05
$\Delta\epsilon_4$		1.27		1.38
ϵ_0	10.65	10.63	11.53	11.96

The contributions of individual phonon modes to the static dielectric tensor are indicated. The tensors are diagonal and have different components parallel (∥) and perpendicular (⊥) to the c-axis. The phonon mode contributions to $\epsilon_0^\parallel$ come from the three IR-active A_{2u} modes, while the contributions to $\epsilon_0^\perp$ come from the four IR-active E_u modes.

decomposition of the static dielectric tensor can further be analysed using the mode-effective charge vectors and the oscillator strength tensors, defined by equations (4.5.12) and (4.5.14), respectively. In table 4.5.11, we present, for each IR-active mode, the magnitude of its mode-effective charge vectors (this vector is parallel and perpendicular to the tetragonal axis for A_{2u} and E_u modes, respectively), as well as the relevant component of the oscillator strength tensor (the parallel–parallel component for A_{2u} modes, and the perpendicular–perpendicular component for E_u modes).

Table 4.5.11. Components of mode-effective charge vectors Z_m^* and oscillator strength tensor S_m for each of the IR-active modes for $HfSiO_4$ and $ZrSiO_4$.

| | HfSiO$_4$ | | ZrSiO$_4$ | |
	Z_m^*	S_m	Z_m^*	S_m
$A_{2u}(1)$	6.85	7.39	7.68	10.06
$A_{2u}(2)$	3.78	4.24	2.76	2.64
$A_{2u}(3)$	6.60	11.22	6.71	11.50
$E_u(1)$	5.93	4.05	6.79	5.91
$E_u(2)$	2.94	1.70	3.51	2.71
$E_u(3)$	1.69	0.91	0.28	0.12
$E_u(4)$	7.21	14.02	7.37	14.69

The description of the reported vector and tensor components corresponding to the two types of mode is given in the text. The components of the mode-effective charge vectors are given in units of $|e|$, where e is the electronic charge. The oscillator strengths are given in 10^{-4} atomic units (1 a.u. $= 0.342\,036\,m^3\,s^{-2}$).

For each symmetry representation (A_{2u} and E_u), the lowest- and highest-frequency modes exhibit the largest effective charges and the largest oscillator strengths. Despite their similar oscillator strengths, the modes of lowest frequency contribute much more to the static dielectric constant than the modes of highest frequency, the frequency factor in equation (4.5.13) playing a crucial role. The second-lowest-frequency modes are moderately strong, while the third E_u modes have a negligible IR activity.

Similarly to what was observed when comparing hafnia and zirconia (see the 'Dielectric permittivity tensors' section), the oscillator strengths and the mode-effective charges are smaller for $HfSiO_4$ than for $ZrSiO_4$ (except for the $A_{2u}(2)$ and $E_u(3)$ modes). The origin of this difference can be traced back to the Born effective charges and the eigendisplacements. Indeed, as discussed in the 'Born effective charge tensors' section, the Born effective charges of M = (Hf, Zr) and Si atoms are smaller in $HfSiO_4$ than in $ZrSiO_4$. Moreover, due to their heavier weight, the displacements of Hf atoms are smaller than those of Zr atoms.

Coming back to the contributions to the static dielectric constant reported in table 4.5.10, we observe that most of the contributions for $HfSiO_4$ are smaller than those for $ZrSiO_4$ (except those of the $A_{2u}(2)$ and $E_u(3)$ modes). In the cases where the oscillator strengths are smaller for hafnon than for zircon, the behaviour of the phonon frequencies will also influence the contribution to ϵ_0. Indeed, when the corresponding frequency in $HfSiO_4$ is higher than in $ZrSiO_4$, the contribution to the static dielectric constant is further increased (as for the $E_u(2)$ modes). In contrast, when the frequency is lower, the difference in the contribution is lower than in the oscillator strength. For instance, for the $A_{2u}(1)$ mode, the oscillator strength S_m in hafnon is 30% smaller than in zircon, but the corresponding contribution to ϵ_0 is only 20% smaller, since the associated frequency is about 10% larger. Finally, for the $A_{2u}(2)$ and $E_u(3)$ modes, the frequencies are very similar (these modes essentially involve displacements of the Si atoms and of some of the O atoms) and therefore the oscillator strength governs the trend of the contributions to the static dielectric constant (larger in hafnon than in zircon).

Amorphous silicates

The dielectric properties of Zr and Hf amorphous silicates constitute an issue of great practical importance. Early experimental measurements suggested a supralinear dependence of the static dielectric constant ϵ_0 on the M = (Hf, Zr) concentration [64]. While several phenomenological theories addressed this behaviour [65, 66], more recent data appear to favour instead a close to linear dependence [67, 68]. In a recent paper [69], we have used DFT simulations to shed some light on this particularly relevant issue by investigating how the permittivity of Zr silicates is affected by the underlying nanoscopic structure.

In tackling this technological problem, we face the more general issue of predicting the dielectric properties of amorphous alloys from first principles. Brute force analysis of numerous large supercells is beyond present computational capabilities. To overcome this difficulty, we establish a relationship between the dielectric properties of Zr silicates and their underlying nanoscopic structure. Using DFT, we compute optical and static dielectric constants for various model structures of Zr silicates, both ordered and disordered. We introduce a scheme which relates the dielectric constants to the local bonding of Si and Zr atoms. This scheme is based on the definition of parameters characteristic of the basic SUs formed by Si and Zr atoms and their nearest neighbours.

Applied to amorphous Zr silicates, our scheme provides a good description of measured dielectric constants, both optical [68, 70] and static [67, 68], and reveals the important contribution of ZrO_6 SUs to the static dielectric constant. In a very similar way, our scheme can also be used to investigate Hf silicates. Only here we briefly indicate how the two systems compare.

We consider model structures of $(ZrO_2)_x(SiO_2)_{1-x}$, nine crystalline and one amorphous, with x ranging from 0 to 0.5, and describe them in terms of cation-centred SUs. We start with three different SiO_2 polymorphs ($x = 0$):

[C_0] α-cristobalite with four SiO_4 SUs per unit cell

[Q_0] α-quartz with three SiO_4 SUs

[S_0] stishovite with two SiO_6 SUs

By substituting one of the Si atoms by a Zr atom for each of these models, we generate three new crystal structures:

[C_1] Zr^{Si} in α-cristobalite with three SiO_4 and one ZrO_4 SUs per unit cell ($x = 0.25$)

[Q_1] Zr^{Si} in α-quartz with two SiO_4 and one ZrO_4 SUs ($x = 0.33$)

[S_1] Zr^{Si} in stishovite with one SiO_6 and one ZrO_6 SUs

We also consider zircon, as well as two other structures generated by replacing Zr by Si:

[Z_2] zircon that contains two SiO_4 and two ZrO_8 SUs per unit cell ($x = 0.5$)

[Z_1] Si^{Zr} in zircon with two SiO_4, one SiO_6 and one ZrO_8 SUs ($x = 0.25$)

[Z_0] fully Si-substituted zircon with two SiO_4 and SiO_6 SUs ($x = 0$)

Finally, the amorphous structure [A] is generated using classical molecular dynamics with empirical potentials [69]. In this study, only a single disordered structure could be afforded because of the noticeable computational cost associated.

The atomic coordinates and the cell parameters of all our model structures are fully relaxed within the LDA to DFT. The calculated optical and static dielectric constants for our model structures are given in table 4.5.12. Due to the well-known limitations of the LDA, the theoretical values are larger than the experimental ones (when available) by about 10%.

In order to understand how the optical dielectric constant (ϵ_∞) depends on the underlying atomic nanostructure, we consider the electronic polarizability $\bar{\alpha}$ which is related to ϵ_∞ by the Clausius–Mosotti relation [66, 68]:

$$\frac{\epsilon_\infty - 1}{\epsilon_\infty + 2} = \frac{4\pi\,\bar{\alpha}}{3\,\bar{V}} \tag{4.5.16}$$

where $\bar{V}$ is the average SU volume. The polarizability $\bar{\alpha}$ can be taken as a local and additive quantity, in contrast with ϵ_∞. Therefore, we define α_i values for each SU i, where $i \equiv SiO_n$ (with $n = 4$ or 6) or ZrO_n (with $n = 4, 6$ or 8), such that:

$$\bar{\alpha} = \sum_i x_i \alpha_i, \tag{4.5.17}$$

where x_i is the molecular fraction. In table 4.5.13, we report the five α_i values that we obtain by solving in a least-squares sense the over-determined system based on the calculated ϵ_∞ values for the nine crystalline models. The optical dielectric constants derived from these α_i values using equations (4.5.16) and (4.5.17) compare well with those calculated from first principles, showing average and maximal errors smaller than 1 and 2.5%, respectively. For the amorphous model, which was not used to determine the α_i values, the derived value $\epsilon_\infty = 3.25$ is in excellent agreement with the first-principles

Table 4.5.12. Composition (x), optical (ϵ_∞) and static (ϵ_0) dielectric constants, volume ($\bar{V}$) in bohr^3, polarizability $\bar{\alpha}$ in bohr^3, characteristic dynamical charge ($\bar{Z}$), and characteristic force constant ($\bar{C}$) in hartree/bohr^2 for the various model systems.

Model	x	ϵ_∞	ϵ_0	$\bar{V}$	$\bar{\alpha}$	$\bar{Z}$	$\bar{C}$
C_0	0.00	2.38	4.30	264.77	19.92	4.21	0.4391
C_1	0.25	2.76	5.25	273.21	24.12	4.59	0.3895
Q_0	0.00	2.54	4.83	240.34	19.46	4.28	0.4169
Q_1	0.33	2.91	5.84	275.28	25.56	4.85	0.3661
S_0	0.00	3.36	10.33	153.74	16.16	4.81	0.2716
S_1	0.50	4.44	24.20	201.88	25.74	6.14	0.1188
Z_0	0.00	3.37	10.11	167.80	17.68	4.76	0.2512
Z_1	0.25	3.94	18.36	189.74	22.42	5.29	0.1287
Z_2	0.50	4.13	11.81	213.28	26.00	5.58	0.2385
A	0.15	3.24	8.92	213.12	21.75	4.83	0.2424

The reported dielectric constants correspond to orientational averages.

Table 4.5.13. Polarizability (α in bohr^3), characteristic dynamical charge (Z) and characteristic force constant (C in hartree/bohr^2) for various SUs.

	SiO$_4$	SiO$_6$	ZrO$_4$	ZrO$_6$	ZrO$_8$
α	19.68	16.14	37.37	35.35	32.69
Z	4.29	4.92	5.66	7.16	6.73
C	0.3597	0.2176	0.4202	0.0817	0.1153

result $\epsilon_\infty = 3.24$. These results justify *a posteriori* the use of equations (4.5.16) and (4.5.17) to model the optical dielectric constant.

For the static dielectric constant (ϵ_0), the phonon contributions preclude a description in terms of a single local and additive quantity as the electronic polarizability. To overcome this difficulty, we focus on the difference between dielectric constants ($\Delta\epsilon$):

$$\Delta\epsilon = \epsilon_0 - \epsilon_\infty = \frac{4\pi}{\Omega_0}\sum_m \frac{S_m}{\omega_m^2} = \frac{4\pi\bar{Z}^2}{\bar{V}\,\bar{C}}, \tag{4.5.18}$$

where ω_m and S_m are the frequency and the oscillator strength of the mth mode. The volume of the primitive unit cell Ω_0 is related to the volume $\bar{V}$ and to the number of SUs $\bar{N}$ by $\Omega_0 = \bar{N}\bar{V}$. The characteristic dynamical charge $\bar{Z}$ and characteristic force constant $\bar{C}$ are defined by:

$$\bar{Z}^2 = \frac{1}{\bar{N}}\sum_\kappa Z_\kappa^2 \quad \text{and} \quad \bar{C}^{-1} = \frac{1}{\bar{N}}\sum_m \frac{S_m}{\omega_m^2\bar{Z}^2}, \tag{4.5.19}$$

where Z_κ are the atomic Born effective charges.

The variation of $\Delta\epsilon$ due to a Si $\rightarrow$ Zr substitution has been analysed in detail in [69], where the contribution from sixfold-coordinated atoms has been highlighted. In fact, these configurations resemble those in ABO$_3$ perovskites. The enhancement of $\Delta\epsilon$ originates from very low-frequency modes in which the cations (A or B) move in opposition with the O atoms while carrying opposite effective charges.

By analogy with the polarizability, we define Z_i and C_i values for each SU such that:

$$\bar{Z}^2 = \sum_i x_i Z_i^2 \quad \text{and} \quad \bar{C}^{-1} = \sum_i x_i C_i^{-1}, \tag{4.5.20}$$

though the locality and the additivity of these parameters is not guaranteed *a priori*. We determine the optimal values Z_i and C_i in the same way as for α_i (table 4.5.13).

For the nine crystalline models, the values of $\Delta\epsilon$ obtained by introducing these parameters in equations (4.5.18) and (4.5.20) match quite well with those calculated from first principles [69], though the agreement is not as impressive as for ϵ_∞. Differences result primarily from the determination of $\bar{C}$. By contrast, the values of $\bar{Z}$ given by equation (4.5.20) agree very well with those computed from first principles, showing an average and maximal error smaller than 2 and 3%, respectively. *A posteriori*, $\bar{C}$ appears to be less local and additive. In fact, it can be demonstrated that the locality of $\bar{C}$ is closely related to the dynamical charge neutrality of the SUs [69].

For the amorphous model, which was not used to determine the Z_i and C_i values, the agreement between the model and the first-principles $\Delta\epsilon$ is excellent with an error smaller than 1% [69]. Indeed, our scheme is more accurate for disordered systems, where the localization of vibrational modes is enhanced and the dynamical charge neutrality appears better respected.

For Zr silicates of known composition in terms of SUs, the parameters in table 4.5.13 fully determine the dielectric constants. Several points are noteworthy. First, the three parameters of Zr-centred SUs all contribute to enhancing the dielectric constants over those of Si-centred ones of corresponding coordination[1]. This is clearly at the origin of the increase of ϵ_∞ and ϵ_0 with increasing Zr concentration. Second, while the polarizability α_i of a given SU (Si or Zr centred) steadily decreases with increasing coordination, such a regular behaviour is not observed for the parameters Z_i and C_i determining $\Delta\epsilon$. In fact, Z_i and C_i concurrently vary to enhance the contribution of ZrO_6 units, which are the SUs giving the largest contribution to $\Delta\epsilon$ in amorphous Zr silicates.

Based on the scheme given by equations (4.5.16)–(4.5.20), we can now estimate ϵ_∞ and ϵ_0 for amorphous $(ZrO_2)_x(SiO_2)_{1-x}$ as a function of Zr composition ($0 < x < 0.5$). Using measured densities for Zr silicates [70], we first obtain ϵ_∞ as a function of x. As shown in figure 4.5.4, our theoretical values[2] are in excellent agreement with available experimental data [68, 70].

To apply our scheme for $\Delta\epsilon$, we need additional information on the cationic coordination. We take the Si atoms to be fourfold coordinated. The coordination of Zr atoms is less well determined. Recent EXAFS measurements [65] indicate that the average Zr coordination grows from about four to about eight for Zr concentrations increasing from $x \sim 0$ to $x \sim 0.5$. In figure 4.5.4, we report calculated ϵ_0 for amorphous $(ZrO_2)_x(SiO_2)_{1-x}$ as a function of x, together with the available experimental data [64, 67, 68, 71].

---

[1] In table 4.5.13, the value of C for SiO_4 apparently leads to a higher contribution to $\Delta\epsilon$ than that for ZrO_4. This is an artifact of the approach we used to determine the Z_i and C_i.

[2] Because the various Zr-centred units have close α values compared to SiO_4 (table 4.5.13), the effect of Zr coordination on ϵ_∞ is negligible.

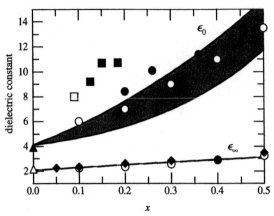

Figure 4.5.4. Dielectric constants (ϵ_∞ and ϵ_0) as a function of composition x for amorphous $(ZrO_2)_x(SiO_2)_{1-x}$. The grey region corresponds to results derived from our model scheme and reflects the indetermination of the number of ZrO_6 units. The upper curve delimiting the band corresponds to structures entirely composed of ZrO_6 units, while the lower curve represents a smooth transition from a structure composed of ZrO_4 units at $x = 0$ to one composed of ZrO_8 units at $x = 0.5$, without the occurrence of any ZrO_6 units. The references for the experimental data are: ◆ [70], ● [67], ○ [68], ■ [64], □ [71], ▲ [72] and △ [73].

The theoretical results are given in the form of a band reflecting the indetermination of the coordination of Zr atoms. We modelled the dielectric constant in terms of suitable distributions of three representative SUs (ZrO_4, ZrO_6 and ZrO_8). The upper curve delimiting the band in figure 4.5.4 corresponds to structures entirely composed of ZrO_6 units. The lower curve is for amorphous systems which do not contain any ZrO_6 units. The average Zr coordination varies linearly from four to eight between $x = 0$ and $x = 0.5$, with concentrations of ZrO_4 and ZrO_8 SUs varying at most quadratically. Note that the upper part of the band agrees well with the recent experimental data [67, 68]. The earlier data [64, 71] cannot be explained. Figure 4.5.4 shows that, for a sufficient number of ZrO_6 units, values of ϵ_0 at intermediate x can indeed be larger than estimated from a linear interpolation between SiO_2 and $ZrSiO_4$. However, in accord with recent experiments [67, 68], our theory indicates that the extent of this effect is more limited than previously assumed [64, 65].

Our scheme could also be applied to Hf silicates which are very similar to Zr silicates. In this respect, the comparison between $HfSiO_4$ and $ZrSiO_4$ carried out in the 'Crystalline silicates' section is very useful. Translated in terms of the quantities defined in this section, we get $\bar{V} = 220.13$, $\bar{a} = 25.74$, $\bar{Z} = 5.50$ and $\bar{C} = 0.2581$ for hafnon to be compared with $\bar{V} = 213.28$, $\bar{a} = 26.00$, $\bar{Z} = 5.58$ and $\bar{C} = 0.2385$ for zircon. Basically, all these parameters show a similar trend that leads to smaller dielectric permittivities (both electronic and static) for hafnon than for zircon. For amorphous Hf silicates, we expect the same kind of consideration to apply.

Conclusions

Using DFT, we have investigated the structural, electronic, dynamical and dielectric properties for a series of high-κ materials belonging to the Hf–Si–O and Zr–Si–O systems. We have considered hafnia and zirconia (the crystalline oxides), hafnon and zircon (the crystalline silicates) and finally the amorphous silicates.

In all the investigated systems, the parameters of the relaxed atomic structures are found to be in very good agreement with experimental ones (when available). The phonon frequencies at the centre of the Brillouin zone, the Born effective charge tensors, and the dielectric permittivity tensors have been obtained using density-functional perturbation theory.

For the crystalline systems, we have found an excellent agreement between the calculated phonon frequencies and their corresponding experimental values. For hafnia and zirconia, both the cubic and the tetragonal phases have been investigated. For t-ZrO$_2$, we have proposed new symmetry assignments that meet all the arguments discussed in the literature. Our assignments solve the apparent contradictions of previous works, clarifying some important issues. We have also illustrated the relationship between the phonon modes of the cubic and the tetragonal phases. In all the cases, the differences between the Hf- and Zr-based systems have been analysed in detail and interpreted in terms of structural changes, the mass ratio and variations of interatomic force constants.

An important anisotropy was observed in the Born effective charge tensors. For some directions, these effective charges are found to be larger than the nominal ionic charge, indicating a mixed covalent–ionic bonding between M $=$ (Hf, Zr) and O atoms, and between Si and O atoms. We have also discussed the effective charges focusing on the changes between the systems containing hafnium and those containing zirconium.

The electronic and static dielectric permittivity constants have been computed, and a detailed analysis of the contributions of individual vibrational modes has been performed, including the computation of mode-effective charges and oscillator strengths. For the tetragonal systems (t-HfO$_2$, t-ZrO$_2$, HfSiO$_4$ and ZrSiO$_4$), we observed, for directions both parallel and perpendicular to the tetragonal axis, that a single mode contributes for more than 60% of the full ionic contribution. Our first-principles approach allows us to obtain the corresponding eigenvectors, showing clearly that the displacement is characterized by Zr and O atoms moving in opposite directions. In the silicates, the displacement of Si atoms in these modes is less than half those of the other species, inducing a substantial distortion of the SiO_4 tetrahedra in contradiction to what was previously thought. For all systems, the modifications related to the presence of Zr rather than Hf have been rationalized in terms of the difference in mass between these atoms, variations of interatomic force constants and changes in structural parameters (minor effect).

For hafnon and zircon, we have also calculated the electronic DOS in which the contributions from Hf 5s and 5p, Zr 4s and 4p and O 2s and 2p are clearly distinguishable, although the spread of the latter indicates hybridization with atomic M = (Hf, Zr) and Si orbitals.

Finally, we have investigated the dielectric properties of amorphous silicates. We have provided a simple scheme which relates the optical and static dielectric constants of Zr silicates to their underlying nanoscopic structure. Our theory supports recent experiments which find a close to linear dependence of ϵ_0 on the Zr fraction x, and shows that higher dielectric constants can be achieved by increasing the concentration of ZrO_6 SUs. For Hf silicates, we expect a very similar behaviour based on the comparison made throughout this chapter.

Acknowledgments

We thank F. Detraux and A. Bongiorno who took an active part in the research leading to the results presented in this chapter. We are also grateful to R.B. van Dover for providing us his results prior to publication. Support is acknowledged from the FNRS-Belgium (G.-M.R. and X.G.), the Swiss FNS under grant No 620-57850.99 (A.P.), the FRFC project (No 2.4556.99), the Belgian PAI-5/1/1 and the Swiss Center for Scientific Computing.

References

[1] Hohenberg P and Kohn W 1964 *Phys. Rev.* **136** B864–B871
[2] Kohn W and Sham L J 1965 *Phys. Rev.* **140** A1133–A1138
[3] Pickett W 1989 *Comput. Phys. Rep.* **9** 115–197
[4] Payne M C, Teter M P, Allan D C, Arias T A and Joannopoulos J D 1992 *Rev. Mod. Phys.* **64** 1045–1097
[5] Baroni S, de Gironcoli S, Dal Corso A and Giannozzi P 2001 *Rev. Mod. Phys.* **73** 515–562
[6] Gonze X 1997 *Phys. Rev. B* **55** 10337–10354
[7] Gonze X and Lee C 1997 *Phys. Rev. B* **55** 10355–10368
[8] Gonze X *et al* 2002 *Comput. Mater. Sci.* **25** 478–492 http://www.abinit.org
[9] Perdew J P and Wang Y 1992 *Phys. Rev. B* **45** 13244–13249
[10] Ceperley D M and Alder B J 1980 *Phys. Rev. Lett.* **45** 566–569
[11] Troullier N and Martins J L 1991 *Phys. Rev. B* **43** 1993–2006
[12] Teter M P 1993 *Phys. Rev. B* **48** 5031–5041

[13] Weast R C (ed) 1985 *Handbook for Chemistry and Physics* 65th edn (Boca Raton, FL: CRC Press) p 165
[14] Ruh R and Corfield P W R 1970 *J. Am. Ceram. Soc.* **53** 126–129
[15] Little E J and Jones M M 1960 *J. Chem. Educ.* **37** 231
[16] Duran P and Pascual C 1984 *J. Mater. Sci.* **19** 1178–1184
[17] Teufer G 1962 *Acta Cryst.* **15** 1187
[18] Ruh R, Garrett H J, Domagala R F and Tallen N M 1968 *J. Am. Ceram. Soc.* **51** 23
[19] Aldebert P and Traverse J P 1985 *J. Am. Ceram. Soc.* **68** 34–40
[20] Perry C H, Liu D W and Ingel R P 1985 *J. Am. Ceram. Soc.* **68** C184–C187
[21] Lawson S 1995 *J. Eur. Ceram. Soc.* **15** 485–502
[22] Hunter O Jr, Scheidecker R W and Tojo S 1979 *Ceram. Int.* **5** 137
[23] Královík B, Chang E K and Louie S G 1998 *Phys. Rev. B* **57** 7027–7036
[24] Rignanese G-M, Detraux F, Gonze X and Pasquarello A 2001 *Phys. Rev. B* 13430 **64** 1–7
[25] Demkov A A 2001 *Phys. Status Solidi B* **226** 57–67
[26] Zhao X and Vanderbilt D 2002 *Phys. Rev. B* 233106 **65** 1–4
[27] Zhao X and Vanderbilt D 2002 *Phys. Rev. B* 075105 **65** 1–10
[28] Wang J, Li H P and Stivens R 1992 *J. Mater. Sci.* **27** 5397–5430
[29] Zhong W, King-Smith D and Vanderbilt D 1994 *Phys. Rev. Lett.* **72** 3618–3621
[30] Lee C and Gonze X 1994 *Phys. Rev. Lett.* **72** 1686–1689
[31] Lee C, Ghosez Ph and Gonze X 1994 *Phys. Rev. B* **50** 13379–13387
[32] Gonze X, Charlier J-C, Allan D C and Teter M P 1994 *Phys. Rev. B* **50** 13035–13038
[33] Giannozzi P, de Gironcoli S, Pavone P and Baroni S 1991 *Phys. Rev. B* **43** 7231–7242
[34] Feinberg A and Perry C H 1981 *J. Phys. Chem. Solids* **42** 513–518
[35] Ishigame M and Sakurai T 1977 *J. Am. Ceram. Soc.* **60** 367–369
[36] Kim D-J, Jung H-J and Yang I-S 1993 *J. Am. Ceram. Soc.* **76** 2106–2108
[37] Hirata T, Asari E and Kitajima M 1994 *J. Solid State Chem.* **110** 201–207
[38] Kjerulf-Jensen N, Berg R W and Poulsen F W 1996 *Proc. 2nd European Solid Oxide Fuel Cell Forum* (Oberrohrdorf, Switzerland: U. Bossel) pp 647–656
[39] Pecharromán C, Ocaña M and Serna C J 1996 *J. Appl. Phys.* **80** 3479–3483
[40] Mirgorodsky A P, Smirnov M B and Quintard P E 1997 *Phys. Rev. B* **55** 19–22
[41] Mirgorodsky A P, Smirnov M B and Quintard P E 1999 *J. Phys. Chem. Solids* **60** 985–992
[42] Bouvier P and Lucazeau G 2000 *J. Phys. Chem. Solids* **61** 569–578
[43] Liu D W, Perry C H and Ingel R P 1988 *J. Appl. Phys.* **64** 1413–1417
[44] Philippi C M and Mazdiyasni K S 1971 *J. Am. Ceram. Soc.* **54** 254–258
[45] Pecharromán C and Iglesias J E 1994 *Phys. Rev. B* **49** 7137–7147
[46] Kukli K, Ihanus J, Ritala M and Leskela M 1996 *Appl. Phys. Lett.* **68** 3737–3739
[47] Gusev E P, Cartier E, Buchanan D A, Gribelyuk M, Copel M, Okorn-Schmidt H and D'Emic C 2001 *Microelectron. Eng.* **59** 341–349
[48] Wood D L and Nassau K 1982 *Appl. Opt.* **12** 2978–2981
[49] French R H, Glass S J, Ohuchi F S, Xu Y-N and Ching W Y 1994 *Phys. Rev. B* **49** 5133–5142
[50] Lanagan M T, Yamamoto J K, Bhalla A and Sankar S G 1989 *Mater. Lett.* **7** 437–440
[51] Dwivedi A and Cormack A N 1990 *Philos. Mag.* **61** 1–22
[52] Negita K 1989 *Acta Metall.* **37** 313–317
[53] Negita K and Takao H 1989 *J. Phys. Chem. Solids* **50** 325–331
[54] Speer J A and Cooper B J 1982 *Am. Mineral.* **67** 804–808

[55] Rignanese G-M, Gonze X and Pasquarello A 2001 *Phys. Rev. B* 104305 **63** 1–7

[56] Mursic Z, Vogt T, Boysen H and Frey F 1992 *J. Appl. Crystallogr.* **25** 519–523

[57] Ghosez Ph, Michenaud J-P and Gonze X 1998 *Phys. Rev. B* **58** 6224–6240

[58] Gonze X, Allan D C and Teter M P 1992 *Phys. Rev. Lett.* **68** 3603–3606

[59] Nicola J H and Rutt H N 1974 *J. Phys. C: Solid State Phys.* **7** 1381–1386

[60] Hoskin P W O and Rodgers K A 1996 *Eur. J. Solid State Inorg. Chem.* **23** 1111–1121

[61] Dawson P, Hargreave M M and Wilkinson G R 1971 *J. Phys. C: Solid State Phys.* **4** 240–256

[62] Gervais F, Piriou B and Cabannes F 1973 *J. Phys. Chem. Solids* **34** 1785–1796

[63] Pecharromán C, Ocaña M, Tartaj P and Serna C J 1994 *Mater. Res. Bull.* **29** 417–426

[64] Wilk G D and Wallace R M 2000 *Appl. Phys. Lett.* **76** 112–114; Wilk G D, Wallace R M and Anthony J M 2000 *J. Appl. Phys.* **87** 484–492

[65] Lucovsky G and Rayner G B Jr 2000 *Appl. Phys. Lett.* **77** 2912–2914

[66] Kurtz H A and Devine R A B 2001 *Appl. Phys. Lett.* **79** 2342–2344

[67] Qi W-J, Nieh R, Dharmarajan E, Lee B H, Jeon Y, Kang L, Onishi K and Lee J C 2000 *Appl. Phys. Lett.* **77** 1704–1706

[68] van Dover R B, Manchanda L, Green M L, Wilk G, Garfunkel E and Busch B unpublished

[69] Rignanese G M, Detraux F, Gonze X, Bongiorno A and Pasquarello A 2002 *Phys. Rev. Lett.* 117601 **89** 1–4

[70] Nogami M 1985 *J. Non-Cryst. Solids* **69** 415–423

[71] Misra V unpublished

[72] Varshneya A K 1994 *Fundamental of Inorganic Glasses* (San Diego, CA: Academic) p 364

[73] Weast R C (ed) 1972 *Handbook of Chemistry and Physics*, 52nd edn (Cleveland, OH: Chemical Rubber Co.) p E-204

Chapter 4.6

Defect generation under electrical stress: experimental characterization and modelling

Michel Houssa

Introduction

The scaling of the SiO_2 gate layer thickness in advanced generations of complementary metal–oxide–semiconductor (CMOS) processes is reaching its limits, both from the point of view of leakage current limitations as well as intrinsic reliability concerns (see the introductory chapter of this book). By using a thicker gate insulating layer with a higher dielectric constant than SiO_2 (3.9), the leakage current flowing through the device is expected to decrease, and the reliability of the gate dielectric is expected to improve. Numerous recent works have indeed reported leakage current reductions in high-κ based MOS devices, as compared to SiO_2 layers with equivalent electrical thicknesses [1–6]. However, defect generation in these materials under electrical stress, which is closely related to the reliability of the devices, has not yet been extensively studied.

The purpose of this chapter is to discuss the generation of defects in MOS structures with very thin $SiON/ZrO_2$ gate stacks. Defect build-up in these devices is investigated by monitoring the variations of the current density and the capacitance–voltage (C–V) characteristics of the structures during constant gate voltage stress experiments.

A polarity effect on the defect generation is reported, namely interface defects, positive and negative charges and bulk neutral traps are observed under electron injection from the TiN gate, while negative charge build-up is only observed under injection of electrons from the Si substrate. These results are discussed within the hydrogen release model [7–10]. This model assumes the liberation of hydrogen species at the anode by the injected electrons,

467

resulting in interface defect generation, followed by the random hopping of the hydrogen species in the gate dielectric, leading to bulk defect generation. This model successfully explains the generation of defects in MOS structures with thin SiO_2 layers under electrical stress [11, 12], as well as under irradiation [10].

The hydrogen release model is developed and extended to the case of MOS structures with very thin high-κ gate dielectric stacks. It is shown that the kinetics for interface and bulk defect generation under gate injection can be well reproduced by numerical simulations. Comparison between simulated and measured kinetics allows us to estimate the position of the defect centroid generated in the gate stack. Possible atomic structures for these defects are then proposed based on these results.

Experimental details

The devices investigated in this work are MOS capacitors with $80\,\mu m \times 80\,\mu m$ or $100\,\mu m \times 100\,\mu m$ gate area. (100) n- and p-type 200 mm wafers were first cleaned using the IMEC clean process. An ultra-thin thermal oxynitride layer (SiON) was grown on the substrate, using a two-step process: (1) growth of a SiO_2 layer in pure (dry) O_2 at 650°C for 10 min, and (2) furnace anneal of the layer in NO gas at 650°C for 10 min. The SiON layer thickness was estimated to be 1.5 nm from spectroscopic ellipsometry and high-resolution cross-sectional transmission electron microscopy measurements (HRTEM). The nitrogen content was estimated to be about 10 at.% from secondary ion mass spectrometry measurements. A ZrO_2 layer was next deposited on the Si/SiON substrate, using atomic layer deposition (ALD), with $ZrCl_4$ and H_2O sources (see the chapter of Ritala in the present book). The thickness of this layer was estimated from HRTEM measurements to be 3 nm. A 100 nm TiN layer was next sputtered on the gate stack to form the gate electrode. MOS capacitors were then patterned using a wet lithography process. The samples were finally annealed in N_2/H_2 at 400°C for 30 min. A schematic illustration of the devices analysed in this work is shown in figure 4.6.1, together with an HRTEM picture of the $SiON/ZrO_2$ gate stack.

The $C-V$ characteristics of the structures were measured using an HP 4275 A multi-frequency *LCR* meter. The equivalent oxide thickness of the gate stack, estimated by comparing the $C-V$ characteristics with numerical simulations, taking into account quantum mechanical corrections, is estimated to be 1.8 nm.

The measurements of the $I-V$ characteristics of the structures, and constant gate voltage stress experiments were performed with an HP 4156A or Keithley 4200 semiconductor parameter analysers. The generation of defects during the electrical stress of the devices was monitored by recording

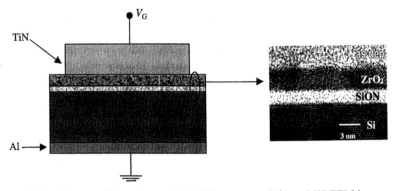

Figure 4.6.1. Schematic illustration of the MOS structures (left) and HRTEM image of a 1.5 nm SiON/3 nm ZrO$_2$ gate stack (right).

the time-variation of the gate current, as well as by measuring their C–V characteristics periodically during the stress. The structures were stressed under accumulation conditions, i.e. with a positive gate voltage for n-type substrates, resulting in the injection of electrons from the Si substrate, see figure 4.6.2 (left), and with a negative gate bias for the p-type structures, resulting in the injection of electrons from the TiN gate, as illustrated in figure 4.6.2 (right).

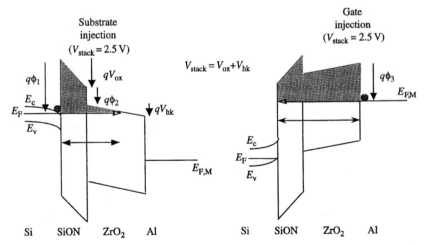

Figure 4.6.2. Schematic energy band diagram of a Si/SiON/ZrO$_2$/TiN structure under substrate injection (left) and gate injection (right) of electrons. The shaded area represents the area defined by the tunnelling distance and the energy barrier height. Due to the asymmetry of the energy band structure, the tunnelling current flowing through the gate stack is higher under substrate injection than under gate injection.

Results

The current density flowing through the structures is shown in figure 4.6.3 as a function of the voltage across the gate dielectric stack, V_{stack}, for the case of gate injection and substrate injection, respectively. It has been shown previously that the dominant charge carrier transport mechanism (at room temperature) in these devices is the tunnelling of electrons through the SiON/ZrO$_2$ gate stack [13]. The lower current density observed under gate injection can be explained by the asymmetry of the energy band diagram of the structure [13, 14] (see also the chapter of Autran *et al* in the present book). Indeed, within the Wentzel–Kramers–Brillouin (WKB) approximation, the electron tunnelling probability is inversely proportional to the area defined by the tunnelling distance and the energy barrier height. It is evident from figure 4.6.2 that this area is much larger under gate injection as compared to substrate injection; hence, the tunnelling current is lower in the former case. It should be also remarked that the tunnelling current flowing through a MOS device with a 1.8 nm SiO$_2$ layer is about 0.1 A cm^{-2}. The structures investigated in this work thus present a leakage current reduction of three to four orders of magnitude as compared to a SiO$_2$ layer with equivalent electrical thickness.

The time-dependence of the gate current density variation $\Delta J_G(t) = J_G(t) - J_G(0)$ observed during constant gate voltage stress of the capacitors is presented in figure.4.6.4, under substrate and gate injection of electrons, respectively [15, 16]. A polarity effect on the defect generation is clearly observed. Under substrate injection, figure 4.6.4(*a*), one observes that the

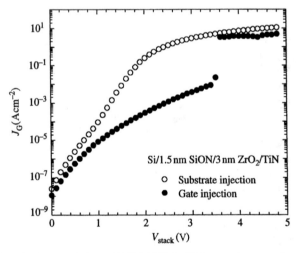

Figure 4.6.3. *I–V* characteristics of Si/SiON/ZrO$_2$/TiN capacitors, under substrate and gate injection, respectively.

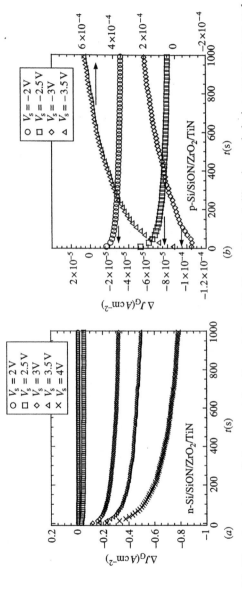

Figure 4.6.4. Time-dependence of the current-density variation $\Delta J_G(t) = J_G(t) - J_G(0)$, observed during constant gate voltage stress of Si/SiON/ZrO$_2$/TiN capacitors, (a) under substrate injection and (b) gate injection, respectively.

current density decreases with time during the electrical stress, for values of the stress voltage V_s between 2 and 4 V and for stress times up to 1000 s. We argue below that this decrease in the current density can be attributed to the trapping of electrons in bulk neutral traps present in the ZrO_2 layer. Under gate injection, figure 4.6.4(*b*), one observes a similar decrease in $\Delta J_G(t)$ for $|V_s|$ smaller than 3 V. For $|V_s| \geq 3$ V, $\Delta J_G(t)$ increases with time. This increase in $\Delta J_G(t)$ can be attributed to the *generation* of bulk neutral traps in the gate stack, leading to the so-called stress-induced leakage current (SILC), which likely arises from the trap-assisted tunnelling mechanism through neutral defects generated during the electrical stress [8] (see below).

The $C-V$ characteristics of MOS capacitors (80 µm × 80 µm area) with n- and p-type Si substrates, recorded at 10 kHz, before and after constant gate voltage stress at $|V_G| = 3.6$ V, are shown in figure 4.6.5(*a*) and (*b*), respectively. Under substrate injection, a very small shift of the $C-V$ curves towards more positive voltages is observed, indicating the generation of negative charge in the gate stack [17]. On the other hand, under gate injection, a much larger shift of the $C-V$ curves to more negative gate voltages is observed as the stress time is increased, indicating the generation of positive charge in the gate stack [17]. Notice that the curves are also stretched out along the voltage axis and present a characteristic bump near the flat-band voltage, suggesting that defects are also generated at the Si/SiON interface during injection of electrons from the TiN gate [18]. We will first discuss the generation of negative charge under substrate and gate injection.

Negative charge trapping

The tunnelling current flowing through a MOS structure depends on the electric field at the cathode, F_c. When charged defects are present in the gate dielectric layer, the electric field is shielded by the trapped charge, as illustrated in figure 4.6.6, resulting in the variation of the cathode electric field as given by [19]

$$\Delta F_c = \frac{-qN_t}{\varepsilon_0 \varepsilon_{ins}} \left(1 - \frac{x_t}{t_{ins}}\right) \tag{4.6.1}$$

where N_t is the trap density, ε_0 the permittivity of free space, ε_{ins} the relative dielectric constant of the insulating layer, t_{ins} the insulator thickness and x_t the charge centroid from the cathode. This variation of the cathode electric field leads to the decrease of the gate current during the electrical stress [19], as observed in figure 4.6.4. By analysing the decrease of the gate current induced by the negative charge trapping under substrate and gate injection, the defect density and its centroid can be estimated. The saturated value of the current density variation ΔJ_{sat} obtained from figures 4.6.4(*a*) and (*b*) is shown in figure 4.6.7 as a function of the stress voltage. The solid lines are fits to the data obtained from the simulation of the tunnelling current through

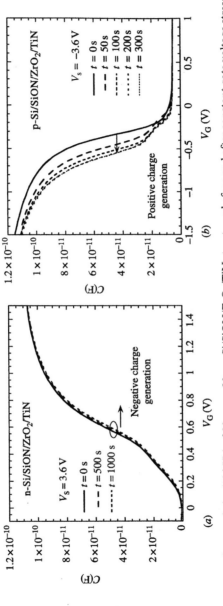

Figure 4.6.5. $C-V$ characteristics (10 kHz) of 80 μm × 80 μm Si/SiON/ZrO$_2$/TiN structures before and after constant gate voltage stress at $|V_s|$ = 3.6 V, under (*a*) substrate and (*b*) gate injection, respectively.

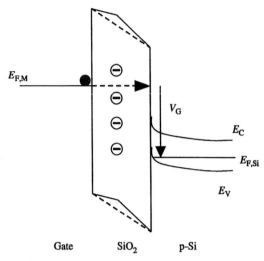

Figure 4.6.6. Schematic energy band diagram of a MOS structure, illustrating the change in electric field across the dielectric layer induced by a sheet of negatively charged defects.

the gate stack within the WKB approximation, under substrate and gate injection, respectively [13], and including equation (4.6.1) in the calculations. From these fits, N_t and x_t are found to be $3 \times 10^{12} \, \text{cm}^{-2}$ and 1 nm (away from the TiN/ZrO$_2$ interface), respectively. This analysis thus suggests that the decrease of the gate current observed under substrate and gate injection is due to the trapping of electrons in the ZrO$_2$ layer, close to the TiN interface.

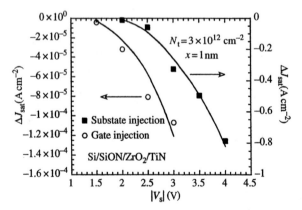

Figure 4.6.7. Saturated current density ΔJ_{sat} as a function of stress voltage $|V_s|$ under substrate and gate injection. The solid lines are obtained from simulations of the tunnelling current through the gate stack, considering the energy band diagram presented in figure 4.2.2, and including equation (4.6.1) in the computation.

The shift of the flat-band voltage induced by the negative charge can be calculated from the expression [17]

$$|\Delta V_{FB}| = \frac{qN_t}{\varepsilon_0 \varepsilon_{ins}} x_t. \tag{4.6.2}$$

For $N_t = 3 \times 10^{12}$ cm^{-2} and $x_t = 1$ nm, the corresponding flat-band voltage shift is 27 meV. This value is close to the one obtained from the results shown in figure 4.6.5(*a*), i.e. 20 meV. It should be pointed out that under gate injection, the positive charge generation is dominant over electron trapping, and a net shift of the C–V curves towards negative gate voltages is only observed in this case. This results from the fact that the positive charge is located close to the Si/SiON interface (see below), while the negative charge is located close to the ZrO$_2$/TiN interface, and, consequently, does not influence much of the flat-band voltage of the structure.

Neutral defect and positive charge generation under gate injection

The current density variation ΔJ_G observed under gate injection for $|V_s| > 3$ V is shown as a function of stress time in figure 4.6.8(*a*). The increase in the current density is more important as the stress voltage increases. The I–V characteristics of a similar capacitor, measured periodically during the stress at -3.5 V, are shown in figure 4.6.8(*b*). One observes an increase in the current density at low voltages (between 0 and -1 V). The time variation of the current and the I–V characteristics of the structures are very similar to those observed in ultra-thin SiO$_2$ layers, and is called the SILC [20–22]. It has been suggested that SILC arises from the trap-assisted tunnelling mechanism through bulk neutral defects generated in the gate dielectric layer during the electrical stress [8, 20–22], as illustrated in figure 4.6.9.

On the other hand, the density ΔN_p of positive charge generated during the electrical stress, under gate injection, and estimated from the flat-band voltage shifts observed in figure 4.6.5(*b*) is presented as a function of the stress time t in figure 4.6.10 for different values of V_s. These defect densities were obtained using equation (4.6.2), assuming that the positive charge is located at about 7 Å from the Si/SiON interface (see below). One observes that ΔN_p increases rapidly with time and saturates after 200–300 s. The saturation value $\Delta N_{p,sat}$ increases with the stress voltage V_s and ranges from about 5×10^{11} to 2.2×10^{12} cm^{-2}.

The hydrogen release model

The generation of interface defects, bulk neutral traps, and positive charge under gate injection can be explained by the hydrogen release mechanism [7–12], which is schematically illustrated in figure 4.6.11, for the case of a

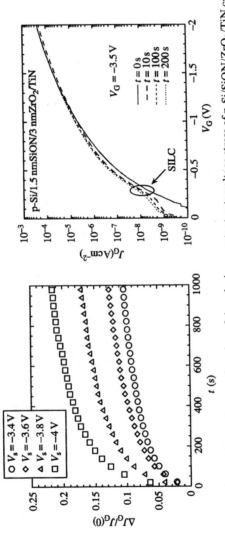

Figure 4.6.8. (*a*) Current density variation ΔJ_G as a function of time during constant gate voltage stress of p-Si/SiON/ZrO$_2$/TiN capacitors, under gate injection. (*b*) C–V characteristics of a capacitor stressed at −3.5 V for different times.

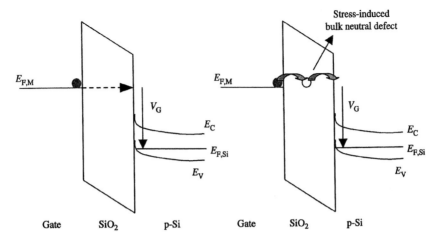

Figure 4.6.9. Schematic illustration of the trap-assisted tunnelling mechanism, responsible for the SILC observed in ultra-thin SiO_2 layers.

p-Si/SiON/ZrO$_2$/TiN capacitor, stressed under a constant gate voltage (injection of electrons from the gate). One assumes that electrons arriving at the anode with sufficient energy can release hydrogen (in atomic or proton form) close to the Si/SiON interface, generating defects at this interface, cf step (1) in figure 4.6.11. Electrons are supposed to be transported ballistically through the gate stack, i.e. the energy released at the anode is equal to qV_s. The liberated hydrogen species are then transported in the gate dielectric, where they can be trapped in the oxide networks, forming hydrogen-induced positive charge, see step (2) in figure 4.6.11, as well as bulk neutral traps, see step (3) in figure 4.6.11.

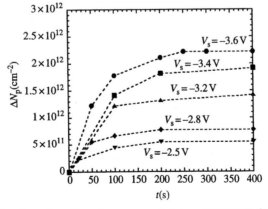

Figure 4.6.10. Density of positive charge generated in p-Si/SiON/ZrO$_2$/TiN structures during constant gate voltage stresses.

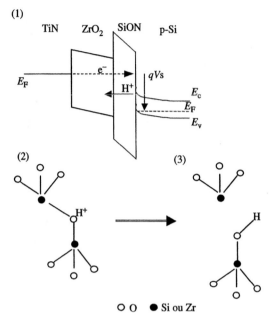

Figure 4.6.11. Schematic illustration of the hydrogen release mechanism in a Si/SiON/ZrO$_2$/TiN structure, stressed under gate injection.

Before discussing in detail the simulations of defect generation kinetics obtained within this model, let us first mention a few results which support the liberation of hydrogen during the electrical stress.

(a) (Indirect) experimental evidence for the release of hydrogen during the electrical stress of the devices, under gate injection, is based on the measurements of the inversion capacitance of the structures during the electrical stress. The relative capacitance at inversion $[C_{inv}(t) - C_{inv}(0)]/C_{inv}(0)$ of p-Si/SiON/ZrO$_2$/TiN capacitors is shown in figure 4.6.12 as a function of time during constant gate voltage stress at -3.6 V. It is evident that the inversion capacitance decreases with stress time. This decrease can be attributed to the deactivation of boron acceptors in the p-type substrate due to the release of hydrogen during the electrical stress, resulting in the formation of B–H (electrically inactive) complexes [23–25].

(b) During the ALD process, H$_2$O is used as a source for the deposition of the ZrO$_2$ layer. One can then expect H$_2$O to react with the interfacial oxynitride layer, producing H-containing electron traps. Upon electron injection, these defects can release hydrogen species (H and/or H$^+$). Therefore, it is reasonable to assume that H can be released close to the Si/SiON interface by the injected electrons under gate injection.

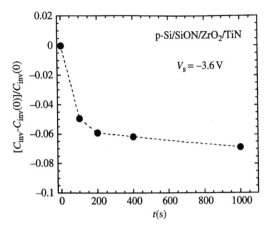

Figure 4.6.12. Normalized variation of the inversion capacitance of p-Si/SiON/ZrO$_2$/TiN capacitors as a function of the stress time during constant gate voltage stress at -3.6 V.

Being liberated, hydrogen will be transported in the gate dielectric, resulting in the generation of hydrogen-induced neutral traps like SiH [9], SiOH [10], and ZrOH [26, 27] as well as positively charged centres like [Si$_2$ = OH]$^+$ [28] (see below). These H-induced defects may then be responsible for the current density increase as well as the shifts of the C–V curves observed under gate injection.

By contrast, one expects much less hydrogen to be present at the ZrO$_2$/TiN interface, because of the cracking of OH bonds and H$_2$O at the metal/insulator interface induced by the post-metallization anneal in N$_2$/H$_2$ [29]. Consequently, under substrate injection, much less hydrogen can be potentially released at the anode, and the hydrogen-induced neutral traps and positive charge are not detected from the current density variation and shifts of the C–V curves under substrate injection.

These considerations are confirmed by hydrogen concentration profile measurements obtained from nuclear reaction analysis measurements, as shown in figure 4.6.13. It appears clearly from these results that hydrogen is located mostly close to the Si/SiO$_2$ interface after the post-metallization anneal of the structure in N$_2$/H$_2$.

(c) Cartier and DiMaria [30] proposed that the cross-section σ_H for the release of hydrogen by electron impact ionization varies like $\sigma_H \propto (E - E_{th})^2$, where E is the energy of electrons arriving at the anode and E_{th} is the threshold energy for the release of hydrogen. In figure 4.6.14, $\Delta N_{p,sat}$ and $[J_G(1000\,s) - J_G(0)]/J_G(0)$ (measured under gate injection for $|V_s| > 3$ V) are plotted as a function of $(E - E_{th})^2$, where $E = |qV_s|$ and $E_{th} = 1.6$ eV. It is assumed here that the current

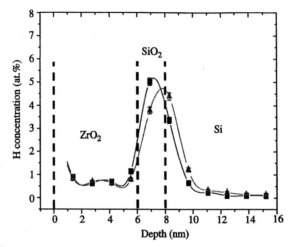

Figure 4.6.13. Hydrogen concentration profile of Si/SiO$_2$/ZrO$_2$/TiN structures obtained from nuclear reaction analysis measurements.

density increase ΔJ_G reflects the variation in the bulk electron trap density ΔN_{ot} generated during the electrical stress, i.e., $\Delta J_G(t)/J_G(0) \propto \Delta_{Not}(t)/N_{ot}(0)$ [8]. It is evident that the positive charge and bulk electron trap density also varies like $(E - E_{th})^2$. These latter results strongly suggest that both type of defect may be induced by the release of hydrogen at the anode and the trapping of hydrogen in the oxide networks.

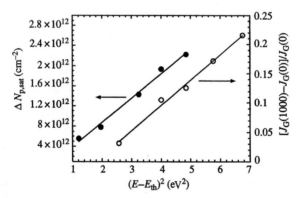

Figure 4.6.14. Saturation value of the positive charge, ΔN_{psat}, and variation of the normalized current density after constant voltage stress for 1000 s of p-Si/SiON/ZrO$_2$/TiN capacitors as a function of the square of the electron energy E arriving at the anode (with respect to the threshold energy E_{th} for hydrogen release).

Modelling

Based on these observations, we developed a model for defect generation in the Si/SiON/ZrO$_2$/TiN structures under gate injection, taking into account the release of hydrogen at the (100)Si/SiON interface, as well as the random hopping (dispersive) transport of hydrogen in the gate dielectric stack [16, 31]. We will first discuss the generation of defects at the (100)Si/SiON interface by electron impact, resulting in the release of hydrogen in the structure.

Defect generation at the (100)Si/SiON interface

The most important electrically active defect identified at the (100)Si/SiO$_2$ interface (essentially by electron-spin resonance) is the trivalent Si dangling bond, Si$_3$≡Si$^{\cdot}$, the so-called P$_{b0}$ centre [32] (see also the chapter by Stesmans and Afanas'ev in this book). The atomic structure of this defect corresponds to a Si atom at the Si/SiO$_2$ interface, backbonded to three Si atoms from the substrate, and presenting an unpaired electron in a dangling sp^3-like orbital, pointing in the [111] direction (see figure 4.6.15). This defect is produced inherently at the interface during the thermal oxidation of silicon, as a result of mismatch between the crystalline Si substrate and the amorphous SiO$_2$ layer; the density of P$_{b0}$ centres at the (100)Si/SiO$_2$ interface is typically of the order of 10^{12} cm^{-2}, for oxidation temperatures in the range 800–950°C [33].

The P$_{b0}$ centre is an amphoteric defect, i.e., it can trap holes or electrons from the Si substrate, depending on the position of the Fermi level in the Si band-gap. The energy levels corresponding to this defect are distributed in the Si band-gap, as illustrated in figure 4.6.16 [32]. This distribution can be approximated by two gaussian distributions, with mean energy levels at 0.3 eV above the valence band and 0.3 eV below the conduction band, respectively. In order to passivate (electrically) this defect, MOS structures

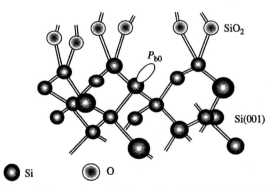

Figure 4.6.15. Schematic illustration of a P$_{b0}$ centre at the (100)Si/SiO$_2$ interface.

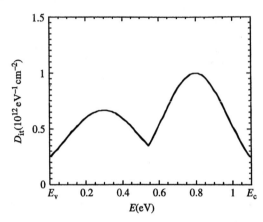

Figure 4.6.16. Distribution of the energy levels of the P_{b0} centres in the Si band-gap.

are generally annealed in a hydrogen-containing atmosphere (usually N_2/H_2). During the anneal, the P_{b0} centres chemically react with H_2, forming passivated $Si_3\equiv SiH$ centres, so-called $P_{b0}H$ [32, 33].

When charge carriers are injected into the structure, they can induce the dissociation of the Si–H bonds, provided the energy of the impacting carriers is higher than the Si–H binding energy. The electrical stress then induces the re-activation of the P_{b0} centres, as well as the release of hydrogen (in atomic or proton form) close to the Si/SiO_2 interface, as illustrated in figure 4.6.17 [32]. The generation of the P_{b0} centres can be monitored by the change of the C–V characteristics of the MOS structure during the electrical stress, as discussed below.

The 10 kHz C–V characteristics of a p-Si/1.5 nm SiON/3 nm ZrO_2/TiN structure (80 μm × 80 μm area) recorded before (circles) and after (squares) constant gate voltage stress at $V_s = -3.6$ V for 400 s, corresponding to the injection of 2×10^{19} electrons cm^{-2} from the TiN gate, are shown in figure 4.6.18(a) [34]. A large 'bump' in the C–V curve is clearly observed between -0.5 and 0 V after electron injection, indicating the generation of defects at the Si/SiON interface [18, 34]. The solid lines in figure 4.6.18(a) are theoretical C–V curves obtained from numerical simulations [18], taking into account the presence of defects at this interface, with a characteristic energetic distribution in the Si band-gap corresponding to the P_{b0} centre (cf figure 4.6.16). A uniform positive charge density was also included in the model, in order to reproduce the shift of the C–V characteristics towards more negative gate voltages after the electrical stress. It is clear from figure 4.6.18(a) that the bump observed in the C–V characteristics after electron injection can be quite well reproduced using such an interface defect energy distribution, and can be thus reasonably attributed to the generation of P_{b0} defects. It should be pointed out that the C–V characteristics measured

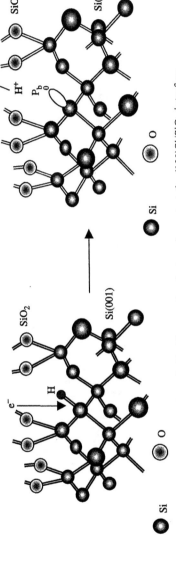

Figure 4.6.17. Illustration of the dissociation of a $P_{b0}H$ centre by electron impact at the (100)Si/SiO$_2$ interface.

484   *Defect generation under electrical stress*

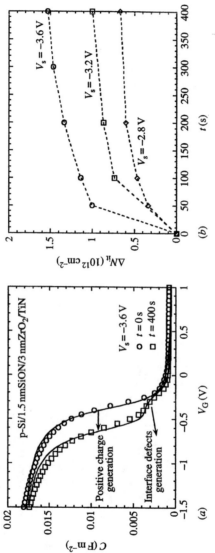

Figure 4.6.18. (*a*) C–V characteristics of a p-Si/SiON/ZrO$_2$/TiN structures, before and after constant V_G stress at -3.6 V. Solid lines are fits to the data. (*b*) Density of interface defects generated during constant V_G stress of p-Si/SiON/ZrO$_2$/TiN structures. Dashed lines guide the eye.

at 10 kHz on p-Si/SiON/ZrO$_2$/TiN capacitors are only sensitive to interface defects in the positive charge state, i.e., located in the lower part of the silicon energy band-gap. The comparison between the measured and calculated C–V characteristics allows us to extract the variation in the density of interface defects ΔN_{it} after electron injection. These densities are presented in figure 4.6.18(b) as a function of time during constant gate voltage stress at -2.8, -3.2 and -3.6 V, respectively. The density of interface defects increases with time and with the stress voltage $|V_{sl}|$, with typical densities in the range 5×10^{11}–1.5×10^{12} cm^{-2}. We will assume here that these defects are P$_{b0}$ centres. It should be remarked, though, that an atomistic identification of such defects is only possible by using electron spin resonance.

The kinetics of P$_{b0}$ defect generation during electron injection from the TiN gate is modelled using the following electro-chemical reaction [34, 35]:

$$\text{Si}_3 \equiv \text{SiH} + e^- \longrightarrow \text{Si}_3 \equiv \text{Si}^{\cdot} + \text{H} + e^-, \qquad (4.6.3)$$

leading to the following first-order differential equations

$$\frac{\mathrm{d}[\text{Si}_3 \equiv \text{Si}^{\cdot}]}{\mathrm{d}t} = k_d[\text{Si}_3 \equiv \text{SiH}][e^-]$$

$$\frac{\mathrm{d}[e^-]}{\mathrm{d}t} = \frac{J_{\text{inj}}}{q} \qquad (4.6.4)$$

where J_{inj} is the injected current density and k_d is the dissociation rate constant, assumed to be given by the Arrhenius equation

$$k_d = k_0 \exp(-E_d/k_B T) \qquad (4.6.5)$$

where k_0 and E_d are related to the attempt frequency and the activation energy for the dissociation of the Si$_3 \equiv$SiH centre, respectively.

It has been recently demonstrated through extensive ESR work [36, 37] that the activation energy E_d is not a single value, but that there exists a (gaussian) spread characterized by the standard deviation σ_{Ed}. The physical origin of this spread is most probably from slight variation in the local strain at the Si/SiO$_2$ interface [36]. Taking into account this gaussian spread of activation energies E_d, the following equation is found for the kinetics of generation of interface defects

$$\frac{[\text{Si}_3 \equiv \text{Si}^{\cdot}]}{N_0} = 1 - \frac{1}{\sqrt{2\pi}\sigma_{Ed}} \int_0^\infty \exp\left(-\frac{(E_d - E_{di})^2}{2\sigma_{Ed}^2}\right)$$

$$\times \exp\left(-\frac{J_{\text{inj}}}{2q} t^2 k_0 \exp\left(-\frac{E_d}{k_B T}\right)\right) \mathrm{d}E_d \qquad (4.6.6)$$

where N_0 is the initial density of $Si_3{\equiv}SiH$ centres and E_{di} the mean activation energy.

The time-dependence of the normalized density of P_{b0} centres calculated from equation (4.6.6) is shown in figure 4.6.19, for $\sigma_{Ed} = 0\,eV$ (corresponding to the case of a single activation energy) and $\sigma_{Ed} = 0.2\,eV$, respectively. The other parameters were fixed to the following typical values [35]: $T = 300\,K$, $J_{inj} = 10^{-2}A\,cm^{-2}$, $E_{di} = 3\,eV$, and $k_0 = 2.5 \times 10^{23}\,cm^2\,s^{-1}$. One can notice a change of curvature of the theoretical plot when the distribution of activation energies is taken into account. Comparing these theoretical curves to the experimental data shown in figure 4.6.18(*b*), it is clear that the distribution of activation energies has to be included in the model in order to explain the kinetics of interface defect generation under electrical stress.

The variation in the density of interface defects during constant gate voltage stress of p-Si/SiON/ZrO$_2$/TiN structures is compared with theoretical curves obtained using equation (4.6.6) in figure 4.6.20. A good agreement between the experimental and theoretical results can be obtained, considering E_{di} as a free parameter (see below), and fixing the other parameters to the following values (for the different values of V_s): $k_0 = 2.5 \times 10^{23}\,cm^2\,s^{-1}$, $\sigma_{Ed} = 0.15\,eV$ and $N_0 = 3.15 \times 10^{12}\,cm^{-2}$. The values of J_{inj} for the different gate voltage stress were obtained from the *I–V* characteristics of the structures presented in figure 4.6.3. It should be noticed that the values obtained for σ_{Ed} and N_0 are close to the one derived from ESR experiments performed after vacuum annealing of Si/SiO$_2$ interfaces [36].

In order to reproduce the stress voltage dependence of ΔN_{it}, we had to assume that E_{di} depends on V_s. The derived values of E_{di} are shown in figure 4.6.21 as a function of the electric field E_{ox} across the interfacial SiON layer, corresponding to different values of V_s. It is found that E_{di} decreases

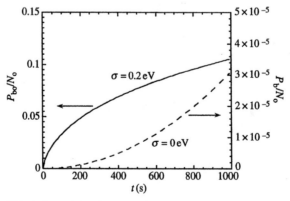

Figure 4.6.19. Simulation of the kinetics of generation of P_{b0} centres during electron injection at the (100)Si/SiO$_2$ interface of a MOS structure.

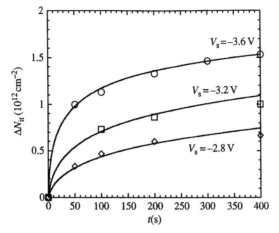

Figure 4.6.20. Density of interface defects generated during constant V_G stress of p-Si/SiON/ZrO$_2$/TiN structures. Solid lines are fits to the data using equation (4.6.6).

linearly with E_{ox}, a behaviour that can be explained as follows. Due to the different electronegativity between Si (1.8) and H (2.1), the Si–H bond presents a permanent dipole moment oriented towards the Si atom. When a negative bias is applied to the gate, an electric field is developed across the SiON layer, and the dipole moment of the Si–H bond tends to align parallel with the electric field, i.e., the Si–H bond tends to flip towards the Si substrate, as illustrated in figure 4.6.22. Since the Si–H dissociation pathway is believed to occur via the flipping of H towards a Si–Si bond next to the

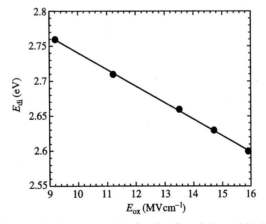

Figure 4.6.21. Mean activation energy E_{di} for the dissociation of P$_{b0}$H centres during electron injection in p-Si/SiON/ZrO$_2$/TiN structures, as a function of the electric field E_{ox} across the SiO$_2$ layer. The solid line is a fit to the data using equations (4.6.7) and (4.6.8).

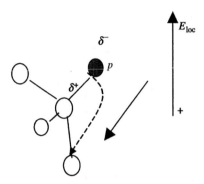

Figure 4.6.22. Schematic drawing of a $P_{b0}H$ centre at the (100)Si/SiO$_2$ interface, illustrating the orientation of the Si–H dipole with respect to the local electric field E_{loc} applied in the SiO$_2$ layer.

dangling bond [38, 39], the alignment of the Si–H dipole with respect to the electric field may favour the dissociation of the H atom, and leads to the reduction of the mean activation energy, given by

$$E_{di} = E_{di}(0) - p\cos(\theta)E_{loc} \qquad (4.6.7)$$

where p is the Si–H permanent dipole moment, θ the angle between the Si–H bond and the electric field ($\theta = 45°$ for the electric field applied normal to the (100)Si surface) and E_{loc} the local electric field, which is related to the field across the oxide by the Lorentz expression [40]

$$E_{loc} = (1 + \chi/3)E_{ox} \qquad (4.6.8)$$

where χ is the electric susceptibility of the gate oxide ($\chi = 2.9$ for SiO$_2$).

The solid line in figure 4.6.21 is a fit to the data using equations (4.6.7) and (4.6.8). From this linear fit, $E_{di}(0)$ is found to be equal to 2.98 eV, i.e. almost the Si–H binding energy [41], and the dipole moment of the Si–H bond at the interface is found to be 1.3×10^{-29} cm. This dipole moment is the product of the Si–H bond length and the effective charge of the dipole δq, where q is the electron charge. Fixing the Si–H bond length at 1.48 Å [39], one then finds $\delta \approx 0.5$. This value is in reasonable agreement with the one calculated by Biswas and Li in bulk hydrogenated amorphous silicon, i.e. $\delta \approx 0.2$–0.3 [42]. It should be remarked that the dipole moment was obtained by fixing the electric susceptibility of the SiON layer at 2.9. Since the presence of nitrogen in SiO$_2$ leads to an increase of the dielectric constant and the electric susceptibility of the material, a somewhat larger value of χ would lead to a lower value of the extracted dipole moment.

Positive charge and neutral trap generation in the SiON/ZrO₂ gate stack

When a $Si_3\equiv SiH$ centre is depassivated by an electron arriving at the (100)Si/SiON interface, an electrically active P_{b0} centre is generated at the interface, and a hydrogen species (atomic hydrogen or proton) is liberated. This hydrogen species, that we will assume to be H^+ in the following, is then transported towards the TiN gate by a random hopping transport in the gate dielectric stack, via Si–O–Si or Zr–O–Zr bonds [12]. The protons can then be trapped in the oxide networks, leading to the generation of bulk neutral and positively charged defects.

The kinetics for defect generation in the gate stack is calculated within a dispersive transport model, i.e. defect generation is assumed to be limited by the random hopping of H^+ and its trapping in the SiO_2/ZrO_2 stack. Let us recall that the hydrogen dispersive transport model was used previously by Brown and Saks to explain the generation of interface states after irradiation of MOS devices [43, 44]. These authors assumed that H^+ was generated during irradiation and was then transported towards the Si/SiO₂ interface, resulting in the depassivation of Si dangling bonds at this interface.

Within this model, the time dependence of the defect density variation $\Delta N(t)$ is given by

$$\Delta N(t) = N(t)_{H^+}\left[1 - \int_0^{l_{SiON}(t)} G_{SiON}(y)\,dy - \int_{l_{SiON}(t)}^{l_{ZrO_2}(t)} G_{ZrO_2}(y)\,dy\right] \quad (4.6.9)$$

where $N(t)_{H^+}$ is the density of H^+ ions generated close to the Si/SiO₂ interface by electron impact. According to the electro-chemical reaction (4.6.3), the kinetics of H^+ generation is equivalent to the kinetics of P_{b0} generation; $N(t)_{H^+}$ is thus calculated using equation (4.6.6). The functions $G_{SiON}(y)$ and $G_{ZrO_2}(y)$ in equation (4.6.9) are related to the probabilities $P_{SiON}(x,t)$ and $P_{ZrO_2}(x,t)$ for finding a hopping ion at a distance x at time t in the SiON and ZrO₂ layer, respectively [44], according to the expression

$$P(x,t) = \frac{G(y)}{\mu t^\alpha} \quad (4.6.10)$$

where μ is the average hopping distance of the H^+ ion, $0 < \alpha < 1$ is a parameter characterizing the dispersive transport of the ion in the material (the smaller the α, the more dispersive the transport process), and y is related to x and t according to

$$y = \frac{x}{\mu t^\alpha} \quad (4.6.11)$$

In equation (4.6.9), the integration limits $l_{SiON}(t)$ and $l_{ZrO_2}(t)$ depend on the position of the hydrogen-induced defect centroid x_c according to the expression

$$l(t) = \frac{x_c}{\mu t^\alpha}.\tag{4.6.12}$$

McLean and Ausman [45] derived an approximate expression for $G(y)$, within the continuous-time random walk model, which describes the random hopping of a particule in a periodic lattice [46–48]. The approximate expression of G reads

$$G(Z) = \begin{cases} (A\theta/y_0)(1 + BZ)^{\alpha-1/2}\exp(-Z) & \text{for } Z > 0 \\ 0 & \text{for } Z < 0 \end{cases}\tag{4.6.13}$$

where $Z = y/y_0$ and $\theta = (1 - \alpha)^{-1}$. The parameters A, B and y_0 depend on the value of α. The parameters were tabulated by McLean and Ausman, and are given in table 4.6.1.

Equations (4.6.9)–(4.6.12) have been used to model the time and voltage dependence of the positive charge, ΔN_p, and the bulk neutral defects responsible for SILC, ΔN_{ot}, observed in the $Si/SiON/ZrO_2/TiN$ structures [16]. Numerical simulations are compared to the experimental data in figure 4.6.23(a) and (b). One can see that the kinetics for the defect build-up can be quite well reproduced by this dispersive H^+ model. The hopping distances were fixed at the distance between nearest neighbour oxygen atoms in the material, i.e. $\mu_{SiO_2} = 2.5\,\text{Å}$ and $\mu_{ZrO_2} = 2.6\,\text{Å}$, assuming that H^+ is hopping between nearby O atoms. The value of α_{SiO_2} was fixed at 0.3, as previously determined from the analysis of radiation-induced [43, 44] and electrically induced [11, 12] defect generation in thin SiO_2 layers.

The time and voltage dependence of ΔN_p could be better reproduced by considering that these defects are generated in the SiON layer, i.e. the third

Table 4.6.1. Values of A, B and y_0 parameters in equation (4.6.13), corresponding to the different values of α [45].

α	A	B	y_0
0	1	0	1
0.1	1.152 18	1.013 88	1.335 15
0.2	1.120 95	2.489 09	1.531 09
0.3	0.959 267	4.835 13	1.665 79
0.4	0.769 309	9.031 30	1.746 83
0.5	0.564 190	—	1.772 45
0.6	0.373 309	32.6538	1.741 66
0.75	0.150 344	121.688	1.586 18
0.9	2.5359×10^{-2}	1749.6	1.291 87
1	0	∞	1

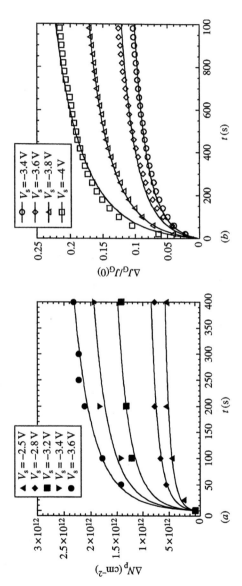

Figure 4.6.23. (*a*) Density of positive charges and (*b*) variation of the gate current as a function of time during constant V_G stress of p-Si/SiON/ZrO$_2$/TiN capacitors. Solid lines are fits to the data using equations (4.6.9)–(4.6.13).

term of equation (4.6.9) was neglected. This is hinted by the observation that the positive charge density increases very fast during the stress and reaches rapidly the saturation value $\Delta N_{p,sat}$, indicating that the distance travelled by H^+ is very small [12, 44]. The position of the positive charge centroid x_{c,SiO_2} is then left as a free parameter, the same value being consistently used for the different stress voltages. We then estimated $x_{c,SiO_2} = 7$ Å, from the Si/SiON interface. The hydrogen-induced positive charge is thus located very close to this interface, suggesting that this defect could be an over-coordinated oxygen centre, i.e. $[Si_2=OH]^+$. Afanas'ev and Stesmans [28] indeed proposed that this positive centre could be stabilized by the strain residing close to the Si/dielectric interface. It should be remarked that due to its location, this positive charge is unlikely to produce the increase in gate current observed in figure 4.6.23(*b*). As a matter of fact, as already discussed above, the current is sensitive to charged defects located close to the cathode [19], i.e. close to the TiN/ZrO$_2$ interface in the case of electron injection from the metal gate. This result is thus consistent with the assumption that the increase in gate current can be attributed to the generation of neutral defects that lead to the trap-assisted tunnelling process.

Concerning the generation of these neutral defects, the results shown in figure 4.6.23(*b*) were better reproduced by assuming that most of these defects are generated in the ZrO$_2$ layer. The build-up of N_{ot} is indeed much slower than the build-up of N_p, see figure 4.6.23(*a*) and (*b*), suggesting that the distance travelled by the H^+ ions is larger for the former defects. In this case, x_{c,SiO_2} was replaced by the SiO$_2$ layer thickness (1.5 nm) and x_{c,ZrO_2} and α_{ZrO_2} were used as free parameters (again, the same set of parameters was used to reproduce the data measured under different voltages). We then estimated $x_{c,ZrO_2} = 11$ Å, from the SiO$_2$/ZrO$_2$ interface, and $\alpha_{ZrO_2} = 0.2$, indicating a highly dispersive transport process of the H^+ ions in the poly-crystalline high-κ material. The hydrogen-induced neutral defect generated in the ZrO$_2$ layer is found to be closer to the SiO$_2$/ZrO$_2$ interface than the TiN/ZrO$_2$ interface; this defect could be a ZrOH centre, as suggested previously [26, 27], this defect being induced by the breaking of bridging oxygen bonds in the oxide networks, and trapping of H^+ at the remaining ZrO sites.

It is worthwhile to add a comment about the atomic identification of defects generated during the electrical stress of gate insulators. In the case of SiO$_2$, electrically detected magnetic resonance (EDMR) techniques have been used to identify paramagnetic defects produced in the gate insulator during electrical stress. It has been shown that bulk defects (still unidentified) are produced during the electrical stress [49]. However, the hydrogen-induced defects discussed here are diamagnetic centres that cannot be detected by electron-spin resonance techniques (like EDMR), the techniques of choice for atomistic identification of defects in solids.

Conclusions

The generation of defects in $Si/SiON/ZrO_2/TiN$ structures, during constant gate voltage stress experiments, was investigated. A polarity dependence for the defect generation was observed, both in the current density variation and shifts of the $C-V$ characteristics during stress time, namely interface defect, negative, positive, and bulk neutral trap generation was observed under electron injection from the metal gate, while a net negative charge was only observed under electron injection from the Si substrate. The analysis of the negative charge build-up, under both polarities, suggested that these defects correspond to the trapping of electrons by traps located in the ZrO_2 layer, at about 1 nm away from the ZrO_2/TiN interface.

The kinetics for defect generation under gate injection could be explained within the hydrogen release model, i.e., assuming that H^+ ions are released at the Si/SiON interface by the injected electrons, producing P_{b0} centres at this interface. The H^+ ions are subsequently randomly hopping in the $SiON/ZrO_2$ gate stack, where they can be trapped in the oxide networks, causing the generation of positively charged and bulk neutral defects. The release of hydrogen species during the electrical stress was hinted by the decrease of the inversion capacitance of the capacitors with p-type substrates, suggesting the formation of B–H inactive centres. Additionally, the injected-electron energy dependence of the density of positive charge and bulk neutral traps was found to be similar to the energy dependence of the hydrogen release cross-section at the Si/SiON interface, suggesting that these defects were induced by the transport and trapping of hydrogen in the gate stack.

Based on these observations, the kinetics for the defect generation was simulated numerically. The interface defect generation was modelled by a first-order electro-chemical reaction, assuming the depassivation of $Si_3{\equiv}SiH$ centres at the (100)Si/SiON interface by the injected electrons. This model included a Gaussian spread for the dissociation energy of these centres, as well as the electric field dependence of the mean dissociation energy induced by the orientation of the permanent dipole moment of the Si–H bond with respect to the applied electric field. The simulated interface defect generation was in very good agreement with the experimental data.

The positive charge and bulk neutral defect generation kinetics were next calculated within the dispersive hydrogen transport model, considering the random hopping of H^+ ions in the gate stack and their subsequent trapping in the oxide networks. By comparing this model to experimental results, we found that the positive charge was located in the SiON layer, within about 7 Å from the Si/SiON interface, suggesting that this charge could be the over-coordinated oxygen centre, $[Si_2{=}OH]^+$. This defect is supposed to be stabilized by the strain residing at the Si/SiON interface. On the other hand, the neutral defects responsible for SILC were found to be mostly generated in the ZrO_2 layer, with the position of their centroid located at about 11 Å from the $SiON/ZrO_2$ interface. This defect could be a neutral ZrOH centre, induced by the breaking of bridging $Zr-O-Zr$ bonds, and the subsequent trapping of the H^+ ions at the remaining ZrO site.

Acknowledgments

The author would like to express his deepest gratitude to Prof. André Stesmans and Dr. Valery Afanas'ev (University of Leuven), Dr. Marc Heyns, Dr. Stefan de Gendt, Dr. Mohamed Naili, Dr. Rick Carter, Dr. Hugo Bender, and Dr. Bert Brijs (IMEC), and Prof. Jean-Luc Autran (University of Provence), for their valuable contribution to this work. Fruitful discussions with Dr. Howard Huff (International Sematech), Prof. Gerry Lucovsky (North Carolina State University), Prof. Bob Wallace (University of North Texas), Prof. Pierre Gentil (Institut National Polytechnique de Grenoble), Prof. Yves Danto (University of Bordeaux I), Dr. Thomas Skotnicki (STMicroelectronics), Dr. François Martin and Dr. Charles Leroux (CEA-LETI), and Dr. Michel Lannoo (L2MP-CNRS) are acknowledged. This work has been financially supported by International Sematech, the MEDEA T653 (ALADIN) project and the RMNT Kappa project.

References

[1] Kwo J, Hong M, Kortan A R, Queeney K T, Chabal Y J, Mannaerts J P, Boone T, Krajewski J J, Sergent A M and Rosamilia J M 2000 *Appl. Phys. Lett.* **77** 130
[2] Wilk G D, Wallace R M and Anthony J M 2001 *J. Appl. Phys.* **89** 5243
[3] Copel M, Cartier E and Ross F M 2001 *Appl. Phys. Lett.* **78** 1607
[4] Copel M, Cartier E, Gusev E P, Guha S, Bojarczuk N and Poppeller M 2001 *Appl. Phys. Lett.* **78** 2670
[5] Ngai T, Qi W J, Sharma R, Fretwell J L, Chen X, Lee J C and Banerjee S K 2001 *Appl. Phys. Lett.* **78** 3085
[6] Misra V, Heuss G P and Zhong H 2001 *Appl. Phys. Lett.* **78** 4166
[7] Arnold D, Cartier E and DiMaria D J 1994 *Phys. Rev. B* **49** 10278
[8] DiMaria D J and Cartier E 1995 *J. Appl. Phys.* **78** 3883
[9] Blöchl P E and Stathis J H 1999 *Phys. Rev. Lett.* **83** 372
[10] Afanas'ev V V and Stesmans A 1999 *J. Electrochem. Soc.* **146** 3409
[11] Houssa M, Stesmans A, Carter R J and Heyns M M 2001 *Appl. Phys. Lett.* **78** 3289
[12] Houssa M, Afanas'ev V V, Stesmans A and Heyns M M 2001 *Microelectron. Eng.* **59** 367
[13] Houssa M, Naili M, Afanas'ev V V, Heyns M M and Stesmans A 2001 *Proceedings of the 2001 VLSI-TSA Conference* (Piscataway, NJ: IEEE) p 196
[14] Zu X, Houssa M, De Gendt S and Heyns M M 2002 *Appl. Phys. Lett.* **80** 1975
[15] Houssa M, Afanas'ev V V, Stesmans A and Heyns M M 2001 *Appl. Phys. Lett.* **79** 3134
[16] Houssa M, Autran J L, Afanas'ev V V, Stesmans A and Heyns M M 2002 *J. Electrochem. Soc.* **149** F181

[17] Sze S M 1981 *Physics of Semiconductor Devices* (New York: Wiley)

[18] Masson P, Autran J L, Houssa M, Garros X and Leroux Ch 2002 *Appl. Phys. Lett.* **81** 3392

[19] DiMaria D J, Cartier E and Arnold D 1993 *J. Appl. Phys.* **73** 3367

[20] Ricco B, Gozzi G and Lanzoni M 1998 *Trans. IEEE Electron Devices* **45** 1554

[21] De Blauwe J, Van Houdt J, Wellekens D, Groeseneken G and Maes H 1998 *IEEE Trans. Electron Devices* **45** 1745

[22] Houssa M, Mertens P W and Heyns M M 1999 *Semicond. Sci. Technol.* **14** 892

[23] Pankove J I, Carlson D E, Berkeyheiser J E and Wance R O 1983 *Phys. Rev. Lett.* **24** 2224

[24] Sah C, Sun J Y and Tzou J J 1983 *Appl. Phys. Lett.* **43** 204

[25] Afanas'ev V V and Stesmans A 2001 *Europhys. Lett.* **53** 233

[26] Houssa M, Naili M, Heyns M M and Stesmans A 2000 *Appl. Phys. Lett.* **77** 1381

[27] Houssa M, Naili M, Heyns M M and Stesmans A 2001 *Japan. J. Appl. Phys.* **40** 2804

[28] Afanas'ev V V and Stesmans A 1998 *Phys. Rev. Lett.* **80** 5176

[29] Afanas'ev V V, de Nijs J M M and Balk P 1995 *J. Non-Cryst. Solids* **187** 248

[30] Cartier E and DiMaria D J 1993 *Microelectron. Eng.* **22** 207

[31] Houssa M, Afanas'ev V V, Stesmans A and Heyns M M 2001 *Semicond. Sci. Technol.* **16** L93

[32] Helms C R and Poindexter E H 1994 *Rep. Prog. Phys.* **57** 791 and references therein

[33] Stesmans A 2000 *J. Appl. Phys.* **88** 489

[34] Houssa M, Autran J L, Stesmans A and Heyns M M 2002 *Appl. Phys. Lett.* **81** 709

[35] Houssa M, Autran J L, Heyns M M and Stesmans A 2003 *Appl. Surf. Sci* **212/213** 749

[36] Stesmans A 2000 *Phys. Rev.* **B61** 8393

[37] Stesmans A 2002 *J. Appl. Phys.* **92** 1317

[38] Van de Walle C G 1994 *Phys. Rev.* **B49** 4579

[39] Van de Walle C G and Street R B 1994 *Phys. Rev.* **B49** 14766

[40] Kittel C 1983 *Introduction to Solid State Physics* 5th edn (Paris: Dunod)

[41] Lide D R ed 1998 *CRC Handbook of Chemistry and Physics* 79th edn (Boca Raton, FL: Chemical Rubber Company)

[42] Biswas R and Li Y P 1999 *Phys. Rev. Lett.* **82** 2512

[43] Saks N S and Brown D B 1989 *IEEE Trans. Nucl. Sci.* **36** 1848

[44] Brown D B and Saks N S 1991 *J. Appl. Phys.* **70** 3734

[45] McLean F B and Ausman G A 1977 *Phys. Rev.* **B15** 1052

[46] Montroll E W and Weiss G H 1965 *J. Math. Phys.* **6** 167

[47] Scher H and Lax M 1973 *Phys. Rev.* **B7** 4491

[48] Scher H and Montroll E W 1975 *Phys. Rev.* **B12** 2455

[49] Stathis J H 1996 *Appl. Phys. Lett.* **68** 1669

SECTION 5

TECHNOLOGICAL ASPECTS

Chapter 5.1

Device integration issues

E W A Young and V Kaushik

CMOS: the low-power building blocks

In the late 1970s, NMOS transistors (n-channel MOSFETs on p-substrate) were the key building blocks in MOS technology. However, since the late 1980s, complementary MOS (CMOS) have become fashionable. For more than 20 years now, CMOS has been the faithful workhorse. The package density of transistors increased over the years, because the MOS transistors (see figure 5.1.1) could easily be scaled down in size without a performance penalty. The key feature of CMOS (see figure 5.1.2) is the ability to limit power dissipation. With the CMOS inverter structure, power is only consumed during switching operations. In between switching operations, power dissipation is limited to source to drain leakage current in the off state (off-state leakage). Until recent times, leakage current through the gate (gate leakage) could be neglected. However, with gate thickness shrinking to a point where direct tunnelling current through the gate becomes large, gate leakage should be taken into account too. Generally speaking, gate leakage becomes a problem when gate leakage $\geq$ off current. Though this depends on the transistor design for a specific application area (see below), generally speaking, gate leakage becomes a concern when it exceeds $1-10\,\mathrm{A\,cm^{-2}}$. This will happen when the equivalent oxide thickness (EOT) drops below 1.5 and approaches 1.0 nm around 2006 [1].

Conventional CMOS

A conventional CMOS process flow consists of a mature set of subsequent process steps. The introduction of high-κ material will have an impact on quite a number of these steps [2].

Conventional CMOS uses NMOS and PMOS transistor structures. In the initial steps of processing, the NMOS and PMOS areas are separated

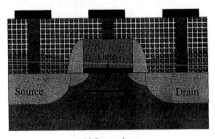

MOS transistor
Leakage current paths

Figure 5.1.1. Schematic representation of a MOS transistor.

(shallow trench isolation, STI) and defined (well implants). The gate stack (dielectric and poly) is deposited next and the poly is subsequently etched to form the gate. Source and drain areas are implanted in a self-aligned process, using the gate as a mask (with additional litho to shield either PMOS or NMOS). After a re-oxidation of the gate area, spacers are formed and subsequent implants are applied followed by thermal activation of these implants by means of rapid thermal anneal (RTA). In the silicidation process, the silicon contact areas are converted into silicides. The final steps in the front-end processing are capping of the structure and the formation of contact holes. High-temperature processes are in the early sequence of steps such as the oxidation steps in the STI formation process and the gate formation. Towards the back end of line (BEOL) of the processing, processing temperatures are kept low to avoid damage to the structure. Source/drain and gate activation anneals are exceptions to this rule. Appropriate activation of the implanted dopants requires high-temperature processing. The materials and structures in the gate stack should be able to withstand this high-temperature process step.

The high-κ deposition options, both MOCVD- and ALD-based processes, are low-temperature processes and match with the STI substrates

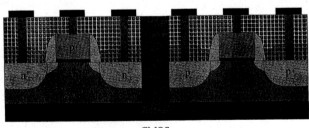

CMOS

Figure 5.1.2. Schematic representation of a CMOS building block.

quite well. There are some issues with the cleaning strategy in particular for SOC applications (see below). Process compatibility of post-high-κ process steps, however, severely limits the materials options and meanwhile urges the use of new processes and integration concepts. Some of the post-high-κ processes that will be highlighted in this chapter are:

- poly-Si deposition on the high-κ;
- gate definition (etch);
- re-oxidation and spacer formation;
- S/D activation.

One CMOS, many applications, one high κ

The CMOS technology can be used in a wide variety of applications. The current ITRS roadmap (figure 5.1.3) distinguishes three main application areas:

(a) high performance (MPU);
(b) low operation power (LOP);
(c) low standby power (LSTP).

Desktop computers are in the high-performance category, while portable equipment, cell phones and the like use low-power chips. The current ITRS roadmap [3] lays out the front-end gate stack requirements for each of the application areas. The more aggressive scaling of the gate thickness (EOT) is in the high-performance applications. Low-power applications tend to follow this EOT scaling trend with a delay of 2–3 years.

In high-performance applications, the high-speed requirement urges the use of relatively high off-state currents. The gate leakage current, even for classical gate oxides, is not expected to exceed the off-state current. In principle, standard gate dielectrics (siliconoxynitrides) should be able to meet the requirements. It is the manufacturability, and to a lesser extent overall heat dissipation, that are the key drivers to replace silicon dioxide gates by low-leakage high-κ in HP applications rather than power consumption due

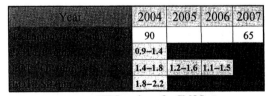

Year	2004	2005	2006	2007
	90			65
0.9–1.4				
1.4–1.8	1.2–1.6	1.1–1.5		
1.8–2.2				

ITRS roadmap for CMOS
Node and gate dielectric thickness (EOT)

Figure 5.1.3. ITRS roadmap for gate dielectrics.

to gate leakage itself. In low-power options, the direct tunnel current through the gate is the driver. The gate leakage will soon exceed S/D off current in these designs and is the dominant factor in battery power loss during standby.

Though the applications might be fairly dissimilar (high performance and low power), it is generally thought that there will be just one high-κ solution. One base material (and deposition tool) should fit high performance as well as low power. In many cases, these process applications run in a single fab, if not on a single wafer (see system-on-chip (SOC)). Needless to say, this high-κ solution should allow scaling and be more than just a one-generation solution [4].

CMOS process integration

The current roadmaps show introduction of high-κ in conventional CMOS with poly gates for nodes ≤ 90 nm. Advanced metal gates will be introduced in a later stage, around nodes ≤ 45 nm (see the section on 'Metal gates'). A suitable gate stack solution should therefore cover poly integration on high-κ. The field of metal gates is still relatively unexplored anyhow.

Tools

In the conventional CMOS flow, the next step after gate dielectric deposition is the poly deposition. It is common practice to do the poly deposition in a stand-alone furnace. This procedure requires a stable dielectric that is not affected by air exposure. However, it is well recognized that high-κ dielectrics do not prevent the underlying silicon from oxidation. Depending on the interface, wafers are susceptible to oxidation even during air exposure at room temperatures. In particular, for thin gates the impact of minor additional oxidation of the silicon on the EOT of the stack is large. In those cases, air exposure should be avoided and a clustered solution is preferred.

Cluster tools are not the preferred choice, especially in industry, because of the relative inflexibility, reliability concerns and the high costs of ownership. Nowadays, a clustered gate/poly stack would raise additional concerns with respect to the throughput in the poly deposition module because of the limited growth rate of RTCVD poly (figure 5.1.4). However, the ITRS roadmap [3] indicates that the poly thickness is decreasing from the current 100 nm (130 nm generation) down to 25 nm (45 nm generation) and, because of that, the throughput is becoming less of an issue over time.

Thermal stability

Activation of source and drain and doping activation represents the highest temperature budget for the gate stack. The requirements on S/D activation,

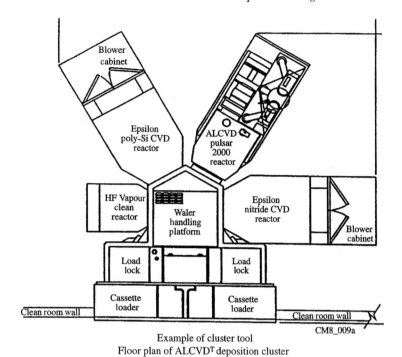

Example of cluster tool
Floor plan of ALCVD^T deposition cluster

Figure 5.1.4. Cluster tool for clean, gate dielectric poly deposition and anneal.

the doping levels and the abruptness of the junction increase from generation to generation. At the current timeframe, development work on high-κ takes a temperature budget of 1000°C for 10 s as a metric to benchmark high-κ temperature stability performance. For the screening of high-κ materials and processes for the 90 nm generation, this metric still suits quite well, but S/D anneals are subjected to development over time.

The thermal budget of the S/D anneal is the subject of elaborate studies [5]. The ITRS roadmap [3] shows a trend towards shallow junction depth and allows some penalty on resistivity as indicated by the boxes in figure 5.1.5.

Table 5.1.1 summarizes the technology trends with respect to anneal conditions. The requirements on the level of dopant activation (resistivity) and junction abruptness (diffusion) urge the use of higher activation temperatures and faster ramp-up and cool-down to limit the thermal budget. It should be noted here that materials selection based on a thermal stability criteria uses 'yesterday's technology to test future materials solutions'. RTA, as we know it today, will not be able to address the needs of technology nodes beyond 90 nm. At this point of time, it is far from clear what will be used in advanced technology nodes. Novel technology such as laser annealing has received a lot of attention. The impact of such technology on the gate stack cannot be predicted. The development of amorphous gate stack materials in

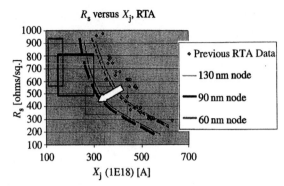

R_s versus X_j, RTA

• Previous RTA Data
— 130 nm node
— 90 nm node
— 60 nm node

Figure 5.1.5. ITRS roadmap source/drain engineering requirements: drain resistivity and junction depth.

particular suffers from the lack of knowledge on the ultimate gate stack thermal processing. Few of the selected candidate materials survive the 1000°C–10 s test without crystallization or phase segregation. Maybe more would pass a laser anneal test. Even more worrying, some of our more promising candidates might turn out to fail in advanced gate processing.

Specifics of high-κ integration

The issues and details specific to the integration of high-κ gate dielectrics into the existing CMOS platforms are now discussed. The key features of a typical conventional CMOS flow are shown in table 5.1.2 below. It is assumed that the starting point for high-κ integration is a set of process conditions for channel and source–drain engineering that is a legacy of the previously existing CMOS platform. The high-κ process flow path can then begin at the pre-gate clean step followed by gate definition, spacer module and source–drain activation steps. The steps prior to the pre-gate clean and the steps occurring after and including silicidation are assumed to be defined or will be discussed elsewhere. If meeting the performance specifications with the high-κ gate dielectric necessitates a change in the channel or source–drain

Table 5.1.1. Trends and expectations in S/D anneal.

S/D activation	
1000°C; 10 s	180 nm
1050°C; 1 s	130 nm
1100°C; spike	90 nm
Zero-time anneal?	60 nm
Laser anneal?	45 nm

Table 5.1.2. Front end processing CMOS.

Shallow trench isolation
Channel implant
Gate stack deposition
Gate electrode patterning
High-κ removal/strip from the S/D areas
Extensions (LDD) and HALO implants
Spacer formation module
HDD + RTA
Silicide co-silicide (Ni available soon)
Back side metal and sinter

conditions, that could become a separate topic in itself, and is beyond the scope of this section.

For convenience, the process flow is divided into several sections. The sections under which the high-κ process integration will be covered are:

(a) gate and electrode depositions;
(b) gate definition—etching and strip modules for electrode and high-κ layers;
(c) spacer module including LDD implants;
(d) source–drain and poly-Si activation module.

An overview of the gate architecture comparing features specific to the gate dielectric choice between SiO_2 and high κ is shown in table 5.1.3. It is evident that some re-engineering of the process modules will be required judging from an integration point of view to produce measurable and functional transistors. Further modifications to meet all the performance specifications such as mobility and reliability will still be required.

Table 5.1.3. Front end process requirements for oxide and high κ.

Process	SiO_2 feature	High κ
Dielectric deposition	Thermal growth	Choice of deposition method
Electrode deposition	Poly-Si for next node	Compatibility and stability
Poly gate etch	Stop on oxide	Stop on high κ
Gate dielectric etch	Stop on silicon	New etch chemistry needed
LDD and halo implants	Well established	Implant through high κ
Re-oxidation	Gate edge modification	Not compatible with EOT
Spacer formation	Well established	Needs re-definition

Gate and electrode deposition

Gate dielectric deposition

Issues of starting surface and deposition techniques have been discussed earlier in this book (see chapters on deposition techniques). The choice of the pre-gate clean surface treatment and post-deposition treatment will define the properties of the interfacial layer, which will have a large impact on transistor properties. Stability of the interfacial layer thickness (as established at the gate deposition) during subsequent transistor processing is an important aspect of the integration scheme. Increase of the interfacial layer or modification of the interfacial layer can occur during any of the following steps:

- post-deposition annealing (PDA);
- etching/gate definition;
- resist removal (oxygen plasma ashing);
- high-κ etch and removal;
- re-oxidation;
- spacer deposition;
- source–drain activation annealing.

Gate electrode deposition

For the initial deployment of high-κ gate dielectric, polysilicon (poly-Si) will be the natural choice of gate electrode. Metal gate options will also need to be explored for subsequent scaling as per ITRS roadmap requirements, and are discussed later. Since n-type or p-type doped poly-Si is typically used as gate electrode in conventional CMOS processing, compatibility of the high-κ material (e.g. HfO_2) with the CMOS poly-Si process module is critical for device fabrication. In specific terms, compatibility refers to the thermal stability and physical integrity leading ultimately to satisfactory electrical performance of the gate stack. This is usually measured by clearly defined performance targets in the ITRS specifications. The thermal stability criterion for the poly-HfO_2 system is suggested by ITRS requirements to be 1000°C for 10 s. This is the thermal budget that is expected to be required to activate the dopants in the source–drain regions and in the poly-Si layer. Electrical performance will be considered satisfactory in a broad sense for the system when the ITRS performance specifications are met for gate and standby leakage, drive current, EOT, yield and reliability. Compatibility between the HfO_2 and poly-Si is, therefore, required during the poly module that includes deposition and activation. Poly-Si can be deposited under a variety of conditions of temperature and pressure resulting in films that are amorphous or polycrystalline with small or large-grained microstructures. Columnar small-grain poly-Si is often preferred for CMOS applications to achieve desirable properties of poly depletion, boron penetration and threshold voltage (V_t) matching [6].

Compatibility issues between HfO_2 and conventional CVD poly-Si have been reported in the literature [7, 8] by using gate leakage and yield as performance metrics for the gate stack. Significant gate leakage degradation and yield loss was seen for poly-Si gates with uncapped HfO_2 films as shown in figure 5.1.6. Yield results were seen to depend on the area of the capacitor structure, enabling the extraction of a defect density [11]. This area dependence suggests a defect-related mechanism for the high gate-leakage yield failures. The problem is aggravated by using high-temperature crystalline poly-Si deposition recipes. The fact that this problem is related to the poly-Si deposition and also to the gate dielectric is illustrated in figure 5.1.7. Figure 5.1.7(*a*) shows that low leakage and ~100% yield is obtained for the same gate dielectric with TiN gate electrodes. Figure 5.1.7(*b*) shows the detection of Si within the HfO_2 film for samples that are as deposited or annealed in N_2 or O_2 prior to poly-Si deposition. The source of the Si is presumably from in-diffusion through the interfacial SiO_2 or the substrate. Despite these issues, however, functional and well-behaved transistors have been fabricated with low-temperature amorphous silicon LPCVD depositions, so that much of the literature on poly-Si–HfO_2 transistors is based on such poly-Si recipes.

Figure 5.1.8 presents a model to explain the observations shown in figures 5.1.6 and 5.1.7. The film has initial weak spots, which are observed in the gate dielectric in the form of etch pits or inhomogeneities after etching with a delineating etch, shown by an SEM micrograph inset [12]. When TiN deposited by PVD or ALD is used as gate electrode, no high leakage failures are seen despite the presence of the weak spots and a near-100% yield is seen.

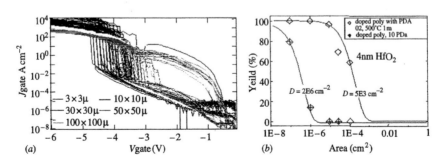

Figure 5.1.6. The gate leakage and yield observed for capacitors using LPCVD poly-Si gate electrodes on uncapped HfO_2 films with post-deposition anneal. (*a*) The area dependence can be observed where low leakage is seen for smaller area capacitors ($<30 \mu$) while larger sizes show higher leakage and lower yield. (*b*) The estimated defect density for un-annealed films (crossed) as $2 \times 10^6 cm^{-2}$ and for annealed films (uncrossed) $5 \times 10^3 cm^{-2}$ [8].

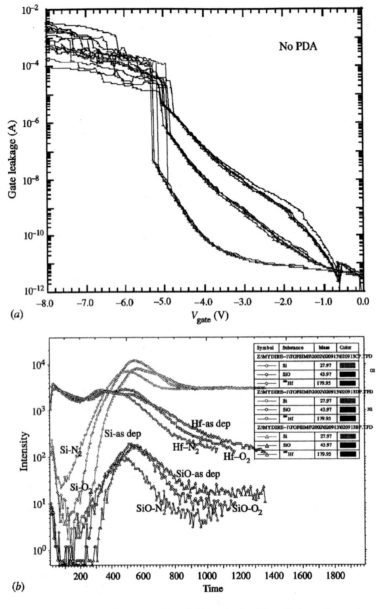

(a)

(b)

Figure 5.1.7. (*a*) Low leakage and ~100% yield is obtained for all sizes of capacitors for TiN gate electrodes on 4 nm HfO_2 gate dielectric. This indicates that the problem is with the poly-Si gate electrode module [8]. (*b*) A TOFSIMS profile through as-deposited versus annealed in N_2 or O_2 HfO_2 films showing the detection of Si within the HfO_2 film after annealing that is higher after N_2 anneal [9].

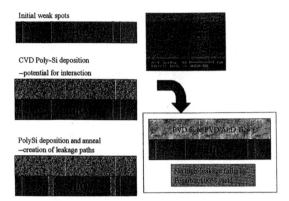

Figure 5.1.8. A schematic representation of the proposed mechanism to explain why some gate electrodes (PVD Si, PVD/ALD TiN) give 100% yield with HfO_2 while CVD poly-Si may show high leakage due to interaction with poly-Si forming conducting paths. The SEM image inset shows an example of the defects seen after etching [9].

When CVD poly-Si is used as gate electrode, there is a potential interaction between the weak spots and the silane source used in CVD poly-Si processes. [7]. This interaction is likely to result in the formation of leakage paths that may be related to Si indiffusion. The problem can be mitigated or eliminated by the use of barriers or capping layers between the poly-Si and the HfO_2, while a longer-term solution would be to improve the quality of the HfO_2 layer, an aspect that will become more important when the thickness of the films is scaled down further.

Gate definition–etching and strip

In this section, we discuss the etch development for poly-Si gate electrodes. The poly-Si patterning in standard CMOS is required to stop on SiO_2. This process is largely unchanged and perhaps somewhat simplified for high-κ gates, since the high-κ layers are difficult to etch with standard etch chemistries used in semiconductor fabrication [13]. Therefore, with a standard poly gate, the poly gate etch recipe only needs minor adjustment with regard to end-point detection and over-etch strategy. However, etch chemistry for the high-κ layer itself is likely to be more of an issue. In this regard, two options are available as shown in figure 5.1.9.

In figure 5.1.9(a) the high-κ layer is removed immediately after gate patterning [14] or poly-Si etching, while in figure 5.1.9(b) the high-κ layer is etched after the spacer is defined [15]. The advantage of the approach of figure 5.1.9(a) is the absence of a high-κ layer in the extension regions adjacent to the channel, such that the LDD or extension implants are not done through the high-κ layer.

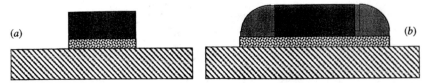

Figure 5.1.9. Panels (*a*) and (*b*) show the two options available for removal of the high-κ layer at source–drain regions. In (*a*), the HfO$_2$ layer is removed immediately after poly definition, while in (*b*) it is removed only after spacer definition.

The disadvantage of this approach is that it imposes stringent requirements on the high-κ etch in its selectivity to the Si below. Conversely, with the approach of figure 5.1.9(*b*), the etch requirements are relaxed, but they will require LDD implants through the high-κ layer. The ultimate choice will depend on overall compatibility and performance requirements.

The next step is high-κ etching. The etch chemistry of HfO$_2$ needs to be selective to Si substrate in order to minimize recess of the substrate. Further, it should be selective to silicon dioxide at the trench isolation regions, and to the capping layer (anti-reflective coating (ARC) or photoresist layer) that covers the poly-Si gate. Although hydrofluoric acid may etch HfO$_2$ films with selectivity to Si, significant loss at field oxide regions will occur. Wet etching studies in the literature [16] have not been very promising for HfO$_2$ layers. The etch selectivity requirements are easier to achieve by dry or plasma etch methods than by chemical or wet etching. Figure 5.1.10 shows a schematic cross-section of a poly-Si line taken after high-κ etch and resist stripping steps. The image shows that with this overetch, some Si recess is present. This recess can be minimized by optimizing the etch/overetch and strip modules to maintain selectivity to Si substrate. It was mentioned earlier that careful attention is required in preserving the thickness and quality of the interfacial layer below the HfO$_2$ layer during transistor fabrication, including

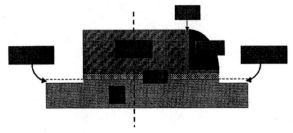

Figure 5.1.10. Schematic cross-section of a poly-Si line after patterning and high-κ removal from adjacent areas. The silicon substrate is seen to be recessed due to exaggerated overetch during high-κ removal, either after gate patterning (left half) or after spacer definition (right half). The Si recess for high-κ removal before spacer (left half) is likely to have more impact on transistor properties.

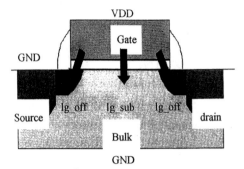

Figure 5.1.11. CMOS showing S/D extensions under channel illustrating the importance of the fringe area for gate leakage.

all the steps after high-κ deposition. This may be a consideration in choosing the etch chemistry.

Since, after the etch, the gate structure is laterally exposed to the ambient, an oxidizing chemistry in the etch or the resist ashing will cause lateral oxidation [17] of the gate dielectric. This should lead to the degradation of the EOT value of the stack, in addition to any other process-induced damage.

At this point, in a conventional poly-Si–SiO$_2$ gate stack, a re-oxidation step is performed. The effect of this step is to grow a thin oxide along the poly sidewalls and increase the gate edge oxide thickness by lateral oxidation. This cannot be performed for a HfO$_2$ stack because the HfO$_2$ layer is quite permeable to oxygen and any poly oxidation will be accompanied by an interfacial layer increase and EOT degradation. It is interesting to note that the absence of re-oxidation has been reported [14] to not to cause a serious degradation to transistor behaviour for high-κ dielectrics. The importance of the fringe area of the high-κ gate oxide should not be underestimated. Leakage currents of S/D to gate as well as channel to gate are equally important in small device structures (see figure 5.1.11).

Spacer module

Typical materials for spacers include oxides or nitrides of silicon. The purpose of the spacer is to provide a lightly doped drain (LDD) or extension region, between the channel and the heavily doped source–drain regions. This helps to reduce hot carrier effects. Prior to spacer formation, the extension (or LDD) implants are carried out. As discussed earlier, the presence of a high-κ layer over the source–drain regions at this point will cause the LDD implant to go through the high κ, creating the possibility of knock-on damage (see figure 5.1.12).

Then the spacer material, e.g. SiO$_2$ or Si$_3$N$_4$, is deposited and patterned by dry etching. In SiO$_2$-based CMOS, the re-oxidized SiO$_2$ in the source–drain

Figure 5.1.12. LDD implant with and without high κ.

regions serves as an etch stop for silicon nitride spacers, thereby providing better selectivity than the Si substrate. However, skipping the re-oxidation step for high-κ gate dielectrics may necessitate some modification to the Si_3N_4 spacer etch procedure. This can be accomplished either by a liner layer to serve as an etch stop (such as deposited SiO_2) or by changing the Si_3N_4 etch to provide better selectivity to silicon substrate.

If deposited SiO_2 is used as a liner prior to Si_3N_4 spacer or as the spacer itself, care must be taken to ensure that there is no lateral re-oxidation of the Si substrate or the poly-Si gate during this step. Lower-temperature SiO_2 deposition options may be helpful in this regard.

Source–drain and poly activation module

After spacer formation, the source–drain implants are carried out, with the same implant often serving to dope the poly-Si gate electrode. For CMOS processing, n- and p-type implants will need to be done by suitable masking steps. Prior to the implant step, the high-κ layer can be removed from the source–drain layers, and additional screen oxides may be used for implanting through, as in conventional SiO_2-based processing. The high-dose implants need thermal annealing steps for activation and for healing any damage to the crystalline Si lattice or for recrystallization of amorphized regions. As noted earlier, current ITRS specifications require 1000°C–10 s of annealing treatment during which the gate stack must be stable. Potential interactions may occur between the HfO_2 film and the gate electrode or between HfO_2 and interfacial oxide or substrate Si. The formation of Zr silicide was reported to cause high leakage and low yield for ZrO_2 gate dielectrics after 1000°C–10 s annealing [17].

The impact of the different choices made in the integration can be eventually seen in the electrical behaviour. An example of this is illustrated in figure 5.1.13 for 3 or 4 nm HfO_2 capacitors with crystalline poly-Si

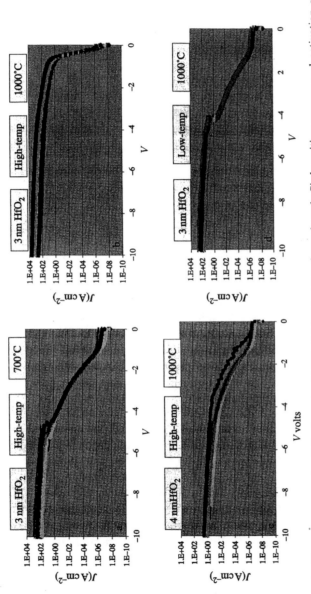

Figure 5.1.13. An illustrative example of the impact of HfO$_2$ thickness, gate electrode poly-Si deposition process and activation anneal on capacitor gate leakage properties. Panel (*a*) shows that for a 3 nm HfO$_2$ layer working devices result with a high-temperature poly-Si process but 700°C activation anneal. (*b*) When the activation anneal is performed at 1000°C shorted devices are obtained. (*c*) To fix this problem, the HfO$_2$ film thickness can be increased at the expense of degraded EOT and spread in leakage, or (*d*) the poly-Si deposition process can be carried out at lower temperatures to improve compatibility [10].

gate electrodes deposited at a higher or lower temperature [10]. Source–drain activation was performed at either 700 or 1000°C. This example tests the compatibility between the gate dielectric deposition conditions, the poly-Si gate electrode deposition conditions and the source–drain activation conditions.

Gate leakage measurements for several devices are shown in figure 5.1.13(*a*) for a 3 nm HfO$_2$ film with the high-temperature poly-Si with 700°C activation. The estimated EOT for this gate stack is ~1.5 nm. Although low leakage with tight distribution and good yield are observed for this condition, the activation anneal is much lower than the specified target. When the same gate stack was annealed at 1000°C, shorted devices resulted, as shown in figure 5.1.13(*b*). To improve the leakage behaviour at 1000°C, we can either increase the thickness of the film as shown in figure 5.1.13(*c*) where a 4 nm film gave no shorts and low leakage, but with some spread across the devices. Another possibility to improve the electrical characteristics is going to a lower-temperature crystalline deposition process for the poly-Si. Figure 5.1.13 shows that with this integration, low leakage and acceptable yield with low spread is achieved for the gate stack.

The above discussion also illustrates another potential problem. Since reducing the HfO$_2$ thickness from 4 to 3 nm necessitated a process change to achieve compatibility with poly-Si, it is reasonable to expect that as the HfO$_2$ thickness continues to be scaled down to meet the EOT targets of 1 nm, re-engineering the poly-Si process or improving dielectric film quality will become increasingly challenging.

Back-end compatibility

State-of-the-art back-end metallization processing has a low temperature budget, i.e. around about 400°C. One of the last steps in the processing is a hydrogen bake to passivate the gate interface with hydrogen. The hydrogen saturates the silicon dangling bonds at the interface. This process is well established and has a large impact on the overall transistor performance. Also, in the case of high κ, this process step turns out to be essential to obtain better performance. Recent findings in the high-κ research area even show that better gate stack performance can be obtained after high-temperature gate passivation in forming gas. Giving the incompatibility of this high-T passivation with the low-T budget requirements of the BEOL processing, the passivation step should be applied prior to the back-end metallization. Moreover, the use of nitride layers to package the high-κ should be well studied in correlation with hydrogen passivation. Nitrides are known to block the hydrogen diffusion and as such can shield the gate area.

SOC compatibility: cleaning strategy

SOC is a sure trend in semiconductor manufacturing processing [18]. Two or even three gate oxide thicknesses on a single chip are on their way to become a commodity. High speed (EOT 1 nm), low power (EOT 2 nm) and I/O (EOT ≥ 3.5 nm) are often combined on a single device without compromise. A typical chip layout is shown in figure 5.1.14. An example of SOC process sequences for a dual-gate-oxidation process is depicted in this figure as well. As indicated, dual (triple) oxides are generated by two (three) subsequent oxidations. First, a thick blanket oxide is grown. In a litho step, this oxide is removed by selective HF etching and subsequent resist stripping. This will leave a 'chemical oxide' on the cleaned parts. The second oxide is thermally grown with this chemical oxide as a starting layer. The thick oxide will increase slightly in thickness too during this second gate oxidation process.

Implementing a high-κ gate dielectric in SOC processing is not straightforward. If we consider an SOC architecture with dual oxide (high κ for high-end functions with EOT and thick standard oxide for I/O application) the use of high κ has its consequence for standard 'oxide' gate architecture and processing as discussed with etch and clean requirements and processing restrictions on post-gate etch processing. Furthermore, high-κ gate deposition is a blanket deposition. As a result, the standard oxide gate dielectrics will have a high-κ layer on top. But most of all, SOC processing limits high-κ pre-deposition clean and surface preparation strategies. The high κ is deposited just after the litho and resist removal step. The chemical oxide created after resist strip matches very well with high-κ deposition technology. But, the chemical oxide after resist strip is rather thick (1 nm). Scaling of this chemical oxide will be required by either etch-back or complete removal and re-growth. HF chemistry-based cleans will be needed here, but care should be taken not to affect the thick standard oxide gate that is already in place (and exposed).

Increasing SOC complexity towards two high-κ gates oxide on chip (one for high performance and one for low leakage) will require even more creative processing solutions. Consequently, for SOC, many process issues remain to be solved and choices will have to be made. Clearly, some combinations of oxide gates and high κ are more realistic than others.

'Metal' gates

In a standard CMOS process, heavily doped poly-Si is used as a 'metal' in the transistor MOS capacitor. The use of poly-Si clearly has a lot of advantages in the CMOS application. The severe disadvantage of a poly gate appears with thinner dielectrics. Though heavily doped, the poly-Si still lacks sufficient charge carriers and shows depletion at the dielectric interface when the transistor channel is in inversion. For high-current transistor operation,

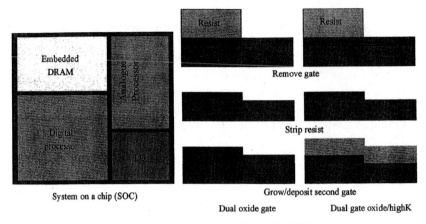

Figure 5.1.14. SOC, and dual-gate technology of high κ for SOC.

charge in the channel in inversion is the key parameter. A higher inversion capacitance (or lower CETinv, capacitor equivalent thickness in inversion) is required. The depletion in the semi-metallic poly-Si limits the capacitance in inversion. The contribution of the poly depletion in the inversion capacitance reduction depends on poly doping. When ITRS roadmap poly doping levels are achieved, poly depletion accounts for about 25% reduction of the inversion capacitance. Figure 5.1.15 shows the poly doping requirements in the ITRS roadmap. These high doping level values might be hard to achieve. Even when the doping level requirements can be met, poly depletion still accounts for a relative important capacitor decrease in sub-nm EOT devices [19].

Figure 5.1.16 illustrates this poly depletion effect. A metal gate stack and poly-gated stack are shown in the figure. In the case of a metal gate stack, accumulation and depletion capacitance develop near mirror-like. In the poly-gated cases, inversion capacitance does not fully develop. The poly depletion depends on the doping level of the poly. To meet the inversion capacitance of a 0.7 nm metal gated dielectric stack with a poly gate stack, the dielectric should be scaled down to compensate for the depletion. At a very high poly doping of $3 \times 10^{20}\,cm^3$, depletion is 0.2 nm and the dielectric thickness should be scaled down 0.2 nm extra. With doping level in the more realistic range of $1 \times 10^{20}\,cm^3$, the depletion effect can be as high as 0.4 nm.

Figure 5.1.15. ITRS roadmap for poly doping requirements.

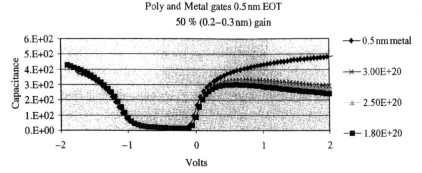

Figure 5.1.16. Calculated *C–V* curves illustrating the advantage of metal gates over poly-Si gates.

The use of metal gates, therefore, would relax requirements of the high κ or can be used to boost transistor performance.

There are other advantages linked with the use of metal gates. In addition to the depletion problem, the poly has a relatively high resistivity. The resistivity limits the drive current. Replacing the poly by metal would help to solve this problem too. Meanwhile, S/D engineering and poly doping activation are currently a single process.

The need to activate the poly provides a boundary condition for the optimization of the S/D anneal conditions. With proper metal gates in place, there is no longer the need to activate the poly, and thermal processing can be optimized for S/D engineering exclusively.

Full metal or metal/poly-Si gates
The semiconductor manufacturing industry is extremely addicted to poly-Si. Indeed, one of the strong points of CMOS processing is the poly-Si gate with its spacer technology and self-aligned implantation of S/D (figure 5.1.12). Replacing poly with metal will introduce a whole lot of new integration issues. To name a few: straight wall metal etch, spacer definition on metal, silicidation of S/D. For the proper work function pinning though, just a few nm of metal is sufficient. A thin-metal-gate approach to fix the work function and eliminate the depletion, combined with a standard (thick) poly-Si layer, will solve a lot of the integration issues. Deposition of thin metals widens the scope of potential deposition technologies. Slow processes like ALD are feasible for thin layers, but less appropriate for thick layers. Gate definition will be on standard poly-Si (for NMOS as well as PMOS; see below). Gate etch will be on standard poly-Si primarily. Etching of such thin metal layers will be much easier. Consequently, thin metal layers with poly-Si might be a first metal gate generation.

This thin-metal/poly-Si approach will solve the depletion problem, but does not tackle the resistivity limitation of the poly gate. The full metal gate

Resistivity	μ·cm
Metal	10
Alloy	50
Nitrides	200
Silicide	20
Poly	200

Figure 5.1.17. Typical values of specific resistivity of front-end compatible conductor materials.

approach can be a final stage of metal gate integration into CMOS. It remains to be seen if there is really a need for full metal gates. First of all, the poly-Si is getting thinner and the novel silicidation processes on small channel devices will consume most, if not all, of the poly-Si anyway. Besides, the advantageous resistivity of metal gates applies for pure (high-melting-point) metals and metal silicides only. Metal alloys and stable compounds such as nitrides are moderate conductors (see figure 5.1.17). Chances to develop a suitable pure metal gate material for PMOS and NMOS are low (see below). A realistic full metal approach consists of a 'metal' stack of a material with appropriate work function and a cladding with low-resistivity material.

Finally, for a full metal (stack) gate approach, replacement gate processing could be applied too. Here, the use of a standard poly gate is maintained throughout the front-end processing. But, after spacer etching and planarization, the poly-Si is removed and replaced by metal 'plugs' or better 'slabs' [20].

Dual or mid-gap work function
The actual requirements of the work function of metal gates (figure 5.1.18) are still subject to debate. Dual-work-function solutions similar to the present dual-work-function poly-Si are mainstream. In a dual-work-function (figure 5.1.19) approach, proper V_t control will require typical 'mid-gap + (0.2–0.3 eV)' (≈ 4.9) for PMOS and 'mid-gap − (0.2–0.3)' (≈ 4.4) for NMOS. If Fermi level pinning [21] is important, PMOS will require a somewhat higher-work-function metal gate that matches the Si

Figure 5.1.18. Metal gate roadmap.

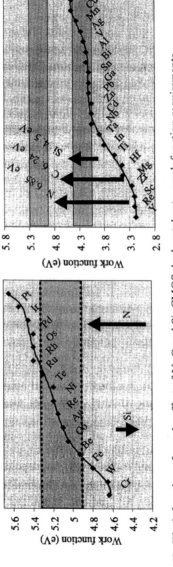

Figure 5.1.19. Work function of metals, effect of N, C and Si; CMOS dual metal gate, work function requirements.

valence band level, 5.2 eV, and NMOS will need conduction band matching, 4.1 eV. Research on metal gates should explore metals within these work function ranges, 4.0–4.4 eV for NMOS, 5.3–4.9 eV for PMOS.

The work function of pure metals is well known [22]. Theory and experimental data are in fair agreement. Figure 5.1.19 gives a summary of metal work function data for PMOS and NMOS transistors.

In the standard CMOS process sequence, the metal gate is formed prior to the SD formation and anneal. Again, thermal stability is the most important and a (1000°C, 10 s) anneal is used as a metric here too. Taking thermal stability requirements into account, few pure metals are candidates. Interesting work in this area currently focuses on work function engineering by mixing metals with silicon, nitrogen and carbon. Nitrogen and carbon will tend to increase the work function; silicon drives the material to mid-gap values. This work function and materials engineering approach offers a wide range of opportunities. TaN and TaSiN for instance are being investigated for use in NMOS [23–25]. Some of the noble metals (Pt, Ir, Ru) attract attention for use in PMOS alongside heavily nitrided TiN [26]. Silicides, Ni silicides in particular, get some attention too. For mid-gap applications, TiN is an attractive candidate, although the work function may vary with composition, deposition technology and thermal history of the material.

Again, interface studies are needed to understand the potential interactions between high-κ material and metal gates. Metal gates are relatively unexplored and many issues are still unresolved. We have just discussed the choice of work function and materials. Many other issues should be considered, for example, tool clustering. Cluster tools might be avoided in case of poly-Si integration; the interfacial oxide is still relatively thick and stable, air exposure tolerable. However, for the aggressively scaled high-κ gate dielectrics that are needed for the 45 nm generation and beyond, air exposure of the high-κ dielectric should preferably be avoided. A high-κ/metal cluster deposition seems thus the way to proceed.

Another serious issue is the technology sequence [24]. Gate stack definition of mid-gap-gated devices is straightforward. Only one gate is required. In the case of a dual-metal-gate approach, the technology becomes rather complex though. Two different gates are needed. The most straightforward processing starts with the deposition of the first metal, subsequently etched back (stopping on the gate dielectric) and the second gate metal is deposited on the first one. It is worthwhile noting here that this procedure cannot take advantage of a cluster solution for both NMOS and PMOS. Air exposure of the naked gate is inevitable for at least one of the transistor types. There might be a need to also etch the gate dielectric layer, deposit a fresh one and the second metal gate, subsequently. Furthermore, there are proposals to use smart materials that can be converted from NMOS to PMOS (or the other way around) by reaction or implantation without the need to fully etch the gate area (figures 5.1.20 and 5.1.21).

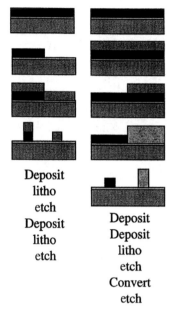

Deposit
litho
etch
Deposit
litho
etch

Deposit
Deposit
litho
etch
Convert
etch

Figure 5.1.20. Technology sequences for dual metal gates.

High-κ dielectrics and metal gates in the fab

Considering all the above, it is clear that a lot of issues remain to be solved before a high-κ dielectric actually finds its way into the fab. It should be emphasized here that many issues have been addressed and solved in the meantime. High κ fulfils its mean promise, leakage current reduction, but it comes with penalties. Without giving a lot of detail here, the performance issues can be depicted in a performance triangle. Thinning the high-κ stack leads to increase of the leakage. Reliability and interface layer thickness are

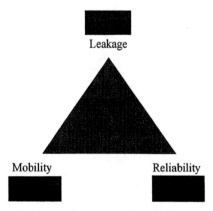

Figure 5.1.21. Diagram showing performance trade-offs for high κ.

linked [27]. Thinning the interfacial oxide layer gives rise to performance reduction, mobility degradation. The work on high-κ gate dielectrics should give insight in the parameter space in this triangle, while there is a need for improvement of the electrical properties of the devices, e.g. mobility performance in particular.

Latest ITRS roadmaps indicate that low- (standby) power applications will be the main drivers for high-κ gate dielectric integration, in order to reduce the gate leakage current. Manufacturability and reliability might turn out to be strong drivers in the high-performance application area.

Acknowledgments

The author would like to thank numerous colleagues at Philips, Motorola and IMEC.

References

[1] Zeitzoff P M, Hutchby J A and Huff H 2002 *Int. J. High Speed Electron. Syst.* **12**, 267–293

[2] Young E W A, Chen J, Cosnier V, Lysagh P, Maes J W, Roozeboom F, Zhao C, Carter R, Richard O and Conard T 2002 *Electrochem. Soc. Symp. Proc. 2002-11*, 125–136. NL-MS-22.352, October 15, 2002

[3] SIA, ITRS 2002, Front end processes chapters

[4] Young E W A 2002 *Electrochem. Soc. Symp*, PV 2002-2 Semiconductor Silicon p 735

[5] Surdeanu R *et al* 2002 *Electrochem. Soc. Symp.* PV 2002-11 Rapid Thermal and Other Short-Time Processing Technologies III

[6] Tuinhout H P, Montree A H, Schmitz J and Stolk P A 1997 *IEDM Tech. Digest* 26.2.1 631

[7] Gilmer D *et al* 2002 Compatibility of polycrystalline silicon gate deposition with HfO_2 and Al_2O_3–HfO_2 gate dielectrics *Appl. Phys. Lett.* **81** 1288

[8] Kaushik V S, DeGendt S, Claes M *et al* 2002 Novel Materials and Processes for Advanced CMOS *Mater. Res. Soc. Proc.* eds J P Maria, S Stemmer, S De Gendt and M Gardner p 745

[9] Kaushik V S, De Gendt S, Caymax M, Young E, Rohr E, Elshocht S V, Delabie A, Claes M, Shi X, Chen J, Carter R, Conard T, Vandervorst W, Schaekers M and Heyns M Compatibility of polysilicon with HfO_2-based gate dielectrics for CMOS applications *203rd ECS Conference Proceedings* vol PV2003_06 *ULSI Process Integration III*

[10] Van Elshocht S, Carter R, Caymax M, Claes M, Conard T, Daté L, De Gendt S, Kaushik V, Kerber A, Kluth J, Lujan G, Pétry J, Pique D, Richard O, Rohr E, Shimamoto Y, Tsai W and Heyns M M 2003 Scalability of MOCVD-deposited hafnium oxide *CMOS Front-End Materials and Process Technology Mater Res. Soc. Proc.* vol 765, eds T-J King, Bin Yu, R J P Lander and S Saito

[11] Hori T 1997 *Gate Dielectrics and MOS ULSIs—Principles, Technologies and Applications* 145 (Berlin: Springer)

[12] Claes M, Witters T, Loriaux G, Van Elshocht S, Delabie A, De Gendt S, Heyns M M and Okorn-Schmidt H F 2002 Open circuit potential analysis as a fast screening method for the quality of high-κ dielectric layers *6th Int. Symp. Ultra-Clean Processing of Silicon Surfaces (UCPSS) (September 2002)*

[13] Christenson K, Schwab B, Wagener T, Rosengren B, Riley D and Barnett J 2002 Selective Wet Etching of High-κ Gate dielectrics *Proceedings of the Sixth International Symposium on Ultra-Clean Processing of Silicon Surfaces* (Oostende, Belgium: September 16–18 2002) p P21

[14] Hobbs C *et al* 2001 80 nm Poly-Si gate CMOS with HfO$_2$ gate dielectric *Int. Electron Devices Meeting 2001 Tech. Dig.* p 651

[15] Kim Y *et al* 2001 Conventional n-channel MOSFET devices using single layer HfO$_2$ and ZrO$_2$ as high-κ gate dielectrics with polysilicon gate electrode *Int. Electron Devices Meeting, 2001. IEDM Tech. Dig.* p 455

[16] Lysaght P S, Chen P J, Bergmann R, Messina T, Murto R W and Huff H R 2002 Experimental observations of the thermal stability of high-κ gate dielectric materials on silicon *J. Non-Cryst. Solids* **303** 54–63

[17] Hobbs C *et al* 2001 Sub-quarter micron Si-gate CMOS with ZrO$_2$ gate dielectric *VLSI Technology International Symposium on Systems and Applications 2001* Proceedings of Technical Papers pp 204–207

[18] Ono A, Fukasaku K, Hirai T, Makabe M, Koyama S, Ikezawa N, Ando K, Suzuki T, Imai K and Nakamura N 2001 *VLSI Symposium (Kyoto)*, TO7A_2

[19] Hauser J R and Ahmed K 1998 Characterization and metrology for ULSI Techn. 1998 Int. Conf. ed D G Seiler *et al* (American Institute of Physics) p 235

[20] Guillomot B *et al IEDM* 2002, 14_01

[21] Yeo YC, Ranada P, Lu Q, Lin R, King T S and Hu C 2001 *VLSI Symp.* (Kyoto) To5A_3

[22] de Boer F R, Boom R, Mattens W C M, Miedema A R and Niessen A K 1988 *Cohesion in Metals* vol 1 (NHPP/Elsevier)

[23] Misra V, Kulkarni M, Heuss G, Zhong H and Lazar H 2000 *Electrochem. Soc. Symp. PV* **9** 291–298

[24] Samavedam S B *et al* 2002 *IEDM* 17_02

[25] Lander R J P, Hooker J C, van Zijl J P, Roozeboom F, Maas M P M, Tamminga Y and Wolters R A M 2002 MRS 2002 spring *Symposium B* MRS Proceedings vol 716, eds J Veteran, D L O'Meara, V Misra and P Ho

[26] Zhong H, Hong S N, Suh Y S, Lazar H, Heuss G and Misra V 2001 *IEDM* 467–470

[27] Kauerauf T, Degraeve R, Cartier E, Govoreanu B, Blomme P, Kaczer B, Pantisano L, Kerber A and Groeseneken G 2002 *IEDM* 20_05

Chapter 5.2

Device architectures for
the nano-CMOS era

Simon Deleonibus

International Technology Roadmap acceleration and issues

Since 1994, the International Technology Roadmap for Semiconductors (ITRS) [1] (figure 5.2.1) has been accelerating the scaling of CMOS devices to lower dimensions continuously despite the difficulties that appear in device optimization.

However, uncertainties about lithography, economics and physical limitations can probably slow down the evolution. For the first time since the introduction of the poly gate in the CMOS device process, showstoppers other than lithography appear to deserve special attention and could require some breakthrough or evolution if we want to continue scaling at the same rate. Design could also be affected by this evolution.

Which are the main showstoppers for CMOS scaling? In this chapter, we focus on the possible solutions and guidelines for research in the next years in order to propose solutions to enhance CMOS performances before we need to skip to alternative devices. In other words, how can we offer a second life to CMOS?

Projections through the roadmap of classical linear scaling meet serious limitations from showstoppers like gate insulator thinning. Because of the continuous increase of device performance, which implies a low threshold voltage, leakage current becomes a main concern. That is why the roadmap distinguishes today three types of product: high performance (HP), low operating power (LOP) and low standby power (LSTP) devices. In the HP case, it is assumed that static power dissipation per device will continuously grow due to the unavoidable leakage current coming from the subthreshold regime. However, dynamic power per device will continuously decrease due to the major contribution of supply voltage scaling (table 5.2.1). By the

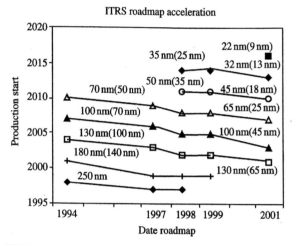

Figure 5.2.1. ITRS roadmap acceleration since 1994 [1]. Example for MPU and ASIC products.

32 nm node, a historical event will happen: the contribution of static power dissipation will become higher than dynamic power contribution! In the following we will analyse the various mechanisms giving rise to leakage current in a MOS device that can impact consumption of final devices. Among them, gate leakage current is already a concern. In the case of LSTP devices, a high-κ gate insulator could be needed earlier than expected in order to limit static consumption (see the 'High doping levels in the channel' section).

Limitation and showstoppers from classical CMOS scaling

Several mechanisms can generate device leakage in ultra-small MOSFETs (figure 5.2.2). One must find a compromise in order to get a good trade-off between saturation current, thus obtaining potentially low enough switching time $\tau = CV/I$ and off state leakage current I_{off}. The mechanisms involved are summarized hereafter:

(1) Classical type

- Drain induced barrier lowering (DIBL) is due to the capacitive coupling between source and drain.
- Short-channel effect due to the charge sharing in the channel in the short-channel devices at low V_{ds}.
- Punch-through between source and drain due to the extension of source space charge to the drain.

Table 5.2.1. High-performance logic technology requirements from 2001 ITRS [1].

Year of production:	2001	2002	2003	2004	2005	2006	2007	2010	2013	2016
DRAM 1/2 pitch (nm)	130	115	100	90	80	70	65	45	32	22
MPU/ASIC 1/2 pitch (nm)	150	130	107	90	80	70	65	50	35	25
MPU printed gate length (nm)	90	75	65	53	45	40	35	25	18	13
MPU physical gate length (nm)	65	53	45	37	32	28	25	18	13	9
Physical gate length high performance (HP) (nm)	65	53	45	37	32	28	25	18	13	9
Equivalent physical oxide thickness for high-performance t_{ox} (EOT) (nm)	1.3–1.6	1.2–1.5	1.1–1.6	0.9–1.4	0.8–1.3	0.7–1.2	0.6–1.1	0.5–0.8	0.4–0.6	0.4–0.5
Gate depletion and quantum effects electrical thickness adjustment factor (nm)	0.8	0.8	0.8	0.8	0.8	0.8				0.5
t_{ox} electrical equivalent (nm)	2.3	2.1	2.0	2.0	1.9	1.9				
Nominal power supply voltage (V_{dd}) (V)	1.2	1.1	1.0	1.0	0.9	0.9	0.7	0.6	0.5	0.4
Nominal high-performance NMOS subthreshold leakage current, $I_{sd,leak}$ (at 25°C) ($\mu A\,\mu m^{-1}$)	0.01	0.03	0.07	0.1	0.3	0.7	1	3	7	10
Nominal high-performance NMOS saturation drive current, I_{dd} (at V_{dd}, at 25°C) ($\mu A\,\mu m^{-1}$)	900	900	900	900	900	900	900	1200	1500	1500
High-performance NMOS device τ ($C_{gate}V_{dd}/I_{dd}$ – NMOS) (ps)	1.6	1.3	1.1	0.99	0.83	0.76	0.68	0.39		
Dynamic power dissipation per ($W/L_{gate}=3$) device $f(C_{gate}3L_{gate})V^2$ (W/device)	9.5×10^{-7}	7.0×10^{-7}	5.4×10^{-7}	4.4×10^{-7}	3.5×10^{-7}	3.0×10^{-7}	2.2×10^{-7}	1.8×10^{-7}	1.3×10^{-7}	7.2×10^{-8}
Static power dissipation per ($W/L_{gate}=3$) device (W/device)	5.6×10^{-9}	6.7×10^{-9}	1.0×10^{-8}	1.1×10^{-8}	2.6×10^{-8}	5.3×10^{-8}	5.3×10^{-8}	9.7×10^{-8}	1.4×10^{-7}	1.1×10^{-7}
MPU clock frequency (GHz)	2.7	3.3	4.0	4.5	5.4	5.8	6.7	11.6	19.6	29.0

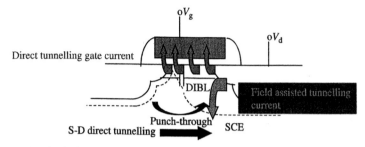

Figure 5.2.2. Physical mechanisms generating leakage in ultra-small MOSFETs.

(2) Quantum and high-field effects

- Direct tunnelling through the gate dielectric.
- Field assisted tunnelling at the drain to channel edge. This effect occurs if electric field is high and tunnelling is enhanced through the thinnest part of the barrier.
- Direct tunnelling from source to drain. This effect will occur in silicon for a thicker barrier than on SiO_2 because the barrier height is lower and the equivalent barrier thickness is higher, due to the higher dielectric constant.

An example of these limitations is given in the following paragraph for 20 nm finished gate length n channel MOSFETs (figure 5.2.3(a)) realized in LETI in the 200 mm diameter Si wafer clean room. Ultra-thin gate silicon dioxide of 1.2 nm is used as the gate insulator for these devices (figure 5.2.3(c)). It operates in direct tunnelling mode [2]. 25 nm devices show a good I_{on}/I_{off} 550/30 $\mu A\,\mu m^{-1}$ trade-off whereas 20 nm devices demonstrate high punch-through current because of the very short channel length [2] (figure 5.2.4).

Field effect is still controllable on 20 nm gate length devices with metallurgical channel length far less than 10 nm: these characteristics can be observed by action on the gate (figure 5.2.4(a)) as well as on the bulk of the devices [2].

Good agreement is observed between experimental $V_T(L_g)$ and $I_{on}(L_g)$ and analytical models dedicated to circuit design [3, 4] (figure 5.2.5(a)). Low-field mobility μ_0 of 130 $cm^2\,V^{-1}\,s^{-1}$ and channel doping of 3.5×10^{18} at. cm^{-3} are extracted [3, 4] from 10 $\mu m \times$ 10 μm transistor characteristics, with precisions of 10 and 20%, respectively, assuming a gate oxide thickness of 1.2 nm. Access resistance of 1650 $\Omega\,\mu m$ and ΔL_{eff} of 13 nm are deduced for short-channel behaviour. That accounts for a channel length $L_{eff} = L_g - \Delta L_{eff} = 7$ nm for the 20 nm gate length devices, in agreement with simulations of figure 5.2.3(b) and 1.37 nm zero-bias depletion width in the extensions.

The performances and the I_{on}/I_{off} ratio of sub-$L_g = 25$ nm gate length devices can be improved at low temperature demonstrating that dispersive

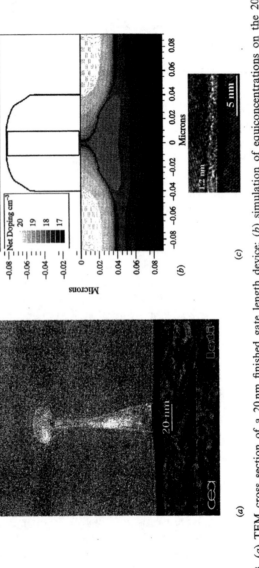

Figure 5.2.3. (*a*) TEM cross section of a 20 nm finished gate length device; (*b*) simulation of equiconcentrations on the 20 nm MOSFET. Metallurgical channel length is 4 nm; (*c*) HR TEM of the 1.2 nm gate oxide [2].

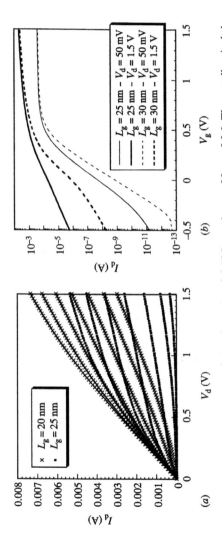

Figure 5.2.4. (*a*) $I_d(V_d)$ characteristics for 25 and 20 nm finished gate length MOS transistors of figure 5.2.3. The metallurgical channel length of the 20 nm device is 4 nm [2]. Gate oxide is 1.2 nm thick SiO_2. (*b*) $I_d(V_g)$ subthreshold characteristics for 2.5 and 30 nm finished gate length MOS devices with the same characteristics as in (*a*) [2].

transport—carrier surface scattering—is the dominant phenomenon at 77 and 293 K [5, 6] (figure 5.2.5(*b*)). One of the most important challenges today is how to take advantage of nonstationary transport in order to enhance drive current. Velocity overshoot and ballistic transport are the most probable mechanisms that will enhance nonstationary transport at the level of sub-100 nm channel lengths. However, the impact of scattering by dopants on transport is not negligible [6].

Superhalo contributes in improving the short-channel effect and DIBL. Demonstration of the Superhalo efficiency to reduce leakage current has been done on 16 nm finished gate length devices (figure 5.2.5(*c*) and (*d*)) to less than $10 \, \mu A \, \mu m^{-1}$ and saturation current of $600 \, \mu A \, \mu m^{-1}$ with gate oxide thickness of 1.2 nm [6].

These devices have been realized by a hybrid DUV e beam lithography [7] and a poly gate etch process highly selective to SiO_2 thanks to an HTO hardmask [2, 6, 8, 9].

Issues in lowering supply voltage

Daily life objects will be transformed in the future in such a way that more and more electronics will be introduced. Furthermore, an increasing number of these objects will be portable. It is assumed that we will enter a nomadic age. That is why very low voltage and low power dissipation will become necessities strongly required by the market. Standby leakage, and also active power dissipation, are the main constraints that we will have to trade off with device performance without relaxing reliability. Microelectronics is living in the regime of a nearly constant field scaling since lower than 5 V supply voltage has been introduced, that is the assumed trend in the most recent version of the ITRS roadmap [1].

However, for sub-0.10 μm devices, the following main issues cannot be avoided.

Direct tunnelling through SiO_2

Direct tunnelling through SiO_2 occurs whenever a thickness less than 2.5 nm is reached, that will increase the contribution of the leakage component to power consumption. Nevertheless, SiO_2 operating in the direct tunnelling current regime has been demonstrated to be usable down to a thickness of 1.4 nm without affecting device reliability [2, 10, 11]. The probability P of tunnelling through the barrier can be expressed by:

$$P = \exp\left[-\frac{4\pi}{h} \sqrt{2me(U_o - E)}T \right] \qquad (5.2.1)$$

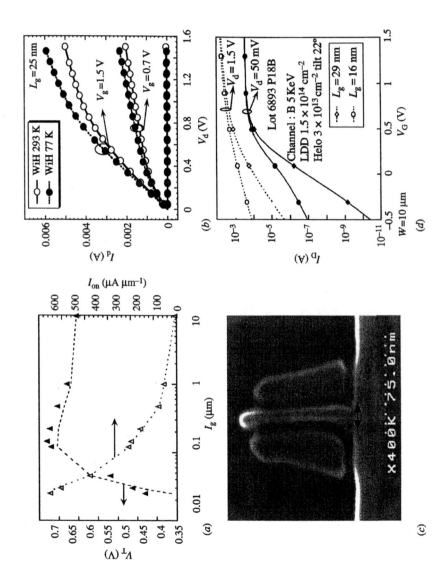

where h is Planck's constant, m is the electron mass, e is the electron charge, U_o is the barrier height, E is the electron energy and T is the barrier thickness.

High doping levels in the channel

High doping levels in the channel reaching more than 5×10^{18} cm^{-3} will enhance field assisted tunnelling reverse current in source and drains up to values of 1 A cm^{-2} under 1 V [12]. The current I through the barrier can be expressed as:

$$I \approx -F^2 \exp\left[-A\frac{E_g^{3/2}}{F}\right] \qquad (5.2.2)$$

where F is the electric field, E_g is the bandgap of the involved semiconductor and A is a constant depending on material characteristics.

Classical small-dimension effects

Classical small-dimension effects are more severe than the fundamental limits of switching due to quantum fluctuations, energy equipartition (statistical mechanics) or thermal fluctuations.

We must give a minimum value to threshold voltage which is related to:

- *Subthreshold inversion* which will increase leakage current up to a few nA μm^{-1} even for an ideal fully depleted silicon on insulator (SOI) (59.87 mV dec^{-1} at 300 K) considering supply voltage of $V_s = 0.50$ V and the relation:

$$S = \frac{kT}{q}(\log 10)\frac{d\left(\frac{qV_G}{kT}\right)}{d\log(I_D)} \approx \frac{kT}{q}(\log 10)\left(1 + \frac{C_w}{C_{ox}}\right) \qquad (5.2.3)$$

C_w and C_{ox} being, respectively, the depletion and the gate insulator capacitance; T is the absolute temperature, k is Boltzmann's constant and q is the electron charge.

Figure 5.2.5. (*a*) $V_T @ V_d = 50$ mV (left-hand scale) and $I_{on} @ V_d = V_g = 1.5$ V (right-hand scale) versus L_g, finished gate length. Measured values (closed and open triangles) and simulated (dashed lines) with models described in [3, 4]. Modelling is in agreement with $L_{eff} = 7$ nm [2]; (*b*) $I_d(V_d)$ characteristics for $V_g = 0$; 0.75 and 1.5 V at 293 and 77 K of $L_g = 25$ nm devices [5]. Source and drain architecture includes BF$_2$ pockets; (*c*) finished gate length is 16 nm. Gate oxide thickness is 1.2 nm; (*d*) subthreshold characteristics of 16 nm finished gate length n channel MOSFET and comparison with 29 nm gate length devices. $V_d = 0.05$ and 1.50 V. BF$_2$ Superhalo is used to reduce the short-channel effect as compared to the devices of figure 5.2.4(*b*) [6].

The limit V_T value would then be 180 mV precluding supply voltage lower than 0.50 V. Otherwise, the limit on leakage should be relaxed. However, a concept like impact ionization MOS (I-MOS) would allow us to reduce the subthreshold slope down to 5 mV dec^{-1} but still performance is an issue [13].

• *The short-channel effect* due to the charge sharing along the transistor channel. This effect is the most severe. It is strongly dependent on the gate/channel coupling capacitance value, the junction depth, the channel length and doping concentration as in the following relation:

$$\Delta V_T = -4\Phi_F \frac{C_w}{C_{ox}} \frac{x_j}{L} \left[\left(1 + 2\frac{W}{x_j} \right)^{1/2} - 1 \right]$$

$$= -4\Phi_F \frac{\varepsilon}{\varepsilon_{ox}} \frac{t_{ox}}{L} \frac{x_j}{W} \left[\left(1 + 2\frac{W}{x_j} \right)^{1/2} - 1 \right]. \qquad (5.2.4)$$

Here the threshold voltage is expressed by:

$$V_T = V_{FB} + 2\Phi_F - \frac{Q_B}{C_{ox}} \qquad (5.2.5)$$

where ΔV_T is the threshold voltage decay, t_{ox} is the gate dielectric thickness, ε and ε_{ox} are the silicon and gate dielectric constants, respectively, L is the channel length, x_j is the drain or source junction depth, W is the space charge region depth, V_T is the threshold voltage, V_{FB} is the flatband voltage, Φ_F is the distance from the Fermi level to the intrinsic Fermi level, Q_B is the gate controlled charge and C_{ox} is the gate insulator capacitance $C_{ox} = \varepsilon_{ox}/t_{ox}$.

If C_{ox} reaches high enough values, the contribution of gate depletion C_{depl} and inversion layer C_{qinv} capacitances in series should not be neglected. In relations (5.2.4) and (5.2.5), C_{ox} will be replaced by C_G, gate to channel capacitance, the equivalent capacitance of C_{ox}, C_{depl} and C_{qinv} in series. C_G can be expressed by:

$$\frac{1}{C_G} = \frac{1}{C_{ox}} + \frac{1}{C_{depl}} + \frac{1}{C_{qinv}}. \qquad (5.2.6)$$

Then gate depletion and quantum confinement in the inversion layer will play an important role in the short-channel effect.

The charge sharing effect is the main limitation to the minimal design rule. It can be of the order of the threshold voltage itself [14] if low threshold voltage values are reached. We must consider that $V_T = V_S/3$ to obtain less than 30% inverter delay degradation (V_S is the supply voltage) [15].

Statistical dopant fluctuations

The effect of dopant fluctuations has already been considered by Schockley in 1961 [16] to evaluate their influence on p–n junction breakdown. Recently, special attention has been paid to this subject because the number of dopants in the channel of a MOSFET tends to decrease with scaling. The random dopant placement in the volume of the MOSFET channel by the ion implantation technique must be considered. Whether the distribution is binomial or poissonian [17–19], dopant fluctuations become very severe for geometries lower than 50 nm if one considers the data of table 5.2.1. Moreover, the discrete nature of dopant distribution can also give rise to device characteristic asymmetry [19]. This consideration as well as the Fowler–Nordheim limitation at high electric field gives more emphasis to the use of the nearly intrinsic SOI thin film as substrate.

In the following, we consider the possible solutions to overcome the physical limitations we could encounter in a classical scaling scenario through the different aspects of MOS devices optimization:

- gate stack and channel/substrate engineering
- source and drain engineering
- gate dielectric engineering.

Gate stack and channel/substrate engineering

Gate and channel. Issues in classical scaling of bulk MOSFET

The gate architecture engineering must be achieved together with the channel engineering as both physical characteristics will affect the nominal threshold voltage value given by the well-known expression for the nMOSFET; if gate depletion and channel quantum effects are taken into account, expression (5.2.5) can be written:

$$V_T = V_{FB} + 2\Phi_F - \frac{Q_B}{C_G} \tag{5.2.7}$$

where

$$V_{FB} = \Phi_{ms} - \frac{Q_{ox}}{C_{ox}} \tag{5.2.8}$$

V_T is the threshold voltage, V_{FB} is the flatband voltage, Φ_F is the distance from the Fermi level to the intrinsic Fermi level, Q_B is the gate controlled charge, C_G is the gate to channel capacitance given by relation (5.2.6), Φ_{ms} is the difference between the extraction potentials of the gate and the semiconductor, Q_{ox} is the oxide charge density and C_{ox} is the unit-area capacitance of the gate insulator.

From the above expressions, V_T depends on the gate material extraction potential as well as the doping concentration in the channel. This a classical issue to deal with in classical scaling of the bulk MOSFET.

The management of low threshold voltage values will be achieved by the following.

Adjusting gate insulator thickness
Addressed in the 'Gate dielectric engineering' section.

Tuning surface doping concentration
Tuning surface doping concentration as low as possible. For bulk devices, the retrograde profile is the conventional answer to achieve maximum bulk concentration—limits punch-through efficiently—together with low surface concentration that maximizes mobility (reduction of scattering by dopants). It is easy to obtain the low threshold voltage values required for low supply voltage. However, excellent localization of the dopant profile is needed to minimize as much as possible junction parasitic capacitance and the body effect. The use of heavy ions such as indium for nMOSFETs and Sb for pMOSFETs has been suggested. Besides the defect annealing issue, the use of In, for example, suffers from the difficulty of carrier freeze-out [20] and activation of these impurities at a concentration higher than 10^{18}cm^{-3} because of the solid limit solubility of these atoms in silicon. Selective Si epitaxy of the channel has also been suggested to achieve almost ideal retrograde profiles [21].

Strained channel engineering
Strained SiGe has been studied for several years as a way to improve hole mobility for bipolar or MOSFET applications [22]. Other solutions like strained $SiGe_xC_y$ based alloy epitaxy or strained Si will increase the channel mobility [23–25]. However, high-quality gate insulator and subthreshold characteristic optimization require modifications of the device vertical structure and low thermal budget [24, 25]. Practically, a Si cap layer of the SiGeC channel is needed to grow high-quality gate insulator on top [25], unless high-κ gate insulator is deposited as reported on strained Si [26]. Carbonated layers introduced by selective epitaxial growth act as a boron diffusion barrier (figure 5.2.6(*a*)) and thus help to improve drastically the short-channel effect [24, 25] (figure 5.2.6(*b*)) and increase low-field mobility.

These architectures will allow a new family of high-frequency operating, low-noise devices that will be needed for future high-volume telecommunications applications.

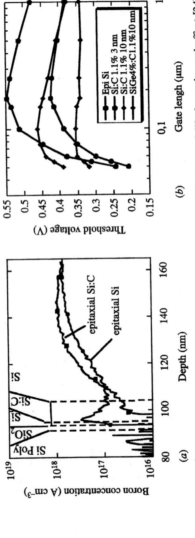

Figure 5.2.6. Effect of introduction of carbonated silicon in channel on (*a*) boron diffusion and (*b*) short-channel effect [24].

Choosing the gate material

An important challenge which will ease the circuit design for low voltage is the symmetry of threshold voltage for n and p channel devices ($V_{TP} = -V_{TN}$), in order to achieve ideal transfer CMOS inverters characteristics, i.e. trade-off between performance and standby leakage. Several alternatives have been envisaged:

- *The use of n+ poly gate for nMOSFET and p+ poly gate for pMOSFET.* That allows theoretically a threshold voltage adjustment by enhancement of the well dopant concentration. This solution suffers from many problems essentially related to boron penetration into SiO_2 coming from the p+ doped gate. The literature is very abundant on this subject. Nitrided SiO_2 by very numerous techniques is a way to *limit without avoiding* this effect. However, the creation of trapping centres in the oxide or at the SiO_2/Si interface will decrease carrier mobility. Practically, gate depletion is preferred as a trade-off.
- *The use of metal gate material.* A metal gate is the best solution to avoid gate depletion. Midgap TiN is available today in the microelectronics industry due to its properties as a barrier in the interconnect schemes. In this case, only one material is needed for n and p MOSFETs. However, the flat band voltage is shifted by half the Si bandgap (i.e. 0.55 eV). It could lead to a too high absolute value of V_T if one does not reduce or compensate the surface concentration. Surface punch-through is thus an issue. In this case, silicon epitaxy can also be used to achieve low surface concentration together with higher bulk concentration to limit punch-through (just as for a polysilicon gate). The use of a midgap gate on bulk or partially depleted SOI will thus be dedicated to supply voltages higher than 1 V. If fully depleted SOI is used then one can tune the silicon film thickness to adjust threshold voltage. However, that will be achievable if the film thickness is controllable.

The important challenge in this field is to implement dual metal gates in order to trade off the best CMOS performance with low leakage current. This is mandatory if one wants to adjust threshold voltage on low-doped fully depleted SOI.

The metal gate integration is not obvious because of the popularity of silicon gate technology coming from the self-alignment of source and drain on the gate. The process sequence has to be modified in order to allow high-temperature source and drain dopant activation.

Several approaches have been proposed. The classical process integration requires the protection of the metal gate material from ion implant as well as oxidation during the dopant activation anneal. TiN is often chosen as a gate material [27], because it is usable as an Al barrier material and as a W adhesion layer. Some attempts at integration with Ta_2O_5 [28, 29] were achieved: however, leakage current was an important issue.

Alternative approaches such as the damascene gate (figure 5.2.7) [30, 31] have been proposed and achieved in order to avoid the source and drain activation temperature issue on the metal gate. These solutions are ideally suited to be integrated with a high-dielectric-constant gate insulator [31]. Moreover, the damascene gate offers the possibility of multi-threshold devices by allowing several gate materials on the same chip. This will be mandatory to optimize low-voltage supply device performances. Moreover, high-frequency optimized devices together with standard CMOS could then be embedded on the same chip.

Gate dielectric engineering

This topic will certainly have the greatest strategic impact on the integration of terabit class devices as much as lithography.

Among the limitations that technology demonstrates and will demonstrate, the gate leakage due to direct tunnelling in standard SiO_2 or SiO_xN_y is one major showstopper (figure 5.2.8) [1]. It will impact directly the static power dissipation P_{stat} according to the relation:

$$P = P_{stat} + P_{dyn} \qquad (5.2.9)$$

$$P_{stat} = V_{dd}I_{off} \quad \text{and} \quad P_{dyn} = CV_{dd}^2 f \qquad (5.2.10)$$

with P being the total power dissipation, P_{stat} being the static power dissipation and P_{dyn} being the dynamic power dissipation. If one considers a circuit with active area of the order of $1\,cm^2$ and gate oxide $t_{ox} = 1.2\,nm$, if I_{off} is due to gate leakage, then considering $V_{dd} = 0.5\,V$, $P_{stat}(0.5\,V) = 5\,W$. We would get $P_{stat}(1.5\,V) = 750\,W$ for a V_{dd} of $1.5\,V$!

This is a major showstopper for scaling of CMOS technology. That is why high κ will be needed in the near future. Besides affecting static power, gate leakage also impacts delay time negatively (figure 5.2.9) [32]. Gate leakage current clearly degrades the power consumption and has an influence on the functionality of logic circuits. SiO_2 dielectrics could be practically used down to a thickness of $1.4\,nm$ corresponding to a leakage of $1\,A\,cm^{-2}$, without device performance and reliability degradation [10]. However, a decrease in performance has been reported if the gate oxide thickness is lower than $1.3\,nm$ [33], suggesting a surface roughness limited mobility process due to the proximity of sub-oxide. Still, the strong band bending due to quantum mechanical corrections affects the lower limit of the supply voltage in the constant field scaling approach [34]. Performance and subthreshold slope optimization will require a $0.5\,nm$ range equivalent SiO_2 thickness gate insulator for sub-$0.5\,V$ supply voltage. A lot of know-how on high-κ dielectrics (HfO_2, ZrO_2, Ta_2O_5, BST, BTO, etc) comes from DRAM device development. However, these solutions still require some research and

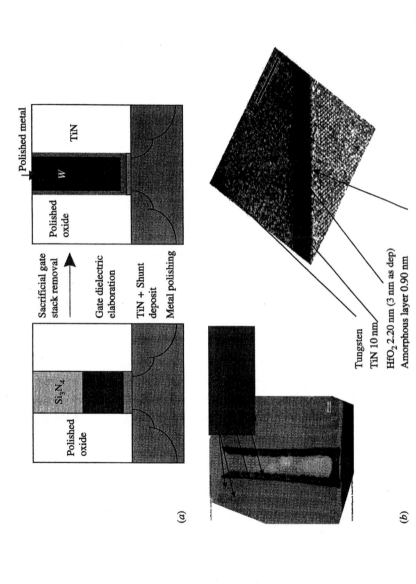

Figure 5.2.7. Damascene metal gate based on (*a*) sacrificial gate architecture developed by LETI and (*b*) SEM cross section of LETI approach [31].

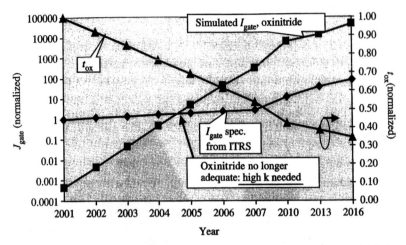

Figure 5.2.8. Gate leakage issue for the ITRS. Once the gate leakage of oxinitride is larger than the expected subthreshold leakage current, there will be a need for high-κ gate dielectric. This issue will come first for LOP devices. From ITRS 2001 [1].

development for MOS applications because a buffer interface compatible with a silicon substrate is needed (figure 5.2.10). On the other hand, solutions compatible with a silicon gate are also investigated to keep compatibility with a standard CMOS process flow: HfSiO$_x$ and ZrSiO$_x$ are given much attention as good candidates [35]. However, these solutions are *dielectric thickness*

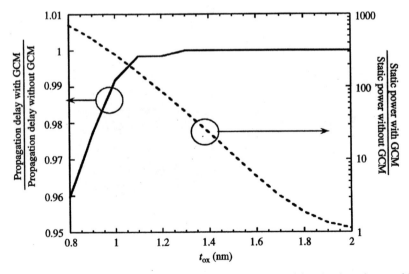

Figure 5.2.9. Propagation delay loss and static power increase with reduction of gate oxide thickness if gate current is taken into account [32]. GCM: gate current model. t_{ox}: electrical oxide thickness (at inversion capacitance).

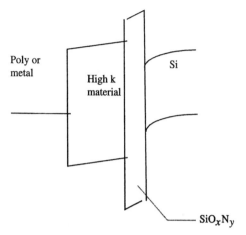

Figure 5.2.10. Energy band diagram of gate/high κ/SiO$_x$N$_y$/Si stack.

budget consuming because a SiO$_x$ interface appears between the high-κ dielectric and the device channel. Furthermore, the high-κ poly gate interface stability is still an issue.

Recently, the lowest leakage current has been reported by using 1.3 nm EOT HfO$_2$ combined with a TiN gate integrated in a damascene process flow [31] (figure 5.2.11(a)). 45 nm functional CMOS devices were obtained. However, electron mobility degradation is reported compared to SiO$_2$ gate dielectric (figure 5.2.11(b)) [31] attributed to the increase of phonon scattering induced by stress.

These materials have a smaller bandgap than SiO$_2$ has and hole trapping is a strong reliability issue [36] (figure 5.2.10 and table 5.2.2). That is why a SiON interface could benefit the leakage current reduction due to the higher bandgap of SiON. Before high-κ materials will be ready for use, silicon nitride or oxinitride will be intermediate solutions [37].

Alternatives to improve CMOS performances and integration density

Silicon on insulator and double gate
The other way to manage low threshold voltage is to use SOI. The fully depleted (FD) architecture is a way to approach a 60 mV dec^{-1} subthreshold slope by using SOI film thickness T_{Si} lower than the space charge region width. In order to obtain the lowest subthreshold slope and acceptable DIBL a practical rule can be used: $T_{Si} \le L_{gate}/4$ [38, 39]. However, the spreading of potential into the buried oxide, due to the coupling with the top gate, will tend to increase the coupling between source and drain and thus impact DIBL negatively. Practically, ultra-thin SOI films are difficult to control. That is why several authors propose partially depleted SOI [39, 40]. Due to complete isolation of the devices on SOI as well as lower junction

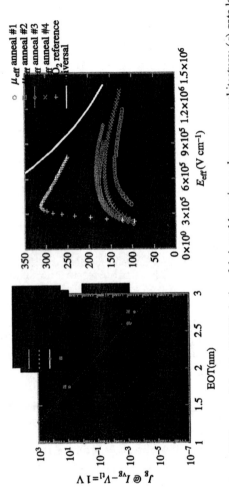

Figure 5.2.11. TiN metal gate/HfO$_2$ gate dielectric CMOS devices fabricated by using a damascene architecture: (*a*) gate leakage, (*b*) n channel effective mobility as a function of effective field [31].

Table 5.2.2. Characteristics of dielectrics—bulk material properties (partly from [36]).

(*)	SiO_2	Al_2O_3	Si_3N_4	Ta_2O_5	BST	HfO_2	ZrO_2
E_c (MV cm^{-1})	10	13	5	5	-10	3.5	
ε_r	3.9	10	7.5	22–27	>103	15–26	14–25
E_g (eV)	9	7–8	5	4.4	4.5	5.68	5.2–7.8

(*) Bulk materials characteristics.

capacitance, improved figures of merit are obtained as compared to bulk (figure 5.2.12). Moreover, the threshold voltage is dependent on Si film thickness whenever the film thickness becomes lower than the space charge region (figure 5.2.13). In this case, V_T given in relation (5.2.5) can be expressed as [39]:

$$V_T = V_{FB} + 2\varphi_F + \frac{qN_A T_{Si}}{2C_{ox}} \qquad (5.2.11)$$

where V_{FB} is the flatband voltage, φ_F is the Fermi potential distance from midgap, N_A is the acceptor concentration, T_{Si} is the silicon thickness and C_{ox} is the gate insulator capacitance.

However, scaling of FD devices encounters some limitations due to the quantum confinement in ultra-thin films and its incidence on the threshold voltage value (figure 5.2.13) [41]: the increase of the fundamental level of the conduction band will increase flat band voltage and V_T consequently.

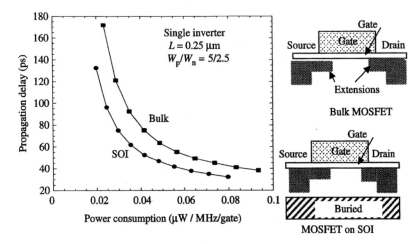

Figure 5.2.12. Dynamic performance comparison between bulk and PD-SOI. Dynamic switching energy of inverter as a function of delay [39].

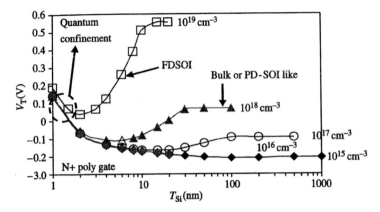

Figure 5.2.13. Evolution of threshold voltage as a function of Si film thickness in SOI MOSFET. Quantum confinement is taken into account for very thin Si layers. Example for n+ poly gate [41]. Thin-film doping ranges from 10^{15} to 10^{19} cm^{-3}.

Recently, the functionality of ultra-small 6 nm gate length devices on 7 nm thin Si film was demonstrated [42] (figure 5.2.14).

Nevertheless, self-heating is an issue on fully isolated devices especially if the insulator does not have good thermal conductivity. This is the case for SiO_2. Replacing SiO_2 by another insulator like Al_2O_3 to circumvent the problem has been proposed as a solution because Al_2O_3 has a thermal conductivity ten times lower than SiO_2 [43, 44] (figure 5.2.15). This is undoubtedly another interesting application of high-κ materials to CMOS provided that parasitic junction capacitance is still acceptable and finely tuned through buried insulator thickness.

The buried oxide offers many advantages for low-power and RF applications which are much more easily manageable on SOI than with bulk technology.

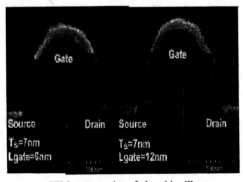

SEM cross-section of ultra thin silicon
channel pFETs with 6 and 12 nm gate lengths

Figure 5.2.14. 6 nm gate length single poly gate on ultra-thin SOI p channel MOSFETs [42].

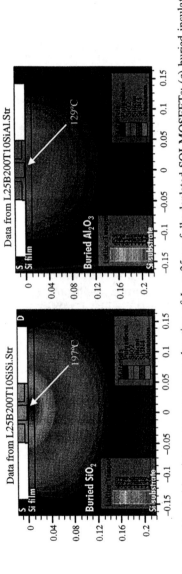

Figure 5.2.15. Impact of buried insulator on temperature elevation of $L_g = 25$ nm fully depleted SOI MOSFETs: (*a*) buried insulator is SiO_2, (*b*) buried insulator is Al_2O_3 with higher thermal conductivity than SiO_2 (factor of 10 difference between materials) [44].

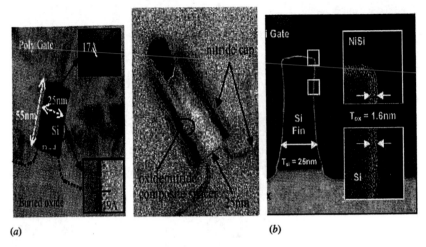

Figure 5.2.16. FinFET (*a*) of the omega type (Ω-FET) [49]; (*b*) with NiSi midgap gate obtained by full poly silicidation [48].

SOI material should allow us to realize attractive devices like multi-gated MOSFETs [45] that will allow further scaling of FD depleted devices which are limited by the quantum confinement issue and DIBL via the coupling of the gate with buried oxide [41]. With multi-gate devices, short-channel effects can be drastically reduced because an ideal subthreshold swing of $60\,\mathrm{mV\,dec^{-1}}$ is obtained as well as high drivability. In these devices, transport occurs by volume inversion due to the coupling of both gates. Consequently, the conditions to control short-channel effects can be relaxed compared to single-gate FD devices [41, 46]. Several process architectures have been proposed in the literature. FinFET is one of the most popular [47–49] (figure 5.2.16(*a*)). These architectures have been proven to be functional down to gate lengths of 10 nm [47]. Double gate devices are for sure the best way to trade off high saturation current and low leakage current [49]: under 0.7 V supply voltage saturation currents of $1300/550\,\mathrm{\mu A\,\mu m^{-1}}$ compatible with $1\,\mathrm{\mu A\,\mu m^{-1}}$ leakage current can be obtained on the n-channel/p-channel $L_g = 25\,\mathrm{nm}$ devices of figure 5.2.16(*a*). This result is the best ever published to date. However, the control of the shape of the fins is still an issue for the mastery of the fabrication process, and the application in the design of high density circuits with large number of devices ($\times10$ million) has to be demonstrated. Other approaches have been proposed like tri-gate FinFET [51] that can relax the gate etching issue due to the high aspect ratio of the fins, or planar PAGODA FET.

The main feature of these devices is to bring a solution to the channel dopant fluctuation problem. Reducing the film thickness to the minimum allows us to use nearly intrinsic Si films because bulk punch-through is no longer a problem. Adjusting V_T to match overdrive with a low supply voltage will require us to adjust the gate workfunction φ_M according to relation (5.2.8) considering

$$\varphi_{MS} = \varphi_M - \varphi_s \qquad (5.2.12)$$

where φ_s is the semiconductor electrochemical potential.

An important motivation for workfunction engineering comes from low-voltage high-performance applications, i.e. that will be emphasized by the introduction of SOI and the related architectures (figure 5.2.16(*b*)) [48]. In the future, it will thus be essential to adjust V_T by tuning very precisely the gate workfunction. If the supply voltage becomes very low, suppressing gate depletion will be mandatory, i.e. metal gate will be necessary. That is why, workfunction engineering on metal gate and high-κ stacks will be a must for low-voltage supply applications. For example, according to figure 5.2.13, adjusting a V_{TN} of 0.165 V on n-channel devices if one considers a supply voltage of 0.50 V will require a material with a workfunction value of 4.37 V (taking into account that the workfunction value of n+ silicon is 4.05 V). The p-channel device will require a gate with a workfunction of 4.765 V that will ensure a V_{TP} of − 0.165 V.

Vertical transistors
Vertical transistors have been initially proposed in order to study ultra-small devices without the need for aggressive lithography. Today, some commercial products have also been proposed successfully—fast read only memories—by Siemens [52] which use RPCVD epitaxy in a pillar structure. More prospective work is done for RF applications [53, 54]. Other approaches using replacement gates have been published [55] (figure 5.2.17): high static performance has been obtained.

Besides the need for integration capability, these approaches have to be improved because:

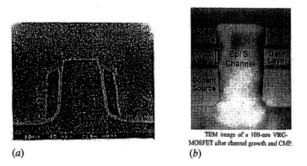

(*a*)   (*b*)

Figure 5.2.17. (*a*) Pillar vertical MOSFET [51], (*b*) vertical replacement gate [57].

- gate/source and drain capacitance is still an important issue. High Miller capacitances can affect dynamic performance despite obtaining good static performance.
- the minimum transistor width is imposed by the contact size unless specific design rules are used.
- the transistor sidewall orientation needs to be perfectly well known as the device channel properties—interface states, mobility, etc—will depend upon these features. No loading dependent sidewall etch can be accepted.

The possibility to wrap around the vertical transistor gate allows us to improve by 25% the integration density as compared to horizontal devices: optimal layout could lead to 12.25F2 area vertical devices instead of 15.75F2 horizontal devices. The static current density per unit area of silicon could be improved by a factor of 5.14 as compared to horizontal devices. However, implementation of minimal gate width devices—i.e. useful to design access transistors in SRAMs—requires us to increase process complexity. Still, short-channel effects are important issues to deal with as the drain depth is imposed by the drain contact layout. Vertical channels are used in other approaches like undercut SOI of the gate all around (GAA) type [56, 57] or in FinFET devices [58]. As already pointed out, large gate to source and drain overlap capacitance is the detrimental aspect of these architectures. Combined with horizontal devices, vertical devices could offer possibilities for 3D integration in the future.

Source and drain engineering

Shallow junctions as well as the introduction of high-κ gate dielectric is one of the most important issues to address to continue CMOS scaling with the rate of Moore's law. Source and drain engineering aims at compromising the reduction of access resistance together with junction depth.

Low-energy ($< 1\,keV$) [33] and heavy molecules (BF_3 [59], $B_{10}H_{14}$ [60], etc) could be the easiest ways to replace boron for the achievement of p+ shallow junctions. Plasma doping is one of the techniques which are investigated as possible alternatives to obtain lower than 25 nm as-implanted p+ junction depths [61]. However, transient enhanced diffusion (TED) is still the limiting process to reach the specified values as final junction depths even if low-energy ion implantation is used. Fast ramp up and down—so-called spike annealing—must be combined with low-energy ion implantation [62] to reduce TED as much as possible, by reducing the role played by extended and dopant defects (figures 5.2.18 and 5.2.19). Excimer laser anneal [63–65] has been reported to demonstrate the best trade-off between activation and junction depth shallowness: highest solid solubility combined with fast processing can be achieved (figure 5.2.19).

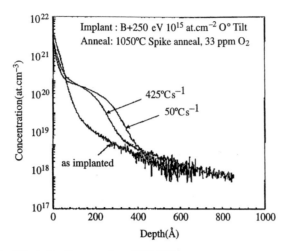

Figure 5.2.18. Effect of spike anneal on TED [61].

Self-aligned silicidation is a necessity with dual Si gate devices in order to shunt $n+$ and $p+$ gates in the dynamic regime. However, a trade-off between source and drain sheet resistance and access resistance must be found due to the dopant consumption during silicidation. In order to achieve low sheet resistance combined with low silicon consumption, monosilicides (NiSi, PtSi) instead of disilicides (TiSi$_2$, CoSi$_2$) will be preferred in the future [66]. Today, elevated source and drain is a very interesting option to minimize source and drain sheet resistance. Still it requires more research to make it a production process due to the criticity of epi/substrate interface prebake.

Practically, extension doping can be increased with the reduction of supply voltage. However, this could have detrimental issues on extension depth and degrade the short-channel effect. Pockets are very helpful to clamp the device punch-through and reduce the short-channel effect. Heavy ions such as gallium or indium have been proposed to limit the SCE as well as junction parasitic capacitance keeping as well good reliability [11] for geometries as low as 50 nm (figure 5.2.20). However, solid solubility of these dopants is an issue to keep a low-thermal-budget process. More generally, the use of pockets can relax the constraint on gate oxide thickness scaling to have a better control of SCE and DIBL (figure 5.2.20).

For smaller geometries, however, boron or BF$_2$ has been extensively studied because In and Ga suffer from solid solubility limit issues (see also figure 5.2.5, 'Limitation and showstoppers coming from classical CMOS scaling' section).

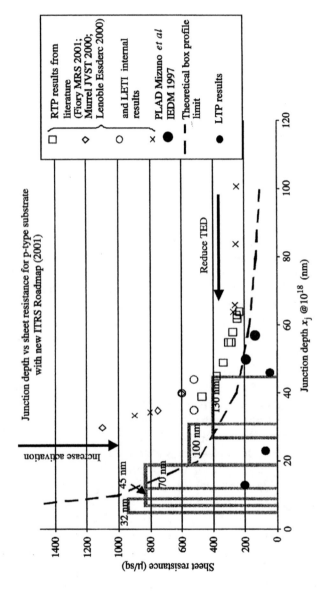

Figure 5.2.19. Trade-off between sheet resistance and junction depth. The 2001 ITRS roadmap goals are reported [1, 64].

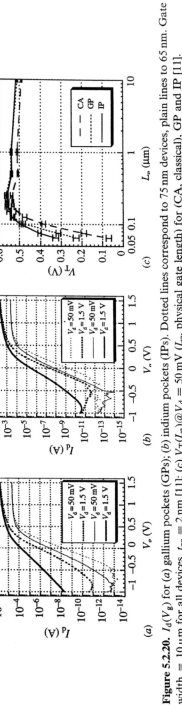

Figure 5.2.20. $I_d(V_g)$ for (*a*) gallium pockets (GPs); (*b*) indium pockets (IPs). Dotted lines correspond to 75 nm devices, plain lines to 65 nm. Gate width = 10 μm for all devices. $t_{ox} = 2$ nm [11]; (*c*) $V_T(L_g)@V_d = 50$ mV (L_g, physical gate length) for (CA, classical), GP and IP [11].

Interconnect issues

Copper integration is widely accepted to replace aluminium combined with low-κ dielectric (figure 5.2.21(a)). For high-speed design, a trade-off has to be realized between the number of interconnect levels and manufacturing cost [67] (figure 5.2.21(b)). The reduction of electromagnetic parasitics [68] and heat dissipation in low-κ dielectrics are new issues to address. A lot of space is free for research on other approaches like local and chip scale optical interconnects.

Alternatives to CMOS or alternative CMOS?

Will there be any alternative devices available to replace CMOS? Many research teams are making efforts on single-electron transistor (SET) operation based on the Coulomb blockade principle (figure 5.2.22(a)) [69]. Demonstration of CMOS inverter operation at 27 K has been achieved by using a vertical pattern dependent oxidation (V-PADOX) process [70] (figure 5.2.22(b)). However, no solution has been found that could compete straightforwardly with CMOS devices. Some teams have pointed out the possibility to achieve memory functional devices by using single-electron trapping by a Coulomb blockade effect for DRAM [71], or nonvolatile applications [72–74]. This effect supposes that the Coulomb energy

$$\frac{e^2}{2C} \qquad (5.2.13)$$

might be much larger than the thermal energy of electrons kT (e is the electron charge and C is the capacitance of the quantum box). This energy is necessary to localize electrons in a Coulomb box provided that tunnelling is the limiting process: implicitly, one has to use very low capacitance and sufficiently high tunnelling resistance. On the other hand, the Coulomb blockade process will be self-limiting by charge repulsion (figure 5.2.22(a)). That means we have to admit a low-speed charge transfer process. However, nonvolatile memory applications can be envisaged by using trapping in nanometre size Si quantum dots (figure 5.2.22(b)) [73]: Al_2O_3 has been chosen as the tunnel insulator with reasonable interface state density (less than 10^{11} cm^{-2}) and can also increase the dot density as compared to other materials (in the range of 10^{12} cm^{-2}). Writing characteristics of the obtained memory cells demonstrate that trapping occurs in the Si dots with low perturbation coming from the interface states [73]. This turns out to be a very interesting application of high-κ materials.

Whether the involved writing or erase mechanisms are due or not to single-electron transfer has been a controversial debate. If the Si dots are randomly distributed in large-area devices then it is very difficult to identify

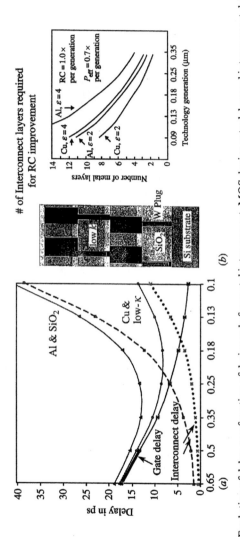

Figure 5.2.21. (*a*) Evolution of delay as a function of design rule for metal interconnect. MOS device gate delay and interconnect delay are reported for aluminium/SiO$_2$ and copper/low-κ dielectric systems; (*b*) number of interconnect layers required for improvement of delay as a function of generation [66].

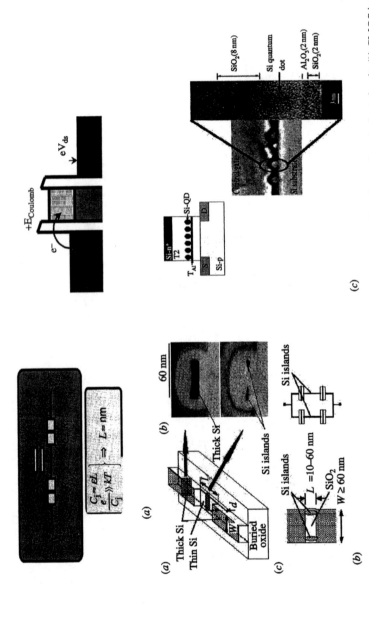

Figure 5.2.22. (*a*) Principle of Coulomb blockade in a potential well surrounded by tunnel barriers and equivalent circuit; (*b*) CMOS inverter built on SET [69]; (*c*) Si quantum dot nonvolatile flash memory cell: schematic cross section of Al$_2$O$_3$ tunnel nucleation insulator and high-resolution TEM cross section [72].

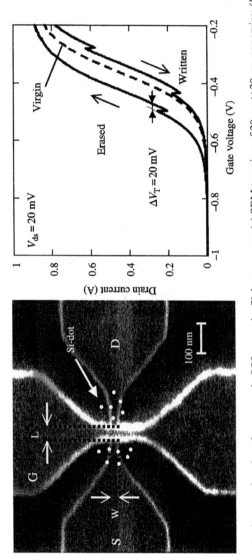

Figure 5.2.23. Single-electron charging phenomena in SOI nanowire Si dot memory: (*a*) SEM top view of 20 nm × 20 nm nanowire; (*b*) writing and erase characteristics of 20 nm × 20 nm (*W* × *L*) devices at room temperature. The spike in $I_d(V_g)$ characteristics is due to trapping or de-trapping of one electron [73].

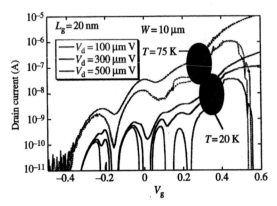

Figure 5.2.24. Drain current oscillations in an $L_g = 20$ nm MOSFET at 75 and 20 K, demonstrating that Coulomb blockade is possible in such devices [6]. See also [74] for 50 nm devices at 4.2 K.

whether the single-electron transfer is occurring or not due to the large number of dots. It is thus very important to use a device of the smallest size possible to get a high sensitivity to single-electron transfer in one dot or a low number of dots. Such a result has been obtained at room temperature on 20 nm × 20 nm nonvolatile memory silicon wire based on silicon quantum dots (figure 5.2.23) [74]: current spikes on the writing or erasing characteristics have been identified as single-electron trapping or de-trapping, respectively.

Coulomb blockade oscillations can be observed if the series access resistance with the quantum well is high enough compared to the resistance quantum

$$\left(\frac{e^2}{h}\right)^{-1}. \tag{5.2.14}$$

This effect has already been reported on 50 nm gate length n channel MOS transistors at 4.2 K [75] making CMOS transistors attractive as single-electron device candidates. As gate length is scaled down to 20 nm, access resistance becomes larger and channel conductance oscillations appear at higher temperatures (here 75 K) (figure 5.2.24) [6].

Conclusions

We have pointed out the main issues to address in order to investigate the limits of CMOS technology. Especially, the demand for low voltage, low power and high performance are the major challenges for nanoelectronic device engineering. The various possible options have been reviewed through the issues to address in gate stack/channel and substrate, source and drain engineering. The most highly strategic issues—besides lithography and interconnect—will be the low-voltage supply management and the necessary gate dielectric engineering in the range of 1 nm or less (equivalent oxide thickness). A few examples were given showing that high-κ materials will play a major role in the future to allow further high-performance CMOS scaling compatible with low power consumption.

Acknowledgments

This work has been carried out within the PLATO Organization teams and tools in the frame of the 'Ultimate CMOS project'. Many thanks to J. Gautier, B. de Salvo, T. Ernst from LETI and B. Guillaumot from STMicroelectronics for their various contributions and helpful discussions.

References

[1] *The International Technology Roadmap for Semiconductors* 2001 edition
[2] Deleonibus S *et al* 2000 *IEEE Electron Device Lett.* **21** 173–175
[3] Enz C *et al* 1995 *Analog Integrated Circuits and Signal Processing* (Dordrecht: Kluwer) pp 83–114
[4] Cheng Y *et al* 1997 *IEEE Trans. Electron Devices* **44** 277–287
[5] Bertrand G *et al* 2000 *Silicon Nanoelectronics Workshop 2000 (Honolulu, HI, June 2000)* pp 10–11
[6] Bertrand G *et al* 2003 *Fourth Workshop on Ultimate Integration of Silicon Proceedings (Udine, Italy, March 2003)* pp 10–13
[7] Tedesco S *et al* 1998 *J. Vac. Sci. Technol. B* **16** 3676–3683
[8] Heitzmann M and Nier M-E 1999 *Tech. Digest Micro and Nanoelectronics Engineering (Rome, Sept. 1999)* pp 23–24
[9] Deleonibus S *et al* 1999 *ESSDERC Tech. Digest 1999 (Leuven, Sept. 1999)* pp 119–126
[10] Iwai H *et al* 1998 *IEDM Tech. Digest 1998 (San Francisco, CA, Dec. 1998)* pp 163–166
[11] Caillat C *et al* 1999 *VLSI Technol. Symp. Tech. Digest 1999 (Kyoto, June 1999)* pp 89–90
[12] Taur Y *et al* 1995 *IBM J. Res. Dev.* **39** 245–260
[13] Gopalakrishnan K *et al* 2002 *IEDM2002 Tech. Digest (San Francisco, CA, Dec. 2002)* pp 289–291
[14] Taur Y *et al* 1997 *IEDM Tech. Digest 1997 (Washington, DC, Dec. 1997)* pp 215–218
[15] Oyamatsu T *et al* 1993 *Tech. Digest VLSI Symp. (June 1993)* pp 89–90
[16] Schockley W 1961 *Solid State Electron.* **2** 35–67
[17] De V *et al* 1999 *Tech. Digest Int. Symp. Low Power Electronics and Design (San Diego, Aug. 1999)* pp 163–168
[18] Nishinohara T *et al* 1992 *IEEE Trans. Electron Devices* **39** 634–639
[19] Wong W *et al* 1993 *IEDM Tech. Digest 1993 (Washington, DC, Dec. 1993)* pp 705–708
[20] Bouillon P and Skotnicki 1998 *IEEE Electron Device Lett.* **19** 19–22
[21] Ohguro T *et al* 1998 *IEEE Trans. Electron Devices* **45** 710–716

[22] Carroll M *et al* 2000 *IEDM Tech. Digest 2000 (San Francisco, CA, Dec. 2000)* pp 145–148

[23] Rim K *et al* 1998 *IEDM Tech. Digest 1998 (San Francisco, CA, Dec. 1998)* pp 707–710

[24] Ernst T *et al* 2002 *VLSI Technol. Symp. 2002 Tech. Digest (Honolulu, HI)* pp 92–93

[25] Ernst T *et al* 2003 *VLSI Technol. Symp. 2003 Tech. Digest (Kyoto)* pp 51–52

[26] Rim K *et al* 2002 *VLSI Technol. Symp. (Honolulu, HI, June 2002)*

[27] Lee J *et al* 1996 *Tech. Digest VLSI Symp. (June 1996)* p 208

[28] Chatterjee A *et al* 1998 *IEDM Tech. Digest 1998 (San Francisco, CA, Dec. 1998)* pp 777–780

[29] Devoivre T *et al* 1999 *Tech. Digest VLSI Symp. (June 1999)* p 131

[30] Yagashita A *et al* 1998 *IEDM Tech. Digest 1998 (San Francisco, CA, Dec. 1998)* pp 785–788

[31] Guillaumot B *et al* 2002 *IEDM 2002 Tech. Digest (San Francisco, CA, Dec. 2002)* pp 335–338

[32] Souil D *et al* 2002 *Third ULIS Workshop 2002 (Munich, March 2002)* pp 139–142

[33] Timp G *et al* 1998 *IEDM Tech. Digest 1998 (San Francisco, CA, Dec. 1998)* pp 615–618

[34] Takagi S *et al* 1998 *IEDM Tech. Digest 1998 (San Francisco, CA, Dec. 1998)* pp 619–622

[35] Lee J *et al* 1999 *IEDM Tech. Digest 1999 (Washington, DC, Dec. 1999)* pp 133–136

[36] Zetterling K 1998 *Abstracts 29th IEEE Semiconductor Interface Specialists Conf. (San Diego, CA, Dec. 1998)*

[37] Song S C *et al* 1998 *IEDM Tech. Digest 1998 (San Francisco, CA, Dec. 1998)* pp 373–376

[38] Ghibaudo G *et al* 1999 *Fall Meeting Electrochem. Soc.*, Abs. no. 1188, Oct 1999

[39] Pelloie J L 1999 *ISSCC Tech. Digest 1999 (San Francisco, CA, Feb. 1999)* p 428

[40] Leobandung L *et al* 1998 *IEDM Tech. Digest 1998 (San Francisco, CA, Dec. 1998)* pp 403–407

[41] Lolivier *et al* 2003 *ECS Spring 2003 (Paris, April 2003)* to be presented

[42] Doris B *et al* 2002 *IEDM 2002 Tech. Digest (San Francisco, CA, Dec. 2002)* pp 267–270

[43] Nakayama H *et al* 2000 *IEEE SOI Conf. 2000 Proc. (Wakefield, MA, Oct. 2000)* pp 128–129

[44] Oshima K *et al* 2002 *SOI Conf. 2002 Tech. Digest (Oct. 2002)* pp 95–96

[45] Wong H S P *et al* 1997 *IEDM Tech. Digest 1997 (Washington, DC, Dec. 1997)* pp 427–430

[46] Allibert F *et al* 2001 *ESSDERC 2001 (Nurnberg, Sept. 2001)*

[47] Yu B *et al* 2002 *IEDM 2002 Tech. Digest (San Francisco, CA, Dec. 2002)* pp 251–253

[48] Kedzierski J *et al* 2002 *IEDM 2002 Tech. Digest (San Francisco, CA, Dec. 2002)* pp 247–250

[49] Yang F L *et al* 2002 *IEDM 2002 Tech. Digest (San Francisco, CA, Dec. 2002)* pp 255–258

[50] Doyle B S *et al* 2003 *VLSI Tech. Symp. 2003 Tech. Digest (Kyoto)* 133–134

[51] Guarini R K W *et al* 2001 *IEDM 2001 Tech. Digest (Washington, DC, Dec. 2001)* pp 425–428

[52] Risch L *et al* 1997 *ESSDERC Tech. Digest (Stuttgart, Sept. 1997)* pp 34–41

[53] Jurczak M *et al* 1998 *ESSDERC Tech. Digest (Bordeaux, Sept. 1998)* pp 172–175

[54] DeMeyer K *et al* 1998 *ESSDERC Tech. Digest (Bordeaux, Sept. 1998)* pp 63–66

[55] Hergenrother J M *et al* 1999 *IEDM Tech. Digest (Washington, DC, Dec. 1999)* pp 75–78

[56] Colinge J-P *et al* 1990 *IEDM Tech. Digest 1990 (San Francisco, CA, Dec. 1990)* pp 595–598

[57] Leobandung L *et al* 1997 *J. Vac. Sci. Technol.* B **15**(6) 2791

[58] Hisamoto D *et al* 1998 *IEDM Tech. Digest (San Francisco, CA, Dec. 1998)* pp 1032–1035

[59] Ha J M *et al* 1998 *IEDM Tech. Digest (San Francisco, CA, Dec. 1998)* pp 639–642

[60] Goto K *et al* 1997 *IEDM Tech. Digest (Washington, DC, Dec. 1997)* pp 471–474

[61] Takase M *et al* 1997 *IEDM Tech. Digest (Washington, DC, Dec. 1997)* pp 475–478

[62] Downey D F *et al* 1998 *Proc. Mater. Res. Soc.* **525** 263

[63] Noguchi T *et al* 1985 *Proc. Mater. Res. Soc.* **146** 35

[64] Tsukamoto H *et al* 1999 *Solid State Electron.* **43** 487

[65] Laviron C *et al* 2001 *Ext. Abst. Second IWJT, IEEE-Cat. No. 01EX541C, 2001: 91-4, Nov. 2001, Tokyo*

[66] Ohguro T 1997 *ECS Symp. on ULSI 1997 (Montreal, CA, Oct. 1997)* p 275

[67] Bohr M 1995 *IEDM Tech. Digest (Washington, DC, Dec. 1995)* pp 241–244

[68] Delorme N *et al* 1997 *ESSDERC Tech. Digest (Stuttgart, Sept. 1997)* pp 125–132

[69] Ahmed H 1998 *FED J.* **9** (Suppl. 2) 15–24

[70] Ono Y *et al* 2001 *IEDM 2001 Tech. Digest (Washington, DC, Dec 2001)* pp 367–370

[71] Tiwari S *et al* 1996 *Appl. Phys. Lett.* **68** 1377

[72] Yano K 1998 *IEDM Tech. Digest (San Francisco, CA, Dec. 1998)* pp 107–110

[73] Fernandes A *et al* 2001 *IEDM 2001 Tech. Digest (Washington, DC, Dec. 2001)* pp 155–158

[74] Molas G et al 2002 *WODIM 2002 (Grenoble, Nov. 2002)*

[75] Specht M *et al* 1999 *IEDM Tech. Digest (Washington, DC, Dec. 1999)* pp 383–341

Chapter 5.3

High-κ transistor characteristics

Jack C Lee and Katsunori Onishi

Introduction

Various high-κ gate dielectrics have been proposed in recent years, and the range of the dielectric constant (κ) is fairly scattered from 7.5 of Si_3N_4 to >100 of ferroelectrics. At this moment, we will categorize the dielectrics into three groups—ultra-high-κ ($\kappa > 100$), moderate high-κ ($4 < \kappa < 10$), and mid-range high-κ ($10 < \kappa < 100$)—and evaluate typical materials for each category.

The ultra-high-κ materials, such as BST ($(Ba,Sr)TiO_3$, $\kappa \sim 300$), are the most advantageous in achieving thinner equivalent oxide thickness (EOT), but have a so-called field-induced barrier lowering (FIBL) effect problem [1]. The physical thickness (T_{phys}) of these materials will be so thick that the cross-section of the dielectric film will be rather a rectangular (with 'high' H/L aspect ratio) than a sheet. Therefore, the channel potential will be controlled by not only the gate electrode but also the source and the drain [1], and the MOSFETs will be difficult to turn off. For example, for an EOT of 10 Å with an ultra-high-κ material with $\kappa \sim 300$, T_{phys} will be ~770 Å or 77 nm, which is comparable to the channel length of the near-future technologies. Although this problem can be relieved by introducing a low-κ interfacial layer [1], this will cancel out the advantages of the ultra-high-κ dielectric. The ultra-high-κ materials cannot be introduced into production unless this FIBL problem is resolved.

Si_3N_4 ($\kappa = 7.5$) and Al_2O_3 ($\kappa = 10$) are well-known candidates for the moderate high-κ materials. These are common materials in the CMOS industry, and can be readily introduced into the production line. On the other hand, several issues remain to be solved for these materials. The dielectric constant κ of Si_3N_4 is not high enough to achieve the advantage of suppressing the gate leakage current significantly. Charge traps due to high nitrogen concentration are also a concern [2]. For Al_2O_3, mobility

560

degradation due to Coulomb scattering from the fixed charges in the dielectric limits the drive current of the MOSFETs [3]. On the whole, these moderate high-κ materials have too many problems to be used for the near-future technologies, and their dielectric constants are not high enough for more advanced technologies.

Considering the disadvantages of ultra- and moderate high-κ materials discussed above, it was found that mid-range high-κ materials were preferable for the gate dielectric application. A variety of mid-range high-κ materials have been reported as possible candidates.

One of the most important material properties for the gate dielectrics is thermal stability in contact with silicon. Ta_2O_5 and TiO_2, for example, have been studied as possible mid-range high-κ candidates. However, these materials were not stable in contact with Si substrate and tended to form low-κ interfacial layers. As a result, these materials required intentional interfacial layers, which cancel out the advantage of their high κ value [4–6].

Hubbard and Schlom [7] have studied thermodynamic stability of binary metal oxides in contact with silicon. They have examined various chemical reactions between binary metal oxides and silicon with respect to the change of Gibbs free energies. Reactions include reduction of the metal oxides, formation of silicide or silicate, and change of the composition of the binary metal oxides. As a result, only BeO, MgO, and ZrO_2 were found to be thermodynamically stable in contact with silicon. According to the theoretical calculations of dielectric constants [8], ZrO_2 is the only high-κ material with $\kappa = 22$ among these three. Although not enough data were available for HfO_2 in [7], HfO_2 is also expected to be thermally stable because of the chemical similarity between Zr and Hf.

ZrO_2 and HfO_2 emerged as promising high-κ dielectrics for ultra-thin gate dielectric application almost at the same time [9, 10]. Their dielectric characteristics measured on MOS capacitors were similar to each other. MOSFET characteristics of these two dielectrics were also found to be well behaved [11, 12]. However, it was found that ZrO_2 was not compatible with the polysilicon gate electrode [13], unlike HfO_2, that exhibited excellent MOSFET characteristics with the polysilicon gate [14]. Although it was demonstrated that nitrogen-incorporated ZrO_2 (ZrON) could show well-behaved $C–V$ characteristics to some extent [15], it was also revealed that ZrON still reacted with the polysilicon and the gate leakage current increased with the polysilicon gate compared to those with metal gate electrodes [16]. In spite of the similar material properties and device characteristics with metal gates, incompatibility with the polysilicon gate was one major disadvantage of ZrO_2 to HfO_2.

HfO_2-based dielectrics have been extensively evaluated and various deposition techniques were demonstrated to be capable of depositing HfO_2, including PVD [10], JVD [17], CVD [18, 19], and atomic layer deposition (ALD) [20, 21]. MOSFET transistors were also demonstrated using these deposition techniques. HfO_2 still presents several issues to be solved before its possible introduction to mass production, such as boron penetration, low crystallization temperature, charge trapping, and low channel mobility. Insufficient reliability data exist in the literature as well. Similar to SiO_2, nitrogen incorporation was found to be effective in suppressing boron penetration. The nitrogen incorporation techniques include a surface nitridation (SN) technique in NH_3 ambient [22], a top nitridation technique [23], Hf oxynitride (HfON) [24], or HfSiON [25]. The crystallization temperature was also improved using various dopings such as HfON [24], HfSiON [25], and HfAlO [17]. Short-channel devices were also demonstrated with the HfO_2-based dielectrics. It was once pointed out that high-κ gate dielectrics were susceptive to oxygen diffusion so that oxidation encroachment from the source and drain sides would be significant [19, 21]. This encroachment deteriorated the gate capacitance and the drive current for the short-channel devices [19, 21]. However, the most recent devices demonstrated comparable performance to that of SiO_2, indicating that the encroachment problem might be overcome [25].

Among the issues related to high-κ gate dielectrics, inadequate mobility compared to SiO_2 is among the most important in order to optimize the MOSFET performance. One approach to solve this issue is to combine the high-κ dielectrics with new substrate materials. Strained silicon was demonstrated to enhance the mobility with ALD-HfO_2 gate dielectric [26]. On the other hand, research has been conducted to enhance the mobility by modifying the dielectric itself. Inserting a thin SiO_2-based intentional interfacial layer was effective in improving the mobility [25, 27]. High-temperature (500–600°C) forming gas (FG) annealing (HT-FGA) was found to be effective in reducing D_{it} of HfO_2 MOSFETs and improving MOSFET mobility as well [27, 28]. In the next section, we will discuss the HfO_2 MOSFET mobility in terms of the effects of HT-FGA.

Another issue that has not been fully addressed is reliability characteristics of high-κ MOSFETs. One of the major reliability issues is bias-temperature instabilities (BTI) due to charge trapping in the dielectric and resulting performance instabilities of MOSFETs. It was observed on HfO_2 MOSFETs regardless of the deposition techniques (e.g. ALD-HfO_2 [20] and PVD-HfO_2 [29, 30]).

HfO_2 MOSFET performance and effects of high-temperature forming gas annealing

In this section, the use of HT-FGA for improving transistor characteristics will be discussed. Although channel mobility is widely used for evaluating the performance of CMOS technologies, it is also well known that the drain current of short-channel devices is degraded from that predicted for long-channel devices using this low-field mobility. In the subsection 'An overview of MOSFET performance', we will relate the mobility to short-channel device performance, and discuss the typical mobility measurements. The subsection 'Experimental results' will describe the performance of HfO$_2$ MOSFETs and the effects of HT-FGA.

An overview of MOSFET performance

Low-field mobility and saturation current of short-channel devices
Performance of MOSFET technology is generally evaluated in terms of the drain current at the saturation condition (i.e. $V_d = V_g = V_{dd}$), and the simplest form of the saturation current for a long-channel device is expressed as [31]

$$I_{dsat} = \mu_{eff}C_{ox}\frac{W}{L}\frac{(V_g - V_t)^2}{2},$$ (5.3.1)

where μ_{eff} is the effective channel carrier mobility, C_{ox} is the gate capacitance at the inversion condition, and W and L are the width and the length of the channel, respectively. I_{dsat} is proportional to μ_{eff}, and μ_{eff} is one of the key parameters to achieve high-performance CMOS technology.

The MOSFET channel mobility is generally degraded compared to the mobility observed in bulk crystalline silicon [31], due to Coulomb scattering from interface and fixed charges of the dielectric and dopants in the substrate, phonon scattering from silicon substrate and dielectrics, and surface roughness at the dielectric/Si substrate interface. On the other hand, Sabnis and Clemens observed on NMOSFETs with a SiO$_2$ gate dielectric a universal relationship between the μ_{eff} and the effective field in the channel E_{eff} [32], which is calculated as

$$E_{eff} = \frac{(\eta Q_{inv} + Q_b)}{\varepsilon_{Si}},$$ (5.3.2)

where η is a dimensionless fitting constant, Q_{inv} is the inversion charge density, Q_b is the substrate charge density, and ε_{Si} is the dielectric constant of the Si substrate. This universal relationship for NMOSFETs is fitted well with an η of 1/2, which agrees with the number calculated for the average field experienced by channel electrons [33]. An η of 1/3–1/2.5 is generally used for SiO$_2$ PMOSFETs [34, 35].

For short-channel devices, on the other hand, equation (5.3.1) is no longer valid, since the velocity of the carriers starts deviating from a linear function of the lateral channel field at high fields. Sodini *et al* modelled the velocity dependence on the lateral field as [36]

$$v = \frac{\mu_{\text{eff}}E}{1 + E/E_{\text{C}}}, \quad E \leq E_{\text{C}} \tag{5.3.3}$$

$$= v_{\text{sat}}, \quad E > E_{\text{C}}, \tag{5.3.4}$$

where E is the lateral electric field, E_{C} is the critical field, at which the carriers are velocity saturated and is equal to $2v_{\text{sat}}/\mu_{\text{eff}}$, where v_{sat} is the saturation velocity at high fields. The velocity approaches $\mu_{\text{eff}}E$ at the low-field limit. By using equation (5.3.3), the saturation current equation for short-channel devices is modified to

$$I_{\text{dsat}} = W v_{\text{sat}} C_{\text{ox}} \frac{(V_g - V_t)^2}{E_C L + (V_g - V_t)}. \tag{5.3.5}$$

It should be noted that at the long-channel limit ($L \to \infty$), equation (5.3.5) approaches equation (5.3.1). On the other hand, for the short-channel limit ($L \to 0$), the equation will approach

$$I_{\text{dsat}} = W v_{\text{sat}} C_{\text{ox}} (V_g - V_t). \tag{5.3.6}$$

With this limit, the saturation current does not depend on the channel length or mobility μ_{eff}, but on the saturation velocity v_{sat}.

While the above discussion on carrier velocity assumes the drift–diffusion approximation under thermal equilibrium, this model may not apply to the very short-channel devices. Laux and Fischetti used a Monte Carlo simulation and found that, at the short-channel limit, some of the carriers did not experience scattering and the average velocity could exceed v_{sat} [37]. This phenomenon is called velocity overshoot.

Lundstrom pointed out from a scattering theory that the maximum drain current at the saturation condition was dominated by the thermal injection velocity v_{T} at the source side [38]. At the short-channel limit, the drain current is still expressed by equation (5.3.6), but v_{sat} will be replaced with v_{T}.

$$I_{\text{dsat}} = W v_{\text{T}} C_{\text{ox}} (V_g - V_t). \tag{5.3.7}$$

This v_{T} is larger than v_{sat} by a factor of 1.5–2 [33]. Lundstrom and co-workers evaluated I_{dsat} of existing technologies and found that it was still far less than the ballistic limit given by equation (5.3.7) [39]. In the Lundstrom theory, the deviation of I_{dsat} from the ballistic limit comes from a backscattering effect at the source edge of the channel, and this backscattering is correlated to the mobility μ_{eff}. With the scattering theory, the saturation current for the intermediate channel length can be written as [38]

$$I_{\text{dsat}} = \left(\frac{C_{\text{ox}}W}{\frac{1}{v_{\text{T}}} + \frac{1}{\mu_n^0 E(0^+)}}\right)(V_g - V_t),
\qquad (5.3.8)$$

where μ_n^0 is the low-field mobility and $E(0^+)$ is the channel electric field.

Having in mind that the ballistic limit of the drain current is given by equation (5.3.7), how close are we to this limit with the most up-to-date technologies? Is it still meaningful to use the long-channel mobility as a figure of merit for transistor performance? Lochtefeld and Antoniadis carefully evaluated the electron velocity at the source side of the NMOSFET channel with $L_{\text{eff}} < 50$ nm [40]. The measured velocity was merely ~40% of the anticipated thermal velocity, and it was therefore confirmed that the mobility had a contribution to the saturation drain current. The same group also examined the correlation between the long-channel mobility and the carrier velocity (v_{eff}) at the source side for a similar technology with $L_{\text{eff}} \sim 45$ nm by intentionally applying a mechanical strain on the wafer [41]. They found that the correlation factor of $\delta v_{\text{eff}}/\delta \mu_{\text{eff}}$ was 0.44–0.46. It was concluded that the low-field mobility was still of crucial importance to carrier velocity and hence to saturated drive current.

Mobility measurements

The effective mobility μ_{eff} at low lateral fields can be defined as

$$\mu_{\text{eff}}(V_g) = \frac{I_d/V_d}{(W/L)Q_{\text{inv}}(V_g)}, \qquad (5.3.9)$$

where Q_{inv} is the inversion charge density. Q_{inv} was conventionally calculated using the charge sheet model,

$$Q_{\text{inv}}(V_g) = C_{\text{ox}}(V_g - V_t). \qquad (5.3.10)$$

The charge sheet model assumes that when V_g is larger than V_t, the surface potential ϕ_S is pinned at $2\psi_B$ and the rest of the gate voltage corresponds to the voltage across the gate oxide. However, this simple equation results in an error in Q_{inv}, particularly in the low-field regime ($V_g \sim V_t$), where the surface potential does not follow the charge sheet model. Instead, the split C–V method is being widely used in order to accurately evaluate Q_{inv} [42].

$$Q_{\text{inv}}(V_g) = \int_{-\infty}^{V_g} C_{\text{gs}}(V_g)\, dV_g \qquad (5.3.11)$$

where C_{gs} is the capacitance between gate and channel. One concern regarding this split C–V method is that C_{gs} tends to be underestimated due to the channel resistance, when C–V is measured at high frequencies. Although we typically measure C–V characteristics at 1 MHz, it was confirmed by measuring the frequency dispersion that the MOSFET $C-V$ was not distorted at this frequency, except a slight increase of the capacitance at $V_g \sim V_{fb}$ due to D_{it} present in the devices.

For the Q_{inv} calculated from equation (5.3.11), the effective field and mobility were calculated using equations (5.3.2) and (5.3.9), respectively. In equation (5.3.2), η values of 1/2 and 1/2.5 were used for N and PMOSFETs, respectively [35]. The measured $\mu_{eff}-E_{eff}$ relationships were compared to the universal curves of SiO_2 MOSFETs. The equations were taken from [35] and can be expressed as

$$\mu_{eff,n} = \frac{630}{1 + (E_{eff}/0.75)^{1.67}},\qquad(5.3.12)$$

and

$$\mu_{eff,p} = \frac{185}{1 + (E_{eff}/0.45)},\qquad(5.3.13)$$

for N and PMOSFETs, respectively.

Experimental results

Experimental details [27, 28]

One of the major challenges in high-κ gate dielectrics is controlling the quality of the interface between the high-κ dielectric and the Si substrate. It was revealed by an XPS analysis that HfO_2 with polysilicon gate electrode formed an interfacial layer between HfO_2 and the Si substrate that has a composition close to SiO_2 [43]. In contrast surface electron mobility in the HfO_2 NMOSFETs was generally inferior to that of SiO_2 [18, 20]. The degraded mobility was often observed along with poor subthreshold swings, indicating that D_{it} was not minimized. In this subsection, we will discuss the carrier mobility in HfO_2 MOSFETs in terms of the effects of HT-FGA ($>400°C$). The HT-FGA was intended to reduce the D_{it} between HfO_2 and Si substrate, and to examine the effect of the D_{it} reduction on the electron mobility. It was found that the HT-FGA was effective in improving the electron mobility. D_{it} reduction was confirmed through electrical characterizations such as $C-V$, subthreshold swings, and charge pumping current measurements.

Devices were fabricated using the conventional self-align MOSFET process, and the process flow is summarized in table 5.3.1. Initially, 400 nm field oxide was grown on 4 inch Si wafers, which have doping concentrations of approximately $3 \times 10^{15}\,cm^{-3}$ for both p- and n-type substrates. The active area was formed by etching the field oxide. After cleaning the wafers with an 'HF-last' dipping, HfO_2 was deposited using reactive dc magnetron sputtering with a modulation technique [10], followed by post-deposition

Table 5.3.1. Process flow of the HfO$_2$ MOS-
FET fabrication.

- Field oxidation and active patterning
- Piranha and HF (1:100) cleaning
- *(Surface annealing in NH$_3$ or NO)*
- Reactive dc sputtering of HfO$_2$
  —Modulation technique with O ambient
- Post-deposition annealing (PDA)
  —500°C 5 min in N$_2$
- Poly-Si deposition and patterning
- Dopant implantation
- S/D oxidation
  —850 or 900°C 1 min
- Low-temperature oxide deposition
- Contact patterning
- Dopant activation
  —900 (PMOS) or 950°C (NMOS) 1 min
- *High-temperature FG anneal*
  —500–600°C 30 min
- Metal (Al) deposition and patterning
- Sintering in FG: 400–450°C 30 min

---

annealing at 500°C for 5 min in N$_2$ ambient. Prior to HfO$_2$ deposition, some selected wafers were annealed in NH$_3$ or NO ambient, in order to examine their effects on the MOSFET performance. Polysilicon (200 nm thick) was deposited using low-pressure CVD at 550 or 580°C in amorphous state. Following polysilicon patterning, phosphorus of 50 keV and BF$_2$ of 60 keV were implanted with a dose of 5×10^{15} cm^{-2} on N and PMOSFETs, respectively. Source and drain (S/D) oxidation at 850 or 900°C for 1 min was then used to anneal out damage incurred during the polysilicon etch and the implantation. Dopant activation was done at 900 or 950°C using rapid thermal annealing. In order to examine the effects of the HT-FGA, selected wafers were annealed in the FG ambient with 4% of H$_2$ in N$_2$ at temperatures of 500 and 600°C for 30 min. This annealing was applied prior to Al interconnect deposition to avoid junction spiking at the S/D regions. After patterning Al interconnect and deposition of backside Al, another FG anneal was applied to lower the contact resistance at relatively lower temperatures (400–450°C).

I–V and *C–V* characteristics were evaluated using an HP4156A precision semiconductor parameter analyser and an HP4194A impedance/gain-phase analyser, respectively, and the measured *C–V* characteristics were simulated to determine the EOT with deduction of the quantum mechanical effect [44].

HfO₂ NMOSFET performance

I_d–V_g characteristics of HfO₂ NMOSFETs with an EOT of 13 Å are shown in figure 5.3.1(*a*) and (*b*) in linear and log scales, respectively, comparing three different FG annealing temperatures. As the FG annealing temperature increased, the drive current increased along with a slight reduction in threshold voltages (figure 5.3.1(*a*)). The subthreshold characteristics are shown in figure 5.3.1(*b*), and the trend of the subthreshold swings is summarized in the inset. The subthreshold swings were decreased as the FG

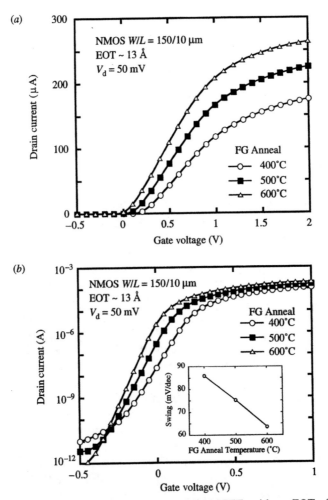

Figure 5.3.1. I_d–V_g characteristics of the HfO₂ NMOSFETs with an EOT of 13 Å, for three different FG annealing temperatures, in linear (*a*) and log (*b*) scales, respectively. Minimum subthreshold swing of 64 mV dec⁻¹ was achieved at the 600°C FG annealing.

annealing temperature increased. Using the 600°C FG annealing, the subthreshold swing was reduced to $64\,\mathrm{mV\,dec^{-1}}$, which is close to the D_{it}-free ideal value (i.e. $60\,\mathrm{mV\,dec^{-1}}$).

I_d–V_d characteristics in figure 5.3.2 confirmed the improvement in the drive current in the saturation region also. The relationship between effective mobility (μ_{eff}) and effective field (E_{eff}) of channel electrons was evaluated for different FG annealing temperatures in figure 5.3.3. The mobility was calculated using the split C–V method [34] using equation (5.3.11). $\eta = 1/2$ was used for NMOSFETs [34, 35]. The universal relationship between μ_{eff} and E_{eff} for SiO₂ NMOSFETs [35] is also shown to compare with the measured data on HfO₂. The mobility was enhanced at the higher FG annealing temperatures for whole range of E_{eff}, although it was inadequate compared to the universal curve.

C–V characteristics of NMOSFETs are compared between 400 and 600°C FG annealing (figure 5.3.4). Capacitance at both accumulation and inversion regions were not degraded by the HT-FGA. There appears to be some slight enhancement in the capacitance for the 600°C FG annealed samples. D_{it} reduction at the 600°C FG annealing was observed as the steeper increases of the capacitance at both flatband and threshold voltages.

Charge pumping current is effective in quantitatively evaluating the D_{it} [45]. Figure 5.3.5(*a*) shows the measured charge pumping currents on the HfO₂ NMOSFETs. Linear dependence of the current on the frequency was observed, and the current was reduced at the higher FG annealing temperatures. D_{it} was calculated using Groeseneken's formula [45], and

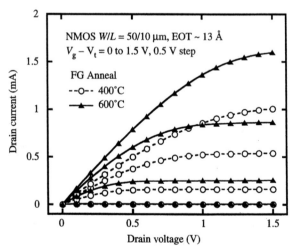

Figure 5.3.2. I_d–V_d characteristics of the HfO₂ NMOSFETs with an EOT of 13 Å for FG annealing at 400 and 600°C. Drain current at the saturation region was also improved at the 600°C FG annealing.

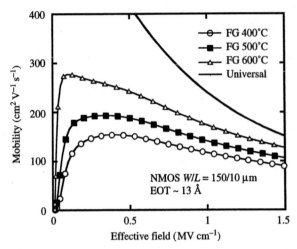

Figure 5.3.3. Effective channel electron mobility (μ_{eff}) of HfO$_2$ NMOSFETs as a function of effective electric field (E_{eff}), for three different FG annealing temperatures.

summarized in figure 5.3.5(*b*). At the 600°C FG annealing, D_{it} was reduced to $3.7 \times 10^{11}\,\text{cm}^{-2}$.

It should be emphasized that all the improvements discussed above were achieved without degrading EOT and gate leakage, as shown in figure 5.3.6. Figure 5.3.7 compares gate leakage currents in NMOS capacitors between 400 (*a*) and 600°C (*b*) for three different measurement temperatures.

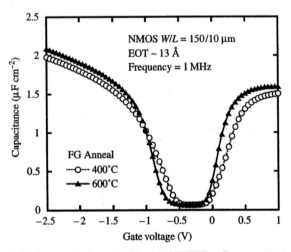

Figure 5.3.4. *C–V* characteristics of HfO$_2$ NMOSFETs. Steeper *C–V* curves in the vicinities of both V_{fb} and V_{t} indicate the reduction in D_{it} for the higher FG annealing temperature.

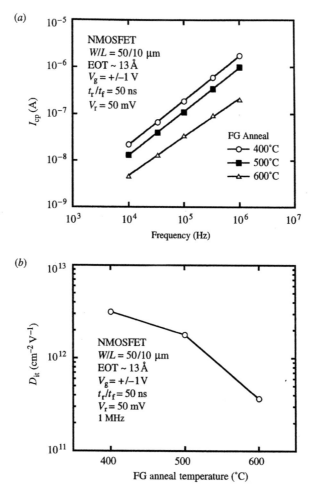

Figure 5.3.5. (*a*) Charge pumping current as a function of gate pulse frequency for three different FG annealing temperatures. (*b*) D_{it} calculated from the charge pumping currents. D_{it} was reduced to $3.7 \times 10^{11}\,\mathrm{cm}^{-2}$ at the FG annealing of 600°C.

The shifts of the minimum points of the gate currents were due to the flatband voltage shifts at the higher temperatures. At high positive gate voltages, the current was limited by the electron generation rate at the p-substrate. Negligible temperature dependence of the leakage current was observed on both samples, except the slight flatband voltage shifts, indicating a tunnelling-type conduction mechanism. The HT-FGA did not change the conduction mechanism. Slightly lower breakdown voltage was observed for the 600°C annealing.

Figure 5.3.8 compares maximum transconductance (Gm_{max}) of HfO$_2$ and SiO$_2$ MOSFETs with various annealing conditions with respect to

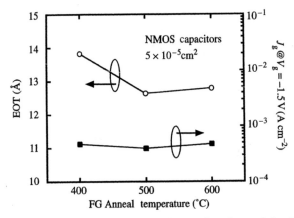

Figure 5.3.6. EOT and gate leakage current (J_g) as functions of the FG annealing temperature. High-temperature FG annealing did not degrade the EOT and the J_g of HfO$_2$ NMOS capacitors.

Gm_{max} for 400°C FG annealing of each dielectric. Unlike HfO$_2$, Gm_{max} of SiO$_2$ was not improved after the HT-FGA. It should be noted that μ_{eff}–E_{eff} relationship of SiO$_2$ MOSFETs was close to the universal curve regardless of the FG annealing temperature (data not shown), indicating that 400°C FG annealing was sufficient for terminating dangling bonds at the interface between SiO$_2$ and the substrate. In figure 5.3.8, Gm_{max} of a HfO$_2$ MOSFET with 600°C N$_2$ annealing was also compared to other FG annealing conditions. The 600°C N$_2$ annealing improved Gm_{max} compared to 400°C FG annealing, but not as much as 600°C FG annealing. It was confirmed that hydrogen was essential for improving the mobility of HfO$_2$ MOSFETs.

Effects of surface treatments

Surface preparation is another process that significantly affects the MOSFET performance. Figure 5.3.9 compares C–V characteristics of HfO$_2$ NMOSFETs with the similar physical HfO$_2$ thickness of ~50 Å for three different surface preparations: i.e., control without any pre-deposition annealing, NH$_3$ annealing at 600°C for 30 s, and NO annealing at 700°C for 30 s. By using the NH$_3$ annealing, an EOT of ~10 Å was achieved, while the NO annealing increased the EOT up to 15 Å. Figure 5.3.10 summarizes electron mobility for these surface preparations. Figure 5.3.10(a) compares mobility at a fixed E_{eff} of 1 MV cm^{-1}, where all the preparations showed mobility improvements at the higher FG annealing temperatures. Figure 5.3.10(b) shows μ_{eff}–E_{eff} relationships for the FG annealing at 600°C. The NO annealing yielded the highest mobility for the whole E_{eff} range. (It should be emphasized that the EOT was the highest for the NO-annealed samples.)

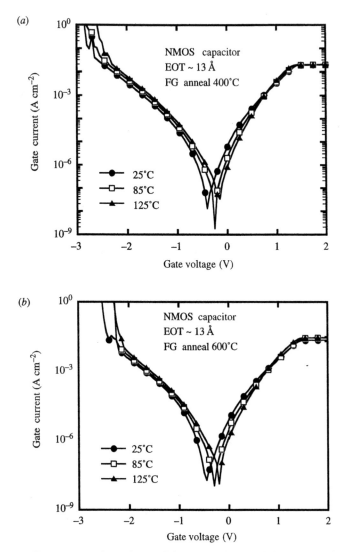

Figure 5.3.7. Temperature dependence of the J_g–V_g characteristics of the HfO₂ NMOS capacitors with an EOT of 13 Å for FG annealing at 400 (*a*) and 600°C (*b*). The high-temperature FG annealing did not change the leakage current, while the breakdown voltage was slightly decreased.

HfO₂ NMOSFETs with TaN metal gate electrodes

HT-FGA was also effective in improving the mobility of HfO₂ NMOSFETs with TaN metal gate electrodes. Figure 5.3.11 shows the μ_{eff}–E_{eff} relationship for these devices. The SN technique by an NH₃ annealing was adopted for these samples. Due to the advantage of the SN technique, an EOT of ~9 Å

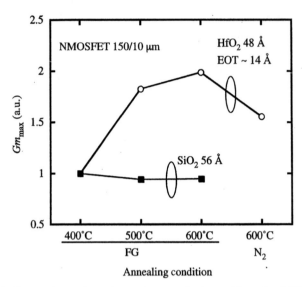

Figure 5.3.8. Comparison of maximum transconductance (Gm_{max}) of HfO$_2$ and SiO$_2$ MOSFETs with various annealing conditions. Unlike HfO$_2$, Gm_{max} of SiO$_2$ was not further improved after the high-temperature FG annealing. 600°C N$_2$ annealing did not improve HfO$_2$ Gm_{max} as much as 600°C FG annealing, indicating that hydrogen was essential for improving the mobility of HfO$_2$ MOSFETs.

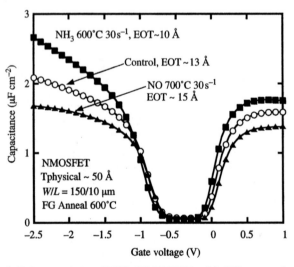

Figure 5.3.9. C–V characteristics of HfO$_2$ NMOSFETs with different surface treatments. For the similar physical thickness of $\sim$50 Å, the NH$_3$ surface annealing has achieved an EOT of $\sim$10 Å. Those without surface annealing and with NO annealing have EOTs of 13 and 15 Å, respectively.

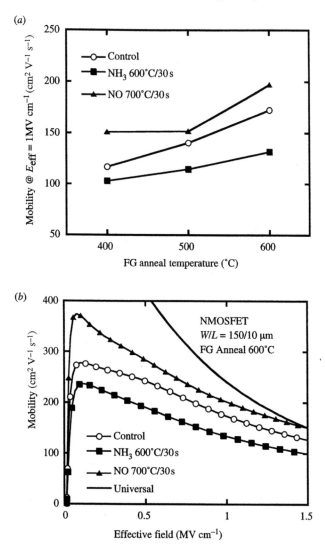

Figure 5.3.10. (*a*) Electron mobility comparison at a fixed E_{eff} of $1\,MV\,cm^{-1}$ for different surface preparations and different FG annealing temperatures. Mobility improvement was observed for all the surface preparations. (*b*) Effective channel electron mobility (μ_{eff}) as a function of effective electric field (E_{eff}) for different surface preparations. EOTs are ~13, 10 and 15 Å for the control, NH₃-annealed and NO-annealed samples, respectively.

was obtained even after transistor fabrication with high-temperature (~950°C) thermal budgets [46]. Similar to the polysilicon gate devices, mobility was improved by using 600°C FG annealing. In fact, peak mobility of ~$400\,cm^2\,V^{-1}\,s^{-1}$ has been obtained for an EOT of ~ 9 Å. These are some of the best data reported for such thin EOTs.

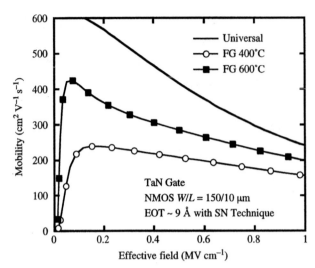

Figure 5.3.11. Effective channel electron mobility (μ_{eff}) of HfO$_2$ NMOSFETs with TaN metal gate electrodes as a function of effective electric field (E_{eff}), for FG annealing temperatures of 400 and 600°C.

HfO$_2$ PMOSFET performance

I_d–V_g characteristics of HfO$_2$ PMOSFETs with an EOT of 14 Å are shown in figures 5.3.12(a) and (b) in linear and log scales, respectively. The HT-FGA similarly improved the I_d–V_g characteristics for the PMOSFETs in both drain currents (figure 5.3.12(a)) and subthreshold swings (figure 5.3.12(b)), while the improvements were not as pronounced as those of the NMOSFETs. The subthreshold swing was reduced down to 63 mV dec^{-1} with the FG annealing at 600°C. The channel hole mobility was improved accordingly as shown in figure 5.3.13, where $\eta = 1/2.5$ was used to calculate the E_{eff} in equation (5.3.2) [35]. Although the mobility saturated to the universal curve of SiO$_2$ at the high E_{eff} for all the FG temperatures, the improvement was clearly observed at the relatively lower E_{eff} region.

Discussion on the high-temperature FG annealing

Electron effective mobility of NMOSFETs with the SiO$_2$ gate dielectric is generally categorized into three regions: i.e., the Coulomb scattering region at the lower E_{eff}, the phonon scattering region at the intermediate E_{eff}, and the surface roughness scattering region at the higher E_{eff} [34]. The behaviour of the mobility degradation in the high-field surface roughness region was attributed to the physical morphology of the interface between SiO$_2$ and Si substrate, and its dependence on E_{eff} was expressed as the universal relationship [47].

(a)

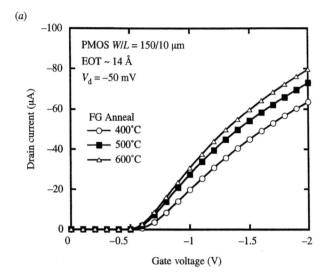

(b)

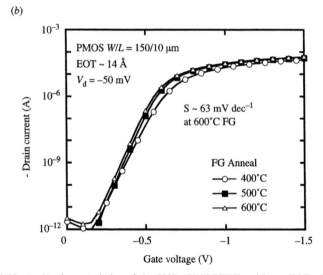

Figure 5.3.12. I_d–V_g characteristics of the HfO₂ PMOSFETs with an EOT of 14 Å, for three different FG annealing temperatures, in linear (a) and log (b) scales, respectively. Similar performance improvements were observed as those for the NMOSFETs.

In figure 5.3.3, the mobility improvement in the HfO₂ NMOSFETs by the HT-FGA was observed for the whole E_{eff} range, but it was more pronounced at the mobility peak in the lower-E_{eff} region. This is due to the reduction in the Coulomb scattering, since the HT-FGA improved D_{it} as discussed above and D_{it} is known to degrade the mobility at the Coulomb scattering region

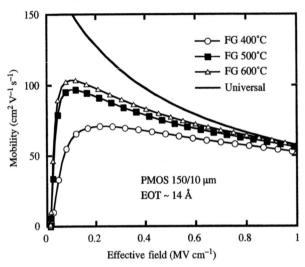

Figure 5.3.13. Effective channel hole mobility (μ_{eff}) of HfO$_2$ PMOSFETs as a function of effective electric field (E_{eff}), for three different FG annealing temperatures.

due to trapped charges [34, 48]. On the other hand, mobility improvement at the higher E_{eff} cannot be attributed to the surface roughness mechanism since the temperatures of the FG annealing used in this study were so low ($\leq 600°C$) that the surface morphology should not be changed. However, the mechanism of the high-E_{eff} mobility degradation is still an open question, due to inconsistent data from different groups. While some groups observed a reasonable correlation between the model and the measurement results [49], some others found a significant discrepancy [50]. In addition, studies on various nitrided SiO$_2$ (SiON) have revealed that the high-field mobility of SiON can be higher than that of the universal curve for SiO$_2$ [51, 52]. Since no clear difference in the interfacial morphology was observed between SiO$_2$ and SiON, D_{it} reduction inside the conduction band of the Si substrate due to nitridation was proposed to explain this improvement [52]. Therefore, a similar mechanism could play some role in improving the mobility at high E_{eff} in HfO$_2$ NMOSFETs.

For the samples studied in this chapter, the optimum mobility at the higher FG annealing temperature is still inadequate compared to that of the SiO$_2$ universal curve. The mechanism is not clear yet, but possible degradation models have recently been proposed, such as wave function penetration [53] and increased remote phonon scattering with high-κ dielectrics [54].

The mobility improvement with the NO-annealed samples is attributed to a thicker and more robust interfacial SiON layer grown by the

NO annealing. This trend can be correlated to the proposed theories [53, 54], since the thicker interfacial layer is effective in both reducing wave function penetration [53] and keeping distance between the channel and the high-κ material [54]. On the other hand, the NH$_3$ annealing degraded the mobility. This is speculated to be due to excess nitrogen concentration in the interfacial layer, since the N concentration in SiON grown by NH$_3$ annealing is higher than that of NO annealing [55], and the higher N concentration compels more Si bonding constraints at the interface [56].

The extent of the mobility improvement in the HfO$_2$ PMOSFETs was relatively limited compared to that of the NMOSFETs, because the mechanism that determines the universality is different. The phonon scattering plays a more significant role to determine the hole mobility [34], which is independent from the interfacial states and therefore is relatively insensitive to the reduction in D_{it}. The hole mobility of the HfO$_2$ PMOSFETs, on the other hand, is closer to the SiO$_2$ universal curve than the electron mobility of the NMOSFETs is. This result also agrees with the model in which the universality of PMOSFET mobility is attributed to the phonon scattering and is therefore insensitive to the dielectric material. A similar trend was observed for various high-κ gate dielectrics, such as Ta$_2$O$_5$ [57, 58] and MOCVD HfO$_2$ [18].

Bias-temperature instabilities of polysilicon gate HfO$_2$ MOSFETs

This section discusses reliability characteristics of HfO$_2$ MOSFETs in terms of BTI due to charge trapping. After reviewing the past studies on BTI and experimental set-ups in the subsections 'Previous studies of bias-temperature instabilities' and 'Bias-temperature instability measurements', respectively, the results will be shown in the subsection 'Experimental results'. Charge pumping measurements were implemented to investigate the nature of the MOSFET instabilities. Potential process solutions were explored as well.

Previous studies of bias-temperature instabilities

It was widely known for the SiO$_2$-based gate dielectrics that high voltage stress on the gate electrode of MOS devices could change flatband (V_{fb}) or threshold (V_t) voltages [59], in particular at elevated temperatures. This phenomenon is called BTI. Much of the past BTI research on the SiO$_2$ dielectric has focused on negative BTI (NBTI), since it caused more severe degradation than positive BTI (PBTI) did [60]. During the NBTI stress, generations of trapped charges (N_{ot}) and interfacial states (N_{it}) were observed [59]. Ogawa *et al* developed a theory on NBTI according to a diffusion–reaction model [61], and succeeded in explaining their experimental results [62], in terms of time evolution, temperature dependence, and thickness (t_{ox}) dependence of the generation rates of N_{it} and N_{ot}. In the diffusion–reaction model, diffusion of hydrogen-related species is responsible for the generation of N_{ot} and N_{it}. Meanwhile, the degradation due to PMOS NBTI remained a minor concern compared to TDDB or HCI until SiO_2 was scaled down to the 30 Å regime. Kimizuka *et al* revealed that at the thickness range thinner than 35 Å NBTI could place a more severe limitation to the SiO_2 scaling than NMOS HCI [63]. They also pointed out that the NBTI degradation could be enhanced with nitrogen incorporation into SiO_2, although nitrogen is effective in suppressing dopant penetration and scaling the oxide thickness [2]. Due to t_{ox}^{-1} dependence of N_{it} generation [61], V_t degradation caused by N_{it} generation became more significant with the thinner SiO_2 than that of N_{ot}. A more recent study of PMOS NBTI on 26 Å SiO_2 revealed that its V_t instability was accounted for only by the N_{it} generation without considering N_{ot} [64].

It is generally recognized that HfO_2 has a significantly larger number of charge traps than that in SiO_2. Gusev *et al* have investigated the traps on HfO_2 deposited by ALD [20], and observed a significant number of traps by hot-carrier injection [65]. We, the authors of this chapter, have been evaluating the trapping characteristics on PVD HfO_2 due to BTI. We found that NBTI on HfO_2 PMOSFETs exhibited sufficient lifetime [29]. On the other hand, NBTI lifetime was deteriorated by the introduction of an NH_3 SN technique [29], which had been effectively used in scaling EOT of HfO_2 and suppressing boron penetration [22, 46]. We also pointed out that unlike SiO_2, NMOS PBTI could be a potential scaling limit of HfO_2 [30].

Bias-temperature instability measurements

I–V and C–V characteristics were measured using an HP4156A precision semiconductor parameter analyser and an HP4194A impedance/gain-phase analyser, respectively, and the measured C–V characteristics were simulated to determine the EOT with the deduction of quantum mechanical effect [44]. BTI measurements were carried out using HP4156A for both stressing and measurements. I–V characteristics were periodically monitored while maintaining the stress temperature to avoid any change in N_{ot} and N_{it} during changing the temperature. Charge pumping current measurement was extensively used to evaluate D_{it} during BTI stressing [45]. An HP8115A pulse generator supplied gate pulses for the charge pumping measurements. Gate pulses of ± 1 V were used at the room temperature, while they were lowered to ± 0.8 V at the elevated temperature (125°C) in order not to damage the devices during the charge pumping measurements.

Experimental results

Positive bias-temperature instability of HfO$_2$ NMOSFET
The time evolution of V_t instability in NMOS PBTI is shown in figure 5.3.14. At relatively low voltages, V_t monotonically increased, while it showed 'turn-around' behaviours at higher voltages. The turn-around behaviours were more pronounced at higher temperatures. The device lifetime should be extrapolated for the positive V_t shift, since the turn-around disappeared at the lower voltages, which would be used for the actual operation. NMOSFET lifetime defined at $\Delta V_t = 50\,\mathrm{mV}$ was extrapolated in figure 5.3.15, using the E-model for the data from lower voltages only. V_g for 10-year lifetime was below 0.3 V. In the same figure, another criterion of $\Delta Gm = -5\%$ was also compared; ΔV_t exhibited more severe degradation, and consequently determined the device lifetime.

Trapped oxide charge density (ΔN_{ot}) calculated from ΔV_t for various stressing voltages is plotted as a function of injected charge density (Q_{inj}) in figure 5.3.16. ΔN_{ot} was found to be a unique function of Q_{inj} (i.e. independent of stress voltage). The ΔN_{ot} values from these PVD HfO$_2$ are amongst the lowest reported for HfO$_2$ [20].

Effects of high-temperature forming gas annealing
It is generally believed that hydrogen is responsible for generation of trapped charges and interfacial states [61, 66, 67]. Although HT-FGA was shown to improve the device characteristics [27, 28], deteriorations in charge trapping characteristics could be a concern. Figure 5.3.17 shows that the

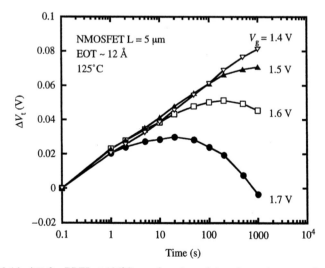

Figure 5.3.14. ΔV_t for PBTI at 125°C as a function of time for various stressing voltages. Turn-around of ΔV_t was observed at higher voltages.

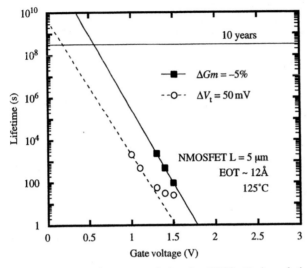

Figure 5.3.15. NMOSFET lifetime extrapolation for PBTI. V_t degradation of 50 mV dominates the lifetime compared to Gm degradation of -5%. V_g for 10 year lifetime by the E-model was smaller than 0.3 V.

charge trapping behaviour was not different between 400 and 600°C FGA devices. In order to further understand the trapping behaviours, charge pumping current (I_{cp}) measurement was implemented to monitor the D_{it} evolution. Figure 5.3.18 shows examples of the I_{cp} as a function of time

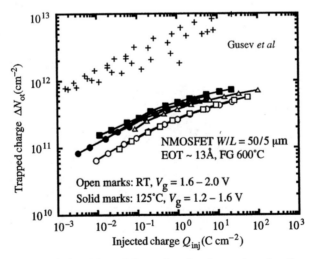

Figure 5.3.16. ΔN_{ot} calculated from ΔV_t was found to be a unique function of Q_{inj}. ΔN_{ot} was enhanced at the higher temperatures.

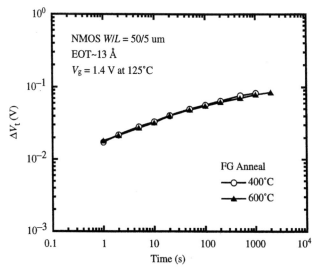

Figure 5.3.17. ΔV_t due to charge trapping was independent from the FG annealing temperature.

during PBTI for the two devices with different FGA temperatures. Although the initial current was an order of magnitude different between 400 and 600°C due to the HT-FGA effect, the increase of I_{cp} from the initial current (ΔI_{cp}) remained similar, indicating D_{it} generation (ΔD_{it}) was also similar.

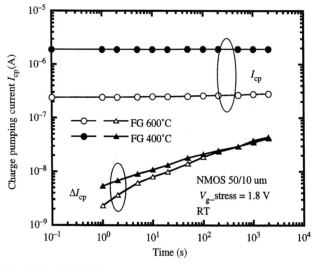

Figure 5.3.18. Although initial I_{cp} was one order different between 400 and 600°C FG annealing, the increase of I_{cp} from the initial current (ΔI_{cp}) due to PBTI was similar.

Generations of trapped charges and interfacial states

ΔD_{it} was calculated from the increase in I_{cp} [45], and it was compared to ΔN_{ot} calculated from ΔV_t (figure 5.3.19(a) and (b)). Although ΔN_{ot} and ΔD_{it} do not have the same units, ΔD_{it} can be correlated to the increase of the interfacial

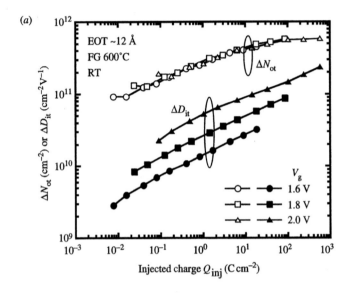

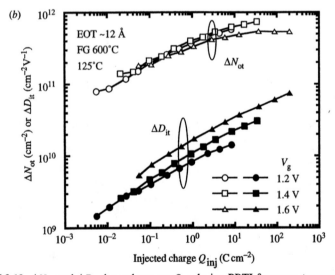

Figure 5.3.19. ΔN_{ot} and ΔD_{it} dependence on Q_{inj} during PBTI for room temperature (a) and 125°C (b). ΔD_{it} was much smaller than ΔN_{ot} and could account for only ≤20% of ΔV_t at most. The difference was more pronounced at lower voltages and higher temperature. A "turn-around" effect was observed at higher stressing voltages at 125°C.

charge density ΔN_{it} as $\Delta N_{it} < \Delta D_{it}E_g$, where E_g is the silicon band gap of 1.12 eV. Since ΔN_{it} at V_t depends on the charged states of ΔD_{it} at $\phi_S = 2\psi_B$, it is very likely smaller than $\Delta D_{it}E_g$. In figure 5.3.19, ΔD_{it} is much smaller than ΔN_{ot}, and the difference is more pronounced at the higher temperature (figure 5.3.19(*b*)). It was also observed that ΔD_{it} was not a unique function of Q_{inj}. Unlike NBTI for SiO$_2$ [64], ΔD_{it} can account only for $\leq 20\%$ of ΔV_t. These results (i.e. the insignificance of ΔD_{it}) suggest that V_t or V_{fb} shifts are not very different from mid-gap voltage (V_{mg}) shift, and these voltage shifts can therefore be used as monitors for trapped charges (ΔN_{ot}) [68].

One possible solution to reduce the charge trapping would be to grow a better Si/dielectric interface, as obtained after NO post-depostion anneals. By applying NO annealing prior to HfO$_2$ deposition, a relatively thick SiON interfacial layer is grown, and interfacial quality closer to the SiO$_2$ dielectric can be expected. Figure 5.3.20 shows the initial mobility comparison between control HfO$_2$ and HfO$_2$ with the SiON interface. Mobility enhancement for the entire effective field range was observed for the device with the SiON interface. However, PBTI results in figure 5.3.21 revealed that the NO annealing did not reduce ΔN_{ot} while it suppressed ΔD_{it}, compared to the control sample in figure 5.3.19(*a*). The better interface did not help to reduce the V_t instability. Bulk traps need to be reduced.

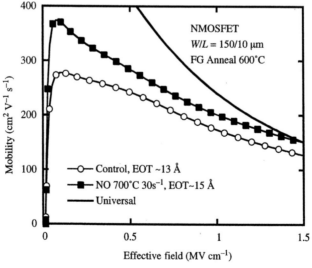

Figure 5.3.20. By using NO annealing prior to HfO$_2$ deposition, NMOS mobility was enhanced compared to the control sample.

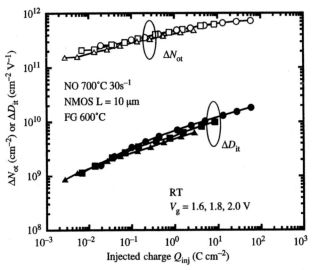

Figure 5.3.21. NO annealing was effective in suppressing ΔD_{it}, compared to figure 5.3.6(*a*), but not in improving PBTI since ΔN_{ot} was not reduced.

Effects of deuterium annealing

D_2 annealing, which is known to improve the SiO_2 MOSFET's immunity against hot-carrier injection [69], is a possible solution. D_2 annealing at 600°C yielded a similar improvement in NMOS mobility to that of 600°C FGA (figure 5.3.22). The D_2 annealing is capable of replacing the HT-FGA

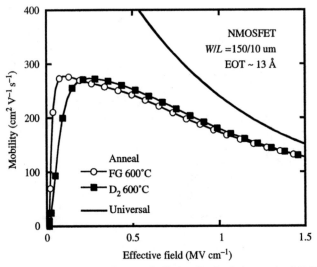

Figure 5.3.22. D_2 annealing at 600°C was similarly effective in improving NMOS mobility as FG annealing at the same temperature.

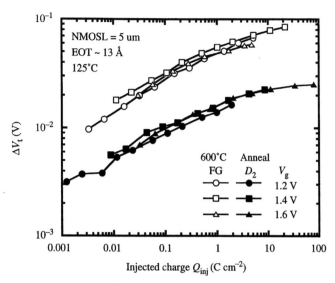

Figure 5.3.23. D$_2$ annealing was effective in suppressing ΔV_t due to charge trapping.

as a technique for enhancing the initial performance of the HfO$_2$ MOSFETs.
Figure 5.3.23 compares ΔV_t as a function of Q_{inj} between the D$_2$ annealing
and the HT-FGA. The D$_2$ annealing was shown to be effective in
suppressing V_t shift. It should be emphasized that the V_t shift for the
D$_2$-annealed samples never exceeded 50 mV. At lower gate voltages, ΔV_t was
small during the stressing time used in this experiment (e.g. ≤ 5000 s). At
higher voltages, a turn-around effect was observed. As a result, lifetime
extrapolation for $\Delta V_t = 50$ mV could not be performed on the D$_2$-annealed
samples. Figure 5.3.24 compares both ΔN_{ot} and ΔD_{it} between the D$_2$
annealing and the HT-FGA. The D$_2$ annealing was effective in reducing ΔD_{it}
as well as ΔN_{ot}.

Negative bias-temperature instability of HfO$_2$ PMOSFET
A similar lifetime definition of ΔV_t of -50 mV was applied for PMOS NBTI,
and the lifetime dependence on V_g is compared between different FGA
temperatures in figure 5.3.25. A 10 year lifetime can be obtained up to
$V_g = -1.4$ V (EOT ~ 14 Å), regardless of the FGA temperature. Unlike
NMOS PBTI, PMOS NBTI is not a serious concern for HfO$_2$, unless
nitrogen concentration is enhanced at the interface [29]. Figure 5.3.26
compares ΔN_{ot} and ΔD_{it} for PMOS NBTI. A slightly larger contribution of
ΔD_{it} to ΔN_{ot} ($\sim 30\%$) compared to NMOS PBTI was observed for all the
stress voltages at both room temperature and 125°C.

PMOS NBTI of HfO$_2$ devices was strongly affected by process
conditions. It is known for SiO$_2$-based gate dielectrics that PMOS NBTI is
degraded if nitrogen is introduced into the dielectric [2]. While the SN

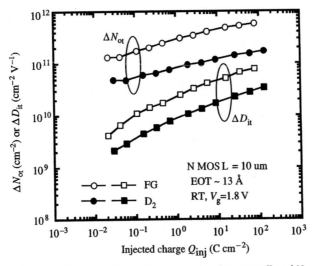

Figure 5.3.24. D_2 annealing was effective in suppressing ΔD_{it} as well as ΔN_{ot}.

technique by an NH_3 annealing was found to be effective in reducing EOT and suppressing boron penetration of HfO_2, there was concern that PMOS NBTI could be degraded. Figure 5.3.27 compared NBTI lifetime between those PMOS devices with and without the SN technique. The NBTI lifetime was clearly degraded with the technique, very likely due to high nitrogen concentration at the interfacial layer.

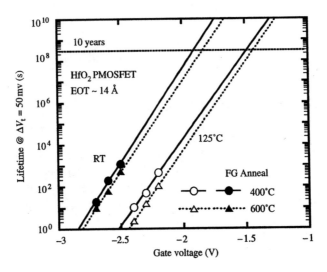

Figure 5.3.25. Lifetime of PMOS NBTI was not significantly altered by the high-temperature FG annealing.

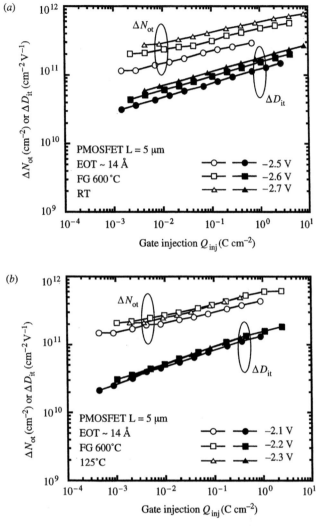

Figure 5.3.26. For HfO$_2$ PMOS NBTI, ΔD_{it} contribution to ΔN_{ot} counted ~30% for all the stress voltages at both room temperature (*a*) and 125°C (*b*).

Discussion on bias-temperature instabilities

As shown in the case of the NO-annealed samples, the V_t instability due to PBTI on HfO$_2$ NMOSFETs was not changed by improving the interfacial quality. Charge traps in the bulk of HfO$_2$ are responsible for the instability.

The larger number of charge traps was generally observed on high-κ dielectrics [20, 70], and Houssa *et al* evaluated the traps on ALD-deposited ZrO$_2$ by analysing the current as a function of time (I–t) during constant V_G

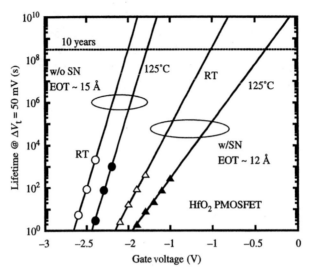

Figure 5.3.27. NBTI lifetime was significantly degraded with the SN technique, very likely due to high nitrogen concentration at the interfacial layer.

stress [70]. They calculated the cross-section of the traps from the charge build-up term of the I–t characteristics, and found that the cross-section of ZrO_2 was much smaller ($\sim 2 \times 10^{-18}\,\mathrm{cm}^2$) than that of SiO_2 ($\sim 1.5 \times 10^{-16}\,\mathrm{cm}^2$). They attributed the small cross-section to the nature of the neutral traps, and explained this by generation of ZrOH trapping centres. H^+ protons generated at the anode by energetic electrons are swept though the dielectric [66] and break the bridging O bonds to form ZrOH centres.

Considering the chemical similarity between Hf and Zr, the charge trapping we have observed on HfO_2 MOSFETs can be correlated to the neutral HfOH centres. In fact, the reduced ΔN_{ot} observed for the D_2 annealing suggested that the trapping reaction was hydrogen related. HT-FGA did not degrade the charge traps on the other hand, although it was expected that the technique introduced hydrogen into the film. It is speculated that the number of HfOH centres are proportional to the number of H^+ protons rather than that of H_2 molecules. Therefore, it must be proportional to the number of energetic electrons, which is independent of the HT-FGA condition.

Neither did HT-FGA degrade D_{it} generation, whose evolution during the electrical stress is modelled by the following rate equation [61].

$$\frac{\partial}{\partial t} N_{it}(t) = G\{N_D - N_{it}(t)\} - SN_{it}(t)[H_2], \qquad (5.3.14)$$

where N_D is the density of hydrogen terminated trivalent Si bonds, $Si_3\equiv Si-H$, and N_{it} is the density of the interfacial states, $Si_3\equiv Si^\cdot$. Although

HT-FGA enhances the first term $(G\{N_D - N_{it}(t)\})$ of equation (5.3.14) by increasing N_D and reducing $N_{it}(0)$, it also increases [H$_2$] of the second term, resulting in the similar time evolution of the N_{it}. The improvement of D_{it} generation achieved with D$_2$ annealing is speculated to be due to suppression of the reaction, and this can be incorporated in equation (5.3.14) by changing the rate constant G. Similar improvement in PMOS NBTI of ultra-thin SiO$_2$ by D$_2$ annealing was observed previously [2].

Another striking difference of HfO$_2$ BTI from that of SiO$_2$ is insignificance of the PMOS NBTI compared to the NMOS PBTI. One possible explanation is the difference in the current amount between N and PMOSFETs at the operating conditions [29]. Figure 5.3.28 shows the band diagrams for both N and PMSOFETs at the operating conditions. In the case of NMOSFETs, the majority of the current is supplied from the channel as electron flows. On the other hand, electron supply from the gate conduction band in PMOSFETs is limited by the generation rate of minority electrons in the p$^+$ gate [71], and therefore the gate current of PMOSFETs is lower. The dominant carriers in the PMOSFET gate current are holes from the channel, whereas valence band electrons from the p$^+$ gate also contribute as the

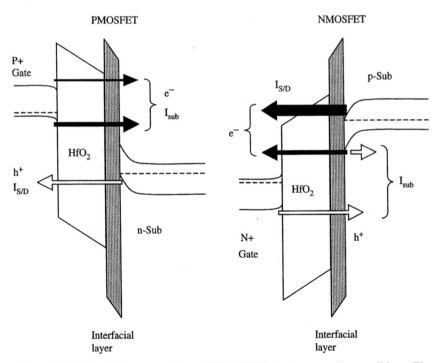

Figure 5.3.28. Band diagrams of P and NMOSFETs during operation conditions. The P+ gate of the PMOSFET supplies insufficient conduction band electrons.

substrate current [29]. As a result, electrons available for charge traps in NMOSFETs are more abundant than those holes in PMOSFETs.

Summary

High-temperature FG annealing was found to significantly improve the drive current or the surface electron mobility of HfO_2 NMOSFETs. The improvement was observed along with the decrease in D_{it} through the characterizations of the subthreshold swings, the $C-V$ characteristics, and the charge pumping currents. The improved mobility and subthreshold swings are advantageous in achieving large on current while suppressing off current, and give more flexibility in V_t adjustment. The mobility enhancement was observed for all the E_{eff} range, and was possibly attributed to the D_{it} reduction. The improvement was further enhanced with the surface NO annealing, while it increased the EOT value. The NH_3 SN was effective in scaling EOT, but the mobility was in general inferior to those of control HfO_2 or the NO-annealed samples. However, by combining the advantage of nitrogen incorporation and hydrogen annealing, high peak mobility ($\sim 400 \, cm^2 \, V^{-1} \, s^{-1}$) with thin EOT ($\sim 9 \, \text{Å}$) has been obtained for TaN gate HfO_2 NMOSFETs. Similar improvements were also obtained on HfO_2 PMOSFETs using high-temperature FG annealing.

A large V_t shift observed on the HfO_2 NMOSFETs after PBTI stress can be a scaling limit of the HfO_2 dielectric. Charge pumping measurements implemented in the PBTI analysis revealed that the ΔV_t was primarily caused by the charge traps ΔN_{ot} in the HfO_2 dielectric, not by the interfacial degradation ΔD_{it}. The PBTI degradation problem was not improved by changing the interfacial layer to the more robust SiON layer grown by NO annealing. D_2 annealing was found to be an excellent method to improve BTI immunity as well as improving mobility of HfO_2 MOSFETs. Unlike SiO_2 devices, NBTI on HfO_2 PMOSFETs was found to be a lesser scaling constraint compared to NMOS PBTI. HT-FGA did not deteriorate either NMOS PBTI or PMOS NBTI.

References

[1] Cheng B, Cao M, Rao R, Inani A, Voorde P V, Greene W M, Stork J M C, Yu Z, Zeitzoff P M and Woo J C S 1999 *IEEE Trans. Electron Devices* **ED-46** 1537–1544
[2] Kimizuka N, Yamaguchi K, Imai K, Iizuka T, Liu C T, Keller R C and Horiuchi T 2000 *Symp. VLSI Tech. Dig.* 92–93
[3] Torii K, Shimamoto Y, Saito S, Tonomura O, Hiratani M, Manabe Y, Caymax M and Maes J W 2002 *Symp. VLSI Tech. Dig.* 188–189
[4] Park D, King Y-C, Lu Q, King T-J, Hu C, Kalnitsky A, Tay S-P and Cheng C-C 1998 *IEEE Electron Device Lett.* **19** 441–443

[5] Luan H F, Lee S J, Lee C H, Song S C, Mao Y L, Senzaki Y, Roberts D and Kwong D L 1999 *IEDM Tech. Dig.* 141–144

[6] Lee B H, Jeon Y, Zawadzki K, Qi W-J and Lee J C 1999 *Appl. Phys. Lett.* **74** 3143–3145

[7] Hubbard K J and Schlom D G 1996 *J. Mater. Res.* **11** 2757–2776

[8] Shannon R D 1993 *J. Appl. Phys.* **73** 348–366

[9] Qi W-J, Nieh R, Lee B H, Kang L, Jeon Y, Onishi K, Ngai T, Banerjee S and Lee J C 1999 *IEDM Tech. Dig.* 145–148

[10] Lee B H, Kang L, Qi W-J, Nieh R, Jeon Y, Onishi K and Lee J C 1999 *IEDM Tech. Dig.* 133–136

[11] Qi W-J, Nieh R, Lee B H, Onishi K, Kang L, Jeon Y, Lee J C, Kaushik V, Neuyen B-Y, Prabhu L, Eigenbeiser K and Finder J 2000 *Symp. VLSI Tech. Dig.* 40–41

[12] Lee B H, Choi R, Kang L, Gopalan S, Nieh R, Onishi K and Lee J C 2000 *IEDM Tech. Dig.* 39–42

[13] Lee C H, Luan H F, Bai W P, Lee S J, Jeon T S, Senzaki Y, Roberts D and Kwong D L 2000 *IEDM Tech. Dig.* 27–30

[14] Kang L, Jeon Y, Onishi K, Lee B H, Qi W-J, Nieh R, Gopalan S and Lee J C 2000 *Symp. VLSI Tech. Dig.* 44–45

[15] Koyama M, Suguro K, Yoshiki M, Kamimuta Y, Koike M, Ohse M, Hongo C and Nishiyama A 2001 *IEDM Tech. Dig.* 459–462

[16] Nieh R, Krishnan S, Cho H-J, Kang C S, Gopalan S, Onishi K, Choi R and Lee J C 2002 *Symp. VLSI Tech. Dig.* 186–187

[17] Zhu W, Ma T P, Tamagawa T, Di Y, Kim J, Carruthers R, Gibson M and Furukawa T 2001 *IEDM Tech. Dig.* 463–466

[18] Lee S J, Luan H F, Lee C H, Jeon T S, Bai W P, Senzaki Y, Roberts D and Kwong D L 2001 *Symp. VLSI Tech. Dig.* 133–134

[19] Hobbs C *et al* 2001 *IEDM Tech. Dig.* 651–654

[20] Gusev E P *et al* 2001 *IEDM Tech. Dig.* 451–454

[21] Kim Y *et al* 2001 *IEDM Tech. Dig.* 455–458

[22] Onishi K, Kang L, Choi R, Dharmarajan E, Gopalan S, Jeon Y, Kang C S, Lee B H, Nieh R and Lee J C 2001 *Symp. VLSI Tech. Dig.* 131–132

[23] Cho H-J, Kang C S, Onishi K, Gopalan S, Nieh R, Dharmarajan E and Lee J C 2001 *IEDM Tech. Dig.* 655–658

[24] Kang C S, Cho H-J, Onishi K, Choi R, Nieh R, Gopalan S, Krishnan S, Han J H and Lee J 2002 *Symp. VLSI Tech. Dig.* 146–147

[25] Rotondaro A L P *et al* 2002 *Symp. VLSI Tech. Dig.* 148–149

[26] Rim K *et al* 2002 *Symp. VLSI Tech. Dig.* 12–13

[27] Onishi K, Kang C S, Choi R, Cho H-J, Gopalan S, Nieh R, Krishnan S and Lee J C 2002 *Symp. VLSI Tech. Dig.* 22–23

[28] Onishi K, Kang C S, Choi R, Cho H-J, Gopalan S, Nieh R, Krishnan S and Lee J C 2002 *IEEE Trans. Electron Devices* **50** 1517–1524

[29] Onishi K, Kang C S, Choi R, Cho H-J, Gopalan S, Nieh R, Dharmarajan E and Lee J C 2001 *IEDM Tech. Dig.* 659–662

[30] Onishi K, Kang C S, Choi R, Cho H-J, Gopalan S, Nieh R, Krishnan S and Lee J C 2002 *Proc. IRPS* 419–420

[31] Sze S M 1981 *Physics of Semiconductor Devices* 2nd edn (New York, Wiley)

[32] Sabnis A G and Clemens J T 1979 *IEDM Tech. Dig.* 18–21

[33] Taur Y and Ning T H 1997 *Fundamentals of Modern VLSI Devices* (Cambridge: Cambridge University Press)

[34] Takagi S-I, Toriumi A, Iwase M and Tango H 1994 *IEEE Trans. Electron Devices* **ED-41** 2357–2362

[35] Chen K, Wann H C, Duster J, Yoshida M, Ko P K and Hu C 1996 *Solid-State Electron.* **39** 1515–1518

[36] Sodini C G, Ko P-K and Moll J L 1984 *IEEE Trans. Electron Devices* **ED-31** 1386–1393

[37] Laux S E and Fischetti M V 1988 *IEEE Electron Device Lett.* **9** 467–469

[38] Lundstrom M 1997 *IEEE Electron Device Lett.* **18** 361–363

[39] Assad F, Ren Z, Datta S, Lundstrom M and Bendix P 1999 *IEDM Tech. Dig.* 547–550

[40] Lochtefeld A and Antoniadis D A 2001 *IEEE Electron Device Lett.* **22** 95–97

[41] Lochtefeld A and Antoniadis D A 2001 *IEEE Electron Device Lett.* **22** 591–593

[42] Koomen J 1973 *Solid-State Electron.* **16** 801–810

[43] Kang L 2000 *PhD Dissertation* University of Texas at Austin

[44] Hauser J *CVC © 1996 NCSU Software Version 3.0* (Raleigh: Department of Electrical and Computer Engineering North Carolina State University)

[45] Groeseneken G, Maes H E, Beltrán N and Keersmaecker R F 1984 *IEEE Trans. Electron Devices* **ED-31** 42–53

[46] Choi R, Kang C S, Lee B H, Onishi K, Nieh R, Gopalan S, Dharmarajan E and Lee J C 2001 *Symp. VLSI Tech. Dig.* 15–16

[47] Sun S C and Plummer J D 1980 *IEEE Trans. Electron Devices* **ED-27** 1497–1508

[48] Matsuoka T, Taguchi S, Khosru Q D M, Taniguchi K and Hamaguchi C 1995 *J. Appl. Phys.* **78** 3252–3257

[49] Cheng Y C and Sullivan E A 1973 *Surface Sci.* **34** 717–731

[50] Fang S J, Lin H C, Snyder J P, Helms C R and Yamanaka T 1996 In: Massoud H Z, Poindexter E H and Helms C R, eds, *The Physics and Chemistry of SiO₂ and the Si–SiO₂ Interface—3* (Pennington, NJ: Electrochemical Society) **96-1**, pp 329–337

[51] Assaderaghi F, Sinitsky D, Bokor J, Ko P K, Gaw H and Hu C 1997 *IEEE Trans. Electron Devices* **ED-44** 664–671

[52] Hori T 1990 *IEEE Trans. Electron Devices* **ED-37** 2058–2069

[53] Polishchuk I and Hu C 2001 *Symp. VLSI Tech. Dig.* 51–52

[54] Fischetti M V, Neumayer D A and Cartier E A 2001 *J. Appl. Phys.* **90** 4587–4608

[55] Hori T 1997 *Gate Dielectrics and MOS ULSIs* (New York: Springer)

[56] Lucovsky G, Wu Y, Niimi H, Misra V and Phillips J C 1999 *Appl. Phys. Lett.* **74** 2005–2007

[57] Chatterjee A *et al* 1998 *IEDM Tech. Dig.* 777–780

[58] Devoivre T, Papadas C and Setton M 1999 *Symp. VLSI Tech. Dig.* 131–132

[59] Deal B E, Sklar M, Grove A S and Snow E H 1967 *J. Electrochem. Soc.* **114** 266–274

[60] Shiono N and Yashiro T 1979 *Jpn. J. Appl. Phys.* **18** 1087–1095

[61] Ogawa S and Shiono N 1995 *Phys. Rev.* **B51** 4218–4230

[62] Ogawa S, Shimaya M and Shiono N 1995 *J. Appl. Phys.* **77** 1137–1148

[63] Kimizuka N, Yamamoto T, Mogami T, Yamaguchi K, Imai K and Horiuchi T 1999 *Symp. VLSI Tech. Dig.* 73–74

[64] Reddy V K, Krishnan A T, Marshall A, Rodriguez J, Natarajan S, Rost T A and Krishnan S 2002 *Proc. IRPS* 248–254

[65] Kumar A, Ning T H, Fischetti M V and Gusev E 2002 *Symp. VLSI Tech. Dig.* 152–153

[66] DiMaria D J, Cartier E and Arnold D 1993 *J. Appl. Phys.* **73** 3367–3384

[67] Afanas'ev V V, de Nijs J M M and Balk P 1995 *J. Appl. Phys.* **78** 6481–6490

[68] McWhorter P J and Winokur P S 1986 *Appl. Phys. Lett.* **48** 133–135

[69] Hess K, Kizilyalli I C and Lyding J W 1998 *IEEE Trans. Electron Devices* **ED-45** 406–416

[70] Houssa M, Stesmans A, Naili M and Heyns M M 2000 *Appl. Phys. Lett.* **77** 1381–1383

[71] Shi Y, Ma T P, Prasad S and Dhanda S 1998 *IEEE Trans. Electron Devices* **ED-45** 2355–2360

Appendix

Material	Relative static dielectric constant	Refractive index	Enthalpy of formation (eV/O atom)	Energy band gap (eV)	Conduction band offset with silicon (eV)	Valence band offset with silicon[a] (eV)
SiO_2	3.9	1.46	−4.68	8.9	3.1	4.7
Si_3N_4	7.4	2.1	/	5.1	2.2	1.8
Al_2O_3	[9, 11]	1.79, 1.87	−5.76	6.2[b], 8.8[c]	2.2[b], 2.8[c], 2.4[d]	2.9, 4.9
Gd_2O_3	[9, 14]	1.88	−6.28	5.2	–	–
Yb_2O_3	[10, 12]	1.55	−6.26	–	–	–
Dy_2O_3	[11, 13]	2.09	−6.44	–	–	–
Nb_2O_5	[11, 14]	2.4	−3.94	4.0	–	–
Y_2O_3	[12, 18]	1.7	−4.93	5.9	2.2[d]	2.6[d]
$Hf_xSi_{1-x}O_y$	[3.9[e], 26[f], 11[g]]	[1.46[e], 2.45[f]]	[−4.68[e], −5.77[f]]	[8.9[e], 5.6[f]]	[3.1[e], 1.3[f]]	[4.7[e], 3.2[f]]
$Zr_xSi_{1-x}O_y$	[3.9[e], 25[f], 12[g]]	[1.46[e], 2.24[f]]	[−4.68[e], −5.66[f]]	[8.9[e], 5.5[f]]	[3.1[e], 1.6[f]]	[4.7[e], 2.8[f]]
$Al_xZr_{1-x}O_2$	[9[e], 25[f], 14[h]]	[1.79[e], 2.24[f]]	[−5.76[e], −5.66[f]]	[8.8[e], 5.5[f]]	[2.8[e], 1.6[f]]	[4.7[e], 2.8[f]]
La_2O_3	[21, 30]	2.0	–	6.0	2.3[d]	2.6[d]
ZrO_2	[14, 25]	2.0, 2.24	−5.66	5.5, 5.8	2.0[b], 1.6[d]	2.4, 3.3[d]
HfO_2	[15, 26]	2.24, 2.45	−5.77	5.6, 5.9	2.0[b], 1.3[d]	2.5, 3.4[d]
Ta_2O_5	25	2.0	−2.09	4.4	0.77[i], 0.3[d]	2.55, 3.0[d]
TiO_2	[50, 80]	2.0	−4.86	3.5, 4.5	1.0[j]	1.4, 2.4
$SrTiO_3$	200	2.47	–	3.2	0.4[d]	1.7

[a] Calculated from the conduction band offsets and energy band gap values.
[b] Experimental, Afanas'ev *et al.*
[c] Experimental, Ludeke *et al.*
[d] Calculated, Robertson *et al.*
[e] Corresponding value for $x = 0$.
[f] Corresponding value for $x = 1$.
[g] Typical value for $x = 0.35$.
[h] Typical value for $x = 0.5$.
[i] Experimental, Lai *et al.*
[j] Experimental, Campbell *et al.*

Index